INTRODUCTION
TO ELECTRONICS

INTRODUCTION TO ELECTRONICS

H. ALEX ROMANOWITZ
RUSSELL E. PUCKETT

Department of Electrical Engineering, University of Kentucky

JOHN WILEY & SONS, INC.   New York, London, Sydney

To Mildred and Dorothy

Preface

The role of electronics in everyday activities is becoming commonplace. Scientists and engineers probe the ocean depths, analyze the atomic nucleus, monitor the physiological behavior of the human body, and venture into the far reaches of space with tools made available by electronics. Many household appliances owe their existence to electronic circuits—adjustable-speed hand tools, food blenders, automatic icemakers, radio and television sets—to name just a few. Electronic control of manufacturing machines by computers is a growing application of fundamental principles in everyday activities.

Semiconductor devices and vacuum tubes are used in many and varied applications of electronic citcuits. The physical principles that govern their behavior, however, are common to all applications. Similarly, the basic concepts of the operating circuits that use these devices are fundamental, whether the circuit be far out in space, inside a giant computer, or carried in one's pocket as a radio receiver. These varied uses of electronic circuits have in common the basic devices and operating characteristics of their individual circuit elements.

This book was written to meet the need for a clear and understandable presentation of these fundamental principles and concepts of electronics. It is an enlargement of the textbook by H. Alex Romanowitz, *Fundamentals of Semiconductor and Tube Electronics*. Algebra and trigonometry are used to discuss the behavior of fundamental electronic devices and circuits; a few derivations and explanations, in which elementary calculus is used, are included in the Appendix.

Chapter 1 is a review of electric circuit analysis, embodying the principles required for an understanding of succeeding chapters. (For students who have already adequately covered the material elsewhere, study may begin with

Chapter 2.) The succeeding chapters may be categorized into two areas: Chapters 2 to 12 present fundamental electronic devices and circuits, and Chapters 13 to 18 cover specific applications of circuits in electronic systems.

Sufficient material is included for a comprehensive first course in electronics, embracing both vacuum-tube and semiconductor devices and circuits. There are more than 550 problems in this book, making it possible for an instructor to choose the desired coverage or alternate problem assignments from year to year. The problems include elementary calculations, those that reinforce important topics of the text, and some that require comprehensive analysis of circuit behavior to challenge the best students. Throughout the text, numerical examples are worked out, and there is a liberal sprinkling of drill problems for student use. Answers to all drill problems and to odd-numbered chapter problems are included.

A *Solutions Manual* comprising detailed solutions to all problems in the book is available for use with it. The *Laboratory Manual*, designed to accompany the textbook, includes more than thirty experiments that have been tested and have proved practical. Selections may be made from them according to the desired coverage and the time assigned for laboratory instruction.

We acknowledge the adaptation of material from the writings of many authors—too numerous to mention individually. The cooperation of manufacturers of electronic equipment, who have furnished photographs and pertinent text material, is greatly appreciated. We also thank Edward S. Berry, Professor of Electronic Technology at Cogswell Polytechnical College, San Francisco, California, for his valuable criticisms and suggestions and the staff of Wiley for its editorial help.

We are grateful for the encouragement and forbearance of our wives, without whose assistance this work could not have been completed.

Lexington, Kentucky
January, 1968

H. Alex Romanowitz
Russell E. Puckett

Contents

CHAPTER THREE—CRYSTAL DIODES   97

CHAPTER FOUR—DIODE TUBES   143

CHAPTER FIVE—RECTIFIERS, POWER SUPPLIES,
AND FILTERS   170

Table of Useful Constants

Charge of an electron, q_e	1.602×10^{-19} C
Mass of an electron, m_e	9.108×10^{-31} kg
Mass of a proton	1.672×10^{-27} kg
Velocity of light	2.9976×10^8 m/s
Planck's quantum of action ($\hbar$)	6.624×10^{-34} J-s
One electronvolt of energy	1.602×10^{-19} J
Boltzmann's gas constant	1.3803×10^{-23} J/°K
Angstrom unit, Å, AU	10^{-10} m
Permittivity of free space, ϵ_0	8.854×10^{-12} F/m
Permeability of free space, μ_0	$4\pi \times 10^{-7}$ H/m
Base, system of natural logarithms, ϵ	2.71828
$\log_{10} \epsilon$	0.4343
$\log_{10} \pi$	0.4971
$\ln \pi$	1.4473

SYMBOLS USED IN CIRCUIT DIAGRAMS

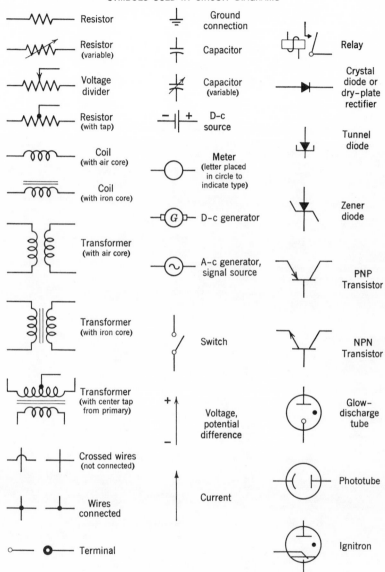

Resistor

Resistor (variable)

Voltage divider

Resistor (with tap)

Coil (with air core)

Coil (with iron core)

Transformer (with air core)

Transformer (with iron core)

Transformer (with center tap from primary)

Crossed wires (not connected)

Wires connected

Terminal

Ground connection

Capacitor

Capacitor (variable)

D-c source

Meter (letter placed in circle to indicate type)

D-c generator

A-c generator, signal source

Switch

Voltage, potential difference

Current

Relay

Crystal diode or dry-plate rectifier

Tunnel diode

Zener diode

PNP Transistor

NPN Transistor

Glow-discharge tube

Phototube

Ignitron

INTRODUCTION TO ELECTRONICS

CHAPTER ONE

Circuit Analysis

A working knowledge of the fundamental principles of electric-circuit analysis is an essential part of the preparation of a student who wishes to profit substantially from a formal study of electronics.

The content of this chapter is presented for review by those who need it. It should be useful to others who may depend on a general course in physics for background.

1.1 Ohm's Law

The amount of current that flows in a conductor at constant temperature depends on its resistance and the voltage (potential difference) applied to it. If the instantaneous value of the voltage remains constant, the amount of current is given by the simple equation

$$I = \frac{E}{R} \qquad (1\text{-}1)$$

which is the mathematical expression for *Ohm's law*. In words,

$$\text{Amperes} = \text{volts} \div \text{ohms}$$

Precise definitions of these terms are available in textbooks on electric-circuit theory. The student should be sure he understands not only the definition of every term he encounters when studying a textbook, intensively, but also the *units* in which the quantities are expressed and their relations to other units.*

* Suggestion: The prefixes kilo, mega, milli, micro should be understood. Also giga $(G = 10^9)$, tera $(T = 10^{12})$, nano $(n = 10^{-9})$, and pico $(p = 10^{-12})$ are now standard prefixes. Examples: nanosecond, picofarad, gigahertz.

Ohm's law applies to a single element in a circuit, to a combination of elements which make up part of a circuit, and to a whole circuit if the combination can be represented as having a single value in ohms. In steady direct-current flow, the ohms value is termed *resistance*; when the current is changing, the ohms value is called *impedance*, and it represents the effect of resistance and either *inductance* or *capacitance* or both. Impedance is denoted by the letter Z or z. It will be expressed as a complex number later in this chapter.

It is evident that any one of the quantities represented in Equation (1-1) may be found by simple algebra if the other two are known.

DRILL PROBLEM

D1-1   An electric heater coil operates on a 120-V supply circuit and has a resistance of 8 Ω. (*a*) How much current flows in the heater? (*b*) What must be the resistance of an electric laundry iron that carries 11 A?

1.2   Kirchhoff's Laws

Circuits are common in which unknown quantities exist that cannot be evaluated by Ohm's law alone. An example is shown in Fig. 1-1. The values of currents I_{AB}, I_{BD}, and I_{BC} are to be found.

Kirchhoff's current law says *the total current entering any junction in a circuit, such as the point B, is equal to the total current leaving that junction.* Applying this law, we can write

$$I_{AB} = I_{BD} + I_{BC} \tag{1-2}$$

Kirchhoff's voltage law says *the sum of all the source voltages in a closed path in a circuit is equal to the sum of all the voltage drops in that path.* Applying

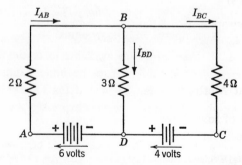

Fig. 1-1   Circuit of example illustrating Kirchhoff's laws.

this law, we can write

$$2I_{AB} + 3I_{BD} = 6 \qquad (1\text{-}3)$$

$$-3I_{BD} + 4I_{BC} = 4 \qquad (1\text{-}4)$$

Equation (1-4) was written for *clockwise summation* around the right-hand loop, in which case $-3I_{BD}$ represents a *rise* in voltage, i.e., a *negative voltage drop*.

Algebraic solution of these three equations gives the following results:

$$I_{AB} = 2\tfrac{1}{13} \text{ A}, \quad I_{BC} = 1\tfrac{6}{13} \text{ A}, \quad I_{BD} = \tfrac{8}{13} \text{ A}$$

The positive value for I_{BD} is due to a correct guess of the direction in which current should flow in the 3-Ω resistance. It is necessary to put arrows on the circuit diagram before writing the equations. Frequently it is necessary to guess the direction of current flow. A minus sign in an answer indicates an incorrect guess. In this example, the current might have been designated I_{DB} with the arrow pointing upward. The answer would then have come out $I_{DB} = -\tfrac{8}{13}$ A.

Examination of Fig. 1-1 will show that the arrow labeled I_{AB} could have been labeled $I_{BD} + I_{BC}$. Kirchhoff's current law applies here, in fact it is the basis of Equation (1-2). This method of labeling current arrows on a circuit diagram eliminates the writing of current equations and thus reduces the number of equations needed to determine current values. The saving of time and work in the analysis of more complicated networks is substantial when this procedure is followed.

The use of determinants in the solution of simultaneous equations of the first degree is recommended. The method is easy to apply and easy to remember for the case of only two unknowns. The procedure for more than two unknowns is explained in section 1.33, and in other references.*

DRILL PROBLEM

D1-2 Redraw the network of Fig. 1-1 and label the current in the 4-Ω resistance $I_{AB} - I_{BD}$ instead of I_{BC}. Write the necessary equations and solve for the current values.

1.3 Mesh-Current Method

A shorter method of solving for the currents in Fig. 1-1 involves labeling the circuit so that each current is *represented as circulating in a closed loop*.

---

* H. Alex Romanowitz, Electrical Fundamentals and Circuit Analysis, Wiley, New York 1966, pp. 110–115.

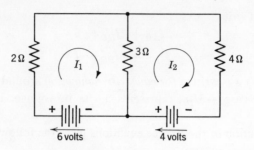

Fig. 1-2   Circuit for mesh-current method of solution.

Figure 1-2 shows the same circuit relabeled for the mesh-current method of solution. After the currents have been evaluated, the following interpretation is given them:

I_1 = *upward* current in the 2-Ω resistance
I_2 = *downward* current in the 4-Ω resistance
$I_2 - I_1$ = *upward current*, or $I_1 - I_2$ = *downward current* in the 3-Ω resistance.

Writing the equations for *voltages*, first *clockwise* around the left-hand loop and then *clockwise* around the right-hand loop, we have

$$2I_1 + 3(I_1 - I_2) = 6 \tag{1-5}$$

$$3(I_2 - I_1) + 4I_2 = 4 \tag{1-6}$$

Kirchhoff current equations are not written, but the principle is used where a circuit element is assigned two currents flowing in opposite directions. The 3-Ω resistor is an example of this.

Solving these equations algebraically gives the same answers as obtained with the Kirchhoff method:

$$I_1 = 2\tfrac{1}{13} \text{ A}, \quad I_2 = 1\tfrac{6}{13} \text{ A}$$

Since I_1 is larger than I_2, the current in the 3-Ω resistor must flow in the direction of I_1, which is downward. It is equal to

$$I_1 - I_2 = 2\tfrac{1}{13} - 1\tfrac{6}{13} = \tfrac{8}{13} \text{ A}$$

1.4   Node-Voltage Method

A widely used method of solving network problems involves *pairs of junctions* in the circuit known as *nodal pairs*. It makes possible the direct solution for a useful voltage value by means of a single equation.

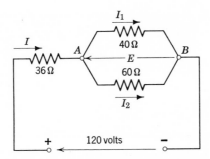

Fig. 1-3   Circuit for node-voltage method of solution.

Consider the circuit of Fig. 1-3. Points A and B form a nodal pair. Observe that the following equation makes use of Kirchhoff's laws.

$$\frac{E}{40} + \frac{E}{60} = \frac{120 - E}{36} \tag{1-7}$$

Solving this gives $E = 48$ V.

It is now a simple matter to get the three currents. They are

$$I_1 = \tfrac{48}{40} = 1.2 \text{ A}, \quad I_2 = \tfrac{48}{60} = 0.8 \text{ A}, \quad I = 1.2 + 0.8 = 2 \text{ A}$$

DRILL PROBLEMS

D1-3   Assume an 18-Ω resistance has been connected in parallel with the 36-Ω resistance in Fig. 1-3. Calculate the current in this 18-Ω resistance, using the node-voltage method.

D1-4   Redraw the network of Fig. 1-3 with the three resistances connected as they are. Use loop-current labeling and solve for all three current values by means of the mesh-current method.

1.5   Superposition Theorem

The problem of analyzing a network which contains two or more voltage sources is simplified if the *superposition theorem* is used. This theorem may be stated as follows: *If a network contains two or more voltage sources, the current through any bilateral linear element is the sum of the currents which would be sent through that element by the voltage sources applied separately to the network with all other source voltages removed and the internal resistances of all sources left in the circuit.* It should be stated that the superposition theorem holds for a-c networks as well, but in that case the word

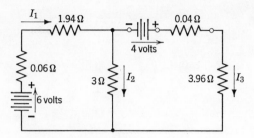

Fig. 1-4   Circuit analyzed with the superposition theorem.

impedances must be used in place of the word *resistances* in the statement of the theorem.

As an example of the use of the superposition theorem, the circuit of Fig. 1-4 will be analyzed. It is desired to determine the current in the 3-Ω resistor. The small resistances (0.04 Ω and 0.06 Ω) represent internal resistances of the batteries.

Use of the superposition theorem requires that *one voltage source at a time* be used when calculations are being made, and that *all other voltage sources be removed but their internal resistances be left where they are* in the network. The procedure will be to use only the 6-V battery first and get a value for I_2, and then to use only the 4-V battery and get another value for I_2. The sum of these two currents should be the correct value of I_2 when both batteries are in the circuit.

Figure 1-5a shows the circuit for the first half of the solution. Note that the 4-V battery has been replaced by only its internal resistance. We shall now solve for the value of I_2', the downward current in the 3-Ω resistor.

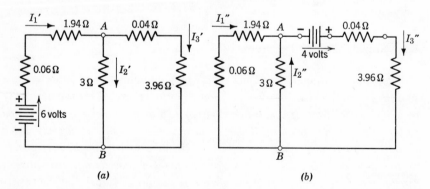

Fig. 1-5   Two circuits used in applying the superposition theorem in analyzing the circuit of Fig. 1-4.

Note that the 3-Ω and the 4-Ω resistors are in parallel. They may be represented by a single resistance R_e connected to points A and B, obtained as follows:

$$1/R_e = \tfrac{1}{3} + \tfrac{1}{4} = \tfrac{7}{12} \text{ mho}$$

$$R_e = \tfrac{12}{7} = 1\tfrac{5}{7} \ \Omega$$

This new unit, when used in place of the 3-Ω and 4-Ω resistances, will be in series with the 2-Ω resistance. The current that will then flow through R_e will be

$$I_{R_e} = \frac{6}{2 + 1\tfrac{5}{7}} = 1\tfrac{8}{13} \text{ A}$$

The voltage across R_e, and therefore between points A and B in Fig. 1-5a, is

$$E_{ab} = 1\tfrac{8}{13} \times 1\tfrac{5}{7} = 2\tfrac{10}{13} \text{ V}$$

The current in the 3-Ω resistor with only the 6-V battery working is then

$$I_2' = 2\tfrac{10}{13} \div 3 = \tfrac{12}{13} \text{ A}$$

Now determine I_2'' of Fig. 1-5b, where only the 4-V battery is working and the 6-V battery has been replaced by its internal resistance of 0.06 Ω. This time the 3-Ω resistance is in parallel with a 2-Ω resistance. R_e between points A and B will now be determined in this case.

$$1/R_e = \tfrac{1}{3} + \tfrac{1}{2} = \tfrac{5}{6} \text{ mho}$$

$$R_e = \tfrac{6}{5} = 1\tfrac{1}{5} \ \Omega$$

The current through R_e when it is connected to points A and B in place of the 2-Ω and 3-Ω resistances is

$$I_{R_e} = \frac{4}{4 + 1\tfrac{1}{5}} = \tfrac{10}{13} \text{ A}$$

The voltage across R_e and therefore between points B and A in Fig. 1.5b is

$$E_{ba} = \tfrac{10}{13} \times 1\tfrac{1}{5} = \tfrac{12}{13} \text{ V}$$

The current in the 3-Ω resistor with only the 4-V battery working is then

$$I_2'' = \tfrac{12}{13} \div 3 = \tfrac{4}{13} \text{ A } \textit{upward}$$

It is important to note here that common sense was used in drawing the arrow for I_2'' pointing upward. It is obvious that when I_3'' comes from the battery and reaches point B it will divide into two parts, one part going upward through 3 Ω to A and the other part going through the other possible path, containing the 2-Ω resistance, to reach point A.

The current I_2 through the 3-Ω resistor in the original circuit (Fig. 1-4) is now considered to be made up of $I_2' = \frac{12}{13}$ A *downward* and $I_2'' = \frac{4}{13}$ A *upward*. The net result must be $I_2 = \frac{8}{13}$ A downward.

The superposition theorem is especially useful in the analysis of vacuum-tube circuits at an advanced level. It requires that the elements left in the circuit, such as resistances, coils, and condensers, be *passive* units, i.e., incapable of generating voltages of their own, and that they be *linear*, i.e., the voltages across them are directly proportional to the currents in them. The elements must also be *bilateral*, i.e., the resistance (or impedance) is the same for both directions of current flow.

1.6   Reciprocity Theorem

In any network composed of linear bilateral impedances, if a source voltage E is applied between any two terminals and the current I is measured in any branch, their ratio, E/I, will be equal to the ratio obtained if the positions of E and I are interchanged. This is a very useful theorem in advanced analysis. The ratio of E between two points (at the input, so to speak) to I in another branch of the circuit (which may be considered the output branch) is called the *transfer impedance* in a-c circuit analysis. Furthermore, it will be shown later how to convert any complex network into an equivalent T type of circuit made up of only three branches and generally considered to be a four-terminal network. The two input terminals and the two output terminals of the T are the same terminals used for input and output in the original circuit. Because of this "equivalent T" convenience, a network theorem may be proved using the T and it naturally follows that the theorem applies to *any linear bilateral network*.

The reciprocity theorem will now be illustrated by means of the T circuit of Fig. 1-6. An input voltage $E = 54$ V is applied to the terminals 1 and 3.

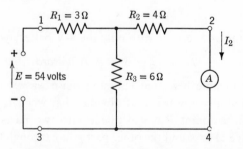

Fig. 1-6   T network used in illustrating the reciprocity theorem.

An ammeter of entirely negligible resistance is connected to the output terminals 2 and 4. The procedure is first to calculate the current I_2 and get the ratio E/I_2. Then interchange E and the ammeter and see if the ratio of E to the current measured at terminals 1 and 3 is the same.

The input current at terminal 1 is calculated by first getting the resistance (R_e) equivalent to R_2 and R_3 in parallel and adding this to R_1. E divided by this total resistance (R_T) is the output current.

$$1/R_e = \tfrac{1}{6} + \tfrac{1}{4} = \tfrac{5}{12}$$

$$R_e = \tfrac{12}{5} = 2.4\,\Omega$$

$$R_T = 3 + 2.4 = 5.4\,\Omega$$

$$I_1 = 54/5.4 = 10\,\text{A}$$

The voltage drop across R_e is $10 \times 2.4 = 24$ V

$$I_2 = \tfrac{24}{4} = 6\,\text{A}$$

$$E/I_2 = \tfrac{54}{6} = 9\,\Omega$$

Now interchange E and A, and follow the same procedure to get the new output current. The new R_e is $2\,\Omega$ as obtained from

$$1/R_e = \tfrac{1}{3} + \tfrac{1}{6} = \tfrac{3}{6}\,\text{mho}$$

Total resistance $= 2 + 4 = 6\,\Omega$

Input current $= \tfrac{54}{6} = 9\,\text{A}$

Voltage drop across the new $R_e = 9 \times 2 = 18$ V

The new value of output current is

$$I_2 = \tfrac{18}{3} = 6\,\text{A}$$

This gives the same ratio as before,

$$E/I_2 = \tfrac{54}{6} = 9\,\Omega$$

Observe that the unit of this ratio is *ohms* because it is obtained by dividing volts by amperes. In an *all-resistance circuit* the ratio could be called the *transfer resistance*, but it is better to call it the *transfer impedance* because it is used much more often in a-c circuits where impedances are encountered very much more frequently than resistances only.

Note that the reciprocity theorem merely states that, if the input voltage and the ammeter (with zero resistance) are interchanged, the *ratio* of E_{in} to

I_{out} will be unchanged. Actually, a value of E different from 54 V could have been used after the interchange was made, but the *ratio* of input voltage to output current would not have changed. It can easily be seen that if 108 V had been used, the ammeter current would have been 12 A.

1.7   Thévenin's Theorem

This theorem states that *the current in any impedance Z_L, connected to two terminals of a network, is the same as would be obtained if Z_L were supplied from a source voltage E' in series with an impedance Z', E' being the open-circuit voltage at the terminals from which Z_L has been removed and Z' being the impedance that would be measured at these terminals after all generators have been removed and each has been replaced by its internal impedance.*

The four-terminal T network of Fig. 1-7 will be used to illustrate the theorem. The impedances are pure resistances and the load impedance is 10 Ω. The theorem states that the *equivalent circuit* has an *input voltage E'* equal to the voltage that will exist at terminals 2 and 4 of the actual circuit after the load impedance is disconnected. This is what is meant by the open-circuit voltage at the load terminals.

With terminals 2 and 4 open, their potential difference will be the voltage drop across the 5-Ω resistor. The current through this resistor is $100 \div 25 = 4$ A. The voltage drop is $4 \times 5 = 20$ V. This is E' in the equivalent circuit.

The value of Z' will now be found. The source E in the original circuit has no internal impedance, so we simply remove the 100 V and connect terminals 1 and 3 together; this means inserting zero resistance. If we now connect an ohmmeter to the open terminals 2 and 4, the ohmmeter "would see" 4 Ω in series with a simple parallel circuit made up of a parallel branch of 20 Ω and 5 Ω. The resistance, which will be the series impedance Z' in the

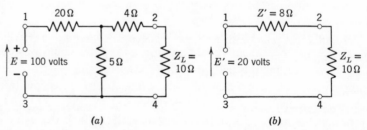

Fig. 1-7   Illustrating Thévenin's theorem. (*a*) Actual circuit; (*b*) equivalent circuit.

equivalent circuit, will be

$$Z' = 4 + \frac{20 \times 5}{20 + 5} = 8 \ \Omega$$

The load current in the equivalent circuit will now be calculated.

$$I_L' = \frac{E'}{Z' + Z_A} = \frac{20}{18} = 1\tfrac{1}{9} \ A$$

It will now be shown that, *so far as load current is concerned*, this equivalent circuit has correctly taken the place of the original circuit. Looking into the original circuit at the input terminals 1 and 3, we see a 20-Ω resistance in series with a parallel pair made up of 5 Ω and 14 Ω. They may be combined into a single resistance R across terminals 1 and 3:

$$R = 20 + \frac{5 \times 14}{5 + 14} = 23\tfrac{13}{19} \ \Omega$$

The input current to the actual circuit is

$$I = \frac{100}{23\tfrac{13}{19}} = 4\tfrac{2}{9} \ A$$

After passing through the 20-Ω resistance this current divides, part going through the 5-Ω resistance and the remainder through the 14-Ω resistance and the load. The load current is conveniently calculated as follows:

$$I_L = 4\tfrac{2}{9} \times \frac{5}{5 + 14} = 1\tfrac{1}{9} \ A$$

This shows that the Thévenin equivalent circuit produces the same current in Z_L as that in the original circuit.

In a complex network, any two terminals may be considered *output terminals* and the impedance connected to them would then be called the load impedance. Any other two terminals could be chosen as the *input terminals*, although the two terminals of a branch having a *voltage source* is usually chosen in a practical situation.

An important point concerning the Thévenin equivalent circuit is that the circuit is used effectively to represent a situation in which an electronic device provides a constant voltage. A *necessary condition* for a constant voltage source (practically constant is good enough, in many applications) is that *the internal impedance of the source be small compared to the impedance of the load*. This condition is not necessary, however, for Thévenin's theorem to hold.

1.8   Norton's Theorem (Short-Circuit Theorem)

Norton's theorem is based on the ability of a generator to supply *constant current* to a network. Certain kinds of electronic tubes can supply a current that remains practically constant over a wide range of anode voltage on the tube. It will be found that Norton's theorem is useful in setting up an equivalent circuit in analyzing the performance of this type of tube.

For an application of Norton's theorem, refer again to Fig. 1-7a. We shall set up a constant-current circuit which will be equivalent to this circuit *as far as load current is concerned.*

The first step is to imagine the output terminals 2 and 4 short-circuited, which means the resistance between them would be reduced to zero. The input current to the whole circuit (at terminal 1) is

$$I_{\text{in}} = \frac{100}{20 + \dfrac{5 \times 4}{5 + 4}} = 4.5 \text{ A}$$

The current through the short circuit is

$$I_{SC} = \frac{5}{4 + 5} \times 4.5 = 2.5 \text{ A}$$

The next step is to determine the impedance Z' looking back toward the source from the output terminals 2 and 4, as was done in the Thévenin theorem case. The result will be $Z' = 8\ \Omega$ as before.

Figure 1-8 shows the equivalent circuit when Norton's theorem is applied to the actual circuit of Fig. 1-7a. The calculated short-circuit current is supplied by a *constant-current generator*, not shown, and the calculated impedance Z' *is connected in parallel with the load impedance*, instead of in series as it was in the Thévenin case. The load current in this equivalent

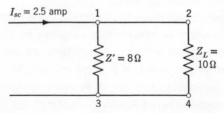

Fig. 1-8   The Norton-theorem equivalent of the circuit of Fig. 1-7a.

circuit should be the same as in the original circuit. The following simple calculation shows that it is:

$$I_L = 2.5 \times \frac{8}{8 + 10} = 1\tfrac{1}{9}\,\text{A}$$

Norton's theorem may now be stated: *The current in any impedance,* Z_L, *between two terminals in a network is the same as it will be when* Z_L *is connected in parallel with another impedance,* Z', *and the pair supplied from a constant-current generator, provided the generator current is equal to the current which flows between the two terminals when they are short-circuited in the original circuit and the parallel impedance* Z' *is equal to the impedance of the network looking back from the two terminals.*

Norton's theorem may well be stated in terms of conductances, since the impedance in the equivalent circuit is placed in parallel with the source and the load: A linear two-terminal network may be replaced, as far as any load connected between the two terminals is concerned, with a source delivering a constant current I' and a conductance G' in parallel with the generator and the two load terminals. I' is the current that would flow through a short circuit between the two terminals in the original circuit, and G' is equal to the conductance measured between the two terminals after all the EMF's are made zero. It must be remembered that current equals voltage times conductance, which means that the voltage drop across a conductance equals the current divided by the conductance.

DRILL PROBLEMS

D1-5 (*a*) Use Thévenin's theorem to solve for the current in the 10-Ω "load resistance" in Fig. 1-9. (*b*) Verify your answer by solving for the current using Norton's theorem.

D1-6 Consider the 6-Ω resistance in Fig. 1-9 to be the "load." Apply Thévenin's theorem, removing this resistor to get the equivalent circuit, and solve for the voltage across it when it is in the circuit. Verify your result using Norton's theorem.

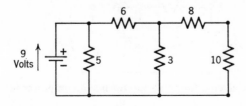

Fig. 1-9   Resistance values are in ohms.

1.9   Maximum Power Transfer Theorem

When electrical energy is to be delivered to a load, there are times when the load should receive *maximum power* rather than maximum voltage or current. One is the driving of a load speaker by an electronic amplifier. Although there are a great many more maximum power transfer situations in a-c circuits than in d-c circuits, only the latter will be discussed at this point. We shall treat the a-c case later, after some work has been done with the representation of voltages, currents, and impedances by complex numbers.

The load resistance R_L in the circuit of Fig. 1-10 will receive maximum power if R_L has *only one particular ohms value* and not any other. Since any four-terminal network may be represented by an equivalent circuit obtained by using Thévenin's theorem, the circuit of Fig. 1-10 will be used in discussing maximum power transfer. The Thévenin equivalent circuit has a supply E' of 72 volts and a series impedance Z' of 64 Ω.

If one used a slide-wire resistance as a load so that its ohms value could be varied, thus varying R_L, a wattmeter showing the power in R_L *would have a maximum reading* only when the slide wire was set at 64 Ω. This can be proved by using simple calculus.* We shall be content with calculating the power transferred to the load for a few values of R_L above and below 64 Ω and show that, if a curve were plotted with power in R_L on the vertical axis and with ohms resistance of R_L on the horizontal axis, the peak of the curve would be at $R_L = 64$ Ω.

The voltage across R_L will always be given by

$$E_L = \frac{R_L}{64 + R_L} \times 72 \text{ V}$$

and the power in R_L by

$$P_L = \frac{E_L{}^2}{R_L} = \frac{72^2 R_L}{(64 + R_L)^2} \tag{1-8}$$

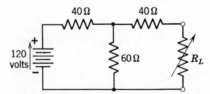

Fig. 1-10   Circuit considered for condition for maximum power into the load R_L.

* Calculus proof is given in the Appendix.

TABLE 1-1

R_L (ohms)	16	36	56	60	62	64	66	68	72	92	112
P_L (watts)	12.96	18.66	20.16	20.22	20.24	20.25	20.24	20.23	20.12	19.59	10.74

Table 1-1 shows how P_L varies with R_L as calculated from Equation (1-8). In a circuit where a generator, with internal impedance R_G, feeds power to a load, with impedance R_L (no net reactance in either), maximum power will be supplied to the load when the load resistance R_L is equal to the generator resistance R_G.

The case when reactances are present is treated in Section 1.28.

1.10 Compensation Theorem

In the analysis of circuits, particularly transmission lines treated in more advanced studies, it is common practice to represent an impedance in a network by a generator. The generator, obviously, must present to its terminals the same instantaneous potential difference that existed at every instant across the replaced impedance, and the generator must not have any internal impedance.

Figure 1-11 illustrates the application. It is the same as Fig. 1-7a with E substituted for the 4-Ω resistor. It is suggested that the student solve for the load current using one of the basic methods described in this chapter.

The compensation theorem may be stated as follows: *Any impedance in a network may be replaced by a generator that has zero internal impedance and a generated voltage which is equal, at every instant, to the potential difference that would have been produced across the replaced impedance by the current flowing through it.*

1.11 T and π Networks

Any complex network can be reduced to an equivalent T by the use of equations involving open-circuit and short-circuit impedances. When the impedances of the three T branches are known, it is an easy matter to calculate the impedances of an equivalent π network by means of three conversion equations. Communication lines are represented by these two types of network, and filters used in such lines actually contain T- and π-shaped sections.

Consider a box with two terminals on one end and two on the other. Inside is a complex network made up of an unknown number of impedances

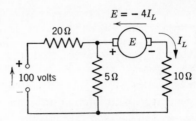

Fig. 1-11   Illustrating the compensation theorem.

connected together in an unknown arrangement. This is referred to as a "black box" in electric-circuit theory. It contains an unknown circuit connected to a *pair of input terminals* and to a *pair of output terminals*. The procedure for setting up an equivalent T circuit, *with all three of its imped-ances completely evaluated*, will now be described.

The impedances Z_1, Z_2, and Z_3 of Fig. 1-12 of the equivalent T that will represent the circuit in the black box, or *any other four-terminal network*, are calculated using the following three equations in which the impedances on the right-hand side of the equality sign are obtained *either by measurement or by calculation*. Calculation would be used when the impedances of the complex network are known.

$$Z_1 = Z_{O1} - \sqrt{Z_{O2}(Z_{O1} - Z_{S1})} \tag{1-9}$$

$$Z_2 = Z_{O2} - \sqrt{Z_{O2}(Z_{O1} - Z_{S1})} \tag{1-10}$$

$$Z_3 = \sqrt{Z_{O2}(Z_{O1} - Z_{S1})} \tag{1-11}$$

In these equations,

Z_1 and Z_2 are the first and second *series* arms of the equivalent T, and Z_3 is the *mid-series shunt* arm, all shown in Fig. 1-12;

Z_{O1} is the impedance looking into the network at the generator end with the load end open (Z_L disconnected);

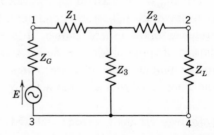

Fig. 1-12   T network with generator and load.

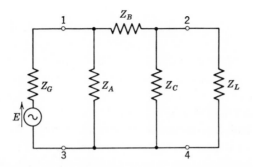

Fig. 1-13 π network with generator and load.

Z_{S1} is the impedance looking into the network at the generator end with the load end short-circuited;

Z_{O2} is the impedance looking into the network at the load end with the generator end open;

Z_{S2} is the impedance looking into the network at the load end with the generator replaced by a short circuit.

These relationships hold for both d-c and a-c circuits.

The same open- and short-circuit impedances values may be used in the following three equations to obtain Z_A, Z_B, and Z_C, the branches of the equivalent π section shown in Fig. 1-13.

$$Z_A = \frac{Z_{S1}Z_{O2}}{Z_{O2} - \sqrt{Z_{O2}(Z_{O1} - Z_{S1})}} \tag{1-12}$$

$$Z_B = \frac{Z_{S1}Z_{O2}}{\sqrt{Z_{O2}(Z_{O1} - Z_{S1})}} \tag{1-13}$$

$$Z_C = \frac{Z_{S1}Z_{O2}}{Z_{O1} - \sqrt{Z_{O2}(Z_{O1} - Z_{S1})}} \tag{1-14}$$

T and π sections are really only *three-terminal* networks instead of four-terminal ones, as is evident from Fig. 1-12 and Fig. 1-13. Note that in both cases terminals 3 and 4 are really *only one-terminal* because there is *zero impedance* between them. If points 3 and 4 were brought together at one common point, the π shape would become triangular and we would have the *delta* connection so familiar to power transmission engineers. Correspondingly, the three arms of the T would be looked upon as forming the *Y* or *star* connection so common in electric power applications.

1.12   Transformation Equations T to π and π to T

There remains to be shown a means for converting a T section to a π section, and vice-versa, by means of special *conversion equations*. These, as well as Equations (1-9) through (1-14), are derived in more advanced textbooks,* to which the student is referred. The derivations are somewhat long but not very difficult to understand. The ability to simplify a complicated circuit by converting a given T section to an equivalent π, or vice-versa, is a very useful tool in the analysis of complex networks. An application of this will be given after the conversion equations are presented.

To convert from T to π, the π-branch impedances are given by

$$Z_A = \frac{Z_1 Z_2 + Z_2 Z_3 + Z_3 Z_1}{Z_2} \tag{1-15}$$

$$Z_B = \frac{Z_1 Z_2 + Z_2 Z_3 + Z_3 Z_1}{Z_3} \tag{1-16}$$

$$Z_C = \frac{Z_1 Z_2 + Z_2 Z_3 + Z_3 Z_1}{Z_1} \tag{1-17}$$

Notice that the same numerator appears in all three equations. To convert from π to T, the T-arm impedances are given by

$$Z_1 = \frac{Z_A Z_B}{Z_A + Z_B + Z_C} \tag{1-18}$$

$$Z_2 = \frac{Z_B Z_C}{Z_A + Z_B + Z_C} \tag{1-19}$$

$$Z_3 = \frac{Z_C Z_A}{Z_A + Z_B + Z_C} \tag{1-20}$$

Notice that the same denominator appears in all three equations. If we were doing a lot of work in these conversions, we would soon catch on to a system of remembering these six equations. If the student would draw the π elements with Z_A and Z_C brought together at one point at the bottom, and then draw the T elements inside and connected to the same points, a pattern for remembering the conversion equations would soon be obvious.

---

* For example, Russell M. Kerchner and George F. Corcoran, *Alternating-Current Circuits*, 4th ed., Wiley, New York, 1960.

1.13   Alternating Voltage, Current, and Frequency

Steady magnetic fields, half of them with north polarity and half with south polarity, rotate with the shaft of an a-c generator. Coils of stationary windings, called the *armature* or *stator*, are mounted very close to the rotating magnetic field *poles* so that magnetic flux first passes through the coil loops. Flux from the north pole passes *into the loops*, and flux of the south poles passes *out of the loops*. The rapid change of flux through the armature winding loops generates a voltage such that its *instantaneous* value e varies as the *sine* of the angle of displacement (θ) of the field poles; that is,

$$e = E_{max} \sin \theta \qquad (1\text{-}21)$$

The angle θ is measured in *electrical degrees* around the periphery of the armature. With only one north and one south pole (one pair) and a single-phase armature winding, there are 360 electrical degrees in one revolution. With two pairs of rotating poles there are 720 electrical degrees in one complete revolution.

When an armature coil has a complete change of flux through it, i.e., the flux has (a) increased from zero to maximum inward, (b) decreased to zero, (c) reversed and increased from zero to maximum in the outward direction, and (d) decreased to zero and is then ready to start increasing in the inward direction again, a full *cycle* of instantaneous voltage values has been generated.

The number of *cycles per second* of the voltage generated by a rotating machine is determined by the number of pairs of poles and the speed. A 4-pole machine (2 north and 2 south poles) generates 2 cycles of voltage per second and must run at 1800 rpm to generate voltage at a frequency of 60 c/s.

The name hertz, abbreviated Hz, is now used instead of cycles per second to denote frequency of voltages and currents. This does not change the basic unit of frequency in electrical engineering, i.e., the reciprocal second, symbolically $1/t$, where t is in seconds.

Electric power frequency in this country is now expressed as 60 Hz, radio broadcast frequencies as 550 kHz to 1,500 kHz (kilohertz), and television frequencies (UHF band) as 300 MHz to 3,000 MHz.

As indicated by Equation (1-21) a plot of instantaneous values of generated voltage throughout a complete cycle has *sine-wave form*. When plotted to scale, the vertical distance from the horizontal, θ, axis to the curve is inter-preted as volts or voltage.

Voltage of sine-wave form sends current of sine-wave form through circuit elements such as resistances, inductances, and capacitances when they have constant impedance. Both voltages and currents have *effective* and *average* values. When they have sine-wave form these values are related to the *instantaneous maximum* (or *crest*) *values* as follows:

$$E_{\text{eff}} = \frac{E_{\max}}{\sqrt{2}} = 0.707 E_{\max} \tag{1-22}$$

Effective values

$$I_{\text{eff}} = \frac{I_{\max}}{\sqrt{2}} = 0.707 I_{\max} \tag{1-23}$$

The *average value* of a voltage, or a current, of sine-wave form is zero, because in one whole cycle there is a *negative instantaneous* value equal in magnitude to every *positive* instantaneous value.

It must be remembered that the relationships above hold only when the voltage or current has sine-wave form, or cosine-wave form, since both of these wave forms have the same *shape* and are merely timed (or *phased*) differently.

The displacement angle θ in Equation (1-21) is usually written as ωt, where ω is the angular velocity in *electrical radians per second*. The instan-taneous voltage would then be written

$$e = E_{\max} \sin \omega t \tag{1-24}$$

1.14   Fourier Series

We frequently encounter voltage or current waves that are not sinusoidal in form. Examples are: the half-sine wave form at the output of a half-wave rectifier, the rectified sine wave at the output of a full-wave rectifier, and the saw-tooth wave form of the horizontal sweep oscillator in a cathode-ray oscilloscope.

These waves are not symmetrical about the horizontal axis. Most non-sinusoidal waves are not. Such waves have *average* values that are called *d-c components* when the waves represent voltage or current. The value of the d-c component can be measured by a d-c meter.

A current or voltage with a nonsinusoidal wave form has a-c components, every one of which may be represented by a sine (or cosine) function of time.

If we let i represent the instantaneous value of a current with a *complex* (nonsinusoidal) *wave form*, the equation describing that current may be written as a *Fourier series* as follows:

$$i = A_0 + A_1 \sin \omega t + B_1 \cos \omega t + A_2 \sin 2\omega t + B_2 \cos 2\omega t$$

$$+ A_3 \sin 3\omega t + B_3 \cos 3\omega t + \cdots + A_n \sin n\omega t + B_n \cos n\omega t \quad (1\text{-}25)$$

A_0 is called the d-c term and it is the average value of the current. The other A's are numerical coefficients in sine-function terms representing the fundamental, second harmonic, third harmonic and all higher harmonics of sine-wave form. The B's are numerical coefficients in cosine-function terms representing the fundamental, second harmonic, third harmonic and all higher harmonics of cosine-wave form.

It is often found that representation of a given complex wave form, to a satisfactory degree of accuracy for engineering analysis, may be done when a relatively few (5 to 11) of the harmonics are evaluated and used. A half-sine wave of current (or voltage) is represented with sufficient accuracy in most practical applications by only four terms of its Fourier series. It happens that A_2, A_3, A_4, B_1, and B_3 are all zero for the half-sine wave. The series is

$$i = 0.318 I_m + 0.500 I_m \sin \omega t - 0.212 I_m \cos 2\omega t$$

$$- 0.0424 I_m \cos 4\omega t - \cdots \text{ amperes} \quad (1\text{-}26)$$

Of course, a half-sine voltage wave would be represented by the same equation after e is substituted for i and E_m for I_m in this equation.

The Fourier series for the output current at the terminals of a full-wave rectifier is

$$i = 0.636 I_m - 0.424 I_m \cos 2\omega t - 0.085 I_m \cos 4\omega t - \cdots \text{ amperes} \quad (1\text{-}27)$$

Substituting voltage symbols for current symbols gives the equation for instantaneous values of a fully rectified sinusoidal voltage wave.

The mathematical procedure by which the d-c term and the coefficients of the component terms are evaluated is explained in reference books on calculus and in books on alternating-current theory.* It should be understood that Equation (1-22) is basic not only for wave shapes containing only curved lines but also for triangular, square, rectangular, and all other wave shapes.

The effective value of a complex wave is obtained by a simple square-root process. For a current wave,

$$I_{\text{eff}} = \left(I_0^2 + \frac{I_{m1}^2 + I_{m2}^2 + I_{m3}^2 + I_{m4}^2 + \cdots + I_{mn}^2}{2} \right)^{\frac{1}{2}} \quad (1\text{-}28)$$

* H. Alex. Romanowitz, *op. cit.*, pp. 501–510.

The I values for the wave of Equation (1-26) are:

$$I_0 = 0.318I_m \qquad I_{m3} = 0$$
$$I_{m1} = 0.500I_m \qquad I_{m4} = -0.0424I_m$$
$$I_{m2} = 0.212I_m$$

I_{m1} is the maximum instantaneous value of the fundamental component of the complex wave represented by the series. I_{m2}, I_{m3}, I_{m4} are the maximum instantaneous values of the corresponding components.

The effective value of a complex *voltage wave* is computed in exactly the same way. It is important to understand that the terms under the radical sign represent the maximum instantaneous values of all of the components of the complex wave. I_{m3} is zero in the above example. Components beyond I_{m4} are small enough to be neglected in the computation for this particular wave form.

If a complex wave is represented by a series that has both sine and cosine terms *of the same frequency*, the terms may be combined into a single term whose instantaneous maximum value is the square root of the sum of the squares of the separate magnitudes. For example, a pair of terms such as $A_1 \sin \omega t + B_1 \cos \omega t$ may be combined into $\sqrt{A_1{}^2 + B_1{}^2}\, \sin(\omega t + \alpha)$ where $\alpha = \tan^{-1}(B_1/A_1)$. If the series represents a current wave, then $I_{m1} = \sqrt{A_1{}^2 + B_1{}^2}$, and $I_{m1}{}^2 = A_1{}^2 + B_1{}^2$ in the equation for the effective value. Therefore, $A_1{}^2$ and $B_1{}^2$ may be used as separate terms of Equation (1-28). We must be sure, therefore, that we include the coefficients of all the sine and cosine terms in the Fourier series up to the point where their magnitudes are small enough to be neglected.

The effective value of the output current of the full wave rectifier represented by Equation (1-27) is

$$I_{\text{eff}} = \left[(0.636I_m)^2 + \frac{(0.424I_m)^2 + (0.085I_m)^2}{2} \right]^{\frac{1}{2}} = 0.706I_m$$

where I_m is the maximum instantaneous value of the whole complex wave.

1.15   Phase Angle; Phasor Diagrams

Consider that a sine-wave voltage E (effective value) is applied to (*a*) a pure resistance, (*b*) a resistance and an inductance in series, or a coil with resistance and inductance, and (*c*) a capacitor and a resistance in series. In the *resistance* the current will be in phase with E (Fig. 1-14*a*). The *coil*

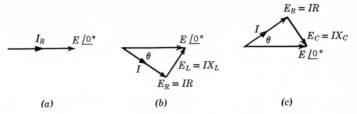

Fig. 1-14   Phasor diagrams for simple circuits.

current will lag E by an angle whose tangent is X_L/R (Fig. 1-14b). The capacitor current will lead E by an angle whose tangent is X_C/R (Fig. 1-14c).

The current in a *pure inductance* would lag 90 degrees behind the voltage applied to the inductance. The current through a *pure capacitance* leads, by 90 degrees, the voltage applied to the capacitance. A lagging current passes through its crest positive, and negative, values *after* the voltage has done so; a leading current passes through these instantaneous values *before* the voltage has done so.

1.16   Time Constant

It is common in electronic circuits for a capacitor to be charged, and discharged, through a resistance, and for a coil and a resistance in series to have increasing or decreasing voltage applied to them.

When a *constant* voltage E is suddenly applied to a *series combination* of *resistance R* and *capacitance C*, current will start flowing immediately and its instantaneous value will be given by*

$$i = \frac{E}{R}\epsilon^{-t/RC} \tag{1-29}$$

in which ϵ is the base of the system of natural logarithms. If a *charged* capacitor, having capacitance C, is suddenly connected to a resistance R, the current at any instant will be given by the same equation. Figure 1-15 shows *current* and voltage plotted against *time* for these conditions. Obviously, the larger the capacitance, the longer it would take to charge the capacitor to full voltage E and the longer it would take to discharge it. In both cases the current will eventually become zero, i.e., stop flowing. At $t = 0$, $i = E/R$, the initial value in both cases.

---

* Derivation is given in the Appendix.

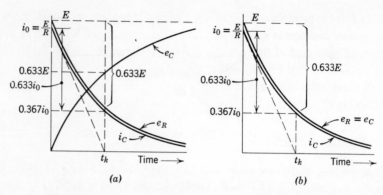

Fig. 1-15 (a) Voltage and current curves for charging of capacitor through a resistor. (b) Voltage and current curves for discharge of capacitor through a resistor.

If $t = RC$ in Equation (1-29), the exponent becomes -1 and

$$i = \frac{E}{R\epsilon} = \frac{E}{2.71828R} = 0.367 \frac{E}{R} \tag{1-30}$$

At $t = \infty$, $i = 0$ and the current has gone through 100 per cent of its change.

Note—and this is *most important*—that at $t = RC$ the current has gone through 0.633, or 63.3 per cent, *of its total change*. RC is called the *time constant* of the circuit. It is the time *in seconds* which is required for charging or discharging current *to go through* 63.3 per cent *of the change* from initial value to final value.

When a constant voltage E is applied to a coil and a resistance R in series, the current is given by*

$$i = \frac{E}{R}(1 - \epsilon^{-(R/L)t}) \tag{1-31}$$

In L/R seconds the current goes through 63.3 per cent of its total change (from zero initial value to E/R final value). L/R is the *time constant*, t_k, in seconds. It is the value required to make the exponent become -1.

If an inductance L, carrying a current, is suddenly connected across a resistance R so that all the energy in the inductance is dissipated in R, the instantaneous current is given by*

$$i = \frac{E}{R}\epsilon^{-(R/L)t} \tag{1-32}$$

* Derivation is given in the Appendix.

in which E is the product of the current and the resistance. As in the previous case, the *time constant* is L/R and it is the time required for the current to go through 63.3 per cent of the total change it will experience in falling from its initial value of E/R at $t = 0$ to zero at $t = \infty$.

In all these cases the current reaches its final value in a matter of seconds or a fraction of a second. Infinite time is only *theoretically* indicated.

1.17 Mutual Inductance

When current flows in a coil which is close enough to a second coil that magnetic flux produced by the current links some of the turns of the second coil, *mutual inductance* is said to exist between the two coils. M_{12} is the symbol for the mutual inductance from coil 1 to coil 2, and M_{21} represents the mutual inductance from coil 2 to coil 1. Note that, if coil 2 is near enough to coil 1 to receive some of its flux, then coil 1 will receive some of the flux produced by the current in coil 2.

Consider that i_1 A in coil 1 produce $N_1\phi_1$ flux linkages in coil 1 and $N_2\phi_{12}$ flux linkages in coil 2. The *self-inductance L_1* of coil 1 is

$$L_1 = \frac{N_1\phi_1}{i_1} \text{ H} \tag{1-33}$$

and the mutual inductance M_{12} from coil 1 to coil 2 is

$$M_{12} = \frac{N_2\phi_{12}}{i_1} \text{ H} \tag{1-34}$$

Let i_2 A in coil 2 produce $N_2\phi_{22}$ flux linkages in coil 2 and $N_1\phi_{21}$ flux linkages in coil 1. The self-inductance of coil 2 is

$$L_2 = \frac{N_2\phi_2}{i_2} \text{ H} \tag{1-35}$$

and the mutual inductance M_{21} from coil 2 to coil 1 is

$$M_{21} = \frac{N_1\phi_{21}}{i_2} \text{ H} \tag{1-36}$$

In these equations,

ϕ_1 is the flux due to i_1;

ϕ_{12} is the part of ϕ_1 which reaches over and links turns of coil 2 also;

ϕ_2 is the flux due to i_2;

ϕ_{21} is the part of ϕ_2 which reaches back and links turns of coil 1 also.

First consider an ideal case in which *all* the flux produced by each current links *all* the turns of both coils. In this situation $\phi_1 = \phi_{12}$ and $\phi_2 = \phi_{21}$, and we may write

$$L_1 L_2 = M_{12} M_{21} = M^2 \tag{1-37}$$

and

$$M = \sqrt{L_1 L_2}\ \text{H} \tag{1-38}$$

which is the theoretical maximum possible mutual inductance.

However, there is always some *leakage flux* produced by each current that does not link *any turns* of the other coil. Nevertheless, it is always true that $M_{12} = M_{21}$ if the magnetic permeabilities of the flux paths remain constant.

The ratio of the actual mutual inductance to the theoretical maximum is called the *coefficient of coupling*, and its symbol is *K*.

$$K = \frac{M}{\sqrt{L_1 L_2}} \tag{1-39}$$

1.18   Mutual Inductance in a Circuit

There is an *EMF of mutual inductance* in each coil which must be taken into account when analyzing a circuit where coupling between coils is present. This is done by writing a term in the voltage equation. In the situation of Fig. 1-16, where r_1 and r_2 represent the resistances of the coils, the voltage equation for the primary circuit may be written

$$E = i_1 r_1 + L_1 \frac{\Delta i_1}{\Delta t} - M \frac{\Delta i_2}{\Delta t} \tag{1-40}$$

The minus sign must be used with the last term because it represents an EMF induced in the primary coil by a flux (ϕ_{21}) *which is increasing in the opposite direction to that of the flux produced by* i_1.

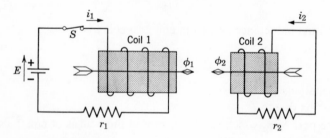

Fig. 1-16   Switch *S* has just been closed. The currents are increasing and their fluxes are bucking each other.

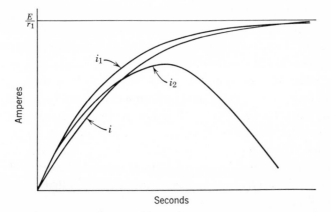

Fig. 1-17   Currents i_1 in primary coil and i_2 in secondary coil of Fig. 1-16. If the secondary (coil 2) were not present, the primary current would behave as shown by the curve (i).

For the secondary circuit the voltage equation may be written

$$0 = i_2 r_2 + L_2 \frac{\Delta i_2}{\Delta t} - M \frac{\Delta i_1}{\Delta t} \qquad (1\text{-}41)$$

Another way to justify the minus signs of the terms with mutual inductance is to recognize that the resulting voltage must be due to the difference of the two inductive effects *because of the opposing fluxes.* Of course, the solution of these two simultaneous equations is beyond the scope of this book. It can be done when the incremental forms are replaced by di_1/dt and di_2/dt and when constant values are used for the applied voltage, the resistances, and the inductances. However, some facts about the general behavior of the currents can be ascertained by considering conditions at the instant the switch is closed. At $t = 0$ both currents are zero and

$$E = L_1 \left.\frac{\Delta i_1}{\Delta t}\right|_{t=0} - M \left.\frac{\Delta i_2}{\Delta t}\right|_{t=0} \qquad (1\text{-}42)$$

$$0 = L_2 \left.\frac{\Delta i_2}{\Delta t}\right|_{t=0} - M \left.\frac{\Delta i_1}{\Delta t}\right|_{t=0} \qquad (1\text{-}43)$$

Solving these simultaneously,

$$\left.\frac{\Delta i_1}{\Delta t}\right|_{t=0} = \frac{EL_2}{L_1 L_2 - M^2} \, \text{A/s} \qquad (1\text{-}44)$$

$$\left.\frac{\Delta i_2}{\Delta t}\right|_{t=0} = \frac{EM}{L_1 L_2 - M^2} \, \text{A/s} \qquad (1\text{-}45)$$

M^2 is always less than the product L_1L_2. Why? Therefore these initial rates of change are positive. It is also evident that the *initial rate of change* of the current in coil 1 is greater than it would be if coil 2 were not present. That is, with $M = 0$ the rate of change in Equation (1-42) becomes E/L_1, a smaller value than when M is present. The curves of Fig. 1-17 show variations of current with time for both the mutual-induction case (i_1 and i_2) and the single-coil case (i) when the secondary coil is not present. The final steady-state value of i_2 is zero. Why? The denominators of Equations (1-44) and (1-45) are equal and may be written $(1 - K^2)L_1L_2$ owing to the relation given in Equation (1-39). Note that the ratio of L_2 to M determines whether the initial rate of change of i_1 is greater than, less than, or equal to that of i_2.

1.19   Measurement of Inductance

The self-inductance of a coil may be measured with an inductance bridge. The mutual inductance of two coils may also be measured with the bridge by the following procedure. Connect the two coils in series and measure their combined inductance. Then reverse the connections to one of the coils and measure their combined inductance again. In one of these two situations the coil fluxes will *aid* each other, and in the other the fluxes will *oppose* each other. When they aid each other, $2M$ is added to the sum of their self-inductances (case A), and, when they oppose (buck) each other, $2M$ is subtracted from the sum. Consequently, using L_A and L_B for the measured inductances in the corresponding cases, we may write

$$L_A = L_1 + L_2 + 2M \tag{1-46}$$

$$L_B = L_1 + L_2 - 2M \tag{1-47}$$

Solving for M,

$$M = \tfrac{1}{4}(L_A - L_B) \tag{1-48}$$

The mutual inductance of two coils depends on the positions they occupy with respect to each other. Care must therefore be taken not to move them when making measurements, especially when reversing the connections.

1.20 ' Alternating-Current Problem Solutions Using Complex Numbers

Several circuits will now be analyzed with the use of complex numbers to represent voltages, currents, and impedance. It may be well to point out here that *these are not vector quantities*, because they do not have *direction* associated with them. They are not representable in *directional component form* as are vector quantities such as *force, velocity, electric field intensity,* and *flux density.*

Consider a coil with 4 Ω resistance and 3 Ω inductive reactance connected to an a-c supply $E = 10 \underline{/0°}$ V effective (root-mean-square) value.

The coil impedance is $Z = 4 + j3$ Ω, and the effective value of the current is

$$I = \frac{E}{Z} = \frac{10 + j0}{4 + j3} \text{ A} \tag{1-49}$$

Rationalizing the denominator and dividing,

$$I = \frac{10 + j0}{4 + j3} \times \frac{4 - j3}{4 - j3} = \frac{40 - j30}{16 + 9} = 1.6 - j1.2 \text{ A}$$

The current value is

$$|I| = \sqrt{1.6^2 + 1.2^2} = \sqrt{4} = 2 \text{ A}$$

The phase angle of the current is the angle θ, where

$$\tan \theta = \frac{-1.2}{1.6} = -0.75$$

$$\theta = -36.9°$$

This shows that the current *lags E* by 36.9°.

The *power factor* is the *cosine of the angle between the voltage and current.*

$$\text{Power factor} = \cos \theta = \frac{1.6}{2} = 0.8 \text{ or } 80\%$$

The *polar form* may be used to advantage in certain places. Since $Z = \sqrt{4^2 + 3^2} = 5\underline{/36.9°}$, we can write

$$I = \frac{E}{Z} = \frac{10\underline{/0°}}{5\underline{/36.9°}} = 2\underline{/-36.9°} \text{ A}$$

Power is obtained by using the formula I^2R or $E_{\text{eff}}I_{\text{eff}} \cos \theta$. In complex number terms, power may be computed by adding the product of the real components to the product of the imaginary components. There happens to be no imaginary component of E in this problem.

$$P = I^2R = 2^2 \times 4 = 16 \text{ W}$$

$$P = EI \cos \theta = 10 \times 2 \times 0.8 = 16 \text{ W}$$

$$P = E_{\text{real}}I_{\text{real}} + E_{\text{imag}}I_{\text{imag}} = 10 \times 1.6 + 0 \times 1.2 = 16 \text{ W}$$

1.21   Series-Parallel Circuit Analysis

In order to illustrate the use of *admittance*, which is the reciprocal of *impedance*, and bring in *conductance* and *susceptance*, we shall solve a series-parallel circuit using the complex number method.

Refer to Fig. 1-18. Calling the admittance of the upper parallel branch Y_3, and using $Y = 1/Z$,

$$Y_3 = \frac{1}{4 + j2} \times \frac{4 - j2}{4 - j2} = \frac{4 - j2}{20} = 0.2 - j0.1 \text{ mho}$$

The *conductance* is 0.2 mho, and the *susceptance* is 0.1 mho. For the lower parallel branch,

$$Y_2 = \frac{1}{1 - j5} \times \frac{1 + j5}{1 + j5} = \frac{1 + j5}{26} = 0.0384 + j0.192 \text{ mho}$$

The conductance is 0.0384 mho and the susceptance is 0.192 mho.

The admittance between points B and C is

$$Y_{BC} = 0.2 - j0.1 + 0.0384 + j0.192 = 0.2384 + j0.092 \text{ mho}$$

The impedance of the parallel branch is

$$Z_{BC} = \frac{1}{Y_{BC}} = \frac{1}{0.2384 + j0.092} \times \frac{0.2384 - j0.092}{0.2384 - j0.092}$$

$$= \frac{0.2384 - j0.092}{0.0568 + 0.00846} = 3.65 - j1.41 = 3.91\underline{/-21.1°} \text{ }\Omega$$

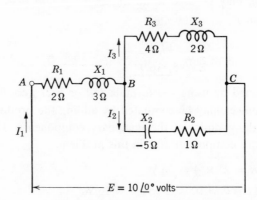

Fig. 1-18   Series-parallel circuit.

The total impedance, Z_T, of the circuit is the sum of Z_{BC} and Z_{AB}, the impedance of the series branch:

$$Z_T = 3.65 - j1.41 + 2 + j3 = 5.65 + j1.59 \ \Omega$$

The input current, I_1, is

$$I_1 = \frac{E_T}{Z_T} = \frac{10 + j0}{5.65 + j1.59} \times \frac{5.65 - j1.59}{5.65 - j1.59} = \frac{56.5 - j15.9}{31.923 + 2.528}$$

$$I_1 = \frac{56.5 - j15.9}{34.45} = 1.64 - j0.461 = 1.71 \underline{/\theta_1} \ \text{A}$$

The phase angle of I_1 is $\theta_1 = \tan^{-1}(-0.461/1.64) = \tan^{-1} - 0.281$, from which $\theta_1 = -15.7°$

$$E_{AB} = I_1 Z_{AB} = (1.64 - j0.461)(2 + j3) = 4.66 + j4 = 6.14 \underline{/40.6°} \ \text{V}$$

$$E_{BC} = I_1 Z_{BC} = (1.64 - j0.461)(3.65 - j1.41) = 5.34 - j4 = 6.67 \underline{/-36.8°} \ \text{V}$$

These are seen to add and give $10 + j0$, the applied voltage E.

The currents in the parallel branches will now be determined:

$$I_2 = \frac{E_{BC}}{Z_2} = \frac{5.34 - j4}{1 - j5} \times \frac{1 + j5}{1 + j5} = \frac{25.34 + j22.7}{26}$$

$$= 0.975 + j0.873 = 1.31 \underline{/41.9°} \ \text{A}$$

$$I_3 = \frac{E_{BC}}{Z_3} = \frac{5.34 - j4}{4 + j2} \times \frac{4 - j2}{4 - j2} = \frac{13.36 - j26.28}{20}$$

$$= 0.668 - j1.334 = 1.49 \underline{/-63.4°} \ \text{A}$$

$$I_2 + I_3 = 1.643 - j0.461$$

which is the value of I_1 above. Figure 1-19 is the phasor diagram for this circuit.

Power is computed, using complex number forms, by multiplying together the real components of current and voltage and adding to this the product of the imaginary components. We shall compute the power in each branch separately, add the values, and compare the sum with the value obtained from the total voltage and input current. The series-branch power is denoted by P_1 and the parallel-branch powers by P_2 and P_3.

$$P_1 = 4.66 \times 1.64 + 4 \times (-0.461) = \underline{5.79 \ \text{W}}$$
$$P_2 = 5.34 \times 0.975 + (-4) \times 0.873 = \underline{1.71 \ \text{W}}$$
$$P_3 = 5.34 \times 0.668 + (-4)(-1.334) = \underline{8.90 \ \text{W}}$$
$$\text{Total power in circuit, } P_T = \overline{\underline{16.40 \ \text{W}}}$$

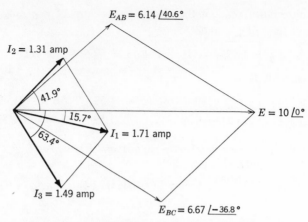

Fig. 1-19   Phasor diagram for circuit of Fig. 1-18.

The total circuit power is also computed by using total voltage and current:

$$P_T = 10 \times 1.64 + 0 \times 0.461 = 16.4 \text{ W}$$

Also, using I^2R,

$$P_T = (1.71)^2 \times 5.65 = 16.4 \text{ W}$$

It should be noticed that $+$ and $-$ signs of the components must be observed.

1.22   Series Resonance

In a series circuit consisting of inductance, capacitance, and resistance, it is found that, when an alternating voltage of *constant magnitude and frequency* is applied, the *current increases* as the *difference* between the inductive reactance and the capacitive reactance is *decreased*. If the capacitive reactance is the larger, it can be reduced by increasing the capacitance until the values of the two reactances in ohms are made equal. By reducing the capacitance, its reactance can be increased to equal a larger inductive reactance.

For fixed values of L and C, a frequency value exists which will produce series resonance, in which case the reactances are equal:

$$2\pi f L = \frac{1}{2\pi f C} \tag{1-50}$$

$$f = \frac{1}{2\pi\sqrt{LC}} \tag{1-51}$$

Obviously, for a fixed frequency a value of L (henrys) or of C (farads) is given by Equation (1-51) if one of them is known.

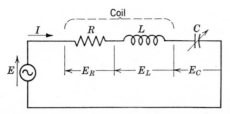

Fig. 1-20   Series circuit in which X_C can be made equal to X_L at a fixed frequency.

EXAMPLE 1-1.   An example of series resonance will be explained with reference to Fig. 1-20. $L = 200\,\text{mH}$, $C = 0.05\,\mu\text{F}$, $R = 100\,\Omega$, $E = 10\,\text{V}$ rms. The frequency of E required for series resonance will be determined, as will values of I, E_R, E_L, and E_C.

Solution.

$$f = 1/(2\pi\sqrt{LC}) = 1/(2\pi\sqrt{0.200 \times 0.05 \times 10^{-6}}) = 1{,}591\,\text{Hz}$$
$$I = E/Z = E/R \quad \text{since } X_L = X_C$$
$$I = 10/100 = 0.1\,\text{A}$$
$$X_C = 1/(2\pi \times 1{,}591 \times 0.05 \times 10^{-6}) = 2{,}000\,\Omega$$
$$X_L = 2\pi \times 1{,}591 \times 0.2 = 2{,}000\,\Omega$$
$$E_R = 0.1 \times 100 = 10\,\text{V}$$
$$E_L = IX_L = 0.1 \times 2{,}000 = 200\,\text{V}$$
$$E_C = IX_C = 0.1 \times 2{,}000 = 200\,\text{V}$$

Notice that, although only 10 V are applied to the circuit, 200 V exist across L and also across C. This is a "resonance rise in voltage" of a 20-to-1 ratio.

The impedance diagram and the phasor diagram are shown in Fig. 1-21.

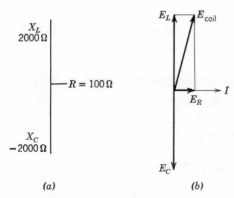

(a)   (b)

Fig. 1-21   Diagrams for circuit of Fig. 1-19. (*a*) Impedance diagram; (*b*) phasor diagram.

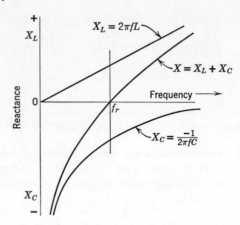

Fig. 1-22   Reactances of an inductance L and a capacitance C plotted against frequency. They are equal in magnitude at f_r, the resonant frequency.

If L and R were values for a coil, the voltage across its terminals would be as shown in the phasor diagram.

When the frequency of the voltage applied to a series circuit composed of L, R, and C is below resonant frequency, the current that flows leads the applied voltage. This means that the total impedance of the circuit is capacitive. When the frequency of the applied voltage is above resonant frequency, the total impedance is inductive; and so for all frequencies greater than resonant frequency the current lags the applied voltage. Since the net reactance of a coil and a capacitor connected in series is given by their algebraic sum,

$$X = X_L + X_C$$

in which X_C is *negative*. It is readily seen in Fig. 1-22 that X will be negative (capacitive) below resonance, and positive (inductive) above resonance. At resonant frequency, X_L and X_C cancel each other, and the circuit is resistive, as has been shown. The current is, in that case, in phase with the applied voltage.

1.23   Sharpness of Resonance

The curves of Fig. 1-23 show the effects of changing the resistance in a series-resonant circuit. When the resistance is increased the current at resonance decreases, as expected, but the curve also loses its sharpness in the region of the peak. The smaller the resistance can be made, the *sharper the peak*, and the *larger* will be the *changes in current* for small changes in frequency. That is, the circuit will *discriminate* in favor of currents whose frequencies are in the neighborhood of the resonant frequency.

The degree of sharpness, or of discrimination, is expressed in terms of half-power points on the current vs. frequency curve. The sharpness of resonance is also denoted by Q, the *figure of merit of the inductance coil* used in a resonant circuit. Q is defined as the *ratio of the coil's inductive reactance to its resistance*:

$$Q = \frac{X_L}{R} \tag{1-52}$$

The current at resonance is given by

$$I_r = \frac{E}{R} \tag{1-53}$$

but

$$R = \frac{X_L}{Q}$$

so that

$$I_r = \frac{E}{X_L} Q \tag{1-54}$$

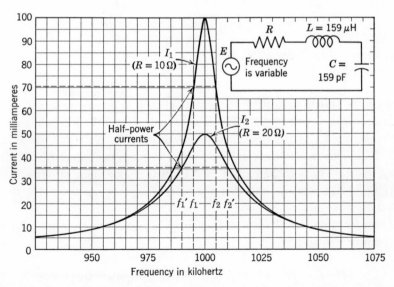

Fig. 1-23　Resonance curves for series R, L, C circuit shown; currents and half-power points when $R = 10\ \Omega$ and when $R = 20\ \Omega$.

1.24   Half-Power Currents and Frequencies

Since power is given by I^2R, one-half the maximum power in the circuit is obtained when I is decreased to $I_{p/2}$ so that

$$(I_{p/2})^2R = \tfrac{1}{2}I_r^2R \tag{1-55}$$

$$I_{p/2} = \frac{I_r}{\sqrt{2}} = 0.707I_r \tag{1-56}$$

Let us now calculate the two values of frequency, at which the current becomes 0.707 of the resonant value. Let $Z_{p/2}$ be the impedance of the circuit at the half-power point:

$$I_{p/2} = \frac{E}{Z_{p/2}} = 0.707\frac{E}{R} \tag{1-57}$$

From this,

$$R = 0.707Z_{p/2} \tag{1-58}$$

This is true when R equals the net reactance $(X_L - X_C)$, as is obvious in a phasor diagram for Fig. 1-23. Since $X = R$ at these points,

$$Z_{p/2} = \sqrt{R^2 + X^2} = \sqrt{2R^2} = \sqrt{2}R$$

$$R = \frac{Z_{p/2}}{\sqrt{2}} = 0.707Z_{p/2} \tag{1-59}$$

It can be shown that, *when the voltage applied to a series-resonant circuit is kept constant while its frequency is increased*, X_L will be increasing *at the same rate as X_C* as the frequency passes through its resonance value. See Fig. 1-22. This means that, during a small increase in frequency, Δf, near the f_r value, ΔX_L is accompanied by ΔX_C of the same value in ohms. Since $\Delta X_L = 2\pi\,\Delta f L$, we can say that the total *change in reactance* due to Δf is $2\,\Delta X_L = 4\pi\,\Delta f L$.

Consider now the change in frequency required to go from f_r to the half-power frequencies:

$$\Delta f = f_r - f_1, \quad \text{where } f_1 \text{ is the } lower \text{ half-power frequency}$$

$$\Delta f = f_2 - f_r, \quad \text{where } f_2 \text{ is the } upper \text{ half-power frequency}$$

Also, since X changes from *zero at f_r* to $X = R$ at these points,

$$4\pi(f_r - f_1)L = R \tag{1-60}$$

$$4\pi(f_2 - f_r)L = R \tag{1-61}$$

$$f_r - f_1 = \frac{R}{4\pi L} \tag{1-62}$$

$$f_2 - f_r = \frac{R}{4\pi L} \tag{1-63}$$

Adding the last two equations gives

$$f_2 - f_1 = \frac{R}{2\pi L} = \Delta f \tag{1-64}$$

This shows that, *the smaller the ratio of R to L, the narrower will be the frequency band between the half-power points*. The *ratio* of this *bandwidth* to the *resonant frequency* is a measure of the sharpness of resonance of the circuit.

$$\frac{\Delta f}{f_r} = \frac{f_2 - f_1}{f_r} = \frac{R}{2\pi f_r L} = \frac{1}{Q} \tag{1-65}$$

It must be pointed out that R is the total resistance of the series-resonant circuit, including the resistance of a generator if one is present.

Obviously, the larger Q is made, the narrower will be the frequency band between the half-power points and the sharper will be the circuit. The circuit effectively rejects currents whose frequencies are outside the "pass band," as $f_2 - f_1$ may be called.

1.25 Parallel Resonance (Antiresonance)

Consider the parallel circuit of Fig. 1-24. A phasor diagram for it is shown in Fig. 1-25. The phasor diagram represents the situation in which the frequency of the applied voltage E has been adjusted so that the input current I to the parallel circuit is *in phase with the applied voltage*. This is the *unity power factor condition*. It will be shown that, to bring this condition about, X_L must be made *practically equal to* X_C in ohms value, and X_L must be *very much larger than* R. It will also be found that under these conditions (i.e., at parallel resonance) the whole circuit will present to the applied voltage source an impedance which is purely resistive and which has a value *much larger* than the resistance R of the inductive branch. Assuming the series resistance of the capacitor to be zero helps to simplify the analysis a great deal; it is negligible in practical circuits.

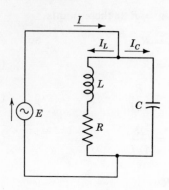

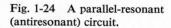

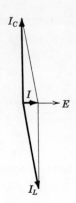

Fig. 1-24   A parallel-resonant   Fig. 1-25   Phasor diagram
(antiresonant) circuit.   for Fig. 1-23.

It was found in the analysis of parallel circuits that the admittances of the parallel branches may be added. The total admittance of the circuit of Fig. 1-24 is

$$Y_T = Y_L + Y_C = \frac{1}{Z_L} + \frac{1}{Z_C} \tag{1-66}$$

$$Y_T = \frac{1}{R + jX_L} + \frac{1}{-jX_C} = \frac{R + jX_L - jX_C}{(R + jX_L)(-jX_C)}$$

$$Z_T = \frac{1}{Y_T} = \frac{X_L X_C - jRX_C}{R + j(X_L - X_C)} \tag{1-67}$$

Rationalizing the denominator,

$$Z_T = \frac{X_L X_C - jRX_C}{R + j(X_L - X_C)} \times \frac{R - j(X_L - X_C)}{R - j(X_L - X_C)}$$

$$= \frac{RX_L X_C - jR^2 X_C - jX_L^2 X_C - RX_L X_C + jX_L X_C^2 + RX_C^2}{R^2 + (X_L - X_C)^2}$$

$$Z_T = \frac{RX_C^2}{R^2 + (X_L - X_C)^2} + j\frac{X_L X_C^2 - X_L^2 X_C - R^2 X_C}{R^2 + (X_L - X_C)^2} \tag{1-68}$$

Proper choice of X_L or X_C, when one of them is known, can be made so that the j term is zero. This will result in the total impedance being *resistive only*; this means *unity power factor* operation. When the reactive component of the total impedance is zero,

$$X_L X_C^2 - X_L^2 X_C - R^2 X_C = 0$$

$$R^2 = (X_C - X_L)X_L \tag{1-69}$$

Although this is precisely the requirement for unity power factor, in high-frequency circuits such as those in radio and television X_L is very much larger than R. This is expressed as

$$X_L \gg R$$

With this in mind, it will now be shown that Equation (1-69) can be reduced for practical applications to $X_C = X_L$. From Equation (1-69),

$$R^2 = X_C X_L - X_L^2$$

$$X_L^2 + R^2 = X_C X_L \qquad (1\text{-}70)$$

If $X_L \gg R$, then R^2 may be dropped in comparison with X_L^2:

$$X_L^2 = X_C X_L$$

$$X_L = X_C \qquad (1\text{-}71)$$

To show the common sense of this procedure, suppose that

$$X_L = 2{,}000 \ \Omega \quad \text{and} \quad R = 25 \ \Omega$$

Using the exact equation (1-69),

$$X_C = \frac{R^2}{X_L} + X_L$$

$$X_C = \frac{625}{2{,}000} + 2{,}000 = 2{,}000.3$$

Therefore, when $X_L \gg R$, X_C may be taken equal to X_L to produce unity power factor in a parallel-resonant circuit.

The impedance of the parallel circuit at *resonant frequency* is a pure resistance given by

$$Z_r = \frac{R X_C^2}{R^2 + (X_L - X_C)^2} = \frac{X_C^2}{R} \qquad (1\text{-}72)$$

since $X_L = X_C$. It is possible to express this impedance in terms of R, L, and C by expressing X_C^2 as $X_L X_C$:

$$Z_r = \frac{X_L X_C}{R} = \frac{2\pi f_r L}{R} \times \frac{1}{2\pi f_r C} = \frac{L}{RC} \ \Omega \qquad (1\text{-}73)$$

The resonant frequency of a parallel circuit with small resistance is given to a high degree of accuracy by the same equation as that used for series resonance, namely,

$$f_r = \frac{1}{2\pi\sqrt{LC}} \qquad (1\text{-}74)$$

1.26   Parallel Resonance with Resistance in Both Branches

It is easily proved in more advanced books that, if there is resistance in *both branches* of a circuit like that in Fig. 1-24 the impedance at parallel-resonant frequency is

$$Z_T = \frac{L}{C(R_L + R_C)} \tag{1-75}$$

where R_L is the resistance in the inductive branch, and R_C is the resistance in the capacitive branch. Note that this equation can be converted to one that contains only X_C or to one that contains only X_L, by first multiplying both numerator and denominator by $2\pi f_r$, thus getting X_L and X_C in the equation, and then writing either $X_L{}^2$ or $X_C{}^2$ in the numerator since they are equal $(1/2\pi f_r C = 2\pi f_r L)$ at resonance. That is, at unity power factor resonance,

$$Z_T = \frac{2\pi f_r L}{2\pi f_r C(R_L + R_C)} = \frac{X_L{}^2}{R_L + R_C} = \frac{X_C{}^2}{R_L + R_C} \tag{1-76}$$

1.27   *Q* of a Parallel-Resonant Circuit

The Q of a parallel-resonant circuit is the *ratio of the reactance of either branch to the total resistance of the two branches.*

$$Q_p = \frac{\omega L}{R} = \frac{1}{\omega CR} \text{ at resonance} \tag{1-77}$$

Since the total impedance Z_T is given by either reactance squared divided by the resistance, it is evident that

$$Z_T = \frac{\omega^2 L^2}{R} = \omega L Q_p = \frac{Q_p}{\omega C} \text{ at resonance} \tag{1-78}$$

In high-frequency circuits (radio frequencies and higher) the resistance of a coil increases with frequency, so that Q is more nearly constant than R. The fundamental equation of *any two-branch parallel circuit* gives the product of the branch impedances divided by their sum as the *equivalent impedance*, which we are now calling Z_T.

$$Z_T = \frac{Z_L Z_C}{Z_L + Z_C} = \frac{Z_L Z_C}{Z_S} \tag{1-79}$$

in which Z_L is the impedance of the L branch and Z_C is the impedance of the C branch, regardless of the resistance values in the branches. Z_S is the *series impedance of the two branches.*

In practical circuits, Q_p is often relatively high (50 or larger), and in such cases it is permissible to neglect the *resistance* values *in the numerator* of Equation (1-79) *but not in the denominator*. When we do this, the numerator becomes L/C, and, since $\omega_r L = 1/\omega_r C$, we can write

$$Z_L Z_C = (\omega_r L)^2 = \frac{1}{(\omega_r C)^2}$$

and the total impedance becomes

$$Z_T = \frac{(\omega_r L)^2}{Z_S} = \frac{1}{Z_S(\omega_r C)^2} \tag{1-80}$$

1.28   Maximum Power to an Alternating-Current Load

It has been seen that in a d-c circuit maximum power is delivered to a load when the load resistance is made equal to the generator resistance. In a-c circuits the load and generator *impedances* must be equal in magnitude and they must be *conjugates of each other*. *The conjugate of an impedance is another impedance of the same magnitude but opposite phase angle.*

$$R_1 + jX_1 \quad \text{and} \quad R_1 - jX_1$$

are conjugate impedances, R_1 having a single real value and X_1 having a single real value.

To prove the maximum power transfer theorem for the a-c case, we recall that power is given by I^2R, no matter what reactance values are present. When a source, or generator, which *develops* or *generates* a voltage E and has an *internal impedance* $Z_G = R_G + jX_G$ is connected to a *receiver circuit* of $Z_R = R_R + jX_R$, the current that flows is

$$I_R = \frac{E_G}{\sqrt{(R_G + R_R)^2 + (X_G + X_R)^2}} \underline{/\theta} \tag{1-81}$$

$$\tan \theta = \frac{X_G + X_R}{R_G + R_R}$$

The power, P_R, delivered to the receiver is

$$P_R = I_R^2 R_R = \frac{E_G^2 R_R}{(R_G + R_R)^2 + (X_G + X_R)^2} \tag{1-82}$$

With R_G and R_R already *fixed and equal*, the *power delivered to the receiver will be maximum when X_G and X_R are equal and of opposite sign*. This makes the *denominator a minimum*, and P_R will be a maximum. The conclusion is that, for a source of a-c power to deliver maximum power to a load, *the complex load impedance must be the conjugate of the generator impedance.* This means that, if $Z_G = R_G + jX_G$, then Z_R must be $R_R - jX_R$, and $R_R = R_G$, $X_R = X_G$. Also, if $Z_G = R_G - jX_G$, then Z_R must be $R_R + jX_R$, where $R_R = R_G$ and $X_R = X_G$.

1.29 Decibels

The *decibel* came into use as the result of a desire for some accurate means of measuring and comparing the power required to produce sounds. The human ear responds to sound intensity in a logarithmic fashion. It readily recognizes the doubling of sound power, but it can hardly distinguish two sounds of the same pitch when their powers differ by as much as 20 per cent.

The common logarithm of the ratio of two powers (they do not have to be sound powers) *is called the bel*. It was named after Dr. Alexander Graham Bell, who invented the telephone. Because the bel represents a relatively large power ratio, the *decibel*, which is *one-tenth of a bel*, is more often used.

The following equation is used to calculate decibels.

$$dB = 10 \log_{10} \frac{P_2}{P_1} \tag{1-83}$$

Suppose we supply a power P_1 to a system, which may be an amplifier or any other transmission network, and get out of it a power P_2. If $P_2 > P_1$, *the dB value obtained from Equation (1-83) comes out positive* and we have that much *dB gain*. If $P_2 < P_1$, the dB value comes out *negative* and we have a *dB loss* (i.e., a negative gain). When $P_2 < P_1$ we may use the expression

$$dB = 10 \log_{10} \frac{P_1}{P_2} \tag{1-84}$$

which will give us the same numerical answer as will Equation (1-83), but there will be no minus sign. We must, of course, remember that we have a loss when the output power (P_2) is less than the input power. All transmission networks not employing electron tubes or other generators have some losses and therefore deliver less power than they take in.

EXAMPLE 1-2. Compare 10 W with 5 W in decibel units.
Solution.

$$dB = 10 \log_{10} \frac{10}{5} = 10 \log_{10} 2 = 3$$

That is, 10 W power is 3 dB above 5 W power. Considering a loss from 10 W input to 5 W output, we have

$$dB = 10 \log_{10} \frac{5}{10} = 10 \log_{10} \tfrac{1}{2}$$

$$= 10(\log_{10} 1 - \log_{10} 2)$$

$$= 10(0 - 0.3) = -3$$

That is, 5 W is 3 dB lower than 10 W. *Every time a power value is halved its decibel value is decreased by 3. Every time a power value is doubled, its decibel value is increased by 3.* We are using 0.3 for the value of $\log_{10} 2$ instead of its five-place value, 0.30103. In practical calculations 0.3 is commonly used.

There is another advantage in using decibels to denote amplification. If one stage of an amplifier multiplies power by 8 and feeds a second stage that multiplies it again by 4, the overall power amplification is 32. That is, we multiply the power amplifications together to get the overall amplification. But when decibels are used they are simply added. The amplification of 8 corresponds to 9 dB, and the amplification of 4 corresponds to 6 dB, so that there is an overall amplification of $9 + 6 = 15$ dB. Checking this by using the overall power amplification of 32, we have

$$dB = 10 \log_{10} 32 = 10 \times 1.5 = 15$$

Zero power does not have a place on a decibel scale, because the power ratio would be either infinite or zero. The logarithm would be either plus or minus infinity.

It is convenient to choose a small value of power and to say that it corresponds to zero decibel or to zero decibel level. Then all other power values may conveniently be expressed in decibels with reference to this level. One *milliwatt* is usually used as a *zero decibel level*. Sometimes 6 mW is used. The power output of a microphone may be as low as 50 dB below 1 mW. Its output would then be rated at -50 dB referred to the 1-mW level. To determine the actual wattage output of such a microphone,

$$-50 = 10 \log_{10} \frac{P_2}{0.001}$$

$$\log_{10} \frac{P_2}{0.001} = -5$$

$$\frac{P_2}{0.001} = 10^{-5}$$

$$P_2 = 1 \times 10^{-8} \text{ W} = 0.01 \ \mu\text{W}$$

EXAMPLE 1-3.   One of the huge hydroelectric generators in the Hoover Dam power station delivers 82,500 kW. How many decibels is this above the 1-mW (0.001-W) level?

Solution.

$$dB = 10 \log_{10} \frac{82.5 \times 10^6}{0.001} = 10 \log_{10} (82.5 \times 10^9)$$

$$= 10(1.916 + 9) = 109.16$$

The amplification of voltage may also be expressed in decibels. One important restriction in doing so, however, is that *the voltages must exist across equal resistances.* Current amplifications may also be expressed in decibels *if the currents flow through equal resistances.*

If P_1 and P_2 exist in two equal resistances (R_1 and R_2), we may write

$$dB = 10 \log_{10} \frac{P_2}{P_1} = 10 \log_{10} \frac{E_2^2/R}{E_1^2/R}$$

$$= 10 \log_{10} \left(\frac{E_2}{E_1}\right)^2$$

which may be written

$$dB = 20 \log_{10} \frac{E_2}{E_1} \qquad (1\text{-}85)$$

For current amplification, we have

$$dB = 10 \log_{10} \frac{P_2}{P_1} = 10 \log_{10} \frac{I_2^2 R}{I_1^2 R}$$

$$= 20 \log_{10} \frac{I_2}{I_1} \qquad (1\text{-}86)$$

It is again emphasized that before using Equations (1-85) and (1-86) the student should be sure that the voltages are across, or the currents are in, equal resistances. These relations also hold for impedances, but with the corresponding restriction that the magnitudes and angles of the two impedances must be equal. That means that the impedances must be equal.

1.30   Power Level Measurement

We are now ready to discuss the use of the multimeter to measure power level. The a-c voltmeter circuit is altered slightly when the selector switch is moved to the OUTPUT position, thus converting the instrument to an "output meter."

The output meter measures the audio-frequency power level in a resistor when the positive and negative terminals are connected across it. If the resistor has a resistance of 600 Ω and happens to be carrying (dissipating) 0.001 W, the voltage across it will satisfy the equation $E^2/600 \, \Omega = 0.001$, from which $E = 0.774$ V. This will cause the meter needle to deflect to the 0-dB mark on the decibel scale. One milliwatt is the accepted reference level for the decibel scale; therefore that power level is 0 dB on the decibel scale. The decibel scale was placed on the meter face so that its divisions line up with the proper voltage readings that correspond to the specified decibel values when those voltages exist across 600 Ω. Therefore, when the voltage across 600 Ω is 0.774 V, the power in it is 0.001 W. The reason 600 Ω is used is that it is the input resistance of commonly used transmission lines in audio-frequency work. Six-hundred-ohm lines are also used, particularly in telephone practice.

The reading on the decibel scale corresponding to 2.38 V across 600 Ω will now be computed.

$$dB = 20 \log_{10} \frac{E_x}{0.774} \qquad (1\text{-}87)$$

$$dB = 20 \log_{10} \frac{2.38}{0.774} = 20 \log_{10} 3.07 = 9.74$$

On the actual instrument face, the 9.74-dB mark on the decibel scale is in line with the 2.38-V point on the 0–2.5 V scale. Any voltage value (E_x) on the 0–2.5 scale has a corresponding decibel value given by Equation (1-87). When other voltage scales of the meter are used, e.g., when using the 0–10 scale, a constant decibel value must be added to the reading on the decibel scale. The amount to be added is designated for each voltage scale by the pointer of the scale-changing switch.

It is interesting to study how to determine the decibel value to be added, when scales other than the 2.5-V scale are used. For example, when the 0–10 V scale is used, 12 dB must be added to the reading taken from the meter's single decibel scale. That figure is arrived at as follows.

First consider the meter to be set for use on the 2.5-V scale and that *the pointer indicates 0 dB*. On 600 Ω, this means 0.774 V. Furthermore, the pointer is located 31 per cent of the way upscale from zero: $0.774/2.5 = 0.31$.

How much voltage would have to be present across 600 Ω to give the same percentage of full-scale deflection when the 10-V scale is used? Obviously, 3.1 V. Now, how many decibels above 1 mW is the power in the 600-Ω resistor which has 3.1 V across it?

$$\frac{(3.1)^2}{600} = 0.016 \text{ W}$$

$$dB = 10 \log_{10} \frac{0.016}{0.001} = 12$$

That is, 12 dB must be added when the meter is used on the 10-V scale and set for decibel measurement. A shorter method of calculation would be to use Equation (1-87):

$$dB = 20 \log_{10} \frac{3.1}{0.774} = 12$$

This equation is applicable because the voltages were across the same resistance (600 Ω). The student should be able to show by calculation that 26 dB must be added to readings when the 50-V scale is used, and 40 dB must be added when the 250-V scale is used.

1.31 Volume Unit

In radio and television broadcasting, power level is expressed in terms of *volume units* (vu) and measured by a *volume level meter*. The zero reference is 1 mW. Accordingly,

$$vu = 10 \log_{10} \frac{P}{0.001} \text{ vu} \tag{1-88}$$

in which P is the amount of power measured, in watts.

The similarity of this equation to Equation (1-83) indicates that a power level at any *positive number of volume units above 1 mW* is a power level of the same number of *decibels above 1 mW, if 1 mW is used as a 0 dB reference*. For example, if P in Equation (1-88) is 0.01 W,

$$vu = 10 \log_{10} \frac{0.01}{0.001} = 10 \text{ vu}$$

A power level of 10 vu is evidently equal to 10 dB on the decibel scale *where 0 dB = 0.001 W*. A voltmeter calibrated to read vu will thus have its zero point at the position of the pointer when the meter is across a 600-Ω load carrying 1 mW of power. The *voltage* indication is evidently calculated as follows:

$$\frac{E^2}{600} = 0.001$$

$$E = \sqrt{0.6} = 0.774 \text{ V}$$

As mentioned earlier, 3 dB and 6 dB are commonly used values. When power is reduced one-half, the new level is spoken of as 3 dB down or 3 vu down. When it is reduced to one-fourth, its former value is 6 dB or 6 vu down. A doubling of power by an amplifier is a 3-dB (or 3-vu) gain, while a power gain of 6 dB (or 6 vu) means a quadrupling of power.

1.32   Determinants

The use of determinants in solving simultaneous equations is convenient and often a time saver. The chance of error is reduced, especially when there are three or more unknowns to be evaluated.

Determinant solutions are applicable to linear algebraic equations, which are those having variables (generally currents and voltages in our work) expressed in first powers only. That is, there are no squares, cubes, etc., or fractional exponents.

Rather than undertake the chore of learning rules by which *any number* of simultaneous equations may be solved by the use of determinants, we shall start with two equations with two unknowns. The *coefficients* of the unknowns are expressed in *second-order determinant* form.

Equations (1-5) and (1-6) in Section 1.3 may be written thus:

$$5I_1 - 3I_2 = 6$$

$$-3I_1 + 7I_2 = 4$$

A determinant form involving only the coefficients of I_1 and I_2 and the numbers on the right may be written for either I_1 or I_2. A determinant, represented by D, which always serves as a denominator, contains the coefficients of the variables. In this example,

$$D = \begin{vmatrix} 5 & -3 \\ -3 & 7 \end{vmatrix}$$

The arrows are not written after the procedure for evaluation D is learned, but they will be helpful here.

To evaluate D, we multiply *top left* by *bottom right* and subtract from this the product of *bottom left* and *top right*. The *denominator* of each expression used in evaluating an unknown current becomes

$$D = (5)(7) - (-3)(-3) = 26$$

Let us first evaluate I_1. To do so we write a form similar to the denominator determinant and use it as a numerator. The complete form, from which we evaluate I_1 becomes

$$I_1 = \frac{\begin{vmatrix} 6 & -3 \\ 4 & 7 \end{vmatrix}}{\begin{vmatrix} 5 & -3 \\ -3 & 7 \end{vmatrix}} = \frac{(6)(7) - (4)(-3)}{26} = \frac{54}{26} = 2\tfrac{1}{13} \text{ A}$$

Observe that the two constants (6 and 4) to the right of the equals signs are used instead of the two coefficients of I_1 and they take the places where those coefficients were written in D. The other elements of the numerator determinant for I_1 are the coefficients of I_2 and they occupy the same places as they do in D.

Now we use 26 for D when solving for I_2

$$I_2 = \frac{\begin{vmatrix} 5 & 6 \\ -3 & 4 \end{vmatrix}}{26} = \frac{20 + (18)}{26} = \frac{38}{26} = 1\tfrac{6}{13} \text{ A}$$

The convenience of solution by determinants is evident. Obviously no algebraic symbols are employed, just numerics.

1.33 Determinants with Three or More Unknowns

Before we examine the directions for solving third- and higher-order determinants, let us evaluate a determinant obtained from three equations containing three unknowns. It will be the denominator determinant obtained from the following equations.

$$2I_1 - 3I_2 + 2I_3 = 7$$

$$4I_1 + 5I_2 + I_3 = 45$$

$$-I_1 + 4I_2 - 3I_3 = -7$$

The denominator determinant that will be used to solve for the three currents represented in these equations by I_1, I_2, and I_3 is

$$D = \begin{vmatrix} 2 & -3 & 2 \\ 4 & 5 & 1 \\ -1 & 4 & -3 \end{vmatrix}$$

A process called expanding into minors, along the first row, gives the following form:

$$D = 2 \begin{vmatrix} 5 & 1 \\ 4 & -3 \end{vmatrix} - (-3) \begin{vmatrix} 4 & 1 \\ -1 & -3 \end{vmatrix} + 2 \begin{vmatrix} 4 & 5 \\ -1 & 4 \end{vmatrix}$$

Note that alternate $+$ and $-$ signs are used *in front of* the top-row numbers as these become coefficients of the minors.

From what we know about second-order determinant solutions,

$$D = 2(-15 - 4) + 3(-12 + 1) + 2(16 + 5) = -29$$

Now we undertake the task of writing the determinant to be put into the numerator of the expression that will give us the value of I_1. We use the same procedure we followed in the second-order case. We write the determinant for the numerator as we wrote one for D, *except* we replace the coefficients of I_1 in D with the constants to the right of the equals signs. This gives us

$$I_1 = \frac{1}{D} \begin{vmatrix} 7 & -3 & 2 \\ 45 & 5 & 1 \\ -7 & 4 & -3 \end{vmatrix}$$

$$= \frac{1}{-29} \left[7 \begin{vmatrix} 5 & 1 \\ 4 & -3 \end{vmatrix} - (-3) \begin{vmatrix} 45 & 1 \\ -7 & -3 \end{vmatrix} + 2 \begin{vmatrix} 45 & 5 \\ -7 & 4 \end{vmatrix} \right]$$

$$I_1 = \frac{1}{-29} [7(-15 - 4) + 3(-135 + 7) + 2(180 + 35)] = \frac{-87}{-29} = 3 \text{ A}$$

Expressing I_2 in determinant form, we replace the coefficients of I_2 in D with the constants to the right of the equal signs:

$$I_2 = \frac{1}{D} \begin{vmatrix} 2 & 7 & 2 \\ 4 & 45 & 1 \\ -1 & -7 & -3 \end{vmatrix}$$

$$= \frac{1}{-29} \left[2 \begin{vmatrix} 45 & 1 \\ -7 & -3 \end{vmatrix} - 7 \begin{vmatrix} 4 & 1 \\ -1 & -3 \end{vmatrix} + 2 \begin{vmatrix} 4 & 45 \\ -1 & -7 \end{vmatrix} \right]$$

$$I_2 = \frac{1}{-29} [2(-135 + 7) - 7(-12 + 1) + 2(-28 + 45)] = \frac{-145}{-29} = 5 \text{ A}$$

Expressing I_3 in determinant form, we replace the coefficients of I_3 in D

with the constants to the right of the equal signs:

$$I_3 = \frac{1}{D} \begin{vmatrix} 2 & -3 & 7 \\ 4 & 5 & 45 \\ -1 & 4 & -7 \end{vmatrix}$$

$$= \frac{1}{-29} \left[2 \begin{vmatrix} 5 & 45 \\ 4 & -7 \end{vmatrix} - (-3) \begin{vmatrix} 4 & 45 \\ -1 & -7 \end{vmatrix} + 7 \begin{vmatrix} 4 & 5 \\ -1 & 4 \end{vmatrix} \right]$$

$$I_3 = \frac{1}{-29} [2(-35 - 180) + 3(-28 + 45) + 7(16 + 5)] = \frac{-232}{-29} = 8 \text{ A}$$

We have seen by this example how a third-order determinant is converted into a series of second-order determinants by expanding across the top row into three minors.

The element of a row, which is written as if it were a coefficient of a minor, is called a *cofactor*; e.g., in the expansion of the last determinant D, the cofactors are 2, -3, and 7. The signs that must be used in front of the cofactors may be determined from $(-1)^{i+k}$ where i represents the number of the row in which the element appears and k represents the number of the column. Using this equation, an array of signs which may be used as a guide may be set up thus

$$
\begin{array}{llcccc}
(i = 1) & + & - & + & - & . & . \\
(i = 2) & - & + & - & + & . & . \\
(i = 3) & + & - & + & - & . & . \\
(i = 4) & - & + & - & + & . & . \\
& . & . & . & . & . & . \\
\end{array}
$$

$$(k = 1)(k = 2)(k = 3)(k = 4)$$

As a check, take $i = 2, k = 3$. $(-1)^{2+3} = -1$ which indicates that the minus sign located in the second row and the third column is correct.

We have shown the basic procedure for solving third-order determinants, but we must add that expansion into minors may be done downward *along a column* also. In doing this (as well as in expanding along a row) we may find it helpful to imagine a line drawn through the column and another drawn through the whole row containing the number when it is used as the coefficient of a minor.

Let us expand the denominator determinant D again, but use the middle column rather than the top row:

$$D = -(-3)\begin{vmatrix} 4 & 1 \\ -1 & -3 \end{vmatrix} + 5\begin{vmatrix} 2 & 2 \\ -1 & -3 \end{vmatrix} - 4\begin{vmatrix} 2 & 2 \\ 4 & 1 \end{vmatrix}$$

$$D = 3(-12 + 1) + 5(-6 + 2) - 4(2 - 8) = -29$$

Whenever a determinant contains one or more zeros as elements, it is desirable to expand along the row or column that contains the most zeros. This eliminates the minors that have the zeros as cofactors. Zero appears as an element when one of the variables is missing in an equation. The first equation in a problem at the end of this section will have zero at the place where the coefficient of I_3 is to be placed.

Let us expand a determinant containing zeros, first along a row and then downward along a column.

$$D = \begin{vmatrix} 20 & 0 & -15 \\ -10 & 0 & 5 \\ 25 & 15 & 10 \end{vmatrix}$$

Along the first row, remembering the sign sequence is $+, -, +,$

$$D = 20\begin{vmatrix} 0 & 5 \\ 15 & 10 \end{vmatrix} + (-15)\begin{vmatrix} -10 & 0 \\ 25 & 15 \end{vmatrix} = 20(-75) - 15(-150) = 750$$

Along the middle column, remembering the sign sequence is $-, +, -,$

$$D = -15\begin{vmatrix} 20 & -15 \\ -10 & 5 \end{vmatrix} = -15(100 - 150) = 750$$

We should add that whenever it is possible to divide a number which is a common factor into the *elements of a whole row or a whole column*, it should be done because this makes the elements smaller for computation. The devisor becomes a factor in the answer. For example, we may divide every row of the determinant by 5, and thus obtain 125 as a factor:

$$D = (5)(5)(5)\begin{vmatrix} 4 & 0 & -3 \\ -2 & 0 & 1 \\ 5 & 3 & 2 \end{vmatrix} = 125\left[-3\begin{vmatrix} 4 & -3 \\ -2 & 1 \end{vmatrix}\right]$$

$$= -375(4 - 6) = 750$$

DRILL PROBLEM

D1-7   Solve for I_1, I_2, and I_3 using determinants:

$$12I_1 + 8I_2 = 28$$
$$10I_1 - 5I_2 + 15I_3 = 45$$
$$9I_1 + 6I_2 - 3I_3 = 12$$

SUGGESTED REFERENCES

1. William H. Timbie and Francis J. Ricker, *Basic Electricity for Communications*, 2nd ed., Wiley, New York, 1958.

2. George V. Mueller, *Introduction to Electrical Engineering*, 3rd ed., McGraw-Hill, New York, 1957.

3. W. L. Everitt and G. E. Anner, *Communication Engineering*, 3rd ed., McGraw-Hill, New York, 1956.

PROBLEMS

1-1   Three resistances are connected in series across 120 V. $R_1 = 12\ \Omega$, $R_2 = 48\ \Omega$, $R_3 = 60\ \Omega$. (*a*) How much current does each resistance carry? (*b*) How much current is in the whole circuit?

1-2   What are the voltage drops across each of the resistances in Problem 1-1?

1-3   Six resistances are connected as shown in the network of Fig. P1-3 and supplied with direct current from the source at a voltage $E = 120$ V. (*a*) Working with resistance values only, calculate the equivalent resistance of, and total current taken by the whole network. (*b*) Calculate the conductance of each resistance unit

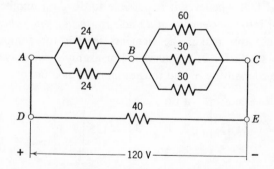

Fig. P1-3   Resistances are in ohms.

and, using these values, calculate the total conductance of the network. Compute the total current using this value and check with your answer to part (*a*). (*c*) How much current flows in the 60-Ω resistor?

1-4 Assume the three resistances of Problem 1-1 are connected in parallel across 120 V. (*a*) Calculate the amount of current in each resistance. (*b*) How much current does the whole circuit take from the supply source?

1-5 A voltage-dropping resistor connected in series with the heater coils of five radio tubes permits the supply of their current directly from a 120-V line. Each of four heater coils operates at 6.3 V and the fifth at 12.6 V. Draw the circuit and calculate the ohms value of the resistor. Assume heater current is 0.3 A.

1-6 Two resistances, R_1 and R_2, are connected in series across 120 V. The resistance of R_2 is five times the resistance of R_1. How much voltage drop exists across R_1?

1-7 Assume that when a third resistance, R_3, is connected in series with R_1 and R_2 of Problem 1-6 the voltage across R_1 is reduced to 10 V. Compute the voltage across R_3.

1-8 Fig. P1-8 shows a voltage divider made of uniformly wound turns of resistance wire so that the ohms values of R_1 and R_2 are directly proportional to the lengths *AB* and *BC* along the surface. The total resistance from *A* to *C* is 18,000 Ω. (*a*) What is the voltage E_2 if the slider makes contact at a point 60 per cent of the distance from *A* to *C*? (*b*) Compute R_1 and R_2 in ohms. (*c*) Assume that a load resistance of 3,600 Ω is connected to points *B* and *C*. Calculate the currents in R_1 and R_2 and in the load, also the load voltage. (*d*) If R_1 and R_2 are to be replaced by fixed resistors, and the load permanently connected as in (*c*) what should their wattage capacities be for safe operation?

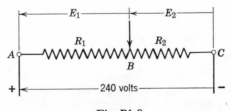

Fig. P1-8

1-9 Two resistors $R_1 = 80\text{ k}\Omega$, $R_2 = 160\text{ k}\Omega$, are connected across 240 V. (*a*) What is the voltage across R_1? (*b*) Assume that a voltmeter with 1,000 Ω/V sensitivity, (its resistance is 250,000 when set on the 0–250-V scale) is used to measure the voltage across R_1. What will be its reading? (*c*) Repeat (*b*) assuming a 20,000 Ω/V meter is used on its 250-V scale. (*d*) Compare the meter readings in (*b*) and (*c*) with the true voltage as obtained in part (*a*) expressing the errors in volts and in per cent of the true value.

1-10 Calculate the potential difference between points A and B in Fig. P1-10. Then assume that a wire with zero resistance is used to connect A and B. How much current would flow in the wire? What is its direction?

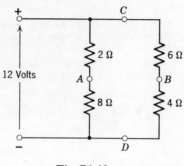

Fig. P1-10

1-11 Solve for the currents in the three branches of Fig. P1-11 using (*a*) Kirchhoff's voltage law with only two unknown current symbols. (*b*) Mesh-current notation. (*c*) Node-pair voltages.

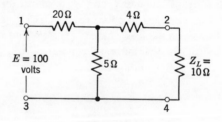

Fig. P1-11

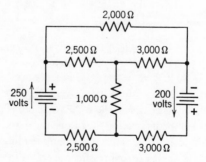

Fig. P1-12

1-12 Determine the current in the 1,000-Ω resistor of Fig. P1-12, using mesh-current notation.

1-13 Solve for the current in the shunt resistor in Fig. P1-13, using the superposition theorem.

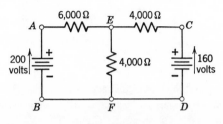

Fig. P1-13

1-14 Use another method of solution to check your answer to Problem 1-13.

1-15 The ammeter in Fig. P1-15 has negligible resistance. Show that the reciprocity theorem holds for the circuit.

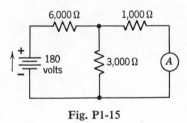

Fig. P1-15

1-16 In the circuit of Fig. P1-16, determine the value to which the variable load resistance *R* should be adjusted for the following conditions, and also calculate the maximum value of the following quantities: (*a*) load current; (*b*) voltage at terminals 2 and 4; (*c*) load power.

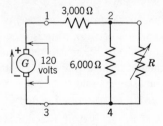

Fig. P1-16

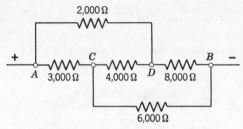

Fig. P1-17   *A* is 180 V above *B*.

1-17 Solve for the current in the 4,000 Ω resistance in Fig. P1-17 using Thévenin's theorem. Assume that resistance to be temporarily disconnected. An open-circuit voltage would then appear between points *C* and *D*.

1-18 Solve for the current requested in Problem 1-17 using Norton's theorem.

1-19 Use the mesh method of solution to find the current requested in Problem 1-17.

1-20 Solve for the line current in Fig. P1-20, using the mesh-current notation.

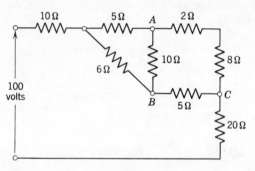

Fig. P1-20

1-21 What is the power in the 8-Ω resistor in Fig. P1-20?

1-22 In Fig. P1-20, transform the circuit connected to points *A*, *B*, and *C* into an equivalent *T* circuit. Then reduce the whole network to a series circuit. Solve for the line current and check with the answer to Problem 1-20.

1-23 A sealed box has four terminals on its lid, each of which is presumed to be connected to a junction point in a network of resistances inside the box. Compute values for, and draw an equivalent *T* circuit to represent, the unknown network, using the following measurements: input impedance with output open, 10,000 Ω; input impedance with output shorted, 6,000 Ω; output impedance with input open, 9,000 Ω; output impedance with input shorted, 5,400 Ω.

1-24 Compute the values of the equivalent π-section circuit from the data given for the network in the box of Problem 1-23.

1-25 Two currents differ in phase by 90°. Their maximum instantaneous values are 12 A and 5 A. (*a*) What is the maximum instantaneous value of their sum? (*b*) What is the sine of the angle between the phasor representing their sum and that representing the 12-A current?

1-26 Two voltages are represented by $E_1 = 86.6 + j50$, $E_2 = 50 - j86.6$ V. Express in complex number form: (*a*) $E_1 + E_2$; (*b*) $E_1 - E_2$.

1-27 Express the sum and difference voltages of Problem 1-26 in polar form: $E \angle \theta$.

1-28 The sum of E_1 and E_2 of Problem 1-26 is applied to a reactance $X = 12 + j5$ Ω. How much current flows, and what is its phase angle with respect to the phase of the applied voltage?

1-29 A 60-Hz alternating voltage having a sine-wave form with a maximum value of 100 V sends a sine wave of current through a resistor. The maximum value of the current is 1 A. (*a*) Plot the two sine waves on a single time axis, using different vertical scales for voltage and current. (*b*) The power at any instant is equal to the product of the instantaneous voltage and current. Plot enough points representing instantaneous power to permit the drawing of a smooth curve showing how the instantaneous power varies throughout the current and voltage cycle. (*c*) The average power is one-half the maximum value in this case. Read its value on the power scale. What is the frequency of the power curve?

1-30 A direct current of constant value flows through a 25-Ω resistor. The average rate of dissipation of electric energy is 100 W. (*a*) Calculate the maximum instantaneous value of an alternating current of sine-wave form which will produce the same dissipation. (*b*) Calculate the maximum instantaneous power. (*c*) What is the effective value of the alternating current?

1-31 At a junction in an a-c circuit a single current I divides into two separate currents denoted by I_1 and I_2, such that $I_1 = 10$ A and $I_2 = 5$ A. I_1 lags I_2 by 30°. Using the cosine law, calculate the value of I.

1-32 Solve Problem 1-31, using complex numbers to represent the currents. Assume I_2 to have zero phase angle. What is the phase angle of I_1 in this case?

1-33 Three currents at a junction are represented by $I_1 = 15 \angle 0°$, $I_2 = 10 \angle 60°$, $I_3 = 20 \angle 90°$, all expressed in amperes. Calculate their sum.

1-34 Compute the phase angle of the current which is the sum of the currents in Problem 1-33.

1-35 A 10-turn air-core coil in which all the flux that is produced links all the turns is found to have 0.01 Wb of flux when the current is 2 A. What is the inductance of the coil in mH?

1-36 A 20-turn coil, near the coil of Problem 1-35 is linked by 0.005 Wb of flux when the current in the 10-turn coil is 2 A. What is the mutual inductance of the pair?

1-37 Refer to the circuit of Fig. P1-37. $E = 100$ V, $L = 0.04$ H, $R = 1,000\ \Omega$. Determine (*a*) the initial and final values of current, e_L', E_R at $t = 0$; (*b*) the initial rate of rise of the current; (*c*) the time constant and value of i, e_R, e_L' at the end of the t_k time interval.

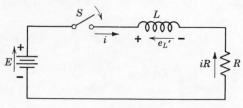

Fig. P1-37

1-38 Compute the initial rate of rise of current and the time constant for the circuit of Problem 1-37, assuming $L = 0.01, 0.08, 0.125$ H, keeping $R = 1,000\ \Omega$

1-39 Sketch current-vs.-time curves for Problems 1-37 and 1-38.

1-40 Assume $R = 500, 1,500, 2,000\ \Omega$ and keep L constant in Problem 1-37. (*a*) Does changing the value of R affect the slope of the current-time curve at $t = 0$? (*b*) Compute the time constant for each value of R. (*c*) Compute the final value of the current in each case.

1-41 Sketch current vs. time curves for Problems 1-37 and 1-40.

1-42 A coil has an inductance of 50 mH. What is its inductive reactance at (*a*) 1 kHz; (*b*) 1 mHz; (*c*) 60 Hz.

1-43 Two coils $L_1 = 0.16$ H and $L_2 = 0.04$ H are so located that their mutual inductance is 0.06 H. What is the coefficient of coupling?

1-44 If a constant d-c voltage of 25 V is suddenly applied to the primary coil ($L_1 = 0.16$ H) of Problem 1-43. (*a*) What will be the initial rate of rise of the current in each coil? (*b*) What would be the initial rate of rise of the current in L_1 if the secondary coil (L_2) were open-circuited before the voltage is applied? (*c*) Account for the difference between the two answers for L_1.

1-45 An alternating voltage of effective value 120 V, 60 Hz, is impressed upon the terminals of a coil that has 0.2 H inductance and 50 Ω resistance. Determine the (*a*) current and its phase angle; (*b*) power (two ways); (*c*) power factor; (*d*) voltages across R and L; (*e*) two component values of current, in-phase and quadrature; (*f*) circuit impedance; (*g*) phasor diagram.

1-46 A motor takes 22 A from a 240-V line. A wattmeter in the line reads 4,490 W. What is the power factor at which the motor is operating?

1-47 (*a*) A reactor is needed for a 1,000-Hz telephone line. The inductive reactance is to be 314 Ω. What should be its inductance? (*b*) The resistance of the wire of the reactor is 32 Ω. How much current will it take from a 48-V, 796-Hz line? Calculate the power loss.

1-48 In a particular series circuit it is desirable to have a power factor of 50 per cent. The frequency is 1 kHz and the resistance and inductance are 1,000 Ω and 100 mH. How much inductance should be added in series?

1-49 A series circuit has $R = 400\ \Omega$, $L = 10$ mH, $C = 0.1\ \mu$F. The frequency is 7,960 Hz. What is the power factor?

1-50 How would you change the series circuit of Problem 1-49 to make the power factor unity? Compute required values.

1-51 What would be the power factor of the new circuit if a pure inductance of 175 mH were added in parallel to a series circuit containing 100 Ω resistance and 100 mH inductance on a 1-kHz line?

1-52 Three 0.1 μF capacitors are to be connected first in series, then in parallel, and then in series-parallel. Calculate the equivalent capacitance for each case.

1-53 A coil has 20 Ω resistance and 3.18 mH inductance. A 3.18 μF capacitor is connected in parallel with it. (*a*) What is the impedance of the parallel circuit at resonant frequency? (Note that ωL is only 1.58 times R.) (*b*) Recalculate the input impedance using $R = 3\ \Omega$. Compare the result with that given by $Z_r = L/(RC)$ which holds when $\omega_r L$ is 10 or more times R.

1-54 A generator with an EMF of 10 V and an internal impedance of $100 + j20\ \Omega$ supplies current at resonant frequency to the parallel circuit of Problem 1-53(*a*) where $R = 20\ \Omega$. What is the terminal voltage of the generator?

1-55 (*a*) What is the resonant frequency of a series circuit in which $R = 10\ \Omega$, $L = 159\ \mu$H, $C = 159$ pF? (*b*) What is the Q of the circuit?

1-56 One volt at resonant frequency is applied to the circuit of Problem 1-55. Compute the voltages across R, L, and C. Compare them with the applied voltage. Is the voltage across C equal to Q times the input voltage? Is E_L also equal to QX_L?

1-57 An 8 μF capacitor with 250 V at its terminals is suddenly discharged through a 10,000-Ω resistor. Determine the (*a*) energy that was stored in the capacitor; (*b*) initial and final values of current in the resistor; (*c*) time required for the current to go through 63.3 per cent of its total change; what will be its value at that instant? (*d*) capacitor current and voltage values at $t = 0.04$ s; (*e*) initial and final rates of change of current; (*f*) time required for the current to decrease to one-half of its initial value; (*g*) time required for the voltage to decrease to 1 V; (*h*) energy lost in the resistor during the complete discharge of the capacitor.

1-58 A coil with an inductance $L = 159$ mH and a resistance R_L of 40 Ω is connected to a generator that has a constant terminal voltage (*E*) of 1 V at a frequency of 1 kHz. See Fig. P1-58. (*a*) Compute the current I_L through the coil and its phase angle. (*b*) Compute the current through a 0.159-μF capacitor C that is connected to the same generator by closing switch S. (*c*) Draw the phasor diagram and show the phasor of the total input current. Calculate its value. (*d*) Determine the Q of the coil and the power consumed in the 40-Ω resistance of the coil. (*e*) Determine the equivalent impedance of the circuit by dividing the applied voltage E by the input current.

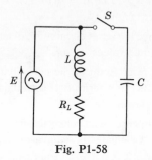

Fig. P1-58

1-59   A 0.02-μF capacitor is in series with a 10-Ω resistor. What is the impedance of the pair at 8 kHz?

1-60   How much inductance is required to produce series resonance in the circuit of Problem 1-59?

1-61   What are the half-power frequencies of the resonant circuit of Problem 1-60, assuming the capacitor has been added. What is the Q of the circuit?

1-62   A coil has an inductance of 100 μH and 5 Ω resistance. What value of capacitance is needed to produce parallel resonance at 1.5 mHz?

1-63   The coil of Problem 1-62 and a parallel capacitor resonate at 1.5 mHz. (*a*) What is the resonant impedance? (*b*) What is the Q of the circuit? (*c*) How much current will the circuit take when 1 V at resonant frequency is applied?

1-64   Compute the current in the capacitor in Problem 1-63 and compare it with the current from the 1-V source. Is the capacitor current equal to the Q times the source current?

1-65   A constant d-c voltage of 100 V is suddenly applied to a 5,000-Ω resistance that is in series with a 2-μF capacitor. What is the (*a*) time constant; (*b*) initial value of current; (*c*) final value of current?

1-66   In Problem 1-65, determine the (*a*) initial rate of change of current; (*Hint*: examine Fig. 1-15a); (*b*) current at the end of the time-constant interval; (*c*) final voltage on the capacitor.

1-67   A constant d-c voltage of 500 V is suddenly applied to a 1,000-Ω resistor in series with a 2-μF capacitor. At the instant the current is 0.20 A, how much energy

($\frac{1}{2}CE_c^2$) is stored in the capacitor, how much charge is on the plates, and what is the rate at which energy is being supplied to the electric field, i.e., what is the power?

1-68 A coil having a resistance of 50 Ω and an inductance of 1.5 H is connected in series with a capacitor having 10 μF capacitance. A voltage of 100 V, 60 H is applied to the circuit. Determine the current and the voltages in the circuit. Draw the impedance diagram and the phasor diagram.

1-69 A voltage of $20 + j30$ V is applied to an impedance of $2 + j5$ Ω. Determine the current and the angle at which it lags the voltage.

1-70 A voltage of $20 + j30$ V is applied to an impedance of $2 - j5$ Ω. Solve for the current and its phase angle with respect to the horizontal axis (where $\omega t = 0$).

1-71 Sketch the voltage and current phasors of Problem 1-69, and those of Problem 1-70. What is the angle of the current in Problem 1-70 with respect to its voltage?

1-72 A capacitor, a coil, and a resistor are connected in parallel. The capacitive reactance is 5 Ω, the coil reactance is 10 Ω and its resistance is 2 Ω, and the resistor has 8 Ω resistance. A voltage of $10 + j0$ is applied to the parallel circuit. (*a*) Determine the current in each branch of the circuit. (*b*) Obtain the complex expression for the equivalent impedance of the circuit. (*d*) Draw the phasor diagram, showing the applied voltage and branch currents.

1-73 Compute the power in the circuit of Problem 1-69 by (*a*) $EI \cos \theta$; (*b*) $I^2 R$; (*c*) complex number method.

1-74 A coil having 12-Ω reactance and 16-Ω resistance is in parallel with a capacitance whose reactance is 20 Ω. In series with this pair is a resistance of 10 Ω. Compute the total impedance of the circuit.

1-75 The circuit of Problem 1-74 is supplied with 120 V rms. Compute all three currents and construct the phasor diagram.

1-76 (*a*) What is the power factor of the whole circuit of Problem 1-74? (*b*) What would you connect in series with the 10-Ω resistance to produce unity power factor? (*c*) How much current would the circuit then take with 120 V applied? (*d*) How much voltage would then exist across the parallel part of the circuit?

1-77 Solve for the following in the circuit of Fig. P1-77: (*a*) I_1, I_2, I, total power loss; (*b*) voltage across parallel part; power loss in each part (two ways); (*c*) power factor of whole circuit; draw the phasor diagram.

1-78 A variable capacitor may be set at any value from 100 pF to 500 pF. Calculate the limits of the frequency band in which this capacitor may be made to resonate with a 50-mH inductance.

1-79 A coil of 10-H inductance and 500-Ω resistance is in series with a 1-μF capacitor. Ten volts at resonant frequency are applied. (*a*) What is the resonant frequency? (*b*) Calculate the current, E_L, and E_C. (*c*) If the applied voltage is held

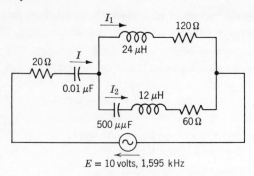

$$E = 10 \text{ volts}, \ 1{,}595 \text{ kHz}$$

Fig. P1-77

constant but its frequency varied, at what values will the circuit receive half as much power as at resonance?

1-80   A series circuit has $L = 100 \text{ mH}$, $C = 100 \text{ pF}$, $R = 100 \ \Omega$. (*a*) Plot a curve for this circuit, showing how the current varies with frequency above and below resonance. Use any convenient value of applied voltage of constant magnitude. Show half-power points. (*b*) Calculate the bandwidth, the Q of the circuit, and the power input at resonance.

1-81   Plot current curves for $R = 200 \ \Omega$ and $R = 50 \ \Omega$ in Problem 1-80. Discuss the results.

1-82   Calculate the per cent frequency discrimination (ratio of bandwidth to resonant frequency) of the circuits of Problems 1-80 and 1-81 for each value of R used.

1-83   It is desired to have 25,000 Ω resistance as the input impedance of a parallel-resonant circuit. A 250-μH inductance with 12 Ω of resistance is available for use. Specify a circuit element which would make this possible at 3.19 MHz.

1-84   A high-frequency generator produces alternating current at 1.5 MHz. Its internal impedance is 10,000 Ω pure resistance. Design a parallel-resonant circuit which would receive maximum power from this generator, using a variable capacitor that has limits of 25 pF and 250 pF. Use an L value which will require that the capacitor be tuned near the middle of its range.

1-85   In Problem 1-84, compute the maximum power received if the generated voltage is 50 V.

1-86   An energy source with $100 + j100 \ \Omega$ internal impedance has an internally generated voltage of 100 V at a frequency of 350 kHz. It feeds a *series circuit* containing a coil ($L = 815 \ \mu$H, $R_L = 100 \ \Omega$) and a capacitor bank consisting of three 80 pF units which may be grouped three ways. Calculate the following values for parallel connection of the capacitor: (*a*) current; (*b*) load power; (*c*) terminal voltage of the source; (*d*) voltage across the capacitors.

1-87 Repeat Problem 1-86 for series connection of capacitors.

1-88 Repeat Problem 1-86 for series-parallel connection of capacitors.

1-89 A generator with an internal impedance $Z = 120 + j50\ \Omega$ is to deliver maximum power to a load. What should be the load impedance? Specify the values of the load elements if the frequency is 400 Hz.

1-90 A parallel circuit has two branches as follows: $L = 1,000\ \mu H$ is in series with $R_1 = 30\ \Omega$, and $C = 2,000\ pF$ is in series with $R_2 = 20\ \Omega$. (*a*) What is the total impedance at resonant frequency? (*b*) Compare $R_1 + R_2$ with X_L and with X_C at resonance.

1-91 In order to tune a radio receiver for reception of a 1,200 kHz signal, what should be the microfarad setting of a variable capacitor to form a parallel-resonant circuit with an inductance of 175 μH?

1-92 The inductance of Problem 1-91 has 20 Ω resistance. What is the resonant impedance of the circuit?

1-93 An antenna supplies 2 μA of current to the resonant circuit of Problem 1-92. What is the voltage across the capacitor?

1-94 An amplifier delivers 0.5 W of power when its input is 1 mW. What is the decibel gain?

1-95 A telephone line with $600 + j0\ \Omega$ input impedance receives 10 mA of current. The current in a terminating impedance of $600 + j0\ \Omega$ is 50 μA. Compute the decibel loss using the current values and check using power values.

1-96 Two equations were obtained in analyzing a network: $5I_1 - 4I_2 = 20$; $4I_1 - 6I_2 = 30$. Solve for the currents using determinants.

1-97 Three equations were obtained in analyzing a network: $2I_1 - 3I_2 + 2I_3 = 7$; $4I_1 + 5I_2 + I_3 = 45$; $-I_1 + 4I_2 - 3I_3 = -7$. Solve for the currents using determinants.

1-98 Solve the following equations for x, y, and z using determinants: $9x + 6y = 36$; $10x - 6y + 15z = 51$; $9x + 6y - 3z = 30$.

1-99 A complex current wave has a d-c component of 5 A, and first, second, and third harmonics whose maximum instantaneous values are 8, 4, and 2 A, respectively. What is the effective value of the current?

1-100 A complex voltage wave is represented by $e = 50 \sin 377t + 25 \cos 377t$. Express this voltage by a single term that has an absolute value and involves only the sine function of an angle. Draw a phasor diagram that represents the voltage as a single quantity with two components.

CHAPTER TWO

Behavior of Semiconductors

The first use of a semiconductor in electronics applications dates back to the beginning of the present century (about 1905), when a crystal of galena (lead sulfide) was used to detect signals in radio receiving sets. A fine pointed wire was pressed against the surface of the crystal to form a *detector*. The wire making point contact was called a *cat whisker*. The detector worked because it conducted current much more readily in one direction than in the opposite direction. This is the requirement of a detector of high-frequency currents used in communications and control circuits. By allowing currents to flow in only one direction through the detector (e.g., in the positive direction) not only is the *average* current per cycle always in the positive direction, but the current also contains low-frequency variations that convey intelligence.

Semiconductors are used extensively in the form of point-contact and junction diodes, as copper-oxide and selenium rectifiers, and as transistors. The general behavior of semiconductors is treated here to provide a good background for an understanding of their operation as crystal diodes and, later, as transistors. The junction-type diode is superior in many ways to the point-contact diode. It will be explained in some detail.

2.1 Atomic Arrangement in Materials

Matter, in general, is composed of atoms that are made up of electrons rotating about a positively charged nucleus which contains enough charge to make the complete atom electrically neutral. The Bohr theory of atomic structure states that the electrons rotate in orbits about the nucleus in the same manner as the planets circle the sun. Rotating electrons can exist only in certain allowed, discrete shells concentric with the nucleus. This is true of atoms, whether they are bound together as in solids and liquids or are in the form of a gas. The distinguishing chemical properties of different materials arise because they contain different kinds of atoms. Each kind of atom has a nucleus carrying the exact amount of positive charge required to neutralize the total negative charge of the orbiting electrons of that atom. Its electrons are arranged in *energy levels* that are unique for the element.

In the Bohr model of an atom, the energy shells are drawn concentric with the nucleus and named K shell (nearest to the nucleus) to Q shell (farthest from the nucleus). The potential energy of an electron (energy of position) increases with distance out from the nucleus. The electrons in the outermost shell of an atom are called *valence electrons*, and it is these that are important in determining the chemical properties of a material. They can also pass easily from one atom to another and this movement constitutes current in the material. The passage of electrons from one atom to another, in a conductor, each moving into a vacant place just left by another electron, is regarded as the mechanism of current. Valence electrons exist in the *valence band* of energy. Those in the highest-energy level are at the greatest distance from the nucleus.

Hydrogen, which has the simplest atom of all elements, has only 1 electron rotating around its nucleus, and this is in the first (K) shell. Copper, the most widely used of all conductors, has 29 electrons arranged in shells about the nucleus: 2 in the K shell, 8 in the L shell, 18 in the M shell, and 1 in the N shell. Silicon and germanium each have 4 valence electrons, which accounts for their behavior as semiconductors, as will be seen.

When a material forms into a solid, the atoms most often are arranged in a regular geometric array, thus defining a crystal structure. Such a structure has symmetry in three dimensions, with the atoms arranged in definite *unit cells*, each of which looks the same throughout space in the *crystal lattice* of the solid. Two examples of crystal arrangements are shown in Fig. 2-1.

Since we shall be primarily concerned with crystalline solids, some of their special characteristics are of interest.

1. The valence electrons of an isolated atom describe a radial dimension for the atom. Consequently, we should expect the atoms in a crystal to be separated by about twice this distance.

2. The outer orbits of neighboring atoms overlap at the atomic separation expected in (1). This produces a distortion of the outer orbits, so that it

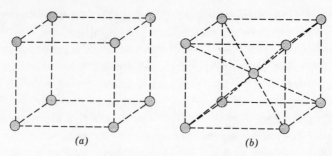

Fig. 2-1   Two examples of atoms arranged in a solid crystal. (*a*) Simplest cubic lattice; (*b*) body-centered cubic lattice.

becomes difficult to say which electron goes with which atom. In this way, some valence electrons are shared by several neighboring atoms.

3. The valence electrons will take on a systematic motion and thus join each atom to its neighbors while maintaining each atom electrically neutral.

2.2   Pattern of Energy Levels

The energy levels of atoms begin with energy values in the K shell for hydrogen and extend through discrete levels called L, M, N, O, P, and Q, for the 103 known elements. There exists a very significant situation concerning the maximum number of electrons that each level can accommodate, owing to a unique principle in physics called *Pauli's principle*.

It has been determined that the K shell can contain only 2 electrons, the L shell, 8 electrons, and other shells: M, 18; N, 32; O, 50; P, 72; Q, 98. However, the largest known number in the O shell is 31, in the P shell 10, and in the Q shell 2. Hydrogen, which is the first one in the table of elements, has already been mentioned. Helium, the second element, has only 2 electrons rotating about its nucleus and they are both in the K shell. Lithium, number three, has 3 electrons, but only 2 of them are in the K shell and the third is in the L shell. There are only 2 electrons in the K shell of each of the remaining 100 elements. Neon has 8 in the L shell and 2 in the K shell as expected. Silicon, the fourteenth element, has 2 (K), 8 (L), 4 (M). Germanium, element 32, has 2 (K), 8 (L), 18 (M), 4 (N).

Immediately following the first element (in the periodic table) that accommodates 8 electrons in one shell (the L shell), the next element in the table starts the next shell with the one added electron that distinguishes it from its predecessor. The L shell of every element in the table beyond and including neon has only 8 electrons. As soon as the M shell of an element has acquired

8 electrons (the first one to do this is argon), the very next element, potassium, starts one electron in the N shell. This process continues on to element 86, radon, which has 8 electrons in the P shell.

It has been explained that as soon as 8 electrons are contained in the outermost shell, further electron orbits must be established in the next outer shell whether or not empty places (energy levels) remain in that shell. The stable condition represented by this so-called "octet" theory is important in the explanation of the behavior of semiconductor materials used in electronic devices.

2.3   Conductors, Insulators, and Semiconductors

Certain elements and their solid-state crystals are classed as conductors, semiconductors, or insulators, based on their relative ability to carry electric current. The crystal structure of highly conductive materials, like copper, gold, silver, and aluminum, is such that the outer electrons are shared by *all* of the atoms in the material. These electrons are actually free to wander from atom to atom, over a very wide temperature range. In most metals, each atom supplies one such free electron, making the number of free electrons per cubic centimeter of the order of 10^{23}. These highly conductive materials have one valence electron, which can readily accept enough energy to free it from its parent atom and enable it to move about in the material. Thus copper is classed as a good conductor of electricity. Silver is even better than copper as a conductor, because its valence electron is in the fifth (O) energy shell, far removed from the binding influence of its nucleus. Copper is more generally used, because it is less expensive than other metals that are better conductors.

Whether a material behaves as a conductor, semiconductor, or insulator can be explained by the permissible energy levels of the atoms making up the material. There is a *forbidden region*, or *forbidden energy gap*, between each of the allowed energy bands from K to Q. Further, an energy gap (forbidden region) may exist between the energy of the outermost valence electrons and the energy needed for conduction from one atom to another. When some outside influence such as an electric field or addition of heat is applied, a valence electron may acquire sufficient energy to jump through the forbidden region and on into the *conduction band* of energy. Here it will have enough energy to free itself from any influence of its positive nucleus and become a carrier of electricity, ready to take the place of another electron that just left its own atom in the conductor in the same manner.

Energy can be given to electrons only in discrete amounts (*quanta*), according to the Bohr theory. Hence, there are certain energy values that an electron simply cannot acquire. As a rough example, if a man were climbing a vertical ladder that had rungs 1 ft apart, he could stand with both feet in a stable position only at 1-ft intervals above the base. He could not stand $1\frac{1}{2}$ ft or $2\frac{1}{4}$ ft above the base nor at any height involving a fraction of a foot. The discrete quantum of energy that can be acquired by an electron is given by *hf*, where *h* is Planck's constant (6.624×10^{-34} J-s) and *f* is the frequency of the energy in cycles/second or hertz (Hz).

Figure 2-2 shows relative energy bands in conductors, semiconductors, and insulators. In conductors, there is no gap between the valence band and the conduction band. Some researchers are convinced that they overlap.

In contrast to the action of conductors, insulators are poor carriers of electricity. The structure of solid insulators is such that almost all of the electrons remain bound to their parent atoms. The forbidden energy gap of these materials is very wide, so the conduction band remains almost empty. Few electrons can acquire enough energy to cross the forbidden region. Sulfur is a good insulator. Although it has 6 valence electrons in the *M* band, it has a very wide energy gap. It is very difficult to impart to these electrons sufficient energy for them to bridge the forbidden gap and enter the conduction band. This is why sulfur is an insulator.

Semiconductors, on the other hand, are neither good conductors nor good insulators, at normal room temperatures. More precisely, they are insulators at very low temperatures and reasonably good conductors at high temperatures. The elements germanium and silicon are the most important semiconductors used in electronics. Their use in crystal diodes almost excludes other semiconductors, although other materials are finding increasing use. However, these two elements appear to be the only ones suitable for the fabrication of transistors.

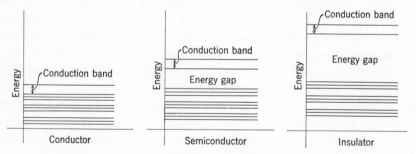

Fig. 2-2   Relative magnitudes of energies in shells and gaps.

The element germanium has 4 electrons in the valence band of its atom. Its conductivity lies in the range somewhere between that of insulators and conductors. The conductivity of a good insulator, such as quartz, may be as low as 10^{-16} mho/m, while that of a conductor may be as high as 10^7 to 10^8 for a typical metal. The conductivity of copper is 5.8×10^7 mho/m. Semiconductors have conductivities in the range from 10^{-3} to 10^5 at 300°K (27°C). In terms of energy gaps, conductors may require as little as 0.05 electronvolts (eV) to jump the forbidden region, while insulators require many electronvolts. Germanium and silicon require 0.7 eV and 1.1 eV, respectively. An electronvolt is the energy acquired by 1 electron falling through 1 V of potential difference.

DRILL PROBLEMS

D2-1   Calculate the energy, in joules, an electron will acquire in falling through 1 V of potential difference; 10 V of potential difference.

D2-2   How many joules of energy are represented by 0.7 eV? 1.1 eV?

D2-3   The frequency range of visible light extends from about 4×10^{14} Hz to 7×10^{14} Hz. Calculate the minimum frequency of photons which may excite electrons from the valence to conduction bands in germanium. May energy at visible-light frequencies excite the electrons?

D2-4   Make a similar calculation as in D2-3, for silicon.

2.4   The Germanium Atom

Germanium and silicon are used in the manufacture of crystal diodes. These two elements are also the dominant material in transistors. The atomic structure of germanium is studied here in some detail, largely because of its universal use as a basic semiconductor material.

The germanium nucleus of 32 positive charges is surrounded by 32 electrons revolving in the *K*, *L*, *M*, and *N* shells, containing 2, 8, 18, and 4 electrons, respectively. Each (except the *N* shell) is filled with the maximum number of electrons that it can accommodate. Because the electrical interactions with other atoms involve only the valence electrons, they are the only ones that will be considered in the behavior of semiconductors.

Figure 2-3 represents the Bohr model of a germanium atom in two dimensions. The black dots around the nucleus and inside the first large circle represent electrons in the filled *K*, *L*, and *M* shells. Each of the valence electrons is shown having a unique energy level in the valence band. The forbidden energy gaps occur above and below the valence band.

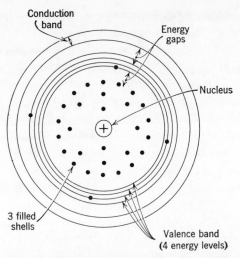

Fig. 2-3   Bohr model of germanium atom, two-dimensional view.

2.5   Covalent Bonding

The 4 valence electrons of a germanium atom are required to make the atom electrically neutral, but they are in excess of those needed for a preferred energy state. Completely filled bands represent the preferred state, which is also the minimum energy state of an atom. As in all physical systems, atoms prefer a state of minimum energy. When an aggregation of atoms are brought together under favorable conditions, they will combine in a state of minimum energy. This state is achieved, for germanium atoms, when each atom shares its valence electrons with 4 of its neighbors. The chemical bonds which attach neighboring atoms are called *covalent bonds*, or *electron-pair bonds*. Under this condition each atom *effectively* has 8 valence electrons and it is closer to a preferred state than when there is no sharing of electrons. Covalent bonding holds the atoms together to form the crystal lattice of germanium.

Covalent bonding in a germanium crystal is shown schematically in Fig. 2-4. Each bar connecting 2 adjacent atoms represents an electron-pair bond.

Covalent bonding action is happening simultaneously to each of the valence electrons of an atom as it encounters a companion valence electron from the atoms above, below, and on the other sides of itself in the crystal. It has been determined that about fifteen times as much energy is needed to free a covalent bonded electron as is required to free an unbonded valence electron.

In addition to nuclear attraction, 2 closely located electrons experience a bonding force that effectively lowers their potential energy, since it makes a still larger energy gap for them to cross before they can enter the conduction band and become free electrons. This bonding force is due to magnetic fields produced by the spinning of electrons on their own axes. When 2 of these electrons come close to each other, their magnetic fields join in additive fashion. This creates a force that tries to keep the electrons from moving farther apart. These bonding effects tend to stabilize the germanium atoms, owing to the fact that each has effectively 8 electrons in its outer orbit.

The germanium atoms arranged in covalent bonds in their crystal structure have important physical properties that make this material useful in electronic devices. The discussion presented here for germanium applies equally well, in a qualitative sense, to silicon and other crystalline materials.

2.6 Adding Impurities to Semiconductors

Germanium atoms arranged in a solid-state crystal will receive enough energy from heat at room temperature to break some of the valence bonds and permit some electrons to act as free-charge carriers in the crystal. Thermal agitation of the electrons results in interactions among them and endows the crystal with electrical conductivity. The conditions in the crystal when an electron escapes from its parent atom are shown in Fig. 2-5. A free electron (minus sign) has produced an imperfection (plus sign) in the crystal. This process is called *thermal ionization* of the crystal.

One electron is now missing from one of the valence bonds, and a *hole*, or vacancy for an electron, exists in the crystal. The missing electron can be easily replaced by another neighbor electron, restoring electrical neutrality

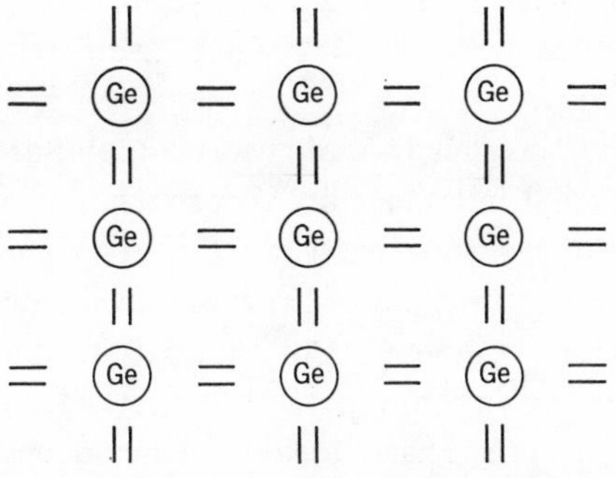

Fig. 2-4 Valence bonds between atoms in a germanium crystal.

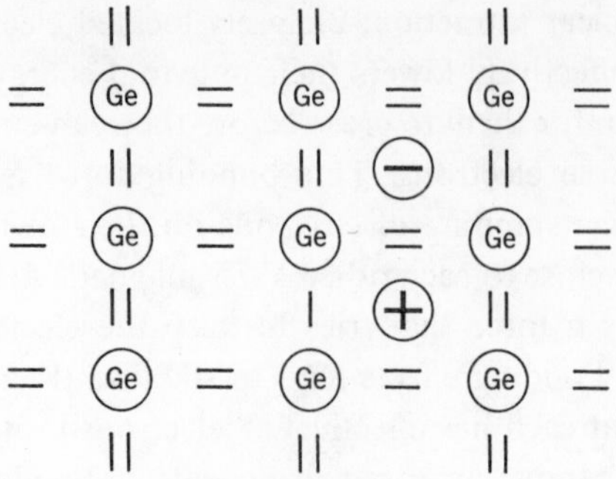

$$\begin{array}{ccccccc}
 & \| & & \| & & \| & \\
= & \text{Ge} & = & \text{Ge} & = & \text{Ge} & = \\
 & \| & & \| & \ominus & \| & \\
= & \text{Ge} & = & \text{Ge} & = & \text{Ge} & = \\
 & \| & & | & \oplus & \| & \\
= & \text{Ge} & = & \text{Ge} & = & \text{Ge} & = \\
 & \| & & \| & & \| &
\end{array}$$

Fig. 2-5   Thermal ionization of a germanium crystal.

to this atom, but the hole has now moved to the neighbor atom which lost its electron. An electron jumping into the hole has effectively moved the hole to a new position in the crystal. Thus, there are now two mobile carriers of current that can move about in the crystal, thermally ejected electrons and the holes they have left behind. The hole carries with it a positive charge equal in magnitude to the electronic charge. If there are enough of these charge carriers available in the crystal, the germanium can act as a fairly good conductor.

Thermal ionization of solid materials is strongly dependent upon the temperature to which they are subjected. Minor variations in the ambient temperature cause radical variations in the number of charge carriers present in the material, thereby varying the electrical conductivity over a wide range. To be useful in electronic devices, the conductivity of the materials should be controllable under any required operating conditions. For this reason, germanium and other semiconductor materials are "doped" by the addition of minute quantities of other elements, called "impurities." The presence of impurity atoms in the crystal allows its conductivity to be increased and controlled by the amount of doping and reduces its dependence on temperature.

Intrinsic germanium crystals (no impurity) contain, as a result of thermal ionization, electrons and holes as current carriers. These crystals are useful in the manufacture of resistors that vary their resistance with temperature changes, but this is about the limit of their usefulness. Semiconductor diodes and transistors require a combination of materials in which the simultaneous existence of an excess of electrons is possible in part of the combination and an excess of holes in the other part. Ideally, semiconductors used in crystal diodes and transistors should contain only one kind of charge carrier.

However, manufacturing limitations do not permit this perfection in the fabrication of semiconductor crystals. In practice, the fraction of impurity atoms can be controlled in the range from about 1 part in 10^5 to 1 part in 10^7.

Impurity atoms are added to intrinsic germanium to make the crystal contain either extra electrons as carriers or extra holes as carriers. An impurity which provides extra electrons is called a *donor* element; one which provides extra holes is called an *acceptor* element.

Donor and acceptor elements are chosen from materials whose atoms have 5 and 3 valence electrons, respectively. When small amounts of these impurity materials are added to germanium, under the proper conditions, they greatly change the electrical properties of the germanium crystal, but its metallurgical properties are not changed perceptibly. When an impurity atom enters the lattice structure of the host crystal, its valence electrons form covalent bonds with adjacent germanium atoms. Donor impurities, having 5 valence electrons, will have 1 electron that is very loosely bound to its parent atom, because it will not be required to complete the covalent bonding with neighboring germanium atoms. Acceptor impurities, on the other hand, do not have enough valence electrons to complete the covalent bonding. An effective hole exists in the lattice because the acceptor atom has only 3 valence electrons. In each case, however, the complete crystal is electrically neutral.

2.7   *N*-Type and *P*-Type Semiconductors

A semiconductor in which donor impurities predominate is called an *N*-type semiconductor, the *N* signifying the negative charge of loosely bound electrons in the material. Acceptor impurities in the material provide holes for electrons, and a semiconductor containing predominantly acceptor atoms is called a *P*-type semiconductor, *P* signifying the positive charge of the holes.

N-type semiconductors use the elements arsenic, antimony, and phosphorus as impurities. Other pentavalent atoms can be used as donor elements, but some are less suitable because of other characteristics which become important in the manufacture of semiconductors. Trivalent atoms, such as those of aluminum, gallium, boron, and indium are used as acceptor elements in the fabrication of *P*-type semiconductors.

If donor and acceptor elements are added to a crystal in equal amounts, the free electrons of donor atoms can fill the holes of acceptor atoms. No current carriers are present except those occurring in thermal ionization. This process of cancelling the effects of donor and acceptor impurities is called *compensation*, and is important in the fabrication of semiconductors. Adding further amounts of either N-type or P-type impurities to a compensated crystal produces an N-type or P-type crystal. Thus, a crystal can be changed during manufacture from an N-type to a P-type and back again, by adding successive amounts of the necessary impurities. It should be remembered that impurity elements make up a very small fraction of the total material, in the range from 1 part in 10^5 to 1 part in 10^7. Their important characteristics are electrical, and their presence hardly affects the metallurgical properties of the crystal.

Figure 2-6 illustrates the lattice pattern of germanium containing impurity elements. A donor impurity supplies an extra electron in the covalent bonding, and an acceptor impurity supplies a hole.

2.8   Energy Levels of N-Type and P-Type Semiconductors

The orbits of electrons in impurity atoms are established either farther from the nucleus or closer to it, in comparison with the distance from the nucleus of corresponding orbits in the germanium atom.

When arsenic, which has a positive charge of 33 in its nucleus, is added as a donor impurity, its electrons rotate in slightly smaller orbits than the electrons of germanium, which has a nuclear charge of 32. The electrons are held closer to the larger positive charge in the arsenic atom. This means that the potential energy level of N-type valence electrons is at a lower level than the level of the valence electrons of "pure" germanium.

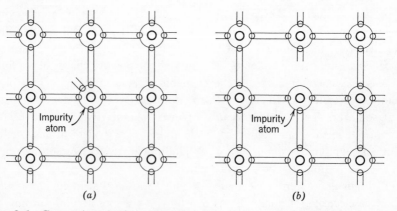

(a)   (b)

Fig. 2-6   Germanium lattice containing impurity atoms. (a) Pentavalent donor impurity with 5 valence electrons. (b) Trivalent acceptor impurity with 3 valence electrons.

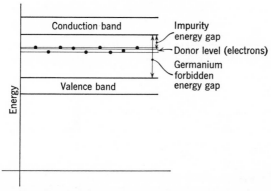

Fig. 2-7   Energy levels in *N*-type germanium. Donor electrons are at high-energy levels.

When indium is added to form *P*-type germanium, its nuclear charge of 49 causes orbital electrons to rotate in smaller orbits than corresponding orbits in pure germanium. But indium has one more permissible energy level than germanium, and its *valence electrons* are rotating in orbits that are *slightly larger* than the orbits of valence electrons of the germanium. This results in the condition that in *P*-type germanium the valence electrons are at a higher energy level than are the valence electrons of a pure (intrinsic) germanium crystal.

This comparison establishes the fact that the potential-energy levels of valence and conduction bands are *higher* in *P*-type semiconductors than in *N*-type. The conduction properties of semiconductors are determined in a large measure by these differences in potential-energy levels. These differences will become important in subsequent discussions of semiconductor devices operating in useful electronic circuits.

The energy levels in *N*-type and *P*-type germanium are shown in Figs. 2-7 and 2-8. Donor electrons are at high-energy levels compared with those of the valence band, and acceptor holes have energy levels closer to those of the valence band.

Some important properties of impurity materials that are commonly used in fabricating germanium and silicon semiconductors are listed in Table 2-1.

DRILL PROBLEMS

D2-5   Calculate the change in ionization energy of intrinsic germanium, when its temperature is increased from 0°K to room temperature (27°C). (See Appendix for temperature conversions.) What is its ionization energy at room temperature?

D2-6   Make a similar calculation as in D2-5, for silicon.

TABLE 2-1 Summary of Properties of Common Impurities Added to Germanium and Silicon Semiconductors.

Element	Atomic Number	Number of Valence Electrons	Function	Majority Carrier	Ionization Energy to Generate Carriers (eV)		Type of Semiconductor
					in Ge[a]	in Si[a,b]	
Boron (B)	5	3	Acceptor	Hole	0.0104	0.045	P-type
Aluminum (Al)	13	3	Acceptor	Hole	0.0102	0.057	P-type
Gallium (Ga)	31	3	Acceptor	Hole	0.0108	0.065	P-type
Indium (In)	49	3	Acceptor	Hole	0.0112	0.16	P-type
Tin (Sn)	50	4	Solder (neutral)	—	—	—	—
Phosphorus (P)	15	5	Donor	Electron	0.0120	0.044	N-type
Arsenic (As)	33	5	Donor	Electron	0.0127	0.049	N-type
Antimony (Sb)	51	5	Donor	Electron	0.0096	0.039	N-type
Silicon (Si)	14	4	Host	(Equal)	$1.205 - 2.8 \times 10^{-4}\,T$		Intrinsic
Germanium (Ge)	32	4	Host	(Equal)	$0.782 - 3.9 \times 10^{-4}\,T$		Intrinsic
					(T in °K)		

[a] R. A. Smith, *Semiconductors*, Cambridge University Press, London, 1959.
[b] E. M. Conwell, *Proc. I.R.E.*, **46**, 1281–1300, 1958.

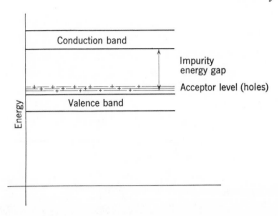

Fig. 2-8   Energy levels in *P*-type germanium. Acceptor holes are at lower energy levels that donor electrons.

2.9   Majority-Carrier Currents

The "missing electron" in covalent bonding of *P*-type germanium constitutes a *hole* which may be neutralized electrically by an electron jumping into it from a nearby bond and thus effectively moving the hole to a new position in the crystal. Energy received from an electric field or from heat could cause the electron to make the jump from the nearby bond to its new position. The motion of the hole, *which acts the same as a positive charge,* constitutes current in the crystal. Similarly, excess electrons in *N*-type germanium are relatively free to move about in the crystal when they receive energy from an external source such as an electric field or heat. Their motion from atom to atom constitutes current, a net motion of electric charge in the material.

Conduction in *P*-type germanium at ordinary temperatures is by means of holes that are created when an acceptor impurity is added. The hole acts as a positive charge, and, since the doped germanium has an excess of these "missing negative charges" in its atoms, the impurity has in effect contributed net positive charge to the conduction process. Holes are in the majority, and in this case they are called *majority carriers* of current. Similarly, electrons supplied by donor impurities in *N*-type germanium are spoken of as the *majority carriers.*

2.10   Minority Carriers

Thermal ionization takes place in doped germanium just as it does in intrinsic germanium. This process yields the same number of holes as electrons, whether it happens in *N*-type or *P*-type germanium. In *N*-type semiconductors, a *minority-carrier* current flows owing to the motion of thermally generated holes. The free electrons arising from thermal ionization receive thermal energy which elevates them into the conduction band, and this increases the *majority* carriers. Holes that are left behind in the valence band may be neutralized by other free electrons falling into the lower potential-energy level occupied by the hole. Others, not eliminated by "recombination" in this way, are driven by an applied electric field toward the negative-potential terminal and constitute a minority-carrier current.

Thermally generated electrons in *P*-type semiconductors may contribute a minority-carrier current. Because of the added thermal energy, electron-hole pairs are generated and the additional free electrons are elevated into the conduction band. Some of them neutralize holes in the crystal, and others are driven by an applied electric field toward the positive terminal of the crystal and constitute a minority-carrier current.

Minority carriers in *N*-type material are holes, while those in *P*-type material are electrons. Minority carriers occur as a result of thermal ionization in the crystal. At very low temperatures few minority carriers are generated. However, as the ambient temperature of the crystal increases, more and more electron-hole pairs are generated. This property of semiconductors becomes important in their application in crystal diodes and transistors operating at high temperatures. Desirable characteristics of these devices are impaired by minority-carrier currents, so it is important to note that crystal diodes and transistors sometimes require extra care in their application to assure that their operating temperature is kept low. More will be said of this in further discussions of their applications.

2.11 Crystal with Impurity Is Electrically Neutral

When thermal ionization generates electrons and holes in the solid, the whole crystal remains electrically neutral, of course, even though some of the individual atoms do not. Even at room temperature the number of free electrons is not nearly so large as it would be in a good conductor. This is true of intrinsic germanium as well as doped germanium, because the germanium atoms taken individually are neutral (not ionized) and the atoms of the added impurity are neutral also. The addition of a donor impurity provides more electrons in the valence band for easy elevation into the conduction band. The addition of an acceptor impurity results in the generation of holes in the valence band. However, no change is made in the net charge in the crystal. As a whole body, viewed macroscopically, the doped crystal is still electrically neutral.

2.12 Atomic Behavior Governed by Quantum Mechanics

The preceding descriptions of atomic behavior and properties of semi-conductors are oversimplified pictures of the situation. We have tried to present some workable understanding of covalent bonding and electron-hole pair generation in terms familiar to the reader. The details of the behavior of semiconductors can be explained more adequately by calling upon modern theory of solids. *Quantum mechanics* has furnished the only satisfactory description. For the purpose of this text, however, the descriptions we have given are sufficient for an understanding of the external behavior of semi-conductors in the form of crystal diodes and transistors.

2.13 *P-N* Junction Diode

When a germanium crystal is doped during manufacture in such a way that a definite boundary exists between the *P*-type and *N*-type semiconductors, a *diode* with very useful properties is made. (The term diode refers to an electronic device having two external electrodes to which electrical connections may be made.) Such an arrangement is called a *P-N junction diode*. It includes the boundary, or *P-N* junction, and some of the *N*-type and *P*-type semi-conductor material on each side of the boundary.

Some of the electrons in the valence band receive thermal energy, and as a result, electrons and holes move about in the material. On the *P* side of the junction there is a high concentration of free holes, and on the *N* side, a high concentration of free electrons. These mobile charge carriers, owing to their motions, spread out and diffuse into opposite sides of the junction. Each kind of charge then finds itself surrounded by charges of opposite sign, and several recombinations take place. Figure 2-9 shows the junction area after diffusion and recombination have occurred. Note that the boundary region soon becomes devoid of holes on the *P* side and of electrons on the *N* side. The result is an accumulation of positive charges at the border on the *N* side and of negative charges on the *P* side. These are referred to as *bound charges* by some authors. In reality, they are positive and negative impurity *ions* bound in the crystal lattice, and they cannot move about. Two equal and opposite charges separated a small distance are called a *dipole*, and in this

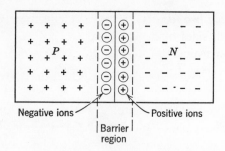

Fig. 2-9   Formation of a dipole layer at a *P-N* junction.

situation a *dipole layer* exists at the *P-N* junction. Such a layer is present in all *P-N* junctions. It is often called a *barrier region* or *depletion region*, because of the absence of free carriers near the junction.

The formation of a barrier region at the junction results in an equilibrium condition, such that free electrons on the *N* side cannot go over to the *P* side because of the opposing forces of its *negative ions*. Free holes on the *P* side cannot go over to the *N* side because of the opposing forces of its *positive* ions.

2.14 Potential Hill at the Barrier Region

Bound charges in the barrier region establish an electric field across the junction, and a potential difference exists because of unlike charges making up the barrier region. The potential distribution inside the diode, with the *P* region taken as a reference potential, is shown as a *potential hill* in Fig. 2-10. The height of the potential hill at the *P-N* junction is determined by an

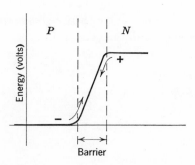

Fig. 2-10   Potential hill through barrier region of reverse-biased *P-N* junction.

equilibrium between two factors: (1) thermal ionization generating electron-hole pairs on both sides of the junction, and (2) the diffusion of carriers across the junction against the potential barrier due to their having acquired sufficient kinetic energy from thermal generation. Under equilibrium conditions with no external circuit for current, these two factors cancel their effects as far as internal current is concerned.

2.15 Forward Biasing of Diodes

If a voltage is applied to the diode as shown in Fig. 2-11, holes in the *P* region move across the junction into the *N* region, and electrons in the *N* region move across the junction into the *P* region. These carriers recombine with opposite charges in the barrier region, lowering the potential hill at the *P-N* junction. External application to the diode of a voltage with this polarity, positive to *P* side, negative to *N* side, is called *forward biasing* of the diode. By lowering the potential hill in this way, easy conduction of current becomes possible in the diode. Forward biasing reduces the strength of the barrier potential. The result is a current (flow of both electrons and holes) that is much greater than that which the resistance of the germanium alone would allow.

When conducting current under forward biasing conditions, semiconductor diodes have only a small voltage drop (a few tenths of a volt) and thus offer very low impedance to current. The internal resistance is almost entirely due to the presence of the potential hill, depending very little on the crystal material on either side of the junction area. A current-limiting resistor such as *R* in Fig. 2-11, is needed to limit the current in the diode.

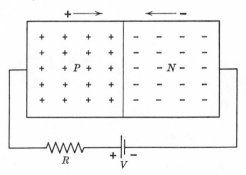

Fig. 2-11   Forward-biased *P-N* junction.

2.16   Reverse Biasing of Diodes

When the positive terminal of the applied voltage is connected to the N side and the negative terminal to the P side, the diode is said to be *reverse-biased*. This condition is illustrated in Fig. 2-12. The barrier region contains no free carriers and it has much higher resistance than the remainder of the crystal on either side of the barrier. In fact, the P and N regions have such comparatively low resistance that they act much like metal contacts between the external connecting leads and the barrier at the common boundary.

The external voltage source attracts electrons and holes away from the junction, producing a wider ion layer at the junction. This results in an increase in the potential hill which charge carriers must overcome in traversing the P-N junction. The ion layers in the barrier region produce an electric field such as an applied battery potential (shown dotted in Fig. 2-12) would produce. This field is in such a direction as to make it still more difficult for *majority carriers* (electrons in the N region) to cross the border from N to P and for holes in the P region to cross from P to N. The current that predominates is due to minority carriers, as will soon be explained.

Remember that the applied bias voltage is trying to produce conduction in the diode. The N side of the crystal, having been made *positive*, is trying to "attract" electrons from the P side through the junction, but N-type semiconductors already have a large excess of electrons. Any migration of electrons toward the positive terminal merely widens the positive-ion layer in the N side.

The P side of the junction has a scarcity of electrons, and although they would willingly go up the potential hill, there are relatively few available to

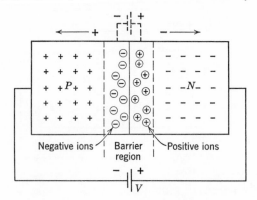

Fig. 2-12   Reverse-biased *P-N* junction.

contribute to current. Of those that receive enough thermal energy to break loose from the valence band in the P material, some may reach the barrier before recombining with a hole that has been created by a similar thermal ionization in the P region. Upon reaching the foot of the potential hill at the barrier, an electron goes up into the N region and becomes a *minority carrier* there. Because its arrival adds one more electron than is needed for equilibrium in the N region, an electron will be delivered to the positive terminal of the biasing battery.

The P side now has an excess hole which was left by this electron. Another electron is accepted from the negative terminal of the biasing battery, and the excess hole is neutralized to restore equilibrium to the P side. Thus, current in the reverse-biased diode is by *minority carriers.*

An electron-hole pair may be generated in the N region in the same way described for the P region. Then a minority carrier (hole) will "slide down" the potential hill at the barrier and "take on" an electron in the P region. The negative terminal of the battery will supply another electron to keep the P region in equilibrium, and because there is now an excess of electrons in the N region, one will go to the positive terminal of the battery.

It is important to note that, while current inside a diode may be composed of both electrons and holes, all *external* current consists of electrons only. This is not too surprising because *conductors* (which have only free electrons for conduction) are used in the external circuitry. The internal material is doped *semiconductor* which contains both holes and electrons.

2.17   Avalanche Breakdown

When reverse bias is increased beyond values for normal operation of the diode, minority-carrier electrons acquire sufficient energy in crossing the boundary region to release valence-band electrons out of the germanium atoms by collisions in the N region. These add greatly to the normal electron flow, rapidly increasing current with increasing reverse-bias voltage, as shown in Fig. 2-13. This phenomenon is called *avalanche breakdown.* Of course, there is simultaneous hole flow across the barrier into the P region. Electrons are taken readily from the negative-bias terminal, and they neutralize the holes. The process of increasing the number of charge carriers by collisions is the same basically as that called *gas amplification* in a gas-type phototube or in a thyratron. This phenomenon is discussed in Chapter 14.

The useful regions of the current-voltage characteristic curve in Fig. 2-13

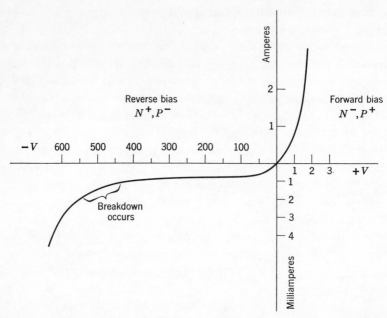

Fig. 2-13   Average current-voltage characteristic of a *P-N* junction germanium diode. Note the change in scales for forward and reverse biasing. (Curve exaggerated to show effects of bias.)

are to the right of the breakdown region. In general, operation in the breakdown mode is to be avoided. The junction diode is a very useful electronic device, however, when operated in the other regions.

The shape of the characteristic may be explained as follows. With forward biasing (*N* side negative) the effect is to lower the potential hill at the barrier. Of course, the battery is now trying to move electrons from the *N* to the *P* side, and they find it easy to surmount the much smaller hill and thus get through the barrier. The current increases in a nonlinear fashion with increasing voltage. The higher the bias voltage, the weaker the barrier, until a voltage is reached at which the barrier has been counterbalanced by the bias battery and disappears. The current-voltage curve from that point on is essentially a straight line and the current is limited only by the external series resistance in the circuit.

With reverse biasing, the current is almost independent of the amount of biasing voltage up to the value necessary to start the breakdown process. The reason is that, at a given temperature, the rate of production of carriers available for migration depends on the concentration of impurity, which is fixed when the crystal is doped. It will be noted that reverse current is much smaller than forward current, typically in the ratio 1:1,000 at the same bias voltage.

The amount of reverse current during breakdown will be limited by the resistance in series with the crystal. This resistance is extremely important and its value is quite critical. Reverse-voltage breakdown must be taken into account in practical applications of diodes. It limits the safe operation of rectifier circuits, and when diodes are used in switching circuits it is desirable to be able to control the amount of voltage that will cause breakdown.

2.18   Zener Effect

Another type of breakdown process that can supply carriers to support conduction when the voltage across the junction is large is called the *Zener effect*. It arises when the barrier potential becomes so large that electrons may be ripped from covalent bonds in the barrier region. Thus, electron-hole pairs are created in the ion layers at the junction. The hole moves quickly into the *P* region under the influence of the electric field in the barrier region and the electron is swept into the *N* region. These moving charges constitute an increase in reverse current in the junction. When the electric field is strong enough to break a single covalent bond, it is sufficiently strong to break many. Once the Zener effect has started, large increases in reverse current may continue with the same applied bias potential. In a diode in which Zener breakdown can occur, its current-voltage curve has a sharp corner at the breakdown region, rather than the smooth transition of Fig. 2-13.

The *intensity* of an electric field is the rate of change of voltage with respect to distance in the direction of the electric flux lines of the field. It is expressed in volts per meter, and this is equivalent to newtons of force per coulomb of charge. The force exerted on an electron is therefore calculated by multiplying the electric-field intensity (in volts per meter) by the charge of the electron, which is -1.602×10^{-19} C.

It has been determined that the field intensity, $\mathcal{E}$, in volts per meter (also newtons of force per coulomb), in the junction region of a germanium diode is given by

$$\mathcal{E} \approx 2 \times 10^5 \sqrt{2E/\rho} \qquad (2\text{-}1)$$

where E is the voltage across a well-defined junction region and ρ is the resistivity of the material on either side of the junction, expressed in ohms-meters. (The resistivity of a material is the reciprocal of its conductivity.) By the units ohm-meter is meant ohms resistance across opposite faces of a meter cube of the material. If the voltage across the junction is assumed to be

10 V and the resistivity is chosen as 0.04 Ω-m, the field intensity is calculated to be about 4,470,000 V/m, which is 4,470 V/mm of junction thickness. This field intensity is strong enough to dislodge a great many electrons from the valence bands and force them across the boundary, thus accounting for large increases in current.

Research has shown that the Zener effect occurs when the diode is heavily doped and when the breakdown occurs at a low voltage—about 10 volts or so. When the breakdown occurs in lightly doped crystals or at high voltages, the mechanism is something other than the Zener effect.

Operation in the breakdown region does no damage to the diode, provided the current is limited to a value set by the manufacturer. This value is determined by the power-dissipating capabilities of the crystal structure and greatly depends on the physical size of the diode. By controlling the doping concentration and other factors, diodes can be designed to have any desired breakdown voltage in the range from about 2 V to over 200 V. Diodes which are designed to operate in the breakdown mode are called *Zener diodes* or *breakdown diodes*.

A crystal diode operating in its Zener region performs many useful and important tasks. The current passing through it may be required to change over a relatively wide range. While this is happening the voltage across the diode will change very little and in some cases by an insignificant amount. We can readily see that this would be a means of holding the voltage between two points practically constant if the Zener diode were connected to the two points. Much more will be said about the operation of Zener diodes when diode applications are discussed.

DRILL PROBLEMS

D2-7 The barrier thickness in an unbiased *P-N* junction is 10^{-4} cm with a potential difference of 0.6 V. Calculate the field intensity $\mathcal{E}$ in the barrier region in units of volts per meter.

D2-8 What force, in newtons, is exerted on the electron by the field in the barrier region?

D2-9 Assuming the *P-N* junction is in a germanium diode, calculate the approximate value of the resistivity of the crystal material.

2.19 Varactor Diodes

In the discussion associated with Fig. 2-12 it is established that a reverse-bias voltage widens the ion layer at the *P-N* junction. The ion layer represents a space charge in the barrier region, since the ions are not free to move but are bound in the crystal. Because the width of the barrier region changes as the voltage applied to the diode changes, a mechanism exists to vary the stored charge with variations in applied voltage. This effect acts as a capacitance and is called the *junction-transition capacitance*.

The stored charge varies with the applied voltage, but not in direct proportion to the voltage, so that the junction-transition capacitance is not a constant but is a function of the applied voltage. The barrier region at the *P-N* junction acts much as a parallel-plate capacitor whose spacing between the plates varies with applied voltage, thereby varying the capacitance. This capacitance is present during forward biasing, but its effect is masked by the large forward current. It becomes important when a reverse bias is applied to the diode.

Since the diode will conduct very little current under reverse bias conditions, the transition capacitance operates as a high-quality capacitor. Its value is dependent on the value of applied voltage and leads to many valuable applications in electronic circuits. A voltage-controlled capacitor can be connected into the resonant circuit controlling a high-frequency oscillator, and the frequency of the oscillator can be controlled by the voltage applied to the diode. Such capacitors can also be used as amplifiers, known as *parametric amplifiers*, because the amplifying action depends on the variation of a circuit parameter. Some junction diodes are especially made to use the transition capacitance in electronic circuits. They are called *varactor diodes*.

2.20   Tunnel Diode

A phenomenon referred to as *tunneling* involves the action of electrons in passing through the junction of a semiconductor diode, when they have received a large increase in energy due to a strong electric field. Recent research has determined that this phenomenon can be built into semiconductor diodes. They are called *tunnel* diodes and have many important features. Among them are the following.

1. Electrical charges move through the junction at speeds approaching the speed of light, rather than at the much slower speeds that occur in ordinary diodes and transistors. Because of this, tunnel diodes give promise of usefulness at extremely high frequences (2 GHz to 10 GHz) in electronic amplifiers, oscillators, and computers.

2. They may operate at temperatures as high as 650°F, whereas conventional silicon diodes stop working at about 400°F and germanium at about 200°F.

3. They resist the damaging effects of nuclear radiation environments which are very harmful to transistors.

4. They exhibit a *negative resistance* characteristic which enables them to amplify and generate power at radio frequencies.

5. Their simplicity permits their fabrication in *micro-modules* and in *integrated circuits* containing complete stages of amplifiers and oscillators formed on a single semiconductor structure not much larger than the letter "o" in the word *semiconductor* in this sentence.

2.21 Negative Resistance of a Tunnel Diode

A tunnel diode exhibits a startling current-voltage characteristic curve compared with that of ordinary junction diodes. The curve of Fig. 2-14 represents the characteristic of a tunnel diode. The shape of the curve in the forward-bias region is quite unusual and it does not behave in the reverse-bias region as a junction diode does.

By doping the semiconductor in increasingly greater amounts, thus producing higher and higher concentrations of charge carriers, the critical voltage causing reverse (avalanche) breakdown has been continuously reduced *even past the zero-voltage point and into the region of small forward-bias voltages.* In this condition, electrons may *tunnel* through the barrier almost instantly, driven by a high field intensity across the junction. The tunnel diode is made by producing an extremely thin barrier region, less than a millionth of an inch, in germanium or silicon. Doping compounds such as gallium arsenide, gallium antimonide, and indium antimonide have been used. Even a forward-bias voltage as small as 0.1 V produces a field intensity

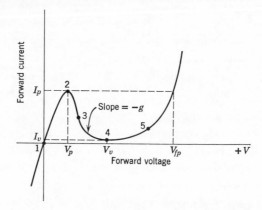

Fig. 2-14   Characteristic curve of a tunnel diode.

of 100,000 V/in. (0.1 V across a millionth of an inch). The current set up by tunneling electrons has been called the *Esaki current* in honor of the scientist who first discovered it. It is shown as the sharp hump in the curve just to the right of the origin in Fig. 2-14.

As forward bias is increased past the hump, current quickly decreases to a low value. Further increase in bias voltage generates a characteristic curve which resumes the conventional shape of a junction diode curve and soon gets steep, as one might expect.

The downward slope of the curve reveals the startling fact that the current *decreases* as the applied voltage *increases* over a certain region. We say that the tunnel diode has *negative resistance* in this region of its characteristics because this is just the opposite of the behavior of normal conductors. This feature of the tunnel diode makes possible its use as an amplifier, an oscillator (a generator of electric voltage and power), and an extremely rapid switching device. The switching function seems obvious when we see from the curve that by reducing the applied voltage a very small amount the current passing through the diode can be made to increase to ten or fifteen times its very small value at the lowest point on the curve.

The tunnel diode is expected to replace the transistor in many of its applications. One thing seems sure—the tunnel diode will be doing jobs at speeds and temperatures at which other semiconductor devices cannot work. Its resistance to damage by nuclear radiation should give it priority in space vehicle and missile circuitry and in applications related to nuclear power production and control.

2.22 The Fermi Level of Energy

The concept of the *Fermi level* of energy in semiconductors will be useful in describing the behavior of a tunnel diode under various biasing values. In a *pure* crystal of semiconductor material, the two most important energy bands are the valence band and the conduction band. Between them is the forbidden region in which no stable electron energies can exist. If the temperature of the crystal is reduced to 0°K (−273°C), all energy levels in the valence band will be occupied and the energy levels in the conduction band will be completely empty.

Now assume that the crystal is brought up to room temperature—about 25°C. The heat, which is a form of energy, taken on by the material will raise the energy of some electrons sufficiently to enable them to break their covalent bonds and move about in the crystal. They are elevated to the conduction band, and each electron thus leaves behind a *vacant energy level* which it formerly occupied in the valence band.

It is desirable to have a particular energy level as a reference to which all electron energies in a solid may be compared. A kind of average value, perhaps midway between the valence and conduction bands, might be a good choice. In practice, it is chosen at an energy value, called the *Fermi level*, for which there is a 50 *per cent probability of occupancy*. For a pure semiconductor crystal, the Fermi level is midway between the valence and conduction bands. However, in doped crystals, donor and acceptor impurities provide energy levels within the forbidden region of a pure, undoped crystal. The Fermi level is shifted slightly by the presence of impurities. In any case, it is the *most probable energy value* that will be found among the electron energies.

2.23   Current-Voltage Characteristic of a Tunnel Diode

The relative positions of the Fermi energy levels in the *P*-type and *N*-type regions of the tunnel diode depend on the amount of bias voltage. At zero bias (point 1 in Fig. 2-14), the Fermi levels are at the same value on the energy scale. This is illustrated in Fig. 2-15*a*. As forward bias is applied and increased, the current increases until point 2 is reached. Here the energy states are represented in Fig. 2-15*b*. The current is due to electrons *tunneling* through the junction from *N* to *P* and filling the vacancies (called empty states) in the *P* region.

At maximum current (point 2) the Fermi level on the *N*-type side is just opposite the top of the valence band on the *P*-type side. The top of a valence band and the bottom of a forbidden gap are at the same energy level. Tunneling ceases when the bottom of the conduction band on the *N* side is opposite the top of the valence band on the *P* side, i.e., when these band boundaries have the same position on the energy scale.

Further increase of the bias potential to that in the region of point 3 results in some of the electrons in the conduction band of the *N* material having energies which are in the forbidden gap of the *P* material. These cannot go across into the *P* region and the current is thereby reduced. At a bias voltage in the region of point 4, practically all the electrons in *N* are at forbidden-gap energy levels, and the current goes to a very low minimum.

Finally, further increase in bias voltage to point 5 and beyond results in normal forward-bias conduction, as occurs in a crystal diode.

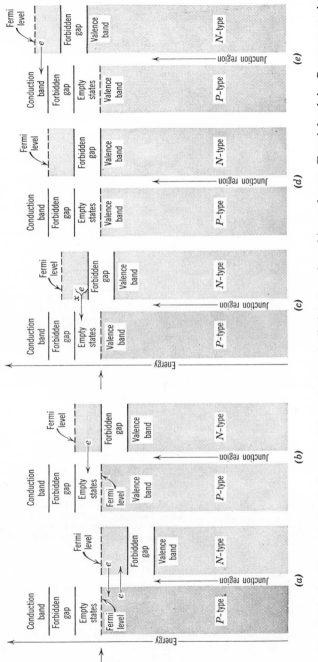

Fig. 2-15   Energy and conduction conditions in tunnel diode at various bias voltages. Fermi level in *P*-type material assumed constant for reference. (*a*) No bias voltage. Electrons (*e*) pass through junction region in both directions. (*b*) Bias increased to voltage value at point 2 in Fig. 2-14. Large current of electrons tunneling across junction to fill empty states. (*c*) Bias increased to voltage value at point 3 in Fig. 2-14. Some electrons in *N* region are opposite forbidden gap. Current decreased. Only electrons in *x* region can tunnel. (*d*) Bias increased to voltage value at point 4 in Fig. 2-14. All electrons are opposite forbidden gap. Very small current flows. (*e*) Bias increased to voltage value at point 5 in Fig. 2-14. Electrons' energies have been raised until they "spill over" into the conduction band of the *P*-type region.

2.24   Circuit Relations of a Tunnel Diode

The current at point 2 of Fig. 2-14 is called the *peak current*, I_p, and it is the value at which negative resistance first occurs. The voltage corresponding to this current is called the peak voltage, E_p. The value of current at which negative resistance (and negative conductance) ceases is called the *valley current*, I_v, and the corresponding voltage is called the valley voltage, E_v. These are the values at point 4, the minimum-current point. The ratio of I_p and I_v is termed the peak-to-valley ratio and the difference $(E_p - E_v)$ is termed the voltage swing when referring to the circuit operation of a tunnel diode. For many practical purposes it is sufficiently accurate to define the negative *conductance* as

$$-g = \frac{2(I_p - I_v)}{E_v - E_p} \tag{2-2}$$

as the average slope in the region between points 2 and 4 in Fig. 2-14.

The equivalent circuit which has been generally accepted for the tunnel diode is shown in Fig. 2-16a and its schematic-circuit symbol in Fig. 2-16b. Inductor L_s represents series inductance of the leads from the *P-N* junction to the external terminals. In some tunnel-diode assemblies, two leads are used

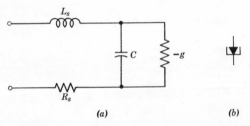

(a)   (b)

Peak-point current, $I_p = 1$ mA
Valley-point current, $I_v = 0.1$ mA
Peak-point voltage, $E_p = 55$ mV
Valley-point voltage, $E_v = 350$ mV
Voltage for forward-peak-point current, $E_{fp} = 500$ mV

Self-resonant frequency $= \dfrac{1}{2\pi}\left[\dfrac{1}{L_sC} - \left(\dfrac{g}{C}\right)^2\right]^{\frac{1}{2}}$;
typical value, 0.95 GHz

Resistive cut-off frequency $= \dfrac{1}{2\pi}\dfrac{|g|}{C}\sqrt{\dfrac{1}{R_s|g|} - 1}$;
typical value, 2.2 GHz

Ratio $I_p/I_v = 10$
Negative conductance $g = 0.0066$ mho
Total capacitance $C = 5$ pF (typical value: substantial variations in production units)
Series inductance $L_S = 6 \times 10^{-9}$ H (typical value, as above)
Series resistance $R_s = 1.5\ \Omega$
Maximum dissipation $= 50$ mW

Fig. 2-16   Equivalent circuit and electrical characteristics (25°C; $\frac{1}{8}$-in. leads) of 1N2939 germanium tunnel diode. (a) Circuit biased in negative conductance region; (b) schematic circuit symbol. (Courtesy General Electric Co., Syracuse, N.Y.)

in parallel to reduce the inductance and resistance. R_s is the series resistance of the *internal leads*, the *ohmic contacts*, and the *semiconductor materials* on both sides of the junction, all lumped together in one value. The capacitor C is the *capacitance of the junction*, and $-g$ is the negative conductance. Typical values of these circuit components are given in the figure.

The ability of the tunnel diode to function well at extremely high frequencies makes it useful in ultrahigh frequency (UHF) applications and in switching operations limited to fractions of a microsecond. Tunneling has been called a majority-carrier effect occurring in fractional picoseconds (10^{-12} s). Its heavy doping renders it practically free from disturbance by nuclear energy fields. Low input-energy requirements, of the order of 1 mW, are favorable for many battery-powered applications such as missile guidance and satellite instrumentation.

SUGGESTED REFERENCES

1. A. K. Jonscher, *Principles of Semiconductor Device Operation*, Wiley, New York, 1960.

2. R. B. Adler *et al.*, *Introduction to Semiconductor Physics*, SEEC Vol. 1, Wiley, New York, 1964.

3. E. J. Angelo, Jr., *Electronic Circuits*, 2nd ed., McGraw-Hill, New York, 1964.

4. Douglas M. Warschauer, *Semiconductors and Transistors*, McGraw-Hill, New York, 1959.

QUESTIONS

2-1 Discuss the physical concept of an isolated atom, using the Bohr model, in terms of a nucleus, electrons, energy levels, and shells.

2-2 Tell all you can about valence electrons.

2-3 What is a quantum of energy? How is it related to Planck's constant? In what units is Planck's constant expressed?

2-4 Why are copper and silver good conductors of electricity?

2-5 Aluminum is a light-weight electrical conductor. Why is it not more universally used to take advantage of its light weight?

2-6 Why is sulfur a good electrical insulator?

2-7   Why are elements like germanium and silicon called semiconductors? How do the Bohr models of these atoms differ from the models of atoms of conductors and insulators?

2-8   What is meant by covalent bonding?

2-9   What is a donor element in a semiconductor crystal, and what purpose does it serve? Name two donor elements.

2-10   What is an acceptor element in a semiconductor crystal, and what purpose does it serve? Name two acceptor elements.

2-11   Describe the concept of a "hole" in semiconductor theory.

2-12   Compare *N*-type and *P*-type germanium.

2-13   What is a *P-N* junction? How is it formed? Why can't a satisfactory *P-N* junction be formed by forcing together pieces of *N*-type and *P*-type crystals?

2-14   Both *N*-type and *P*-type semiconductors have *majority carriers*. Explain what they are.

2-15   Discuss the terms conduction band, valence band, and forbidden band.

2-16   What is a *minority carrier*? Describe minority-carrier current in *N*-type and *P*-type semiconductors.

2-17   Discuss the internal behavior of a *P-N* junction under (*a*) forward bias, and (*b*) reverse bias. Relate this behavior to a plot of current-voltage characteristics of a junction diode.

2-18   Describe the potential hill across the barrier region of a *P-N* junction and its effect on current through the junction.

2-19   What is meant by avalanche breakdown?

2-20   Describe the Zener effect in a semiconductor diode.

2-21   Name the special features of the tunnel diode that make it superior to a conventional semiconductor diode.

2-22   Describe the action of a varactor diode that makes it behave as a voltage-controlled capacitor.

2-23   Discuss the current-voltage characteristic curve of the tunnel diode.

2-24   Why is the factor 2 in Equation (2-2)?

2-25   Why are "holes" restricted to the interior of the crystal diode, whereas electrons are transported in the external circuit to permit hole conduction?

PROBLEMS

2-1   An electron in intrinsic germanium at room temperature must be excited with about 0.7 eV of energy to go from the valence band to the conduction band of energy. How many joules is this?

2-2 How many joules of energy are represented by 1.1 eV?

2-3 Calculate the energy an electron will acquire in falling through 1 V of potential difference. Express the energy in joules and in electronvolts.

2-4 Repeat Problem 2-3 for a potential difference of 100 V.

2-5 Suppose the electron of Problem 2-1 receives the 0.7 eV from a photon that strikes it and gives up its energy. Calculate the frequency of the photon.

2-6 Visible-light frequencies are in the range from about 4×10^{14} to 7×10^{14} Hz. Determine the minimum frequency of photons that may excite electrons from the valence band to conduction bands in germanium at room temperature. Is this frequency in the visible range?

2-7 Repeat Problem 2-6 for silicon at room temperature.

2-8 Using the data of Table 2-1, calculate the change in ionization energy of intrinsic germanium when its temperature is increased from 0°K to room temperature (27°C).

2-9 Repeat Problem 2-8 for intrinsic silicon.

2-10 What is the ionization energy of intrinsic germanium at room temperature (27°C)?

2-11 Which material decreases its ionization energy faster with increases in temperature—intrinsic germanium or intrinsic silicon? Show clearly how you know.

2-12 A germanium *P-N* junction diode has its bias suddenly changed from 1.5 V in the forward direction to 200 V in the reverse direction. Sketch circuit diagrams for these two cases, showing the correct polarity for external bias batteries. Mark the direction of current in the diode in each case.

2-13 If the diode of Problem 2-12 has the characteristic of Fig. 2-13, what are the magnitudes of current before and after the change?

2-14 How much is the current magnified in the diode of Problem 2-13?

2-15 The barrier thickness in an unbiased germanium *P-N* junction is 10^{-4} cm, with a potential difference of 0.6 V as a barrier potential. (*a*) Calculate the field intensity in the barrier region and express it in volts per meter. (*b*) What force (in newtons) is exerted by the field on the electrons in the barrier region?

2-16 In an *N*-type germanium crystal the impurity atom and the excess electron form a hydrogen-like structure. The electric field in which the electron moves can therefore be assumed to be $\mathcal{E} = q/4\pi\epsilon_0 r^2$, where q is the electronic charge, $\epsilon_0 =$ permittivity of free space, and r is the radius of the electron orbit in germanium (about 5.47×10^{-10} m). Estimate the energy necessary to permit the electron of the impurity atom to escape, assuming it is arsenic. Express it in electronvolts.

2-17   Using typical values for the 1N2939 tunnel diode given in Fig. 2-16, calculate the input impedance at 1.59 GHz (1.59×10^9 Hz), using the equivalent circuit for the diode given in the figure.

2-18   Compute the self-resonant frequency of the tunnel diode described in Fig. 2-16, using the given typical values for the circuit parameters.

2-19   The peak-to-valley ratio for the 1N2939 tunnel diode (see Fig. 2-14) is close to 10 and the valley-point current is 0.1 mA. The peak-point voltage, V_p, is 55 mV and the valley-point voltage, V_v, is 350 mV. Assuming the voltampere characteristic of the tunnel diode to be linear between V_p and V_v, find the resistance of the device in this region. Compare your answer with the value specified by the manufacturer.

2-20   The barrier potential across the junction in a typical tunnel diode is estimated to be about 0.8 V. (*a*) What is the minimum bias that should be applied if an electron were to climb over this potential barrier? (*b*) The bias value found in (*a*) is impressed on the 1N2939 tunnel diode. Explain in what way the operation of the tunnel diode will be affected.

CHAPTER THREE

Crystal Diodes

This chapter will extend the discussion of semiconductors and the operation of *P-N* junctions. It is natural that we should want to learn about their fabrication in the construction of crystal diodes and how these devices perform in some typical circuit applications. We shall see how they are used to control voltages, to protect other circuit elements, to act as switches, and to *rectify* alternating current, i.e., obtain a direct current in a load from an alternating current source. Each of these applications is important in a multitude of electronics circuits, from a simple laboratory voltmeter to complex systems of computers and industrial control.

3.1 Preparation of Semiconductor Metal

Depending on the type of semiconductor being formed, the preparation and structure of the metal may produce a nearly perfect single crystal of the material or a polycrystalline state. Crystal diodes and transistors depend for their operation on a single-crystal structure. Its preparation will be discussed here. Since these devices perform as the result of added impurities in the crystal, and any defects in the crystal behave much the same as impurities, the preparation of germanium and silicon metal for transistor and diode manufacture has become a major industry.

In order to have the desired control over the concentration of impurities in *P*-type and *N*-type semiconductors (from 1 part in 10^5 to 10^7), the raw metals are purified so that the impurity concentration is as low as 1 part in 10^{11}. Afterwards, control of the dominant impurity can be obtained by doping the material during the manufacture of single crystals.

Germanium is usually obtained as a byproduct of zinc refining in the United States and from certain coal dusts in Europe. It is available commercially in compound form as germanium dioxide and germanium tetrachloride. The initial purification of silicon typically involves chemical reactions which produce silicon tetrachloride or dioxide. These compounds can be processed to obtain metallic germanium or silicon of relatively high purity. Its purity can be developed further by a process called *zone refining*.

The zone-refining technique for metal purification makes use of the fact that many impurities are more soluble when the metal is in a liquid state. If a short section of a germanium bar is melted and the melted region caused to move slowly from one end of the bar to the other, most of the impurities present in the metal tend to remain dissolved in the liquid state. Thus, as the melted region is moved from one end of the bar to the other, the impurities tend to concentrate at one end of the bar. Such heating of small bars of metal is most easily done by the induction heating process. The heat necessary to produce a narrow molten zone in the bar is accomplished by coils encircling the bar and carrying radio-frequency energy to induce eddy currents in the bar.

A simplified zone-refining apparatus is shown in Fig. 3-1. A bar of low purity germanium or silicon is placed in the apparatus so that the induction

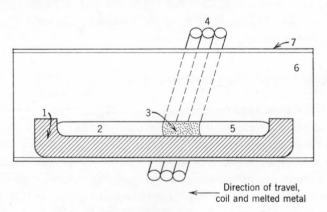

Direction of travel, coil and melted metal

1, Graphite boat
2, Low purity metal
3, Molten zone
4, Induction coil
5, High purity solid metal
6, Inert atmosphere
7, Quartz container

Fig. 3-1   Zone refining apparatus for purifying germanium metal.

heating coil surrounds it at the right end of the figure. As radio-frequency energy is applied, the right end of the bar will melt. Once melting has started, the coil is moved slowly to the left at such a rate that the molten zone moves along the bar at the same rate. In this way, the impurities in the liquid state move to the left end of the bar. As the coil progresses from one end to the other, the purity of the material to the right of the molten zone is increased as it returns to a solid state. Several passes of the bar through the apparatus may be necessary to yield the desired purity of the germanium.

It is important that the metal be protected from other contaminants during the refining process. Graphite and quartz are used in the apparatus to hold the metal, and the container is either filled with an inert gas or held as a vacuum. This prevents the introduction of additional impurities into the melt.

After purification to the degree necessary, the metal is ready for doping and forming into a single crystal. One common method for growing a single crystal is shown in Fig. 3-2. The crucible contains metal at a temperature a few degrees above its melting point. A small piece of single crystal of the metal, previously prepared, is lowered into the liquid metal. This small piece is referred to as a *seed crystal*, because, as the seed crystal is slowly withdrawn from the melt, it permits the formation on it of another metal having the same uniform lattice structure. If the temperature conditions are properly set and controlled, a single crystal of the metal can be *grown* onto the seed crystal until all of the metal is contained.

A more recent technique for both refining and growing single crystals has been introduced. It is called the *floating zone* method, and it is very similar to zone refining except that the graphite boat is eliminated. Instead, clamps at both ends of the bar hold it in a vertical position. Absence of the graphite container helps to reduce possible contamination during the process. The metal in the molten zone is held in place by surface tension of the liquid metal. Doping materials can be added at one end of the bar, and they will be distributed uniformly through the crystal by a single cycle of zone refining. Thus, the metal can be refined and grown into a doped, single crystal in one operation. This technique has been used successfully in producing high-quality silicon semiconductors.

3.2   Forming a *P-N* Junction

During the growth of single crystals of semiconductor metals, appropriate impurities may be added to the melt to form either a *P*-type or an *N*-type

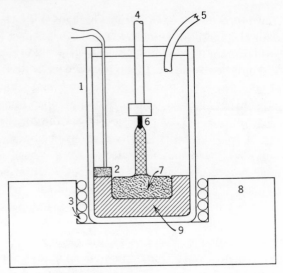

1, Quartz container
2, Thermocouple, to measure temperature
3, Heating coils
4, Pulling rod
5, Gas inlet
6, Seed crystal
7, Molten metal
8, Insulation
9, Graphite crucible

Fig. 3-2   Furnace for growing single-crystal germanium.

semiconductor. Their concentrations can be controlled and the growth of P-N junctions in the crystal permitted with temperature and rate control. Alternate doping with P-type and N-type impurities produces several junctions, each of which has the required change of conductivity across the junction. Each junction will behave as a P-N junction diode. When diodes are formed during crystal growth in this way, they are called *grown-junction diodes*.

The grown-junction technique for forming a P-N junction is used to manufacture crystal diodes and transistors. The fundamental difference in these two devices is the fact that the diode has only one P-N junction, whereas the transistor has *two*, separated by a very narrow region. A separate chapter is devoted to transistors, so only the diode will be discussed here.

There are two major examples of the grown process: rate-grown and grown-diffused. These processes are illustrated in Fig. 3-3 and discussed in the following sections.

In the rate-grown process, molten germanium metal in the crucible contains both donor and acceptor impurities. The donor element is sensitive to the rate of growth, so that the amount of this impurity in the crystal varies as the growing conditions change. While the temperature of the melt is dropping, the crystal grows very rapidly as the metal solidifies. Then heating power is rapidly applied, causing the growth to stop and the crystal starts to remelt. As the metal is allowed to cool again, melting stops and the crystal begins to grow. At the boundary where the growth rate is zero, the acceptor impurity will predominate, forming a *P* region across the germanium crystal. A junction is thus formed between the *N* region and the *P* region, forming a grown-junction diode. Repeating this process several times as the crystal grows will yield *N-P-N* structures which can be used to manufacture transistors.

The grown-diffused process is started with a crystal that is doped to some desired conductivity. *N*-type and *P*-type impurity elements are added to the molten material. The addition of other impurities to the melt while the crystal is growing increases the total impurity concentration. By taking advantage of the different diffusion rates of donor and acceptor elements and by controlling the added amounts of each, a junction area can be formed in the crystal. This method is used to manufacture transistors in silicon, where the higher diffusion of acceptor elements produces a narrow *P* region between two *N* regions. Thus, the product is actually two junctions separated by the

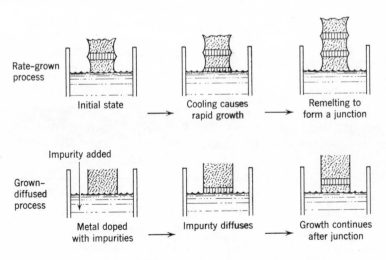

Fig. 3-3   Processes for forming grown-junction diodes.

narrow *P* region, leading to an *N-P-N* transistor structure. Of course, machine processing may be used to slice the junction area in order to form two *P-N* junction diodes.

These processes can be used for making crystal diodes, but other methods are preferable. They are discussed in the following section.

3.3   Impurity Contact Junctions

While the grown processes are suitable for forming a *P-N* junction in a semiconductor crystal, they are better suited for manufacturing transistor structures, which require two junctions spaced so closely together that they interact with each other. A definite single junction is desirable for use in a crystal diode.

Three basic methods for producing single junctions in a semiconductor crystal are: alloy, diffusion, and point-contact. These techniques involve contact between solid forms of semiconductor and impurity, rather than growth from a molten batch of doped metal. These methods are illustrated in Fig. 3-4.

In the alloy process, a wafer of semiconductor doped with the desired impurity has a small pellet of impurity material pressed against it. Heat is applied to melt the pellet and some of the wafer and yields an alloy where the

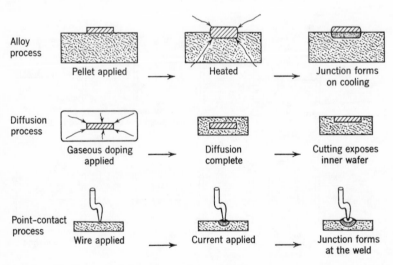

Fig. 3-4   Junction formed by impurity contact methods.

materials meet. After the heat is removed the solution becomes solid again. At the alloy-semiconductor boundary, a heavy concentration of donors or acceptors is formed. The boundary then behaves as a *P-N* junction diode. This method is also employed in the manufacture of transistors, using two pellets each, in multiple graphite molds.

In diffusion processes, a wafer of semiconductor having appropriate doping is heated in a gaseous atmosphere containing the opposite impurity. Control of the temperature and pressure causes some of the gaseous impurity to diffuse into the wafer. When germanium is used as the host material, donor elements diffuse more rapidly than acceptor elements, but either type of impurity may be used to form a diffusion layer on the wafer. After the diffusion cycle is complete, cutting and etching of the wafer yields a *P-N* junction diode.

Early forms of diodes were of the point-contact type. In their manufacture, a thin wafer of *N*-type semiconductor, a few millimeters square and a fraction of a millimeter thick, has a fine pointed wire called a *cat whisker* pressed against one surface. The wire is usually phosphor-bronze, which is quite stiff. Its tip is welded to the crystal surface by passing a heavy current through the wire during a short time interval. Atoms of the phosphorus, an acceptor impurity, diffuse from the wire tip into the *N*-type crystal and form a *P-N* junction in the region close to the point of contact.

Uniformity in characteristics of crystal diodes is desired, whether they are manufactured by one process or another. The ability to manufacture diodes of the same type and characteristics is determined in part by the methods used to form the junction. The rate-grown process, for example, will yield several junctions from the same crystal and they will be similar if proper control of the variables is maintained. A point-contact type of diode, on the other hand, will be affected by the placement of the contact wire as well as by variables associated with the welding process. Uniformity in lead attachment and encapsulation of the finished product can contribute effectively to promote uniformity of the junctions in crystal diodes.

3.4 Packaging Techniques

Ohmic contacts are required for attaching external leads to crystal diodes. Materials which will not produce extraneous *P-N* junctions are used to connect the *N* and *P* regions to external leads. Unless care is taken in forming the ohmic contacts, the diode may not function well at high currents and high temperatures. Some of the diffused junctions have only a small area for attaching contacts. This is a common cause of the formation of spurious junctions.

Connecting leads may be made of aluminum, gold, indium, nickel, or other suitable metals. Gold is especially useful because it can be readily doped either *P* or *N* type.

Alloying, soldering, and welding are used for attaching leads to the crystal

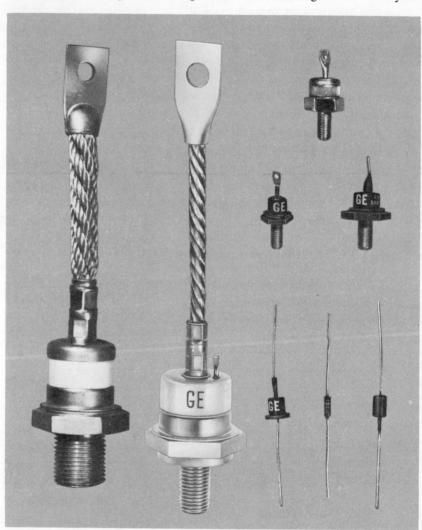

Fig. 3-5 Typical silicon and germanium diodes. (Courtesy of General Electric Co., Syracuse, N.Y.)

structure. Gold and aluminum can be alloyed with germanium and silicon. They are extensively used in the manufacture of crystal diodes and transistors. Fluxless soldering is the preferred method in some cases, particularly for attaching leads to the pellet of alloy-process diodes. Close control of the manufacturing processes is necessary to maintain a balance between good electrical contact and mechanical ruggedness.

Encapsulation is the process which seals the finished diode and its connecting leads into a suitable structure that will permit it to be handled and inserted into a practical circuit. Its primary purpose is to assure reliability of the device when it is used as a circuit component. It protects the junction and its ohmic contacts from mechanical damage and provides a seal against damaging environmental effects. Diodes that are designed to carry high currents and dissipate large amounts of power during operation are mounted on metal-stud conductors and require external leads made of large wire. Low-power diodes are encapsulated in small metal cans or in glass cylinders. Examples of germanium and silicon diodes as they have been encapsulated are shown in Fig. 3-5.

3.5 Diode Ratings

Ratings are the limiting values assigned by a manufacturer to various parameters of a diode and to environmental factors in its operation. These assigned values, if exceeded, may result in permanent damage to the diode or impairment of its performance or life.

The load ratings of semiconductor diodes are based on their ability to dissipate heat losses and thus not exceed the junction temperature limits specified by the manufacturer. Losses are usually given in watts, determined by electrical characteristics such as forward voltage drop during conduction and leakage current in the reverse direction. Thermal characteristics are specified as a certain number of degrees temperature rise per watt at the junction, measured from some reference temperature such as that of the metal stud or of ambient conditions.

Manufacturers usually specify semiconductor ratings according to the symbols and definitions shown in Table 3-1.

3.6 Diode Characteristics

Characteristics of a semiconductor diode are those measurable properties or attributes that are inherent in its design. We have looked previously at the

TABLE 3-1

Term	Symbol	Definition of Rating
Peak reverse voltage or Peak inverse voltage	PRV or PIV	Maximum allowable *instantaneous* reverse voltage that may be applied across the diode under other specified conditions. Voltage may be a sine wave, but a continuous d-c rating, defined below, is not specified. While this rating does not refer to a "breakdown" voltage value, it should not be exceeded on an instantaneous, repetitive basis.
Transient peak reverse voltage	PRV_{trans}	Maximum allowable instantaneous value of reverse voltage that may be applied on a "one-shot," nonrecurring basis. Its duration and other conditions are specified with the rating.
Reverse d-c voltage, for continuous operation	V_{RDC}	Maximum reverse voltage that the diode may block over long periods of time.
Maximum d-c output current	I_F	Maximum d-c forward current which may be allowed under stated conditions of temperature and reverse voltage. This value will be the *average* current in the forward direction, when the load is resistive or inductive. Derating factors are usually applied for capacitive loads and other conditions.

graphical plot of current and voltage variations for a typical crystal diode. Such a display of characteristics is useful in the analysis of circuits containing the diode as a component. However, manufacturers usually present the important characteristics of semiconductor diodes by tabulating typical values at specified operating points, rather than a continuous plot point-by-point.

The characteristics which are usually specified by a manufacturer are given in Table 3-2.

TABLE 3-2

Term	Symbol	Definition of Characteristic
Forward voltage drop	V_F	Value of instantaneous forward voltage drop during conduction of load current, and under any other specified conditions
Reverse (or leakage) current	i_R	Instantaneous value of reverse current at the stated conditions of temperature and voltage

As one might expect, the characteristics of the point-contact diode are similar to those of the junction diode. However, the point-contact diode, because of its design, is inferior to the junction diode in many of its characteristics. For example, it cannot carry nearly as much current because of heat which is generated in its comparatively small active area. Further limitations are: (1) reverse current increases significantly with reverse voltage instead of remaining practically constant, and (2) its peak-inverse-voltage (PIV) rating is much lower than that of a junction diode. Figure 3-6 illustrates the

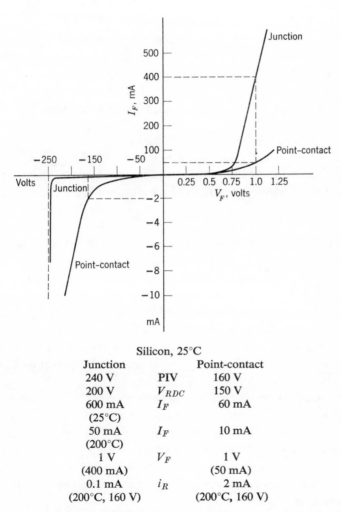

Silicon, 25°C

Junction		Point-contact
240 V	PIV	160 V
200 V	V_{RDC}	150 V
600 mA	I_F	60 mA
(25°C)		
50 mA	I_F	10 mA
(200°C)		
1 V	V_F	1 V
(400 mA)		(50 mA)
0.1 mA	i_R	2 mA
(200°C, 160 V)		(200°C, 160 V)

Fig. 3-6   Comparison of point-contact and junction diode characteristics.

basic differences between a point-contact and a junction diode, by comparing their ratings and characteristics.

The average current-voltage characteristic of a typical point-contact diode closely resembles the portion of the curve for a junction diode in forward conduction, except for the smaller maximum current. On the reverse-bias side of the curve, leakage current starts to increase rapidly with increasing reverse voltage. This illustrates the lack of a small reverse-saturation current, which is so important in many applications of diodes in practical circuits. The unilateral conductivity of semiconductor diodes, the ability to pass current readily in one direction and to effectively block it in the reverse direction, is their most important property. Point-contact diodes are inferior to junction diodes in this respect. Junction diodes can carry larger forward currents and they have higher peak-inverse-voltage ratings.

3.7   Comparison of Germanium and Silicon Diodes

Either germanium or silicon may be used as the host material in a semiconductor crystal diode, but silicon has many favorable properties which make it superior to germanium. In general, however, silicon units cost much more than comparable germanium units, but they can be designed to perform in many applications where germanium units simply cannot meet the specifications.

Silicon can operate at temperatures up to 200°C (392°F), while germanium is limited to about 105°C (221°F). Silicon diodes can be designed to carry currents up to 250 A and to operate under reverse voltages to 1,000 V/cell. Germanium is limited to a maximum of about 400 V PIV per unit, although diodes can be connected in series to obtain higher voltage ratings when circuit conditions require them. When so connected, each diode should be shunted with a resistance of 100 kΩ or more to equalize the diode reverse voltages which would otherwise divide in proportion to the individual diode reverse resistances.

Germanium is superior to silicon in one important property. It has lower forward voltage drop. The lower cost of germanium units for use in low-power installations may make this property an advantage. Typically, silicon diodes exhibit about 1-V forward voltage drop, while germanium units have a forward voltage drop of about 0.6 V when operated under similar conditions. Figure 3-7 illustrates and compares forward characteristics of similar silicon and germanium diodes.

The reverse-bias characteristics of crystal diodes deviate from a nominal

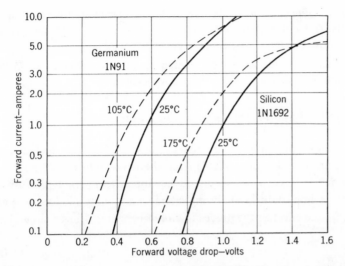

Fig. 3-7   Forward characteristics of silicon and germanium diodes. (After *Rectifier Components Guide*, courtesy of General Electric Co., New York.)

reverse saturation current and depend on both the reverse voltage and junction temperature. Figure 3-8 shows the relative shapes of reverse characteristics and changes that occur because of their temperature dependence.

DRILL PROBLEMS

D3-1   A certain diode has these specifications supplied by the manufacturer: PIV 200 V, V_{RDC} 200 V, I_F (100°C case temperature) 1.5 A, V_F (1.5 A) 1.2 V, i_R (150°C, 200 V) 0.1 mA. Would you expect it to be silicon or germanium?

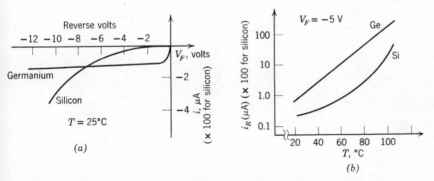

Fig. 3-8   (a) Reverse-bias characteristics for silicon and germanium diodes; and (b) effects of temperature on saturation current. Note the change in current scale for silicon diode.

D3-2   A derating factor is supplied for the diode: "Derate I_F for case temperature above 100°C by the factor -2.5 mA/°C." What is the maximum permissible forward current at 200°C case temperature?

D3-3   What temperature in degrees Fahrenheit is: 25°C, 100°C, 150° C, 200°C? (Temperature conversions are given in the Appendix.)

3.8   Diode Symbols for Circuit Diagrams

The curves representing characteristics of crystal diodes, such as those in Fig. 3-6, are usually not available from the manufacturer, although ratings and characteristics may be furnished in tabular form for typical operating values. It is important in analyzing circuits containing diodes to be able to investigate the effect of the diodes in circuit behavior. Linear circuit analysis

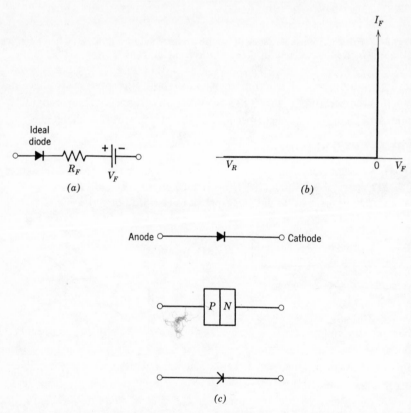

Fig. 3-9   Circuit model and symbols for crystal diode. (*a*) Circuit model. (*b*) Characteristics for ideal diode. (*c*) Symbols.

techniques are easier to work with than are nonlinear techniques. For this reason a crystal diode is often represented in circuit analysis by a linear model which describes its characteristics over the operating region of currents and voltages. A reasonably accurate model for a crystal diode is given in Fig. 3-9, together with various symbols used in circuit diagrams. Notice that the arrow of the symbol points in the direction of forward current.

The ideal diode, whose characteristic is presented in Fig. 3-9b, has no leakage current and no forward voltage drop. The forward voltage drop of an actual diode is taken into account by the internal battery shown in Fig. 3-9a. The internal, linear resistance element (R_F) represents the approximately linear variation of forward current with forward voltage drop. When a crystal diode is operated at reverse voltages below its breakdown potential and within other maximum ratings, this model is sufficiently accurate for analysis of circuits containing diodes. The model more nearly approximates the behavior of junction diodes than point-contact diodes, and the applications which are discussed in subsequent sections are assumed to use junction diodes.

In many of the circuit diagrams used to illustrate typical applications of crystal diodes, the symbol may be that of an ideal diode, but it should be analyzed in circuit behavior as in Fig. 3-9. In some applications, it may not make much difference if an ideal diode is assumed. In such cases the forward voltage drop and reverse current are so small that they can be neglected and the diode can be represented by an ideal element.

DRILL PROBLEMS

D3-4 Determine values for the circuit model of a crystal diode whose characteristic is given by the curve for a junction diode in Fig. 3-6.

D3-5 Under what conditions can this diode be assumed an ideal element?

3.9 Diode as a Rectifier

The characteristic of an ideal diode shows that when it carries forward current, its voltage drop is zero regardless of how large the current may become. Thus, an ideal diode behaves as a short circuit to current in the forward direction. However, it behaves as an open circuit when the voltage drop across the diode is negative. No reverse current passes through the diode regardless of the magnitude of the reverse voltage. Whenever the voltage changes polarity the diode changes from short-circuit to open-circuit behavior, or the reverse.

This ideal model of a diode is, of course, not representative of any physical device but has utility in the analysis of some circuits containing diodes, as we shall see. A more accurate model is given in Fig. 3-10a and its characteristic is shown in Fig. 3-10b. This model will be used to describe the operation of a diode as a rectifier, i.e., to change an alternating voltage from a source to a direct voltage across a load.

This circuit model of a practical diode will permit forward current only when the forward voltage across the diode terminals is greater than the internal voltage drop, V_F. When forward voltage exceeds V_F, forward current is limited by the linear resistance, R_F. For voltages which are less than V_F, no current is permitted in the diode. Thus, in a series circuit containing this diode, an open-circuit condition exists until forward voltage is greater than V_F, and for greater voltages the diode behaves as a linear resistance. The ideal diode, then, is just a *switch*.

In order to illustrate the operation of a crystal diode as a rectifier, we consider the operation of the circuit in Fig. 3-11. This simple combination of circuit elements is called a *half-wave rectifier*. It will be studied in some detail in a later chapter. Here, our interest is in the behavior of the diode rather than the circuit as a whole.

A sinusoidal voltage source, $e_s = E_s \sin \omega t$, is applied to the circuit containing the diode and a resistance load, R_L. When e_s is positive and greater than V_F, it produces a forward current in the diode and R_L. Since this is the forward direction for current in the ideal diode, it acts as a short circuit and

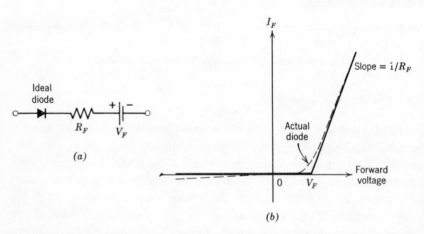

Fig. 3-10   Diode circuit model and its characteristic curve. (*a*) Circuit model. (*b*) Characteristic curve.

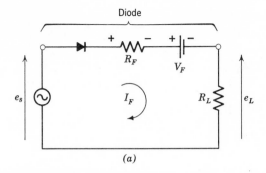

(a)

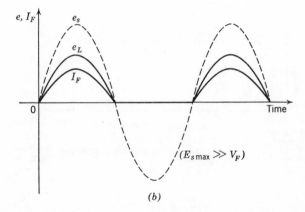

(b)

Fig. 3-11   Diode rectifier circuit and waveforms. (a) Circuit. (b) Waveforms.

the voltage across R_L will be determined by the current. For values of e_s less than V_F, no current will be supplied to the load.

The wave forms of diode current and voltages in the circuit are shown in Fig. 3-11. The load voltage may be expressed as

$$e_L = \frac{R_L}{R_F + R_L}(e_s - V_F), \qquad \text{for } e_s > V_F$$

and   (3-1)

$$= 0, \qquad \text{for } e_s < V_F$$

Current in the circuit may be expressed as

$$I_F = \frac{e_s - V_F}{R_F + R_L}, \qquad \text{for } e_s > V_F$$

$$= 0, \qquad \text{otherwise}$$

(3-2)

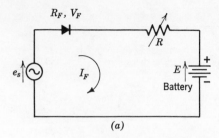

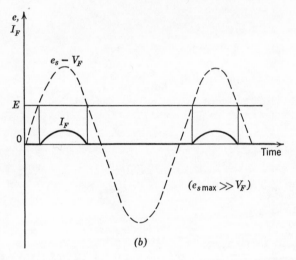

Fig. 3-12   Rectifier circuit as a battery charger. (*a*) Circuit. (*b*) Waveforms.

EXAMPLE.   The diode of Fig. 3-11 has a forward resistance $R_F = 1\ \Omega$ and internal voltage drop $V_F = 1$ V. A load resistance of 27 Ω is used to develop a rectified version of the input sinusoidal voltage, $e_s = 20 \sin \omega t$. We shall determine the magnitudes of diode current and load voltage.

From Equation (3-1) we see that load voltage will exist only when the input voltage is greater than V_F:

$$e_L = \frac{R_L}{R_F + R_L}(e_s - V_F), \qquad \text{for } e_s > V_F$$

$$= \frac{27}{1 + 27}(20 \sin \omega t - 1), \qquad \text{for } e_s > 1\text{ V}$$

When $\sin \omega t = 1$ (at 90°), the load voltage will have an amplitude of

$$e_L = \frac{27}{28}(20 - 1) = \frac{27}{28}(19) = 18.3\text{ V}$$

From Equation (3-2) we see that the diode current will also have its maximum amplitude when $\omega t = 90°$:

$$I_{F,\max} = \frac{20 \sin 90° - 1}{1 + 27} = \frac{19}{28} = 0.678 \text{ A}$$

This circuit may be used to charge a battery from a source of alternating current. The battery would be connected in the circuit in place of the load resistance. In order to accumulate charge in the battery, the diode must permit the source to force forward current into the battery. The action of the circuit as a battery charger may be understood from the wave forms in Fig. 3-12. When $e_s - V_F$ is greater than the battery potential, the diode will allow current to flow into the battery. At all other times the diode prevents current flow and I_F is zero.

A physical diode has maximum ratings that must not be exceeded in its operation. The ratings of importance in the battery charger circuit are the maximum forward current and peak-inverse-voltage. It is necessary to ensure that these two quantities do not exceed the maximum permissible values. The series resistance R shown in the figure can be used to limit the charging current to a desired value. Since the full amplitude of the source voltage is impressed across the diode in the reverse direction, the peak-inverse-voltage rating will determine the maximum permissible amplitude of the source voltage.

DRILL PROBLEMS

D3-6 A germanium diode is rated for $I_F = 100 \text{ mA}$ and a maximum forward current of 350 mA. It is to be used in a half-wave rectifier circuit supplied by a sinusoidal source of 100 V rms. (a) What value of load resistance can be used to receive maximum power without exceeding the diode ratings? (b) Calculate d-c load voltage and current when the load resistance of (a) is used.

D3-7 A silicon power diode, type 1N412A, is used to charge a 12-volt battery from a 36-V rms source in a half-wave rectifier circuit. The average charging current is to be 50 A. Assuming the battery resistance is 0.1 Ω, calculate the value of series current-limiting resistance needed. Assume diode voltage drop is 1 V.

3.10 Diode as a Voltage Limiter (Clipper)

The diode is a useful circuit element in applications requiring voltage levels to remain within prescribed amplitude levels. A circuit which uses a diode as a voltage limiter is illustrated in Fig. 3-13a. This circuit is designed to prevent the output voltage from exceeding the value of the d-c source voltage marked

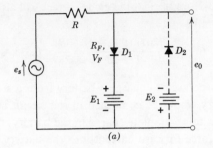

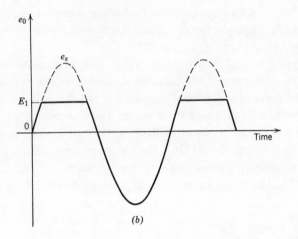

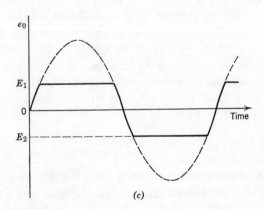

Fig. 3-13   Diode voltage limiter (clipper) circuit and waveforms. (*a*) Circuit.
(*b*) D_1 operative. (*c*) D_1 and D_2 operative.

. E_1. The circuit operation is easy to understand. When the diode is reverse-biased by the external source no current can pass through the diode and the output voltage will follow any variations of the external source. However, when this value reaches the amplitude of $E_1 + V_F$, the diode will be forward biased. Current can then pass through the diode, and the output voltage will be equal to $E_1 + V_F$ during periods of forward biasing. If the internal voltage drop of the diode is very small compared to the d-c source, then the output voltage will be practically limited to the value of E_1 as required.

It is sometimes necessary to limit voltage variations in both positive and negative ranges about some zero reference voltage. The additional circuitry and wave form of the output voltage are shown in Figs. 3-13a and 3-13c. The diode limiter might be used to alter the wave form of a sinusoid as the input voltage. If E_1 and E_2 are chosen equal in magnitude and much smaller than the input amplitude, then the output will be very nearly a square wave of voltage.

DRILL PROBLEM

D3-8 Low-power silicon diodes are used in the circuit of Fig. 3-13 to generate a nearly square wave of voltage from an input sinusoidal source of 100 V rms. (a) Assuming the diodes identical and ideal elements, what must be the magnitude of d-c limiting sources E_1 and E_2, if the square wave is to be 20 V, peak-to-peak? (b) What must be the PIV rating of the diodes? (c) What value of R is required to limit diode current to 100 mA?

3.11   Diode as a Voltage Clamper

In many electronics applications, it is necessary or desirable to eliminate either all positive or all negative amplitudes of input voltage wave forms. The circuit which can perform this alteration of an input wave form is called a *clamper circuit*. The diode voltage limiter can be extended to this application. Simply by making E_1 equal to zero, the positive output voltage will be limited to the internal voltage drop of the diode. The negative voltages will be clamped by D_2 and E_2 at the output terminal. If the other voltages in the circuit are very large compared to the internal diode drop, which is usually less than 1 V, then it can be said that the output voltage is *clamped at zero volts*. The diode limiter might also be said to be a clamper, where its output is clamped to positive E_1 and negative E_2. References, in the literature, to the use of diodes in these applications sometimes employ the terms *voltage limiter* and *voltage clamper* synonymously. (The term "limiter" is also employed for a circuit used for a different purpose in frequency-modulated (FM) communications receivers.) Of course, the action of this particular voltage clamper circuit is similar to that of the diode limiter described, the only difference being the amplitude to which the output voltage is restricted.

3.12   An A-C Voltmeter

A common laboratory multimeter, such as the Simpson Model 270, can be used to measure a-c voltages as well as d-c quantities, even though the indicating meter is a D'Arsonval moving coil type. This type of meter responds only to the *average* current passing through it. If a source of a-c voltage is applied to it, the average current will be zero and the meter will read zero. Then, some arrangement which provides an average current must be built into the multimeter to permit the measurement of a-c voltage. The meter may be calibrated to read effective values.

The diode rectifier circuit of Fig. 3-11 generates a load current, called I_F, whose average value is not zero but is proportional to the a-c voltage at the source. From an analysis of the pulsating load current, by Fourier series or other methods, it can be shown that the current will have an average value given by

$$I_F \approx \frac{E_{s,\max} - V_F}{R_F + R_L}\left(\frac{1}{\pi}\right), \qquad (E_{s,\max} \gg V_F) \tag{3-3}$$

where R_L includes the meter resistance and other series current-limiting resistances. If the voltage to be measured is much larger than the voltage drop of the diode, then the average meter current will be

$$I_F \approx \frac{0.318}{R_F + R_L}\sqrt{2}E_{s,\mathrm{rms}} \tag{3-4}$$

in which $E_{s,\mathrm{rms}}$ is the effective value of the sinusoidal source voltage e_s. Since all quantities are constants except the applied voltage, the meter may be calibrated to read rms values of sine waves. The use of a diode rectifier in this way allows a d-c meter element to be used as an a-c voltmeter.

This technique of using a diode to generate a *current* whose average value can be utilized to represent an a-c voltage is employed in many electronics applications. The simple a-c voltmeter is a very useful application, and is representative of many similar ones. They can be found where it is important to monitor or measure an alternating parameter. Another common example is the output meter on a laboratory oscillator.

It is essential that we recognize the meter scale will give correct readings only for the alternating waveform for which the scale is calibrated, i.e., either for *half-sine waves* or *full sine waves*, depending on the rectifier circuit in the multimeter. Input voltages with other kinds of waveform will generate average currents that correspond to the shapes of the waveforms but will not, in general, have the same values as those of the rectified sine waves. The a-c voltmeter can be calibrated for only one waveform and it will indicate incorrect values if an a-c voltage with a different waveform is applied. When using a measuring instrument it is important to know not only the use for which it is designed but also limitations on its performance that may be imposed by parameter wave shapes.

DRILL PROBLEMS

D3-9 An a-c voltmeter is constructed using a D'Arsonval moving-coil meter and a half-wave rectifier circuit to generate the unidirectional current for the meter. Its calibration is marked on the scale for sine-wave measurements, so that the rms value of a sine wave may be read directly from the scale. The meter is used to measure a recurring triangular wave of voltage having peak amplitudes of 150 V. What value will the meter indicate on its scale? What is the rms value of the triangular wave?

D3-10 The voltmeter of D3-9 is used to measure the rms value of a voltage waveform consisting of a 10-V rms sine wave superimposed on a 16-V d-c source. What value will the meter indicate? What is the rms value of the waveform?

3.13   The Diode Detector or Peak Rectifier

The half-wave rectifier of Fig. 3-11 delivers a pulsating current to the load. It is satisfactory for many applications, such as the battery charger discussed in connection with the circuit. There are many applications, however, where it is necessary to obtain a waveform other than the half-sine wave. In the chapter on rectifiers and power supplies it will be shown how the circuit is used to generate a direct voltage from the pulsating current.

Half-wave rectifier operation will be discussed here to introduce another circuit that behaves in much the same way but with higher-frequency input voltages. This circuit is referred to as a *diode detector*. The name is derived from its ability to detect (identify) information signals that are carried by high-frequency voltage waveforms, such as those transmitted by commercial radio stations.

The basic circuit is shown in Fig. 3-14. It is the same as the rectifier circuit,

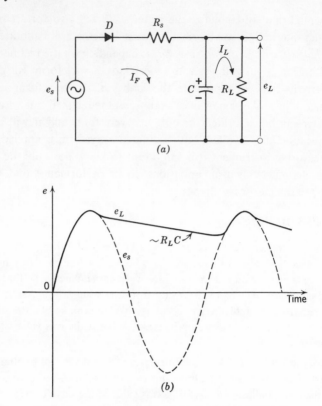

Fig. 3-14   Peak rectifier or diode detector circuit.

except for the additional capacitance, C, connected across the load resistance. The capacitor acts as a reservoir for electric charge, tending to hold the output voltage at a constant value between current pulses. From the waveform of output voltage, though, it can be seen that the time constant $R_L C$ determines how well it does its job. For small load currents the output voltage will be practically equal to the peak value of the input sinusoid. A small load current flows when the resistance is large. This means a long time constant exists. When the circuit generates an output voltage nearly equal to the peak of the input voltage, it is called a *peak rectifier*.

Assume e_s is applied at t equals 0 and the capacitance has no charge initially. Then, as e_s increases to its positive maximum, forward current will deliver charge to the capacitance. If the series resistance R_s is small enough, the voltage appearing across C will be essentially equal to e_s until it reaches its peak. When e_s starts to decrease, the diode will be reverse biased by the charge accumulated on C, and no current will pass back to the input. However, C now has a path for discharge through the load R_L. Its voltage will start to decrease at a rate determined by the time constant $R_L C$. After some time has elapsed the input voltage will change polarity and start its positive increase again. When the value of e_s reaches the potential remaining on C, the diode will again be forward biased and current will recharge the capacitor. The cycle repeats as the input voltage goes through successive periods of its alterations.

In a communications receiver, where it is necessary to detect and separate low-frequency variations caused by changes in the *amplitude* of a high-frequency signal, the circuit behaves as a *diode detector*. In this application it is desired that the time constant of the circuit permits the output voltage to follow the low-frequency variations but not those of the high frequency. These features of the diode-detector circuit will be studied in detail in a subsequent chapter when its use is dictated by the behavior of amplitude-modulation (AM) communications receivers.

DRILL PROBLEMS

D3-11 In the circuit of Fig. 3-14, what effect on circuit operation would result from (*a*) reversing the diode in the circuit? (*b*) doubling the value of C?

D3-12 For an input signal of a 50-Hz sine wave applied to a diode detector, calculate and graphically show the effect of $R_L C$ time constants equal to: (*a*) 0.01 s, (*b*) 0.1 s, (*c*) 1 s. Assume that e_L decreases from its peak amplitude as a starting point.

3.14 Diode as a Logic Circuit Element

Electronic computers have been in use since about 1940 and those of modern design employ thousands of semiconductor diodes which perform logical functions at tremendous speeds. In operation, the diodes change from low-current to high-current conduction states (and the reverse) in a few millionths of a second. Operating as switches in this way, they can be used as circuit elements in the design of functional circuits for logical operations, such as addition and subtraction, in computers. We can learn how such control functions are provided by studying circuits called *switching gates*. They are called gates because in their operation a signal current *may* or *may not* pass through them, depending on conditions existing at the time the signal is applied. Another control signal will determine whether an input signal will appear as an output current or voltage from the gate circuit.

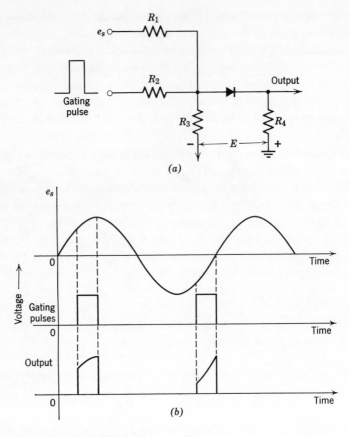

Fig. 3-15   (*a*) Coincidence gating circuit, and (*b*) timing diagram.

The circuit of Fig. 3-15 is called a *coincidence gate*. It will have a desired output when both input signals occur in coincidence. The output signal will be the same as the signal e_s, if the *gating pulse* is large enough to override the reverse bias on the diode. Thus, the gating pulse amplitude and duration can be used to control the operation of the circuit. An output will exist only when the gating pulse occurs, so that a precise control of the output duration in time becomes possible. Such control circuits are common in computer switching circuitry and in time-division multiplexing of several communications channels.

The diode is reverse biased by the negative battery $-E$, because of the voltage divider (R_1, R_3) and the amplitude $E_{s,\mathrm{max}}$. These values are chosen to hold the diode in reverse bias until a gating pulse is applied. If its amplitude is large enough, the diode will be forward biased by the combination of e_s and the gating pulse. The output voltage will follow the wave shape of e_s during the period of the gating pulse. Its amplitude will be determined by the values of voltage and resistance chosen in the design of the circuit. The output voltage cannot take on negative amplitudes, of course, for this would require the diode to conduct heavily in the reverse direction.

In computer circuits it is possible to use diodes to perform certain operations in logic, *involving the thought processes* of AND and OR. Circuits designed to combine signals in these relationships are gates in a general sense and are called AND circuits and OR circuits.

The circuit of Fig. 3-16a is an AND circuit, designed so that its output will be a negative value equal to $-E$ when all inputs are more negative than $-E$. If any one input signal is more positive than $-E$, its associated diode will be forward biased, raising the output voltage at O to equal the input signal. In a practical circuit, the input signals are generated by voltage pulses varying from ground potential to a value more negative than $-E$. The output voltage at O will be equal to $-E$ only when input A AND input B (AND input C, etc.) are negative. Thus, this circuit is referred to as a *negative* AND *circuit*.

The circuit of Fig. 3-16b is called a *negative* OR *circuit*. With the diodes as shown, a negative input *on any terminal* will cause the output to be negative. Thus, input A OR input B (OR input C, etc.) will yield the desired output. Both diodes are in the forward conducting state. Notice that the output cannot have positive amplitudes, because this would require that the diodes conduct heavily in the reverse direction. As an example of the circuit operation, suppose that a negative pulse (with respect to ground) is applied to one input terminal, say input A. This will make diode A carry more current,

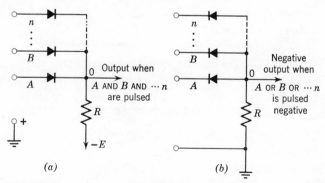

Fig. 3-16   Simple diode gating circuits (*a*) AND circuit, (*b*) OR circuit.

which must pass through the load resistance, lowering the potential of the output terminal O from ground potential to some value below ground potential. This reverses the bias on diode B (and other input diodes), putting it in the reverse-conduction state. If more than one input has a signal impressed on it, the circuit behaves as if only one is active. *Any one input signal* will activate the circuit, whereas in the AND circuit *all inputs* were required to yield an output.

Reversing the diodes and reference polarities in these circuits will yield *positive* AND and OR circuits. Whether an output signal should be positive or negative with *respect to a reference ground* will be determined in an application by other factors in their operation. The theory of logic circuit design is well beyond the scope of this book, but an understanding of diode performance in logic circuits should not be too difficult for the student who masters the foregoing material.

A positive AND logic circuit with two inputs is shown in Fig. 3-17. It is a positive AND circuit because the output will be at its most positive value only when positive signals are applied simultaneously to both inputs (A and B). An output signal in this case will be a voltage of about 12 V above ground potential. In general, the variation of the voltage at the output terminal must be large enough to give the desired effect in a circuit or device connected to the output terminal. Any smaller variation that will not accomplish the desired effect would not be classed as an output signal, practically speaking. The effect desired may be the operation of an electromechanical relay or the magnetization of a bit of metal in the form of a magnetic film or computer memory core.

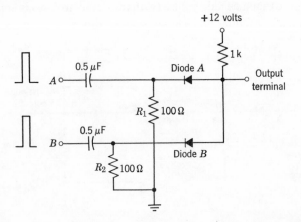

Fig. 3-17   Forward conduction (positive) AND circuit with two inputs.

The circuit is designed so that the diodes operate in the forward-conduction region. Assuming the diodes to have a forward drop of about 1 V, the current through the 1-kΩ resistor causes a drop of about 10 V, making the potential of the output terminal about 2 V above ground. The top terminals of R_1 and R_2 are then at about 1 V above ground. These voltage values are approximate because of the variable forward drop of the diodes and the voltage-dividing action of the resistors in the circuit.

Now suppose that two voltage pulses of short duration (a few microseconds) are applied *simultaneously* to terminals A and B, raising the top terminals of R_1 and R_2 to 12 V above ground. This would *reverse the polarity* across both diodes and thus change their operation to reverse-conduction conditions. The current in the 1-kΩ resistor reduces to practically zero (except for reverse leakage current), causing the output potential to rise to about 12 V, thus providing a positive output voltage or signal.

When only one input signal is applied, the diode in the other input circuit will continue to pass current, but it is not disturbed enough to alter the voltage drop across it or its associated 100-Ω resistor. These two voltages are small compared with the drop across the 1-kΩ load, and their sum will remain at about 2 V as before. If the input signal is applied to input A, for example, only diode A will be changed to the reverse-conduction state. Diode B and its 100-Ω resistor will continue to operate at practically the previous voltage drops, with the result that the output voltage will remain essentially constant at 2 V.

This is more readily seen when we consider that, since the current through diode B and R_2 is essentially unaffected by the signal on input A, their voltage drops do not change appreciably and therefore the output voltage changes very little. Diode B and R_2 are said to *clamp* the output potential to the presignal value, about 2 V in this case. This means there will be no appreciable output signal when only one input is applied. But, as explained above, if both input signals are large enough and applied *in coincidence*, both diodes will be changed from forward to reverse conduction. This will almost completely shut off current in the 1-kΩ resistor, resulting in a substantial rise in the output voltage, from 2 V to 12 V in this case.

The positive OR logic circuit, given in Fig. 3-18, produces an output *when either input circuit* receives a positive signal large enough to reverse bias the diode of the other input circuit. In the circuit as shown, both diodes are forward biased when no input signals are present. Again, assume that the forward drop of the diodes is about 1 V. Notice that, with the 1-kΩ load resistor connected to the negative side of the bias supply, the ground potential

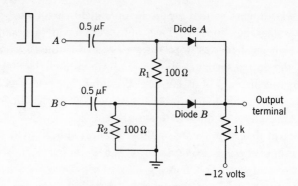

Fig. 3-18   Forward conduction (positive) OR circuit with two inputs.

is positive, making each diode conduct heavily because of its forward biasing. In this condition, the output terminal will be at about -2 V with respect to ground.

Now assume that a positive pulse (with respect to ground) is applied to one of the input terminals, say terminal A. Current will increase in diode A and in the 1-kΩ resistor, raising the potential of the output terminal *from* -2 V to a value *above ground potential*, i.e., a positive value. This reverses the polarity of diode B, putting it in reverse conduction. The *output signal is the potential increase* at the output terminal.

DRILL PROBLEMS

D3-13   Assuming ideal diodes in Fig. 3-16a, $R = 1$ kΩ, $-V = -10$ V, how much current flows in R when the output is negative? What is the value of the output voltage under this condition? How much current flows in R when the output voltage is at ground potential?

D3-14   Assume ideal diodes in Fig. 3-16b, $R = 1$ kΩ, and output voltage to be either ground potential or -10 V. How much current flows in R when the output is at ground potential? If two inputs are simultaneously -10 V, what will be the value of output voltage?

3.15   Switching with Zener Diodes

The phenomenon of *avalanche breakdown* in diodes was discussed in Section 2.18 in relation to the operation of Zener diodes. A device in which a sudden change from low-current conduction to high-current conduction can be produced is obviously a useful tool for switching operations. This is easily accomplished in a semiconductor diode whose characteristic permits avalanche breakdown, called Zener operation. It should be particularly noted that *for Zener operation, the diode must be biased for reverse conduction.*

Figure 3-19 is presented to show the behavior of a silicon Zener diode in the reverse-bias region of its characteristic. The value of voltage at which Zener breakdown occurs is called the *Zener point*, designated E_Z in the diagram.

Zener diodes are available with their avalanche-conduction region at various values of Zener points. For example, one manufacturer makes diodes with Zener points at the following approximate voltage values: -4.7, -5.6, -6.8, -8.2, -10, -12, -15. Referring to Fig. 3-20, these diodes are represented by the curve designated by the diode type number 1Z followed by the E_Z value. For example, 1Z 15 has $E_Z = -15$ V. It is shown that not only does the current increase abruptly when the Zener voltage is approached but also the voltage across the diode remains almost constant even though the current through it is allowed to increase substantially. This feature can be exploited in circuits requiring voltage regulation, as will be discussed later.

The use of a Zener diode in switching circuits (operation in the Zener region of the curve) makes possible extremely fast performance in computer applications. In the handling of information at speeds that require switching at frequencies above 2.5 MHz, diode switching about the zero point (i.e., the origin of the curves) on the characteristic is sometimes too slow, because the "recovery time" of the diode is too long. Note that 1 cycle at a switching frequency of 2.5 MHz has a period of 0.4 s. Thus, to be effective in this

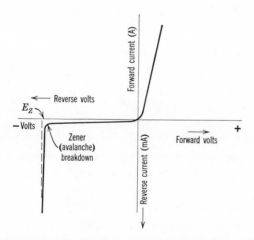

Fig. 3-19 Silicon diode curve showing critical Zener voltage E_Z, and almost constant voltage for large change in current in Zener region.

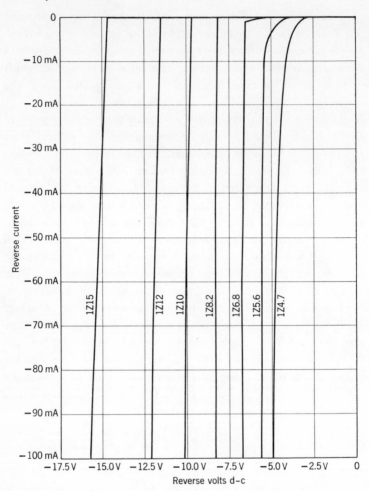

Fig. 3-20   Reverse-current curves of typical Zener diodes. (Courtesy of International Rectifier Corp., El Segundo, Calif.)

application the diode should be able to change its conduction state, i.e., perform its switching function and recover to its original state, and be ready for the next switching action in less than 0.4 s. Zener diode switching accomplishes this action in less than 0.1 s.

Zener-diode switching will be described with reference to Fig. 3-21. To provide the required variation of output-terminal voltage, a selection is made of the Zener diodes to be used and then a choice of steady-state voltage (29 V here) and resistances (R_L, R_1, R_2) is determined.

We shall assume there is no input signal at *A* or *B*. Operation in the avalanche region is normal and not harmful to the diodes. Note that they have reverse-voltage polarity. The output terminal is 7.9 V positive with respect to ground, the potential drop across each diode is 6.8 V, and its left-hand terminal is at 1.1 V above ground. A positive-signal voltage applied to either input terminal, say *A*, raises the potential of the left-hand terminal of the diode sufficiently to cause its current to drop below its Zener-conduction value to the very low normal reverse-current value of the order of 0.5 mA.

While the operation of diode 1 is going on, diode 2 has not changed its state of conduction, and its voltage drop and the voltage across its 100-Ω resistor remain practically constant. This "ties down" the potential of the output terminal to its former value (7.9 V) and so no output signal is produced.

If input signals are applied simultaneously to both inputs (*A* and *B*), both diodes go immediately from avalanche conduction to normal reverse-current conduction (about 0.5 mA each) and the potential at the output terminal rises suddenly to 28+ V and thus a positive output signal is produced.

Figure 3-22 shows an OR circuit, utilizing the reverse-breakdown (Zener) type of operation. With no input signal, both diodes are conducting in the avalanche region. Note that the "bases" of the diode symbols are positive with respect to their "points"; this means that they are biased in the reverse-conduction direction.

A positive input signal at either *A* or *B* will drive the diode of that circuit farther into the avalanche-conduction region and substantially increase the current toward the right through it and downward through the 1-kΩ load resistor. This will abruptly raise the potential of the top point of the 1-kΩ

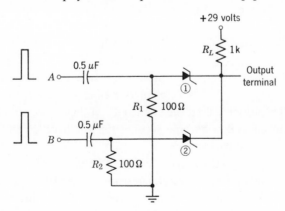

Fig. 3-21   Reverse conduction AND circuit. Zener operation.

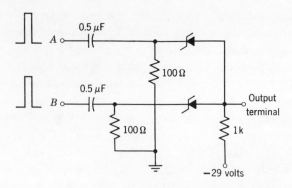

Fig. 3-22   Reverse-conduction OR circuit. Zener operation.

resistor and the sudden rise in potential constitutes a positive output signal. Two input signals occurring simultaneously will merely produce a stronger output signal.

3.16   Voltage Regulation

A very simple voltage regulator circuit using a Zener diode is shown in Fig. 3-23.   Assume nothing is connected to the output terminals.   The current-limiting resistor, R, should be chosen so that the reverse (Zener) current in the diode will not exceed the safe operating value when the fluctuating d-c voltage, applied at the left, is at its maximum value.

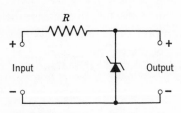

Fig. 3-23   Simple voltage regulator using a Zener diode.

Now, when a load is connected to the output terminals it will take current, but the voltage drop across the diode will not change. Keeping this fact in mind, let us consider two conditions: first, the applied d-c voltage does not change when the output current is flowing, and second, the applied d-c voltage decreases with increase in load current.

With unchanging input voltage, a constant voltage across the diode means a constant voltage across the series resistor, R, and therefore a constant input current. A demand for output current, or for an increase in output current, will be accommodated by the diode without a change in the voltage. With a decreasing input voltage, the current through R will decrease, and the diode will still maintain a constant voltage across the load but it will be carrying less current than in the first condition because it will "give up" some of its current as before.

Two or more diodes in series will provide larger values of regulated output voltage.

3.17   Constant-Current Supply

The simple circuit in Fig. 3-23 has a very important application in the supply of practically constant filament, or heater, current to vacuum tubes. An unregulated d-c voltage input may come from a generator-battery power source of an airplane, a land or water vehicle, or some fixed installation operating from a d-c or rectified source. Series or parallel resistors used to help provide the correct current value for the filaments or heaters of tubes do not prevent fluctuations of the current, and the result is unstable operation of the electronic circuit and shortened tube life.

Suppose that the heaters of one or more tubes were connected in parallel to the output terminals of Fig. 3-23. If they are 6.3-V heaters, a Zener diode with proper operating range, say between −6.2 and −6.5 V, could be connected as a load and the resistance value of R computed to reduce the average input voltage to 6.3 V when both the filament current and Zener diode current are flowing. Then fluctuations in the input voltage would cause the current through R to change, but the diode would accept these changes in current without appreciable change in voltage at its terminals. In practical operation the input voltage to the circuit could change by 12 per cent but the heater voltage would change only about 1.5 per cent, which is an improvement ratio of about 8 to 1.

Where heaters are supplied with alternating current, two diodes may be used as shown in Fig. 3-24. At high instantaneous values of voltage at the transformer secondary terminals S_1 and S_2, one diode is in its Zener region

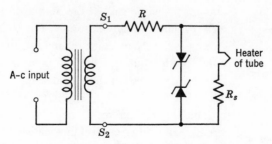

Fig. 3-24   **Zener diodes prevent excessive current in tube heater.**

and the other conducts simultaneously in the forward direction. The forward-conduction characteristic of each of these diodes is such that, at current values used, there is little change in voltage drop with changes in forward current through them. The effect of this application of Zener diodes as voltage regulators is to provide nearly constant heater voltage in the event of substantial variations in line voltage feeding the filament transformer.

Zener diodes have been used across the primary of a filament transformer, and with a current-limiting resistor ahead of them in one of the lines, to provide voltage regulation at the secondary terminals of the transformer.

3.18   Reference Voltage

There are many applications in which it is desirable to maintain a constant voltage between two points in a circuit that is not so simple as the ones just described. This is accomplished in more elaborate voltage-regulator circuits. They require a reference voltage of unchanging magnitude, to which a small error voltage is referred. The same requirements appear in servomechanism circuits, in which the error is amplified and used in such a way as to reduce the error to zero.

A reference voltage is obtainable with a cold-cathode gas diode tube, but the voltage is usually fixed at relatively high values, such as 75, 90, 105, 150 V. Tubes that operate at these voltages are types OA3, OB3, OC3, and OD3, respectively. Their current-voltage characteristics are straight lines essentially perpendicular to the voltage axis. This means that they are similar to the portion of the crystal-diode characteristic that is in the Zener region. Obviously, one disadvantage in using such tubes as reference-voltage elements may be that they must operate at such large voltages.

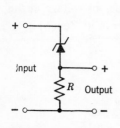

Fig. 3-25   Circuit for reducing range of variation while drawing small values of current.

If the input terminals of the voltage-regulator circuit of Fig. 3-23 are connected to two points whose potential difference is liable to change only a small amount, as is the case at the output of a moderately well-regulated d-c power supply, this circuit will provide a practically constant reference voltage at its output. For this purpose the resistance R in Fig. 3-23 must have a value which will cause the diode to operate in its Zener region. Reference voltages are important in measurement circuits, error-correction circuits, and other control applications.

A useful application of the circuit of Fig. 3-25 is

the case where a small varying component of voltage is superimposed upon a d-c voltage and where it is desirable to reduce materially the d-c voltage but leave the varying component unchanged.

Suppose a voltage E_{d-c}, with a signal voltage e (rms) superimposed upon it is given. A range of currents from I_1 to I_2 in the *Zener* region of the diode characteristic (Fig. 3-19) is available. By choosing the value of R (Fig. 3-25) the current is determined. The diode can be chosen to take a substantial part of E_{d-c}, leaving the remainder for the IR drop. The diode will, of course, be carrying a fluctuating direct current which never reverses. This current produces a varying component of voltage drop across R, which is the signal voltage e originally on E_{d-c}.

As an example, suppose E_{d-c} is 60 V, e is 1 V, and it is desired to have the 1-V signal superimposed upon only 10 V at the output. The diode must have a drop of 50 V, and its current must have a value in the Zener part of the curve. After selecting a diode and then choosing a current value I at the midpoint of the Zener range, R is given by $10 \div I$.

3.19   Crystal Diodes as Power Supply Filter Elements

Voltage regulators using electron tubes are very effective with ordinary loading conditions. But, when the load is of the repeating-transient type, such as currents of on-off nature required in the operation of two or more independent relays, an isolation effect is needed. This can be accomplished with crystal diodes.

Figure 3-26 shows two separate load connections to the output side of a power supply filter. D_1 and D_2 are crystal diodes, and the capacitors C_1 and C_2 are about 30 to 50 μF each.

First, assume that D_1, D_2, C_1, and C_2 are not present and that a load is suddenly and repeatedly connected and disconnected, load 2 for example.

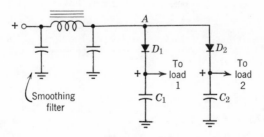

Fig. 3-26   Diodes used to isolate a load from power supply when a separate load is suddenly applied.

Load 1 has been taking steady current, but each time load 2 is turned on the voltage of load 1 will drop and, when it is turned off, the voltage will rise.

Now assume that D_1 and C_1 are present, but D_2 and C_2 are not. A sudden application of load 2 will have very little effect on the voltage of load 1 because C_1 will hold that voltage up largely because it is isolated by D_1, which cannot carry current upward. Therefore a sudden drop in potential at point A will not cause the potential at load 1 to drop. Obviously, the use of D_2 and C_2 is desirable for isolation of load 2 from sudden changes in load 1.

When a diode is used where surge currents (when power is first turned on) are likely to be large, a protective resistor should be inserted in series with the diode to prevent its being damaged. The higher the microfarad values of C_1 and C_2, the larger will be their initial charging currents (surge current). A protective resistor of perhaps 10 Ω would be inserted just above each diode before the connection to the filter output at A is made. We must also be sure that the diode can handle the maximum peak-inverse voltage that will be impressed upon it.

3.20   Relay Delay Circuit

A semiconductor diode may be used to cause a time delay between the operation of two relays. A common requirement is that a particular relay be energized first and deenergized last. The simple d-c circuit of Fig. 3-27, containing one forward-biased diode, will assure that relay 1 pulls in first and drops out last.

Upon closing switch S, the voltage across relay 1 will immediately rise well above the voltage across relay 2. Note that the voltage across the diode will remain comparatively small, but that across C will begin building up with

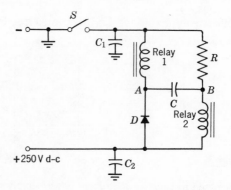

Fig. 3-27   Relay delay circuit. (Courtesy *Electronic Design.*)

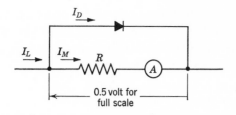

Fig. 3-28   Meter protection by a silicon diode.

the left side positive. Its charging current must pass through R. Soon point A will be substantially higher in potential than point B, and much current will be diverted through the coil of relay 2, causing it to close. The product of the ohms resistance of R and the farad capacitance of C will determine the amount of delay before relay 2 operates. The product RC is called time constant and is discussed in Chapter 1.

When S is opened, the current in relay 2 falls quickly to almost zero and the relay opens. The charge on C, however, must be dissipated and this takes time. C will discharge through relay 1 and R, the discharge current delaying the opening of relay 1 a length of time determined also by the product RC. An important requirement of the diode used is that it must be capable of withstanding the peak-inverse voltage that builds up across the capacitor C.

3.21   Instrument Protection

The delicate moving element of a meter may be protected by a silicon diode arranged as shown in Fig. 3-28. Silicon diodes have a characteristic like the curve in Fig. 3-29. As long as the line current I_L does not exceed the amount that gives full-scale deflection of the meter, the current through

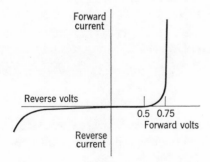

Fig. 3-29   Silicon diode characteristic.

the diode is zero.* At higher values of line current, the voltage drop across the diode increases and it begins to conduct. The diode's characteristic is such that its forward resistance will be small enough to prevent more than 50 per cent overload current going through the meter.

3.22 Power Diodes

Diodes that are expected to handle power, rather than relatively weak signal current or control currents, must withstand a relatively large temperature rise. This occurs in rectifier circuits, and in such applications silicon diodes are superior to those made of germanium.

Diodes having modest power-handling capacities (about $\frac{1}{2}$-W power dissipation) are made in glass envelopes. Typical ones have 400 mA capacity at 25°C and a peak-inverse-voltage rating up to 600 V.

Intermediate-size power diodes are encased in a metal envelope, have maximum ratings of 1 A at 25°C and 1,000 peak-inverse volts.

Silicon diodes with the largest power ratings are stud-mounted so that they may be screwed into a metal base for heat conduction. They will carry in excess of 5 A. Present full-wave, single-phase, bridge rectifiers using these diodes can deliver more than 4 kW at 98 per cent efficiency. As we might surmise, the very low forward resistance and the absence of cathode heater power are responsible for the excellent efficiency of the crystal-diode rectifier.

SUGGESTED REFERENCES

1. A. K. Jonscher, *Principles of Semiconductor Device Operation*, Wiley, New York, 1960.

2. J. D. Ryder, *Electronic Fundamentals and Applications*, 3rd ed., Prentice-Hall, Englewood Cliffs, N.J., 1964.

3. P. E. Gray et al., *Physical Electronics and Circuit Models of Transistors*, SEEC Vol. 2, Wiley, New York, 1964.

QUESTIONS

3-1 Why is it necessary to purify semiconductor metals to 1 part in 10^{11}, when impurity concentrations can be much higher than this in a crystal diode fabricated from the metal?

* Problem 3-26 illustrates this.

3-2 Why is it necessary to "grow" *P-N* junctions, rather than form them as welded joints between *P*-type and *N*-type semiconductors?

3-3 What is the basic principle of "zone refining"?

3-4 Why is zone refining accomplished in an inert atmosphere or in vacuum?

3-5 What characterizes a "single crystal" of semiconductor material?

3-6 Explain the difference between a grown-junction diode and an alloy-junction diode.

3-7 What are the reasons for strict control over manufacturing processes to produce uniform characteristics in any particular type of diode?

3-8 Give some reasons why encapsulation of crystal diodes is desirable.

3-9 Why do manufacturers of semiconductor diodes provide ratings as well as characteristics of their products?

3-10 What are the limitations of a point-contact diode compared with a junction diode?

3-11 Why is a large value of reverse leakage current undesirable?

3-12 Compare the properties of silicon and germanium for use as the host materials in semiconductor crystal diodes.

3-13 Under what conditions is the circuit model for a diode (given in Fig. 3-10) representative of a physical device?

3-14 Why is the circuit of Fig. 3-11 called a "half-wave" rectifier?

3-15 A semiconductor diode is sometimes referred to as a "nonlinear" circuit component. Why?

3-16 Explain the action of a diode in a voltage limiter circuit.

3-17 Discuss the need for two diodes in a voltage limiter used to alter a waveform in both positive and negative ranges of its amplitude.

3-18 The diode voltage limiter circuit is sometimes referred to as a "clipper." Why is this term descriptive of its operation?

3-19 How can a diode be used to calibrate a d-c current-operated meter for measuring a-c voltages?

3-20 What limitations apply to the use of a voltmeter which is calibrated for sine waves?

3-21 Explain the operation of a diode detector circuit, to show the function of each component of the circuit.

3-22. What is the basic function of a "switching gate"?

3-23 How does the AND circuit differ in operation from the action of the OR circuit?

3-24   What characteristics of a semiconductor diode make it especially useful in electronic computers?

3-25   Draw from memory the basic negative AND circuit; also the basic negative OR circuit.

3-26   Discuss the changes necessary to make a negative AND circuit operate as a positive AND circuit; also the OR circuit.

3-27   What is "Zener operation" of a diode?

3-28   Explain the performance of a Zener diode.

3-29   On what part of a semiconductor diode characteristic curve should a circuit operate to provide practically constant current under varying voltage conditions?

3-30   On what part of a Zener diode's curve should it operate to provide practically constant voltage under varying current conditions?

3-31   Explain the operation of a semiconductor diode to provide a reference voltage.

3-32   Explain the action of the load-coupling sections of Fig. 3-26 when the load current of one of them is suddenly changed.

3-33   Discuss the need for surge-current limiting resistors in certain applications of crystal diodes.

3-34   Describe the operation of the delay circuit of Fig. 3-27. Can you think of a possible application of this circuit?

3-35   What hazard is present when operating semiconductor diodes in the forward conducting region? Is this hazard present in Zener operation?

3-36   Why are power-handling diodes physically much larger than those used in small-signal applications?

3-37   What is the purpose of the metal stud mounting of diodes having high power ratings?

PROBLEMS

3-1   A manufacturer specified a certain diode with these values:

PIV	200 V
V_{RDC}	200 V
I_F	1.5 A (at 100°C case temperature)
V_F	1.2 V (at $I_F = 1.5$ A)
i_R	0.1 mA (at 200 V, 150°C)

Would you expect the diode to be constructed of germanium or of silicon? Why?

3-2   The diode of Problem 3-1 has a derating factor for operation above 100°C: "Derate I_F linearly for case temperature above 100°C by the factor -2.5 mA/°C, to a maximum of 200°C." What is the maximum permissible forward current at 200°C case temperature?

3-3   Determine values for the circuit model of the junction diode whose characteristic curve is given in Fig. 3-6.

3-4   Justify or deny that the diode of Problem 3-3 may be represented by a linear resistance in series with an ideal diode. Sketch the equivalent circuit for the diode, for both forward and reverse conduction. What are the values of resistance in the equivalent circuit? Are they constant values for all possible ranges of bias voltages?

3-5   The junction diode whose characteristic is given in Fig. 3-6 is used as a half-wave rectifier for a source of 120-V rms, 60-Hz voltage. The rectifier supplies power to a 1-Ω load. (*a*) Sketch the waveform of load current. (*b*) Calculate the average power developed in the load. (*c*) What possible danger to the diode exists in this circuit?

3-6   Assuming ideal diodes in the circuit of Fig. P3-6, make a graphical plot of *I vs. E*.

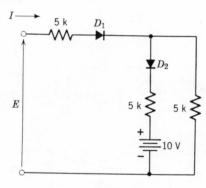

Fig. P3-6

3-7   Repeat Problem 3-6 for D_2 reversed in the circuit.

3-8   Show that a triangular, periodic waveform symmetrical about its average value, has an rms value given by

$$\text{rms} = \frac{1}{(2\sqrt{3})} \text{ times peak-to-peak value}$$

Refer to the waveform in Fig. P3-8. (*Hint:* A general triangular waveform as shown may be used to set up the expressions needed for the calculation.) Note that two sections of the waveform can be analyzed by the methods of analytic geometry, and in general

$$\text{rms} = \left(\frac{1}{T}\int_0^T e^2\, d\theta\right)^{\frac{1}{2}} = \left(\frac{1}{T}\int_0^P e_1{}^2\, d\theta + \frac{1}{T}\int_0^{T-P} e_2{}^2\, d\theta\right)^{\frac{1}{2}}$$

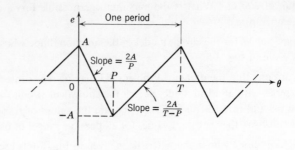

Fig. P3-8

3-9 A battery charger is to be designed to charge a 6-V battery of resistance 0.1 Ω, at a maximum average charging rate of 3 A. Determine the a-c voltage required to supply a half-wave rectifier which uses the diode whose characteristic is given in Fig. P3-9.

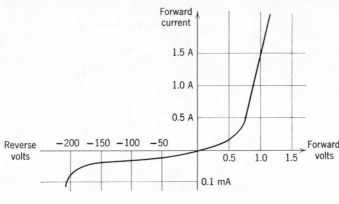

Fig. P3-9

3-10 A half-wave rectifier circuit consists of a junction diode, 10-Ω resistor, and 20-V, 60-Hz signal. (*a*) Assuming ideal characteristics for the diode, find the peak current in the 10-Ω resistor. (*b*) Determine the peak current in the circuit, if the diode is the 1N1692 whose characteristics are given in Fig. 3-7.

3-11 The circuit of Fig. 3-13 is used to generate a nearly square wave of voltage from an input sinusoidal source of 100 V rms. Assuming the diodes to be ideal elements, what must be the magnitudes of d-c limiting sources E_1 and E_2, in order to generate a square wave that is 20 V, peak-to-peak?

3-12 In the circuit of Problem 3-11, what must be the PIV rating of the diodes? What value of R is required to limit diode current to 100 mA?

3-13 An a-c signal of 0.5 V amplitude is superimposed on a 12-V d-c signal. What is the rms value of the combination?

3-14 A diode detector circuit, as in Fig. 3-14, has an input signal of 50 Hz. Calculate the effect of $R_L C$ time constant equal to 0.01 s. Show this by a graphical sketch, assuming e_L decreases from its peak amplitude as a starting point.

3-15 In the circuit of Problem 3-14, what effect on circuit operation would result from reversing the diode connections? What would happen if the value of C were doubled? What would happen if C were to become an open circuit?

3-16 In the circuit of Fig. 3-17 two simultaneous positive voltage pulses are applied to inputs A and B. What should be their values in order to raise the output terminal voltage to 12 V?

3-17 In the circuit of Fig. 3-17, how much current flows in the 1 k resistor before an input pulse is applied? How large an input pulse is needed to raise the output voltage to 4 V above ground?

3-18 In the circuit of Fig. P3-18, $C = 0.02$ μF, forward resistance of the diode = 100 Ω, and reverse resistance = 1 MΩ. Determine the output waveform for the input signal given.

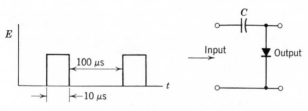

Fig. P3-18

3-19 Zener diodes type 1Z10 are used in the circuit of Fig. 3-21. Determine the output potential when no signal is applied. How large will the output signal be when both signals (at A and B) produce normal reverse conduction? How large must the inputs be to produce conduction?

3-20 The circuit of Fig. 3-22 uses 1Z10 Zener diodes to provide a reverse-conduction OR circuit. What is the potential of the output terminal when no signal is applied? Determine the potential across the 100-Ω resistors. What is its polarity with respect to ground?

3-21 In the circuit of Fig. 3-22, how large a signal is needed at A or B to produce a 5-V output signal? Will the output terminal become more or less negative with respect to ground?

3-22 A simple voltage regulator circuit is required to hold a voltage within the range from 6.5 to 7.0 V, when the input voltage varies between 10 and 20 V. Design a circuit, similar to that of Fig. 3-23, using a Zener diode and a series resistor. (*a*) Choose a diode from the types given in Fig. 3-20, and calculate the required value of series resistor. (*b*) Determine the minimum power rating of the resistor. (*c*) Determine the diode current limits. (*d*) Determine the output voltage.

3-23   Refer to Fig. 3-24. Each diode has reverse voltage applied to it on alternate half-cycles of secondary voltage. Assume the tube heater should not receive more than 8.5 volts at any instant and that the transformer secondary voltage has rms value of 10 volts. Choose two Zener diodes that will work properly in the circuit. Calculate required values of R and R_s. (Neglect diode forward-conduction voltage drops. Why?)

3-24   Refer to the circuit of Fig. 3-27. Assume the capacitors have no initial charge. At the instant the switch is closed, what is the potential difference E_{AB}? To a good approximation, what must be the voltages across the relays?

3-25   In the circuit of Fig. 3-27, assume RC = 5 seconds. Neglect for the moment the inductance and resistance of relay 1. How long after the switch is opened will it take for current in relay 1 to drop to 36.7% of its initial value? How is the current in relay 2 behaving during this time? Would the emf of self-inductance of relay 1 help or hinder the brief continuation of its current? Why does the current in relay 2 drop to zero much more quickly than the current in relay 1?

3-26   The meter element in Fig. 3-28 takes 1 mA at full-scale deflection. What is the resistance of that branch of the circuit? Assuming the silicon diode carries 10 mA when the meter reads full-scale, calculate the diode resistance. What per cent error is made by assuming that the diode current is zero and that the meter current is the line current?

3-27   In the circuit of Fig. 3-28: (*a*) Calculate the voltage drop when the meter is 50 per cent overloaded. (*b*) Calculate the diode resistance when the line current becomes 5 times the full-load meter current, i.e., I_L = 5 mA. Compare it with the value obtained in Problem 3-26.

3-28   A junction diode has a reverse saturation current of 12 μA at reverse-bias voltage up to its Zener breakdown potential. It has a breakdown voltage of 15 V and has negligible ohmic resistance in its breakdown region. A 24-V d-c source is applied to the diode as a reverse bias, in series with a 2-kΩ resistor. Calculate the diode current.

CHAPTER FOUR

Diode Tubes

Although semiconductor diodes and transistors have taken over most of the tasks previously performed by high-vacuum and gas-type tubes in the low-current and low-power areas of electronic applications, there will continue to be a need for tubes for many years to come. This appears to be true especially for many industrial applications where comparatively large amounts of power are involved and where high-frequency performance is not a requirement. Even in the communications field, tubes will be in demand for a long time in high-power radio broadcast systems and television transmission systems. In this chapter we shall find out how diode tubes operate and how they differ from semiconductor diodes in their behavior.

4.1 Construction of Electronic Tubes

Vacuum tubes are made with envelopes (enclosures) of glass, metal, and ceramic materials. Types which are in common use, manufactured in production lots, vary in size from large X-ray tubes, several feet long and approximately a foot in diameter at the bowl, to grain-of-rice size. Almost everyone is familiar with the size and shape of tubes used in radio and television receivers. Television picture tubes have the largest cross-sectional dimensions of all present-day high-vacuum tubes.

Gas-type diodes also are constructed with glass and metal envelopes. They contain an inert gas at low pressure instead of a high vacuum. Electrons and electrically charged gas molecules, called *ions*, are the charge carriers in gas tubes.

High-vacuum and gas-type diodes depend for their operation on the

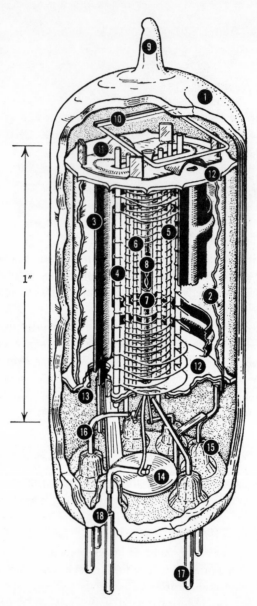

Fig. 4-1 Internal structure of a miniature pentode high-vacuum tube. 1, glass envelope; 2, internal shield; 3, plate; 4, grid 3 (suppressor); 5, grid 2 (screen); 6, grid 1 (control grid); 7, cathode; 8, heater; 9, exhaust tip; 10, getter; 11, spacer shield header; 12, insulating spacer; 13, spacer shield; 14, inter-spin shield; 15, glass button-stem seal; 16, lead wire; 17, base pin; 18, glass-to-metal seal. (Courtesy Radio Corporation of America.)

144

release of electrons from emitting materials within the envelope. The emitter electrode of a tube is called a *cathode*, since it is the negative electrode. The positive, collector electrode is called an *anode*, or *plate* of the tube. These two electrodes make up the internal, electronic structure of a *diode*, a two-electrode device. The emission of electrons from the cathode usually results from the addition of heat energy to the emitting material of the cathode. This requires a *heater* within the envelope in addition to the two active electrodes.

Other tube types, such as *triode*, *tetrode*, and *pentode* have additional electrodes within the envelope. Their designations indicate the number of active electrodes in the tube; for example, a triode is a three-electrode device. A triode consists of a cathode, an anode, and a *control grid*, which is placed between the cathode and anode to control the electron flow in the tube. The tetrode and pentode tubes include additional grid structures to control the flow of electrons. The behavior of these types of tubes will be discussed in subsequent chapters. They have been introduced here to emphasize that a diode is the simplest type.

The internal structure of a miniature vacuum tube with a glass envelope is illustrated in Fig. 4-1. This cut-away view of a pentode shows the internal arrangement of active electrodes as well as the shielding and insulating structures necessary for the proper operation of the tube. The base pins at the bottom of the tube permit its insertion into a socket so that electrical connections may be made from the tube to parts of the circuit.

4.2 Electronic Tube Applications

High-vacuum tubes serve as the most important elements in circuits designed to perform several different functions in electronic applications: (*a*) rectification, changing alternating current to direct current; (*b*) oscillation, the production of electric voltages and currents having desirable frequencies and waveforms; (*c*) amplification, the strengthening of voltages and currents; (*d*) modulation, the process of causing an electric voltage or current, such as a radio signal, to change its waveform in accordance with variations in sound or in another kind of signal; (*e*) demodulation or detection, a kind of rectification devised to recover from a modulated signal a voltage variation similar to the one which was used to generate the modulation; (*f*) electron-beam production, as in the picture tube of a television receiver or an X-ray tube; (*g*) photoelectric action, the change of light energy into electric current as in the phototube or electric eye.

The diode is used in many electronic applications. Its principal service is rectification, which is an electrical valve-like action that results in the flow of current in one direction only. It is interesting that British literature refers to all electronic tubes as "valves."

Gas-type electron tubes are used in: (*a*) switching circuits, especially where more power is needed than semiconductor devices can handle; (*b*) power rectifiers and inverters; (*c*) voltage and current regulators; (*d*) phototube applications.

4.3   Symbols Used in Circuit Diagram

The symbols that are used in circuit diagrams to represent electronic tubes are shown in Fig. 4-2. Notice that some tubes may contain the elements of more than one type within the same envelope; e.g., the twin triode and twin diode-triode. The letters beside the external connections designate the active electrodes or other necessary elements. For example, P designates an anode or plate, H the heater, G the grid, etc. Subscripts on the letters indicate a

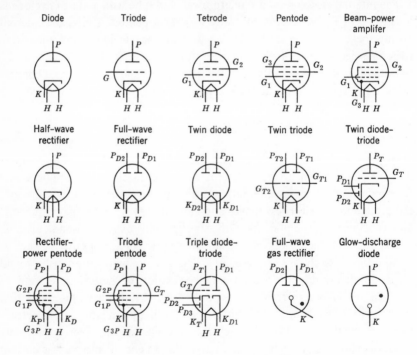

Fig. 4-2   Symbols representing electronic tubes.

distinction among several letters of the same kind. On the symbols for full-wave gas rectifier and glow-discharge diodes a black dot is shown within the envelope. This symbol is used to designate gas-type tubes.

EXERCISE. The student will find it very instructive to procure a receiving-tube manual* and look up the tube types listed below. This will familiarize him with the symbols for various types of tubes, as well as with the manner in which manufacturers provide information about their products. He should make a tabulation of the following information for each tube type: recommended application, type of cathode, rated heater voltage and current, average (d-c) and peak plate current, type of socket required, and pin numbers of filament or heater, plate, cathode, and grids. In multigrid tubes the control grid is the one nearest to the cathode; it is labeled G_1 in Fig. 4-2.

1S4	6AL5	6CM7	6SN7GTB	12AT7	5751
1T4	6AU6	6F6	6V6	12AX7	5814A
1U4	6AX5-GT	6L6	12AL5	117Z3	6005
5U4-GB	6BA6	6SJ7	12AQ5	117Z6GT	6203
5Z4	6CB6	6SK7	6X4	5727	6973

What is, in general, the significance of the number ahead of the letter in the designations of high-vacuum receiving-type tubes? Observe the differences between two tubes which are recommended for the same kind of application.

4.4 High-Vacuum Diode with Thermionic Cathode

An electronic tube that has only an anode (plate) and a cathode as active electrodes is called a diode. A vacuum diode also includes a cathode heater that does not enter directly into the operation of the tube but does influence its behavior. These elements are enclosed in a glass or metal envelope that has been evacuated of air and other gases during manufacture to leave a very high vacuum in the interelectrode space.

Cathodes designed for use in high-vacuum diodes are in two general classes: (1) filamentary, or directly heated, and (2) equipotential, or indirectly heated. In the filamentary construction a wire is heated by passing an electric current through it, and electron emission takes place directly from the wire or a special coating on it. The *filament* is the *cathode element* of the diode when the cathode is directly heated; the heating current passes through the filament wires, from which electrons are emitted. The directly heated cathode structure may appear as in Fig. 4-3a. It is built to have good emission properties and mechanical ruggedness.

* RCA Electronic Components and Devices, Harrison, New Jersey; Tube Department, General Electric Company, Owensboro, Kentucky.

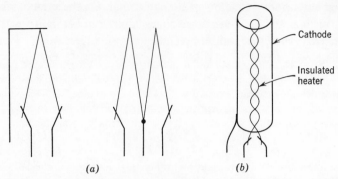

Fig. 4-3   Thermionic cathode construction. (*a*) Directly-heated filament. (*b*) Indirectly-heated cathode.

Directly heated cathodes operate with the best economy of heating power when they are coated with certain metallic oxides. Barium oxide is commonly used with a small amount of strontium oxide added to give the coating mechanical strength and longer life. Filaments are made of materials that may be fabricated to operate over a wide temperature range without losing their electrical and mechanical properties. Nickel and tungsten are widely used for the current-carrying filament, because their metallurgical properties permit a good metallic bond for the oxide coatings.

Some directly heated filaments are made of thoriated-tungsten wire that has been given a special heat treatment. They are not quite so efficient as oxide-coated cathodes, but they are much more efficient than cathodes made of pure tungsten. Pure tungsten must be used when operation is at extremely high temperatures.

Fig. 4-3*b* shows the construction of indirectly heated cathodes. A loop of tungsten wire insulated with a refractory material is inserted into a thin, hollow cylinder, usually made of nickel. The outside of the cylinder is coated with the emitting material. Electric current in the wire heats the cylinder to emitting temperature. Since no heating current is in the cylinder, the cathode is an equipotential surface. This means all points on the cathode are at the same electrical potential. Cathodes of this type exhibit heat storage that reduces the disturbing effect of temperature variations that may occur in the heater.

Indirectly heated cathodes have some important advantages, but because of current limitations their use is restricted to the relatively small-sized tubes. Some of their advantages are: (1) the entire emitting surface is at one potential, so that no potential difference exists along the surface as is the case when direct heating is employed; (2) operation of two or more cathodes at different potentials is possible while the same heating current is being used; (3) alternating current may be used for heating without introducing disturbing effects that may result from stray electric fields due to the continuously changing current and from nonuniform plate-to-cathode potentials; and (4) these cathodes have higher emission efficiency.

4.5 Thermionic Emission; Work Function

Diode tubes depend for their operation on a continuing supply of electrons emitted from the cathode and collected by the anode, or plate. Study of the behavior of diode tubes can be separated into two more or less independent parts: the properties of electron emission from solids and the control of internal current caused by the motion of the released electrons. Both of these parts have features similar to the conduction properties of semiconductor diodes. We shall study them in turn, starting with a consideration of how electrons may be released from the metallic cathodes of diode tubes.

The cathode is a metallic conductor that contains atoms oriented in crystal lattices, electrons bound to the atoms, and free electrons that are not bound to any particular location in the cathode. Cathode emission occurs when free electrons in the metal are caused to escape from the surface of the cathode. The release of electrons from the cathode, called *cathode emission*, is a very important action and it takes place in every type of electronic tube. When electrons are emitted from the cathode because of added heat energy, the process is called *thermionic emission*.

Whether the energy possessed by an electron is sufficient to enable it to escape from the cathode is determined by the amount of work necessary for emission. The amount of work, or energy, given up by an electron as it leaves the surface is called the *surface barrier energy*, and is designated E_B. It consists of two parts, the energy given up in breaking loose from the surface and the work required to overcome other internal forces at greater distances within the cathode. A more detailed explanation of electron energies is usually given in advanced texts.* Only a short description is given here.

When a metal or a metallic compound is heated, the atoms acquire increased energies. Certain electrons become free of the influence of the atoms to which they normally "belong" and move about with increased energies. The kinetic energy of an electron is expressed as $\frac{1}{2}mv^2$ (where m is its mass and

* W. G. Dow, *Fundamentals of Engineering Electronics*, 2nd ed., Wiley, New York, 1952.

v is its velocity), just as the kinetic energy, or energy of motion, of any other moving body is expressed. These high-energy electrons move about within the metal, and some of them approach the surface nearly at right angles. Indeed, it is to be expected that some of them do arrive at the surface of the metal along paths making right angles with the surface.

The energy of such an electron carries it out of the metal for a distance that depends on the kind of metal and the energy of the electron. There is exerted on an electron attempting to escape a holding-back force which is due to a so-called *image charge* within the metal. The image charge is a positive charge equal in magnitude to the negative charge of the electron; it is located as far below the metal surface as the electron is outside the surface (Fig. 4-4).

The force of attraction between this positive charge and the electron tends to prevent the escape of the electron. If its velocity (which determines its kinetic energy) is sufficient to overcome this holding-back force, the electron will escape completely. It will then be free to follow other paths of motion dictated by forces arising from fields outside the region of influence of the image force.

Even at absolute zero temperature many electrons in metals possess an energy near the Fermi level (E_F) of energy. It is only necessary to supply an additional energy, E_W, in order to cause the electron to be emitted from the metal surface. This minimum additional energy is called the *work function*, and has a definite value for any particular metal. The work function may be expressed in ergs or joules, but more often it is expressed in electronvolts or just volts.

The work function depends on the type of material, its surface conditions, and any impurity materials within the metal or on the surface. Because of the dependence of work function on surface conditions, it is not surprising that

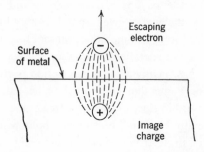

Fig. 4-4   An electron leaving a metal surface. A holding-back force is exerted by the field of the image charge.

TABLE 4-1   Values of Emission Constants

Material	A_0 (A/m²/°K²)	b_0 ($11,600E_W$)	E_W (eV)	Melting Point (°K)
Calcium	60.2×10^4	26,000	2.24	1,083
Molybdenum	60.2×10^4	48,150	4.15	2,895
Tantalum	60.2×10^4	47,600	4.10	3,125
Thorium	60.2×10^4	38,900	3.35	2,118
Tungsten	60.2×10^4	52,700	4.54	3,643
Nickel	26.8×10^4	58,000	5.00	1,725

values reported by different experimenters differ considerably. Those given in Table 4-1 are representative of materials used in the fabrication of thermionic cathodes.

The current composed of escaping electrons passing through the surface of a thermionic cathode can be expressed by an equation of the form

$$J = A_0 T^2 \epsilon^{-(E_B - E_F)(e/kT)} \text{ A/m}^2 \qquad (4\text{-}1)$$

The factor A_0 is a semiempirical constant; T is temperature of the source of electrons, expressed in °K; $E_B - E_F$ is just the work function, E_W; e is the electronic charge; k is Boltzmann's constant; $\epsilon = 2.7182\ldots$, the base of the system of natural logarithms.

This equation is often written in terms of the surface area of emission and a factor containing the work function in the exponential

$$I = A_0 S T^2 \epsilon^{-(b_0/T)} \text{ A} \qquad (4\text{-}2)^*$$

where S is the surface area (m²) and $b_0 = (e/k)E_W = 11,600E_W$, which has been called the temperature equivalent of the work function.

Emission currents are extremely small at room temperature, so that the cathode of a diode tube must be heated to a high temperature to obtain useful emission currents. Only a few materials are suitable for use as thermionic cathodes. Most of those that have a small work function have a low melting point, and most of those with high melting points have large work functions. A small work function is desirable because of the smaller energy needed to cause thermionic emission, but a low melting point prohibits

---

* The form of Equation (4-2), but having the power of T as $\frac{1}{2}$, was developed by O. W. Richardson in 1914. Saul Dushman developed the equation in 1923, having the factor T^2, which is believed to be a better theoretical description of thermionic current. Consequently, references to this equation in the literature cite it as the *Richardson-Dushman equation.* For more information and a derivation of the equation, see Dow, *op. cit.*

operation at the elevated temperatures that accompany a suitable current in the tube. A compromise must be worked out to produce a useful cathode.

Almost all thermionic cathodes are made of nickel coated with barium and strontium oxides, of thoriated tungsten, or of pure tungsten. The work functions of these materials are 1.0, 2.6, and 4.52 eV, respectively. Cathodes made of these materials have widely different emission currents because of the exponential effect of work function in Equation (4-2). Normal operating temperatures for these cathodes are about 750°C, 1,700°C, and 2,100°C, respectively.

EXAMPLE 4-1.   Calculate the emission current density of thorium heated to 1,600°K.

$$J = 60.2 \times 10^4 (1,600)^2 \epsilon^{-38,900/1,600} \text{ A/m}^2$$

We will tackle the exponential factor first:

$$\epsilon^{-38,900/1,600} = \epsilon^{-24.3} = \frac{1}{\epsilon^{24.3}} = \frac{1}{35.7 \times 10^9} = 0.028 \times 10^{-9}$$

Then the emission current density from the thorium cathode is

$$J = 60.2 \times 10^4 (1,600)^2 \times 0.028 \times 10^{-9}$$
$$= 43.2 \text{ A/m}^2$$

of emitter surface area. (This is 4.32 mA/cm² of emitter surface.)

DRILL PROBLEMS

D4-1   Calculate the emission current density of oxide-coated, thoriated tungsten, and pure tungsten cathodes when they are operated at their normal temperatures.

D4-2   What comparative surface areas would be required for each of these cathodes to produce the same current in diode tubes using them?

4.6   Other Forms of Electronic Emission

Electrons may be emitted from the surface of a metal by means other than thermionic emission. For example, electrons passing from the cathode to the plate of a diode tube may have enough energy on arrival at the plate to knock several other electrons from its surface. When electrons are knocked from a material by bombarding particles, the process is called *secondary emission*.

Photoelectric emission is another example of electronic emission from solid materials. Photons of light falling onto a surface may give up enough energy to electrons to enable them to escape from the material. This process is the basis for operation of a photoelectric tube, or electric eye.

Secondary emission is usually undesirable in vacuum tubes, although it is combined with photoelectric emission in a special manner in the *photomultiplier tube*. This device is discussed in some detail in a subsequent chapter. For the purposes of this chapter, secondary and photoelectric emission will be considered negligible or of no importance in the proper functioning of diode tubes.

4.7 Characteristics of the High-Vacuum Diode

It is desirable to understand how the plate current in a diode is affected by cathode temperature. To determine this experimentally, connect the tube in a circuit (as shown in Fig. 4-5) with a particular value of d-c voltage, E_b, between the plate and the cathode.

A change in the heater-cathode current, I_f, will change the temperature of the cathode, and because of this there will be a change in the plate current, I_b, of the diode.

Curve 1 of Fig. 4-6 shows how the plate current of a high-vacuum diode depends on cathode heating current, and therefore on cathode temperature, at a constant value of plate-to-cathode voltage, E_{b1}. When a higher value of constant plate voltage is used, a curve like curve 2 is obtained. The curves coincide for more than half of their length.

Note that no plate current is indicated at low values of heater current. The reason is that the cathode is not hot enough to give off electrons in sufficient quantities to constitute a significant current.

The plate current rises rapidly after the emission temperature has been reached, and soon levels off if the increase in heater current is continued.

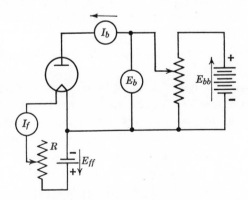

Fig. 4-5   Measurement of diode characteristics.

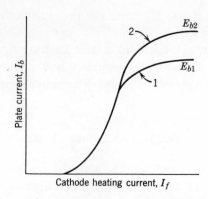

Fig. 4-6   Diode characteristics. Cathode temperature varied, plate voltage having constant values.

The leveling off is due to the fact that electrons repel one another, and so they gang up in the space in front of the cathode and form a *space-charge cloud*. This negative space charge limits the rate of flow of electrons toward the plate to a value determined partly by the electric field between the plate and cathode, and it drives the excess electrons back into the cathode. The plate current is said to be *space-charge-limited*. Naturally, the higher the plate voltage, the stronger will be the electric field in the space between the plate and the space charge; hence more electrons per second will reach the plate, manifesting a larger current in the tube. The plate current is said to be *temperature-limited* when existing conditions make the curves coincide, as in Fig. 4-6. Evidently, in that range of cathode temperature, the higher voltage, E_{b2}, does not cause any more current to flow than does the lower voltage, E_{b1}, because, even at the lower value of plate voltage, the electric field in the space is able to force the electrons toward and into the plate as fast as they are emitted from the cathode. *Temperature saturation* is said to exist where the curves have leveled off, because further increase in cathode temperature does not increase the current.

The other condition, that of constant heater current, is used normally in electron-tube operation. The dependence of plate current on plate voltage is shown in Fig. 4-7 for two constant values of heater current. Again, the lower portions of the curves coincide. In this region the current is limited by space charge, and at voltages corresponding to the lower values of current (up to point A on the graph) the cathode temperatures are high enough to provide electrons as fast as the electric field forces, just off the cathode, can take them away. At plate voltages higher than that corresponding to the

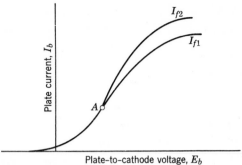

Fig. 4-7 Diode characteristics. Plate voltage varied, cathode temperature having constant values.

current at A, however, the current begins to be temperature-limited and an increase in filament or heater current from I_{f1} to I_{f2} will cause the plate current to increase even though the plate-to-cathode voltage is held constant.

Where the curves begin to level off, the condition known as *voltage saturation* begins. Evidently, when the curves become practically horizontal the only way to get more plate current is to increase the cathode temperature, i.e., increase the filament or heater current.

4.8   The Electronvolt

Consider a vacuum in which an electron has just emerged with zero velocity from a surface. Assume there is also in the vacuum a positive plate, the potential of which is 1 V above the potential of the first surface. The electron will travel to the positive plate and arrive there with a certain velocity and, therefore, with a certain kinetic energy. The amount of the energy is

$$W = Eq_e \qquad J$$

where $E = 1$ V and $q_e = 1.602 \times 10^{-19}$ C of charge (the electronic charge).

The energy of the electron upon arrival is therefore 1.602×10^{-19} J. It is also correct to say that its energy upon arrival at the plate is 1 eV, or 1 V.

If the plate is given a potential of 100 V positive with respect to the first surface, the energy of the electron upon arrival will be 100 eV or 100 V. The energies of positive particles, as well as of negative particles, may be expressed in electronvolts or in volts.

4.9   Potential Distribution in a High-Vacuum Diode

We shall now see how the electric *potential* changes its value at different points along a straight line between the cathode and the plate in a high-vacuum diode having parallel-plane electrodes.

A potential-distribution curve shows the potential with respect to the cathode *at every point along a straight line* extending through the space between the cathode and the other electrode. It is very important to note that Fig. 4-8 is drawn for the special case of a diode having plane-parallel electrodes, and for particular conditions of plate-to-cathode voltage and cathode temperature. The straight line labeled "no space charge" shows the potential distribution when the cathode is not emitting electrons (a cold cathode).

The other curve represents the potential distribution when the temperature of the cathode is high enough to produce a plentiful supply of electrons and the current flow to the anode is steady. This curve is horizontal at the point where it touches the vertical line representing the cathode surface. There is space charge in the region in front of the cathode, and that accounts for the slow rise of potential as the distance to the point of measurement of potential is increased. It is worthy of note here that experiments have actually been performed with parallel-electrode tubes into which probes were inserted to measure the potentials at points in the interelectrode space at selected distances from the cathode. In that way much has been learned about the nature of potential distribution in different types of electron tubes.

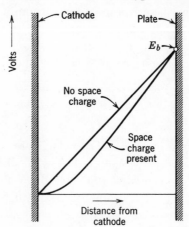

Fig. 4-8   Potential distribution between parallel-plane electrodes in a vacuum.

4.10   The Three-Halves Power Law for Space-Charge-Limited Current

In most hot-cathode vacuum tubes the plate current is limited by space charge. The cathode is operated at such a high temperature that many more electrons per second actually leave it than are needed to maintain the plate current. Obviously, some of them are driven back into the cathode by the negative space-charge cloud in front of it. The actual current that flows is determined by the potentials and depends also on the shapes and dimensions of the electrodes.

An equation that gives the density of plate current, in milliamperes per square meter of cathode surface, in a diode having *plane-parallel electrodes* close together and having large area, was first worked out by Child and is called Child's law:

$$J_b = \frac{2.33 \times 10^{-6} E_b^{\frac{3}{2}}}{d^2}$$

in which J_b = amperes per square meter of the area of the cathode; E_b = volts potential difference between the electrodes; d = meters in perpendicular distance between the electrodes.

In the derivation of this equation several other assumptions were made in addition to the one that the electrodes are large, flat, parallel, equipotential surfaces placed close together. It was assumed that the current is limited by the space charge, that the space between the electrodes is free of gas particles, and that upon emergence from the cathode the electrons have zero initial velocity as they begin their journey to the plate.

In actual high-vacuum tubes the conditions assumed are true only to a limited extent, and so the equation gives only approximate values of plate current. It must also be kept in mind that the equation holds only when the cathode temperature is high enough so that electrons are being emitted more rapidly than they are needed; i.e., temperature saturation exists. In contrast, Dushman's equation holds for the condition that all the electrons emitted pass immediately to the plate; i.e., voltage saturation exists. Remember that voltage saturation means that the plate voltage is sufficiently high so that no space-charge cloud can form in front of the cathode to control the plate current.

4.11   Discussion of Child's Law

Child's law may be expressed in the general form

$$i_b = K e_b{}^n \tag{4-3}$$

in which K is a constant called *perveance*, the value of which depends on the dimensions and shape of the electrodes and their separation; e_b is the voltage between plate and cathode. The exponent n is approximately $\frac{3}{2}$.

This law may be verified experimentally by operating a high-vacuum diode under temperature-saturation conditions (space-charge-limited current) and measuring plate current at each of several values of plate voltage. When the logarithms of plate-current readings are plotted against the logarithms of corresponding plate-voltage readings, the points will lie on a straight line. For each set of readings the temperature must be kept constant. This is accomplished by keeping the filament current constant.

To show how a straight line is obtained, first take the logarithm (to the base 10) of both sides of the general equation:

$$\log i_b = \log K e_b{}^n = \log K + n \log e_b \tag{4-4}$$

The term $\log K$ is a constant; call it C. The quantities $\log i_b$ and $\log e_b$ are variables; call them y and x, respectively. The *form of the equation* is then

$$y = C + nx$$

This is the equation of a straight line the slope of which is n. The slope of a straight-line graph is the tangent of the angle the line makes with the x-axis. We see, then, that if values of $\log i_b$ (the variable corresponding to y) are plotted against corresponding values of $\log e_b$, which we may think of as values of x, we shall get a straight line. Figure 4-9 shows results of a test on a type 5Y3 high-vacuum diode, at two values of filament current. The slope of each line is $n = 1.6$.

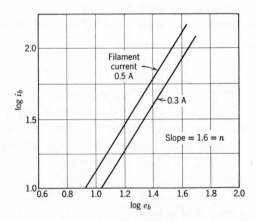

Fig. 4-9   Experimental curves for 5Y3 high-vacuum diode, from which the exponent in the general Child's law equation is determined.

DRILL PROBLEM

D4-3 A vacuum diode was used to determine experimentally its perveance and the value of the exponent of its voltage-current curve. Determine by graphical means the values of K and n for Equation (4-3) from the following data obtained from the tube.

e_b (V)	i_b (mA)
1.0	0.03
5.0	0.35
10.0	0.97
15.0	1.75
20.0	2.75
25.0	4.00
30.0	5.05

4.12 Gas Diodes

It is interesting and instructive to compare the operating currents and voltages of a tube containing a gas (mercury vapor in this case) with those of a high-vacuum tube of very nearly the same physical dimensions. The type 83 mercury-vapor diode and the type 5Y3 high-vacuum diode, both full-wave rectifiers, are more nearly alike in physical size than any two later types.

When these tubes are operated at recommended values of full-load current, the voltage drop in the 5Y3 diode is about 60 V and in the 83 diode it is about 14 V. These figures mean that by using mercury, which readily vaporizes in the tube when the cathode is heated, the amount of current that the tube can handle is greatly increased and at the same time the power lost within the tube during operation is greatly reduced. The power loss, in watts, within the diode is the tube current, in amperes, multiplied by the plate-to-cathode voltage, in volts.

Figure 4-10 shows how the voltage across the tube and the tube current

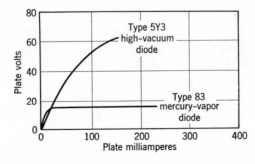

Fig. 4-10 Voltage-current characteristics of small rectifier diodes.

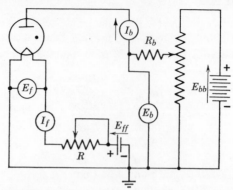

Fig. 4-11   A circuit for measuring characteristics of a gas diode. The resistance R_b should be sufficient to limit the tube current to a safe value when full voltage is used.

are related. These curves are drawn on a common current axis, and so they are characteristic of tubes of the same maximum-current ratings.

Note that the plate voltage of the high-vacuum diode changes with the slightest change in plate current, while the plate voltage of the mercury-vapor diode remains practically constant throughout about 90 per cent of the current range.

The circuit shown in Fig. 4-11 may be used to obtain data for the plotting of these curves. The symbol representing the tube has a dot in it to indicate that it is a gas-type tube.

4.13   Gas Ionization

Although mercury vapor is used in many gas tubes, some employ other gases such as neon, helium, argon, krypton, and xenon—the familiar inert gases studied in high-school chemistry.

When the cathode of a mercury-vapor diode is heated, electrons are given off copiously. Additional mercury vapor is formed by the increase in temperature, thus adding to the mercury vapor already present in the low-pressure envelope. The vapor consists of millions of atoms and molecules of mercury moving about between the cathode and the plate of the tube.

A mercury atom can be ionized (i.e., made to give up its outermost electron and thus become a mercury ion) by being struck with an electron that has an energy of 10.4 V or more. When high positive plate potential is applied to a mercury-vapor tube, the electrons acted upon by the force of the electric field almost instantly acquire energies of at least that value of potential, and ionization of the mercury atoms by electron collision takes place. Some of the electrons released from the atoms by this process acquire higher energies and in turn ionize other mercury atoms when they strike them.

The positive ions move toward the cathode, relatively slowly because of their comparatively large mass, and neutralize the negative space charge caused by the initial electrons. *The plate current thus reaches a value that is determined and controlled only by the resistance that is external to the tube and connected in series with its plate.* The last sentence should be read over and over again because it describes one of the most important characteristics of gaseous-conducting tubes, whether they be diodes, triodes, or any other kind. After you have thought about this for a while, you will perhaps ask yourself the question "What if the external series resistance is small?" The answer is, as you might surmise, that the current will then be high, and if the resistance is too small the tube will be ruined as a result of overheating. Destruction of the cathode, as well as of the plate, may result. Now for a very important conclusion which takes the form of a caution. *Always connect enough resistance in series* with the plate of a gas tube *to limit the current to a safe value.*

The maximum safe value of tube current is usually somewhat greater than the value at which it is desired to operate the tube. In experimental work we often use a variable resistance to adjust the current in a gas tube. If a fixed resistance of proper value is placed in series with the current-adjusting resistance, the tube will be protected from excessive current.

4.14 The Conducting Region in a Gas Diode

The potential distribution within a hot-cathode (thermionic) gas tube, like type 83, is shown in Fig. 4-12. Such a tube is called an *arc-discharge tube*

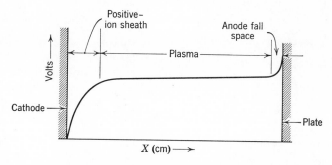

Fig. 4-12 Potential distribution in arc-discharge tube.

or an *arc tube*. Most of the space between the cathode and anode is occupied by what is known as the *plasma*. In the plasma the number of electrons per cubic centimeter is practically the same as the number of ions per cubic centimeter. There is a slight increase in the space potential with respect to the cathode as the plate is approached.

Another important precaution should be observed in the operation of gaseous-conducting tubes, particularly those that employ mercury vapor. Sufficient time should be allowed, before plate voltage is applied, for the cathode to get hot and for the mercury to vaporize. The length of time depends on the tube and its cathode structure. The time is specified by the tube manufacturer and may be as short as 10 s or as long as 10 or more minutes.

If sufficient heating time is not allowed before plate potential is applied, the cathode of the tube may be damaged in the following manner: The insufficient supply of electrons would mean a value of plate current lower than circuit conditions require; this current would produce less than the proper voltage drop in the external resistance connected in series with the tube; the plate-to-cathode voltage would then be above normal; the positively charged gas ions, always accelerated to the cathode, would strike it with too much force, and the cathode coating would be destroyed.

Automatic control is often used to keep the plate-voltage circuit open for sufficient time to allow proper cathode heating. In contrast, the plate voltage of a high-vacuum tube may be applied at the instant the cathode current is turned on, or even before, without damaging the tube.

To balance the passage of electrons from the plasma to the plate, ions move to the cathode, and some diffuse sidewise out of the plasma and reach the walls of the tube. There the ions take on electrons, thus losing their positive charge, and become neutral gas atoms.

The electrons leave the cathode and the cathode region so rapidly, and the positive ions in the cathode region move so slowly, that a layer of ions accumulates about the cathode, forming what is known as a positive-ion sheath. It is across this sheath that most of the potential drop within the tube exists. The relative thickness of the sheath is much exaggerated in Fig. 4-12.

At the anode there is a short anode-fall space across which the potential may either rise a few volts above the plasma potential or fall a few volts below it, depending on the current and the area of the plate.

Larger mercury-vapor diodes, such as the 866A, have been used in rectifier circuits for many years. This type has operating characteristics identical with those shown for the type 83 mercury-vapor diode, except that the current capacity is much larger. Other types of mercury-vapor tube such as the *thyraton* and the *ignitron* are discussed in later chapters where industrial types, or arc-discharge tubes, and their applications will be presented.

4.15   Cold-Cathode Gas Tubes

A diode tube containing an inert gas like argon and having an *unheated* cathode will conduct electricity if the anode is made high enough in potential with respect to the cathode to cause the gas to *ionize*. Ionization is the removal of an electron from a gas atom, which thereby produces a *positive ion* and a *free electron*. Conduction of electricity through the gas in the tube follows ionization, and the tube is said to have a *glow discharge*.

A glow discharge differs somewhat from an arc discharge, principally in the potential difference that exists between the electrodes and in the amount of current that flows. One would expect the current to be comparatively small because the cathode is not heated so that thermionic emission can readily take place. One might also imagine that a powerful electric field near the cathode would be needed to prevent the electrons at the surface from being driven back into the metal by image forces. Actually these conditions do exist in the cold-cathode tube while it is conducting.

Popular types of glow-discharge diodes are the OA3, OB3, OC3, and OD3. They operate at anode-to-cathode voltages of 75, 90, 105, and 150 V, respectively. The currents are in the range of 5 to 40 mA for all four types. These tubes are called voltage-regulator tubes, and they are able to maintain a nearly constant voltage across their terminals throughout the full range of current operation.

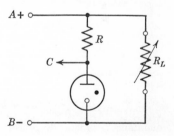

4.16   Applications of Voltage-Regulator Tubes

It is often necessary to have a voltage remain essentially constant, even though other voltages may change owing to various causes. The circuit of Fig. 4-13 shows how a glow-discharge tube may be used to main-

Fig. 4-13   Glow-discharge tube maintains E_{CB} practically constant.

tain a voltage practically constant. Suppose E_{AB} changes as the result of changing the value of R_L or for some other reason. The result will be a change in the current in R_L and also a change in the tube current. But the tube current can change to any value in its range of normal operation (5 to 30 mA) with a change of only about 1 V across its terminals. In practical circuits the change in current in the tube may amount to only 5 to 10 mA, in which case the variation of E_{CB} in the circuit would be of the order of one-third of a volt in 150 V across the tube. This is a very small percentage change, and in most cases it would be allowable.

EXAMPLE 4-2.   Assume that E_{AB} is 350 V and that it changes from 340 to 360 V when R_L is varied to get desirable values of load current. The value of R can be chosen so that the tube current is at midrange value when E_{AB} is at 350 V. The drop across R will be $350 - 150 = 200$ V, and the tube voltage 150. For 20 mA tube current,

$$R = \frac{200}{0.020} = 10,000 \ \Omega$$

The curve of tube current versus tube voltage for the OA2, given in the General Electric tube manual, shows a 2-V decrease when the current increases from 5 to 30 mA. Let us calculate how much the voltage across R would change if the current through it changed by 5 mA.

$$\text{Voltage change} = 0.005 \times 10,000 = 50 \text{ V}$$

Since this change in current will not change the potential drop across the tube more than about $\frac{1}{2}$ V, it is obvious that E_{AB} could change from 340 to 360 V with a rather insignificant change in E_{CB}. It is also obvious that E_{AB} could change between the limits of 325 and 375 V if $\frac{1}{2}$ V change in E_{CB} is allowable.

In some cases current is drawn at C, which produces an additional voltage drop in R. Or, an additional resistance may be required in the input line between A and the junction point of R and R_L. These complicate matters but they do not cancel the usefulness of the voltage regulator tube.

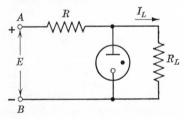

A glow-discharge tube may be used to maintain the current constant in a device. In Fig. 4-14, changes in the input voltage E will change the current in the series resistor R. The voltage drop in R_L and its current will remain practically constant because the tube will accept changes in

Fig. 4-14   Glow-discharge tube maintains I_L practically constant.

current within its operating range. Two or more tubes in parallel will make possible larger changes in E without a change in current in R_L. All the voltage change is taken up by R.

Arc tubes have a practically constant voltage drop for larger current values than glow tubes can handle, and over a wider current range. They must be well protected from excessive current by adequate series resistance.

EXAMPLE 4-3. Assume that a device is to be connected in series with R_L in Fig. 4-14 and that it must carry a constant current of 100 mA. $E = 250$ V. An OB2 tube with a rating of 105 V at 5 to 30 mA is available. (*a*) Determine R_L and R for this circuit. (*b*) Assume R_L constant and determine the permissible range of variation of E during which I_L will stay constant at 100 mA.

Solution. (*a*) Assume an average current of 18 mA in the tube. This means R carries 118 mA with a voltage drop of $250 - 105 = 145$ V.

$$R = 145/0.118 = 1,220 \ \Omega$$
$$R_L = 105/0.100 = 1,050 \ \Omega$$

(*b*) If E were increased to make the tube take its full load of 30 mA at 105 V (theoretical), the current in R would be 130 mA, and its voltage drop would be $0.13 \times 1,220 = 158.6$ V.

$$E = 158.6 + 105 = 263.6 \text{ V}$$

If E were reduced until the tube current decreased to 5 mA, the voltage drop across R would be

$$E_R = 0.105 \times 1,220 = 128.1 \text{ V}$$
$$E = 128.1 + 105 = 233.1 \text{ V}$$

Theoretically, then, E could be raised from 250 V to 263.6 V or dropped from 250 V to 233.1 V, a range of 10.5 V, without causing any change in current in R_L. In actual practice there is at least a slight change in tube voltage when the tube is operated over its full current range.

SUGGESTED REFERENCES

1. W. G. Dow, *Fundamentals of Engineering Electronics*, 2nd ed., Wiley, New York, 1952.

2. E. Milton Boone, *Circuit Theory of Electron Devices*, Wiley, New York, 1953.

3. Paul D. Ankrum, *Principles and Applications of Electron Devices*, International Textbook Co., Scranton, Pa., 1959.

4. George E. Happell and Wilford M. Hesselberth, *Engineering Electronics*, McGraw-Hill, New York, 1953.

QUESTIONS

4-1 A high-vacuum diode is operated at normal values of voltage and current. How does the number of electrons per second emitted by its cathode compare with the number per second arriving at its plate?

4-2 Draw a curve showing the relation between plate current and plate voltage in a high-vacuum diode having a fixed value of filament current. Show plate current on the ordinate (vertical) scale and plate-to-cathode voltage on the abscissa (horizontal) scale. Explain the behavior of the curve in the range of high values of plate voltage.

4-3 On the same axes used for the preceding exercise, show a curve for the same tube operated at a higher value of filament current. Explain the differences in the two curves.

4-4 What is meant by the statement that the current in an electron tube is limited by the space charge?

4-5 Name three advantages of using an indirectly heated cathode instead of a directly heated one.

4-6 Electrons are escaping from a hot-filament wire (the cathode of a diode). What is meant by "work function" of the cathode surface? What is known about an electron that is just barely able to escape from the cathode? What would prevent some electrons from escaping?

4-7 What restrictions were imposed in the derivation of the three-halves power law for a diode? Which of these restrictions are not met in an actual commercial diode?

4-8 What are the important differences between the operating characteristics of high-vacuum diodes and gas diodes?

4-9 Compare high-vacuum and gas diodes in regard to (*a*) steadiness of voltage drop across the tube when external resistance in the plate circuit is varied; (*b*) precautions that should be observed when connecting these diodes in a circuit; (*c*) cathode-temperature requirements.

4-10 An arc-discharge tube is conducting current. Assume a straight line between cathode and anode and perpendicular to both. At every point on that line there will be a particular value of potential with respect to the cathode. Draw a graph showing how this potential varies from point to point, and tell what is happening to electrons and to ions in important regions.

4-11 Compare glow-discharge tubes with an arc discharge tube in regard to (*a*) current capacity; (*b*) voltage drop; (*c*) efficiency of operation (power loss); and (*d*) precautions in use.

4-12 What characteristic of a glow-discharge tube makes it useful as a voltage-stabilizing device? What would be the advantage of using two glow-discharge (voltage-regulator) tubes in series?

4-13 How does a glow-discharge tube make current stabilization possible? When would it be desirable to use two tubes in parallel?

4-14 What is an electronvolt? How many electronvolts of energy will an electron just off the cathode acquire by the time it reaches the plate in a tube where $E_b = 200 \text{ V}$?

PROBLEMS

4-1 An electron that starts from rest at one point and moves to another point 1 V higher in potential than the first may be said to have "risen through a potential difference of 1 V." It is stated in this chapter that the energy of the electron, in such a case, is 1.602×10^{-19} J upon arrival at the second point. Compute the velocity of the electron upon arrival at the second point. Express it in meters per second and in miles per hour.

4-2 With what velocity do electrons arrive at the plate of a high-vacuum diode when it is operated at 200 V between plate and cathode? Assume the electrons leave the cathode with zero velocity. Express their velocity at the plate in meters per second, feet per second, and miles per hour.

4-3 The diode of the preceding problem has a cathode-to-plate distance of 2 mm. Assuming the electrons leave the cathode with zero velocity, how long does it take them to cover the distance between cathode and plate?

4-4 A tungsten-wire filament is used in an electron tube. What value of thermionic-current density is possible if the wire temperature is 2,500°K?

4-5 The cathode of a high-vacuum diode is separated from the anode by a distance of 3 mm (0.003 m). The cathode surface area is 2 sq cm (0.0002 sq m). Plot a curve showing plate current (on the ordinate) and plate voltage (on the abscissa) for voltages from 0 to 250 V. Assume that the conditions required for the application of Child's law are fulfilled.

4-6 In Section 4.5 the thermionic current density was given by a numerical expression containing 60.2×10^4 $(1,600)^2$ multiplied by an exponential factor. Use common logarithms to compute the answer. (The common logarithm of ϵ, the base of the natural system of logarithms, is 0.4343.)

4-7 The screen grid of a vacuum tube must be held at a practically constant voltage of $+90$ V with respect to the cathode, and it will carry a constant current

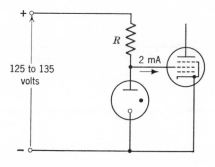

Fig. P4-7

of 2 mA. Design a voltage-regulating circuit using a glow-discharge tube that will do the job when the voltage supplying the circuit varies from 125 to 135 V. A suggested circuit is shown in Fig. P4-7. How can you be sure that the resistor R will not overheat?

4-8   The current in a 500-Ω element is to be held constant at 10 mA even though the voltage input to the control circuit varies between 125 and 135 V. Design a current regulator circuit that will do the job. How much power will the regulator circuit take?

4-9   What velocity must an electron acquire at the surface of the metal molybdenum in order to escape at right angles to the surface? Would it escape if it were traveling with that total velocity in a direction at 45° with the surface? Why? What total velocity must it have in order to escape while traveling in the 45° direction?

4-10   Assume that the tube in the circuit of Fig. 4-14 has a voltage drop of 105 V when carrying 20 mA. $R_L = 1,050 \ \Omega$. (*a*) Calculate the value of R for an input voltage $E_{AB} = 350$ V. (*b*) The input voltage rises an amount to cause the tube current to increase to 30 mA. Assuming that the tube voltage rises to 106 V, compute the rise in input voltage. (*c*) Allowing for a possible maximum tube current of 40 mA, calculate the wattage rating that should be specified for the resistor R.

4-11   The current in the coil shown in Fig. P4-11 must be varied between the limits of 170 mA and 500 mA. It is also necessary to maintain E_{DC} nearly constant at 75 V during the variation of coil current. After the characteristic curve for the tube was

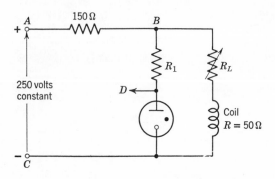

Fig. P4-11

examined, it was decided that its voltage will drop just about 3 V while its current changes from 10 to 30 mA. (*a*) Calculate the value of R_1 and the maximum and minimum settings of R_L. The maximum tube current should not exceed 30 mA. (*b*) Specify wattage capacities for all three resistors.

4-12   A diode is operated at such a cathode temperature that plate currents at various voltages are determined by Child's law. At a plate voltage of 60 V the current is 25 mA. Calculate the value of plate current at a plate voltage of 120 V.

4-13 An electron just off the cathode of a thermionic diode has zero velocity and begins to move toward the anode under the influence of the electric field supplied by 250 V on the plate. (*a*) Calculate the joules of energy of the electron at the instant it arrives at the plate. (*b*) How many electrons per second strike the plate when the current is 200 mA? (*c*) How much power is supplied to the plate by the bombarding electrons?

4-14 A tungsten-wire filament is used in a high-vacuum diode. (*a*) How much thermionic current density is possible when the wire temperature is $2,970°K$? (*b*) How many milliamperes could be emitted from a filament wire 0.010 in. in diameter and 0.5 in. long?

4-15 A diode operates at such a temperature that plate current follows Child's law for the voltages considered in this problem. At a plate voltage of 50 V the plate current is 20 mA. Determine the current if the plate voltage is raised to 150 V.

4-16 A certain diode is known to obey the relation $i_b = K e_b{}^n$. When the plate voltage is 100 V, plate current is 10 mA, and when plate voltage $= 10$ V, plate current $= 0.3$ mA. Determine the perveance of the tube and the value of the exponent n.

4-17 A 500-Ω resistor is connected in series with the plate of a 5Y3 diode. If the cathode is heated to its normal emitting temperature and $E_{bb} = 100$ V, how much plate current will flow?

4-18 A vacuum-tube diode is operated at $E_b = 200$ V and its plate current is 200 mA. Calculate the number of electrons per second striking the plate. Calculate the power (in watts) supplied to the plate by the bombarding electrons.

4-19 Calculate the resistance of a 5Y3 diode when its plate voltage is 50 V. (Refer to Fig. 4-10.)

4-20 A certain diode was used to determine the following data:

E_b (V)	I_b (mA)
1.0	0.05
5.0	0.28
10.0	1.02
15.0	2.24
20.0	3.41
25.0	4.68
30.0	6.71

Determine graphically the values for K and n in the Child's law expression that applies to this tube.

CHAPTER FIVE

Rectifiers, Power Supplies, and Filters

An electronic tube conducts current only when its anode (plate) is electrically positive with respect to its cathode. When an alternating voltage is used to supply the plate current of a tube, the current will flow only during the positive half-cycles because the plate cannot furnish electrons to flow to the cathode during negative half-cycles of the alternating voltage. A transistor behaves in much the same way, requiring a d-c power supply for its operation in useful circuits.

In order that a tube or transistor may conduct current continuously, without being shut off periodically, the power supply must furnish a direct voltage that does not become zero. Because all commercial electricity is available at alternating voltage only, it must be *rectified* in order to supply electronic tubes and transistors with continuous d-c power. This chapter considers the operation of crystal diodes and diode tubes used as rectifiers in *power supplies* needed in electronic circuits. Various rectifier circuits will be studied to see how they convert a-c voltages and currents into d-c voltages and currents. The part that *smoothing filters* play in their operation will be discussed.

5.1 Half-Wave Rectification

We have observed that during each cycle an alternating voltage, as shown by the sine curve of Fig. 5-1, rises to a maximum value, falls to zero and reverses, reaches a maximum in the reverse direction, and then returns to zero. When a voltage with this kind of waveform is applied to a resistor, the current flowing in the resistor will be alternating and of the same waveform as the voltage.

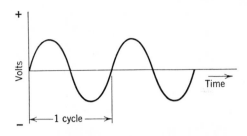

Fig. 5-1   Sine curve representing an alternating voltage.

When an alternating voltage as represented in Fig. 5-1 is applied to a diode rectifying element (crystal diode or diode tube) connected in series with a resistor, the variations in the voltage across the resistor (and the current through it and the diode) are as shown in Fig. 5-2. The waves are drawn on different vertical scales. The diode acts as a half-wave rectifier by allowing current to flow only during the positive half of each alternating-voltage cycle. Current can flow and thus establish voltage across the resistor (Fig. 5-3) only during the positive half-cycles when the plate is more positive than the cathode. During the negative half of each cycle the plate is negative with respect to the cathode and the tube will not conduct because of its reverse bias.

5.2   The Nature of Rectifier Output Voltage

If a d-c milliammeter is connected in the half-wave rectifier circuit of Fig. 5-3 it will show a steady reading even though pulses of current (see Fig. 5-2) are flowing through it. One current pulse occurs in each cycle. It is interesting that the reading of the meter will be only 0.318 of the maximum value to

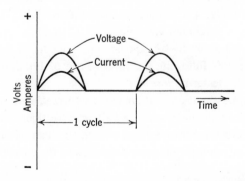

Fig. 5-2   Rectified voltage and current through load resistor.

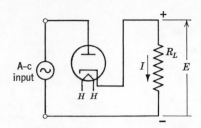

Fig. 5-3 Half-wave rectifier circuit.

which the current rises in each pulse, i.e., if the maximum value of the current supplied by the half-wave rectifier is 10 mA, for example, the d-c milliammeter will read 3.18 mA. This can be verified by mathematical analysis of the current pulse, a half-sine wave.

The mathematical expression representing the complete half-sine wave of voltage during one cycle of the input signal can be determined* by *Fourier analysis* as

$$e = 0.318E_m + 0.5E_m \sin 2\pi ft - 0.212E_m \cos 2\pi(2f)t$$

$$- 0.0424E_m \cos 2\pi(4f)t - \cdots \quad (5\text{-}1)$$

It is important to understand that each term of Equation (5-1) represents a component of voltage and that *all these components exist simultaneously.* They add together from instant to instant, producing the half-sine wave as their total effect.

A half-sine wave of voltage is found to be made up of several components: (1) an average, d-c component having a constant value 0.318 of the maximum value, (2) an a-c component having a maximum value equal to 0.5 of the original half-sine-wave maximum and having the same frequency as the unrectified sine wave, (3) an a-c component of twice the fundamental frequency having a maximum value equal to 0.212 of the original sine-wave maximum, and (4) many high-frequency components of decreasing amplitudes. The three dots following the fourth term of Equation (5-1) indicate additional terms of lesser importance.

Figure 5-4 shows the d-c, fundamental frequency, and second harmonic components in the half-sine wave of a half-wave rectifier. Harmonics higher than the second are not shown because of their small amplitudes. The vertical distance from any point on the horizontal axis to the dashed half-sine curve

* H. Alex Romanowitz, *Electrical Fundamentals and Circuit Analysis*, Wiley, New York, 1966, p. 503.

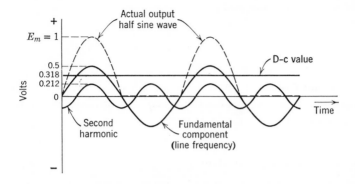

Fig. 5-4   Components of half-sine wave.

can be approximated by adding together the vertical distances from that point to the component graphs. Due regard must be paid to the negative sign when distances are measured downward from the horizontal axis. The contribution made by the fourth harmonic (only one-fifth as large as the second harmonic) is negligible. Contributions from higher harmonics will be correspondingly smaller.

DRILL PROBLEM

D5-1   Show by graphical addition that the Fourier components of a half-sine wave add together to yield the curve as their total effect.

5.3   Output Current and Power of a Half-Wave Rectifier Circuit

It can be shown mathematically that the peak instantaneous value of the current wave through the load resistance is given by

$$I_m = \frac{E_m}{r_p + R_L} \text{ A} \tag{5-2}$$

where E_m is the peak instantaneous value of the a-c input voltage and r_p is the plate resistance of the tube.

The voltage impressed upon R_L is represented by Equation (5-1), the *first term* of which is the *d-c component*. Since $0.318 = 1/\pi$, it may be written E_m/π. The direct current in the load is then

$$I_{\text{d-c}} = \frac{1}{\pi} \frac{E_m}{R} = \frac{I_m}{\pi} \text{ A} \tag{5-3}$$

and the d-c power in the load is

$$P_{\text{d-c}} = I_{\text{d-c}}^2 R_L = \left(\frac{I_m}{\pi}\right)^2 R_L = \frac{E_m^2 R_L}{\pi^2 (r_p + R_L)^2} \quad \text{W} \tag{5-4}$$

Power is lost in the tube because the current must flow through r_p, the tube resistance. This cuts down appreciably the efficiency of conversion from a-c to d-c power. The *effective value* of the *half-sine-wave current* can be shown to be (over the full period)

$$I_{\text{rms}} = \tfrac{1}{2} I_m$$

so that the a-c power in the circuit is given by

$$P_{\text{a-c}} = \left[\frac{E_m}{2(r_p + R_L)}\right]^2 [r_p + R_L] = \frac{E_m^2}{4(r_p + R_L)} \quad \text{W} \tag{5-5}$$

The *efficiency of power conversion* is then obtained by dividing the d-c power by the a-c power. To get the expression for efficiency, we divide Equation (5-4) by equation (5-5) and get

$$\text{Power conversion efficiency} = \frac{R_L}{r_p + R_L} \cdot \frac{4}{\pi^2} (100)\% \tag{5-6}$$

If the tube resistance is zero,

$$\text{Maximum theoretical power conversion efficiency} = \frac{4}{\pi^2} \times 100 = 40.6\%$$

Because r_p is rather large (a thousand or more ohms in high-vacuum tubes) the actual efficiency is much lower than this. For $r_p = R_L$, it is only 20.3 per cent.

The peak value of current which the tube must be able to pass safely is, from Equation (5-3),

$$I_{\text{peak}} = I_m = \pi I_{\text{d-c}} \quad \text{A} \tag{5-7}$$

The *peak-inverse voltage* to which the tube is subjected *is equal to the peak of the a-c voltage applied to the circuit*. This means that when the tube is not conducting (during negative half-cycles) the full voltage is impressed across the tube in reverse polarity. Should the tube conduct (*arc back*) when the polarity is reversed, its rectifying properties would fail and the tube might be destroyed. Data given by manufacturers include the safe value of peak-inverse voltage and also ratings of average and peak current. *Peak-inverse voltage* is defined as the maximum instantaneous voltage applied in the direction opposite that of normal current conduction.

After the full-wave rectifier has been described, a comparison of the advantages of the two types will be made.

DRILL PROBLEM

D5-2 A half-wave rectifier is supplied from a 60-Hz, 120-V rms source. A diode tube having internal resistance of 500 Ω is used to supply d-c power to a load resistance of 2 kΩ. (*a*) Calculate the maximum value of diode current; load current. (*b*) What is the d-c power in the load? (*c*) What is the rms value of the output current? (*d*) Calculate the power conversion efficiency. (*e*) What must be the PIV rating of the tube?

5.4 Full-Wave Rectification

A full-wave rectifier circuit and the waveforms of rectified voltage and current in a resistance load are shown in Figs. 5-5 and 5-6. This circuit is called full-wave because it utilizes the full wave of input voltage to supply power to the load. Each diode is *forward biased* on alternate half-cycles of the input voltage, and each conducts current through the load in the same direction. Thus, the full wave of input voltage is rectified and produces two half-sine waves of load voltage and current for each cycle of input sine wave.

Fourier analysis of the load voltage waveform shows that it may be expressed in this form:

$$e = 0.636E_m - 0.424E_m \cos 2\pi(2f)t - 0.085E_m \cos 2\pi(4f)t - \cdots \quad (5\text{-}8)$$

Note that the d-c component is twice as large as that produced by the half-wave rectifier. This should be expected because there are two half-sine-wave portions of the voltage during each input cycle, and their average value is twice that of one half-sine wave during the same time. A result that might not be expected is the *absence of a component having input frequency*. Only its second and higher harmonic components appear in the load resistance. The second harmonic and high-frequency terms are twice their values in the half-wave case. This rapid reduction in the a-c components (the magnitude

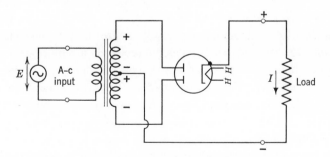

Fig. 5-5   Double-diode full-wave rectifier circuit.

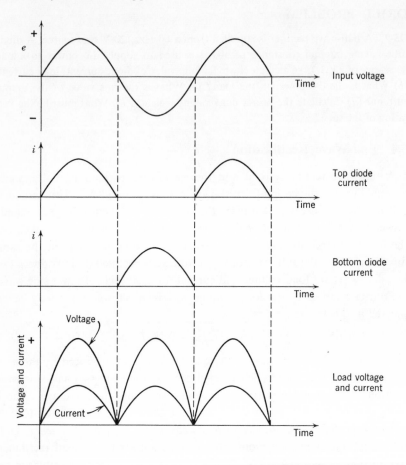

Fig. 5-6 Voltage and current waveforms in full-wave rectifier with resistance load.

of the third term is only $0.085E_m$) helps to improve the power-conversion efficiency and reduces the amount of filtering necessary to get a constant d-c power for the load. It therefore cuts down the size and cost of filters.

If an *unfiltered* rectified sine wave is impressed on a pure resistance load, the load will *simultaneously* carry d-c current and a-c components at the frequencies given by Equation (5-1) or (5-8). The peak value of each component of current will be given by the coefficient of the corresponding sine or cosine term, divided by the total circuit resistance, $r_p + R_L$.

The components of the full-wave rectifier output are shown in Fig. 5-7. Waves for terms of the expression that represent higher harmonics have been neglected.

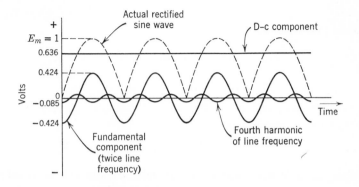

Fig. 5-7 Components of full-wave rectifier output.

5.5 Full-Wave Rectifier Output

Equations for current, power, and efficiency will now be obtained.
As explained above, and also from Equation (5-8),

$$I_{\text{d-c}} = 0.636\frac{E_m}{R} = \frac{2}{\pi}I_m \ \text{A} \tag{5-9}$$

so that the d-c power in the load is given by

$$P_{\text{d-c}} = 4\left(\frac{I_m}{\pi}\right)^2 R_L \ \text{W} \tag{5-10}$$

The effective value of the fully rectified sine wave of current is equal to the
instantaneous maximum divided by $\sqrt{2}$:

$$I_{\text{rms}} = \frac{I_m}{\sqrt{2}} \ \text{A}$$

The a-c power is

$$P_{\text{a-c}} = \frac{I_m^{\ 2}}{2}(r_p + R_L) \ \text{W} \tag{5-11}$$

Dividing Equation (5-10) by Equation (5-11) gives the power conversion
efficiency:

$$\text{Efficiency} = \frac{8}{\pi^2}\left(\frac{R_L}{r_p + R_L}\right) \times 100\%$$

The maximum theoretical power conversion efficiency is again found by
assuming $r_p = 0$:

$$\text{Maximum theoretical efficiency} = \frac{8}{\pi^2} \times 100 = 81.2\%$$

The peak value of current which the tube must be able to pass safely with resistance load is, from Equation (5-9),

$$I_{\text{peak}} = I_m = \frac{\pi}{2} I_{\text{d-c}}$$

The peak-inverse voltage on the tube which is not conducting is twice the peak value of the input a-c voltage, *provided the voltage drop through the conducting tube is assumed to be zero.* Thus

$$\text{Peak-inverse voltage} = 2E_m = 2\sqrt{2}\, E_{\text{rms}} = 2.82 E_{\text{rms}}$$

Although r_p is certainly not zero, the possible peak-inverse voltage applied to each tube on alternate half cycles when the load current is small (and it can be zero if the load circuit is open) should be considered to be twice the instantaneous peak value of the applied a-c voltage.

DRILL PROBLEM

D5-3 Repeat D5-2 for two diodes in a full-wave rectifier circuit, assuming the transformer supplies each diode with 120-V rms input. Compare the results with those for the half-wave rectifier.

5.6   Comparison of Full-Wave and Half-Wave Rectifiers

When compared with the half-wave rectifier, the full-wave rectifier has the following advantages:

1. It is capable of considerably higher efficiency (theoretically double).

2. Its d-c output power is four times as great, for a given load resistance, because the d-c component of load current is twice as large.

3. Output requires much less filtering because the a-c components of output current are, on the whole, much smaller in amplitude.

4. If a transformer must be used, a smaller one may be used for a given power output. This is deduced from Equations (5-4) and (5-10). By equating them for equal power conditions, it can be seen that the required transformer current for full-wave rectification is only one-half that for half-wave rectification.

5. The d-c component of load current produces an alternating magnetic flux in the transformer because it flows in opposite directions in the halves of the secondary windings. This does not polarize the core, whereas it is polarized by the unidirectional flux in the half-wave case. This is another reason for using a smaller transformer in the full-wave circuit.

The peak-inverse voltage applied to the tube in a half-wave rectifier is equal to E_m, the maximum instantaneous value of the input voltage. In the full-wave rectifier, the tubes are subjected to twice this amount of peak-inverse voltage, i.e., $2E_m$.

The half-wave rectifier has an advantage, if the a-c voltage need not be stepped up or down, in that its circuit does not require a center-tapped transformer. Also, in applications which must operate on either alternating or direct current, as in some small radio receivers, the half-wave rectifier is useful whereas the full-wave rectifier is not. Half-wave rectifiers have application also in television receiver circuits.

5.7   Smoothing Filters

It was stated earlier that the plate of an electronic tube should be supplied with a steady, smooth, direct voltage rather than a pulsating voltage. It is possible to reduce greatly the amplitudes of the a-c components in the output voltage of a rectifier by means of one or two sections of a *smoothing filter*. Its action provides a path from the rectifier output to the load for d-c components and blocks the a-c components or short-circuits them around the load. Since the a-c components are eliminated from the load current, only the d-c component produces a voltage across the load.

Smoothing filters used with rectifiers that supply power to electronic tubes and transistors are usually made of iron-core inductances and capacitors. In circuits where the load currents are very small, or if the load current need not be extremely smooth, only capacitors and resistors may be used in the filter. When the load voltage has only a small amount of a-c voltage, its variation about the average (d-c) value is called *ripple voltage* or simply *ripple*.

A filter usually takes one of three forms: (1) shunt-capacitor, (2) L-section, and (3) π-section. A shunt-capacitor filter is just what its name implies, a capacitor connected in parallel with the load resistor. The L-section and π-section filters derive their names from their appearance in a schematic diagram of the circuit. Typical filter circuits are shown in Fig. 5-8.

The tandem L-section filter is called a choke-input filter because the current from the diodes in the rectifier first encounters a *choke coil*, which is an inductance coil wound on an iron core. A choke-input filter often has only one L-section rather than two in tandem as shown. Of course, two sections will do a better filtering job, but the extra cost of circuit elements may be prohibitive in some applications and may be considered an over-design of the circuit.

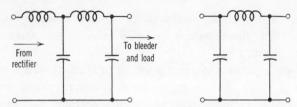

Fig. 5-8   Choke-input and capacitor-input smoothing filters. (*a*) Tandem *L*-section. (*b*) π-section.

The π-section filter is called a capacitor-input filter for obvious reasons. Its operation results in a higher voltage at the load than would be obtained from a choke-input filter. The first (input) capacitor charges up to the peak of the rectifier output voltage and does not discharge very much during the remainder of each cycle while the rectifier input voltage is falling and going through its negative half-cycle. The nature of the voltage at the input and output of the capacitor-input smoothing filter is shown in Fig. 5-9.

5.8   Shunt-Capacitor Filter

A simple filter to remove the a-c ripple components in the output of a rectifier is obtained by shunting a capacitor across the load resistance, as shown in Fig. 5-10 for a half-wave rectifier. If the value of capacitance is chosen so that its reactance at the fundamental frequency is much less than the value of load resistance, the a-c components will have a low-reactance

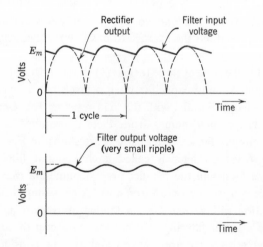

Fig. 5-9   Curves for capacitor-input smoothing filter.

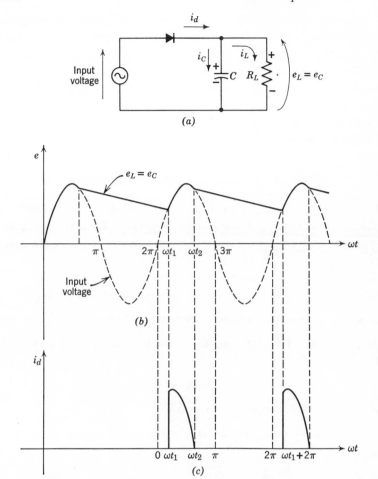

Fig. 5-10   Half-wave rectifier with shunt-capacitor filter and load resistance; circuit waveforms. (*a*) Circuit. (*b*) Voltage waveforms. (*c*) Diode current pulses.

path around the load resistor. Only a small a-c current flows through the load and produces a small ripple voltage.

The action of the capacitor can be explained as follows. During the time when the rectifier output voltage is rising, the capacitor will charge to a voltage equal to the rectifier output. As the rectifier output decreases, the capacitor discharges through the load resistor. The load voltage waveform meanwhile will be an exponential curve determined by the time constant $R_L C$. When the rectifier output again rises to equal the voltage remaining on the capacitor, the diode will again be forward biased and will immediately start conducting a current equal to the existing current in the load resistor. This can easily be understood, because when the rectifier output voltage equals the voltage on the capacitor, the capacitor will be neither charging nor discharging. But the load resistor has some value of current in it. Kirchhoff's current law must hold, so that the diode will pick up at this time whatever current exists in the load resistor.

The diode delivers a current pulse during each cycle to charge the capacitor, and acts as a switch to disconnect the load from the rectifier. If the discharge time constant is long compared with one period of the input voltage, the load voltage will be nearly constant.

The current pulses supplied by the rectifier will have a waveform determined by the time constant of the filter and load resistance. We shall now develop a simplified expression for the diode current pulses in terms of the charging interval, ωt_1 to ωt_2, and the time constant of the filter.

Referring to Fig. 5-10, when the capacitor is charging, the voltage across it equals the input voltage (assuming no forward drop in the diode), or

$$e_C = E_m \sin \omega t, \qquad \text{when } \omega t_1 < \omega t < \omega t_2 \tag{5-12}$$

During this same time interval, the diode current equals the sum of the capacitor and resistor currents,

$$i_d = i_C + i_L$$

Since the load current can be written as

$$i_L = (E_m/R_L) \sin \omega t, \qquad (\omega t_1 < \omega t < \omega t_2)$$

and the capacitor current is

$$i_C = C \frac{de_C}{dt} = \omega C E_m \cos \omega t$$

then the diode current is a pulse having a form

$$i_d = E_m \left[\left(\frac{1}{R_L} \right) \sin \omega t + \omega C \cos \omega t \right], \qquad (\omega t_1 < \omega t < \omega t_2) \tag{5-13}$$

When the input voltage reverses direction, a magnitude will be reached at which the capacitor discharge cannot follow the decreasing voltage. At this point, ωt_2, the diode current will cease and the capacitor current will equal the load current in magnitude, and

$$\left(\frac{1}{R_L} \right) \sin \omega t_2 = -\omega C \cos \omega t_2$$

From this,

$$\omega t_2 = \tan^{-1}(-\omega R_L C) \tag{5-14}$$

Equation (5-13) may be written as

$$i_d = \frac{E_m}{R_L} \sqrt{1 + \omega^2 R_L{}^2 C^2} \sin(\omega t + \phi)$$

where $\phi = \tan^{-1}(\omega R_L C) = \pi - \tan^{-1}(-\omega R_L C) = \pi - \omega t_2$. Making this substitution for ϕ, and from trigonometric identities, yields this simplified expression for the diode current pulses

$$i_d = \frac{E_m}{R_L} \sqrt{1 + \omega^2 R_L{}^2 C^2} \sin(\omega t_2 - \omega t).^* \tag{5-15}$$

It should be noted that Equation (5-15) is valid only during the charging interval, ωt_1 to ωt_2. The diode current is zero at other times.

Since the waveform of diode current depends on the factor $\omega R_L C$, which includes the input frequency and time constant of the filter circuit, it should be expected that the current waveform will be variable and highly dependent on this factor. Figure 5-11 shows some typical magnitudes and waveshapes of diode current.

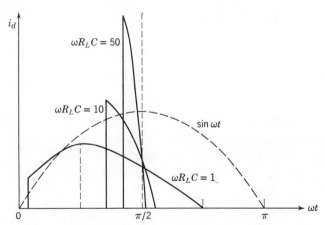

Fig. 5-11   Waveform of diode current pulses for different values of $\omega R_L C$. (Curve for $\omega R_L C = 1$ is exaggerated to show its sine-wave shape.)

---

* After substituting, let $\omega t - \omega t_2 = \alpha$. Show that $\sin(\alpha + \pi) = \sin(-\alpha)$.

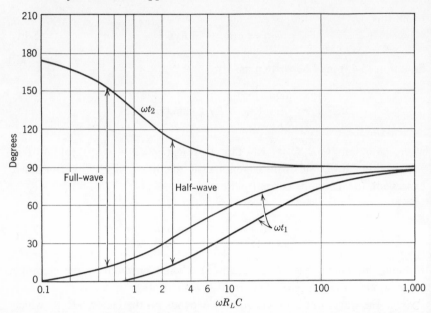

Fig. 5-12   Conduction angles for shunt-capacitor filter on half-wave and full-wave rectifiers.

It can be seen from Fig. 5-11 that the conduction angles for the diode are also functions of the factor $\omega R_L C$. The relationship of $\omega R_L C$ to ωt_1 and to ωt_2 can be calculated, but the procedure is beyond the scope of this book. It can be shown graphically, as in Fig. 5-12, for both the half-wave and full-wave rectifiers.

It should also be noted that the maximum amplitude of the diode current pulse increases rapidly as the factor $\omega R_L C$ increases. This means that when the load voltage has a small ripple the diode current may take on very large amplitudes. Then it is important to design the filter so that the maximum diode current rating will not be exceeded in the operation of the rectifier.

When the conduction angle is small, the inverse voltage on the diode will be approximately twice the peak input voltage. Therefore, PIV rating of the diode must be equal or greater than twice the peak input voltage.

DRILL PROBLEMS

D5-4   A diode having negligible forward resistance is used in a full-wave rectifier whose load resistance is shunted by a large capacitor. Sketch and label the waveform of diode current that has a total conduction angle of 90°.

D5-5   Sketch the waveform of the voltage across the diode during 1 cycle of input voltage to the rectifier of D5-4.

5.9 Bleeder Resistor

Power supplies with smoothing filters are usually terminated in a fixed resistance of high ohmic value, called a *bleeder resistor*, i.e., a resistor of perhaps 10 to 50 kΩ or more is connected permanently to the output terminals of the smoothing filter. Its resistance is usually of such value that it carries about 5 to 10 per cent of the normal full-load current to be furnished by the power supply.

One reason for having a bleeder across the output terminals of a choke-input filter is to prevent the current output of the rectifier from becoming too small. When the load current becomes quite small, the waveform will have a high ripple content. Another reason is to provide a means of discharging the filter capacitors when the power supply is turned off. The capacitors operate at high voltages, and if there is no way for them to lose their charge, it will remain on their plates for a long time after the power supply is turned off. Thus a dangerous voltage would be maintained at the output terminals of the filter. The capacitors cannot discharge through the diodes of the rectifier because of their reverse bias.

Frequently it is necessary to furnish to other parts of an electronic circuit, voltages that are lower than the full output voltage of the filter. These voltages are conveniently obtained by means of taps on the bleeder resistor. It is necessary to take into account the currents flowing through the upper portions of a tapped bleeder and the voltage drop in those sections due to currents and resistances. The power loss that must be accomplished through heat dissipation by the bleeder is an important factor in bleeder (and in voltage-divider) design. (Refer to Problem 5-14.)

5.10 Smoothing-Filter Calculations

It has been noted that the job of the power supply smoothing filter is to prevent the flow of alternating components of current through the load circuit connected to the filter. We may readily observe that, if a filter is capable of eliminating the alternating component of current that has the lowest frequency, the higher-frequency components will be eliminated at the same time. This is readily seen when we recall that the inductive reactance of the choke ($2\pi fL$) increases with frequency, and therefore the higher-frequency components will be diminished even more than the fundamental-frequency component. Likewise the bypassing capacitor will more readily allow the higher-frequency currents to bypass the load because its reactance decreases with increase in frequency.

The ripple factor in a pulsating d-c voltage or current wave is defined as the ratio of the rms value of all the a-c components (taken together) to the d-c component of the wave:

$$\text{Ripple factor} = \frac{\text{rms value of a-c components}}{\text{d-c value}}$$

It is much more convenient, and adequate enough for practical purposes, to use for the ripple factor the ratio of the *rms value* of the *fundamental component* to the d-c component of the wave. Designating the *ripple factor* by ρ the d-c component of the voltage by $E_{\text{d-c}}$, and the rms value of the *fundamental component* of the voltage by E_{r1}, the ripple factor at fundamental frequency is

$$\rho = \frac{E_{r1}}{E_{\text{d-c}}} \tag{5-16}$$

Consider Equation (5-16) to represent the ripple factor at the input to a smoothing filter. E_{r1} is the effective value of the fundamental a-c component of input voltage to the filter and $E_{\text{d-c}}$ is the d-c component of that voltage. Using primes to designate filter *output* quantities, the filter *output* ripple factor is

$$\rho' = \frac{E_{r1}'}{E_{\text{d-c}}'} \tag{5-17}$$

The *smoothing factor* at fundamental frequency is defined as the ratio of the rms value of the fundamental-frequency component of the input voltage at the filter to the rms value of the fundamental-frequency component of the output voltage of the filter. Designating the smoothing factor by S, we have

$$S = \frac{E_{r1}}{E_{r1}'} \tag{5-18}$$

Note that the smoothing factor of a filter may be determined by dividing the ripple factor at the filter input by the ripple factor at the filter output, *if the d-c voltage drop through the filter choke may be neglected.* This may usually be done because it is seldom more than 5 per cent of the input voltage. We then have

$$S = \frac{\rho}{\rho'} = \frac{E_{r1}/E_{\text{d-c}}}{E_{r1}'/E_{\text{d-c}}'} \tag{5-19}$$

Considering

$$E_{d\text{-}c} = E'_{d\text{-}c},$$

$$S = \frac{E_{r1}}{E_{r1}'}$$

which is the definition of smoothing factor given by Equation (5-18).

The smoothing factor is a number usually much larger than unity, whereas the ripple factor is usually less than unity except in a half-wave device.

Let us consider the ripple factor of the output voltage of the full-wave rectifier. The equation of the full-wave rectified voltage (Equation 5-8) shows that $E_{r1} = 0.707 \times 0.424E_m = 0.30E_m$ and $E_{d\text{-}c} = 0.636E_m$. The ripple factor is then

$$\rho = \frac{0.30E_m}{0.636E_m} = 0.472*$$

EXAMPLE 5-1. A smoothing filter connected between a full-wave rectifier and a load filters the tube output so that the load receives a voltage that has a d-c component of 200 V and a fundamental ripple component of only 1 V. Compute the ripple factor of the load voltage and the smoothing factor of the filter.

Solution. The ripple factor of the load voltage is

$$\rho' = \frac{E_r'}{E_{d\text{-}c}'} = \frac{1}{200} = 0.005$$

The smoothing factor of the filter is

$$S = \frac{\rho'}{\rho} = \frac{0.472}{0.005} = 94.4$$

5.11 Inductance and Capacitance Values in Smoothing Filters

It was indicated in Section 5-9 that, when the current delivered by a rectifier with a choke-input filter is too small, the waveform of the load voltage is poor. Incidentally, it will be recalled that, if a voltage is applied to a pure resistance, the current that flows in the resistance will have the same waveform as the voltage. That is, if the load is resistive only and current value is small, the waveform of the load current delivered by the filter having insufficient inductance will also be poor. It is found that there is a minimum value of inductance that the choke should have, i.e., unless the inductance of the choke is at least as large as a certain value, which depends on the load resistance, the current input to the filter will be shut off

---

* Although ρ would be about 2.1 per cent higher than this, if the effect of the harmonics were taken into account, the smaller value is satisfactory for use in design.

during part of each cycle of ripple frequency. The shutting off will be due to the capacitor voltage being too high to allow current flow during part of each cycle. The result will be excessive ripple in the load voltage.

A good rule to follow in the design of choke-input filters is to make the value of the choke inductance (L) at least as large as $R/1,000$, where R is the largest value of operating load resistance. Of course, when the load resistance is largest, the load current will be at its smallest value. It will soon be shown that the lowest value of the inductance of a choke-input filter should be equal to $R_t/1,130$, where R_t is the sum of the load resistance and the resistance of the winding of the choke. When there is insufficient inductance in a single-section choke-input filter, the waveform of the rectifier-tube output current (i.e., filter input) is somewhat as shown in Fig. 5-13. Because the current ceases to flow during part of each cycle, the ripple content is very high. The reason the current stops flowing is that the capacitor voltage remains high and blocks the flow of rectifier current during the early part of each cycle. This happens particularly at light loads (high-load resistance), in which case the load resistance is so high that the capacitor voltage cannot drop fast enough to permit continuous current flow from the rectifier.

Consider for a moment that the filter consists of a capacitor only and no inductance. If there were no load resistor or bleeder, the capacitor could not discharge at all, and, once it acquired a voltage equal to the maximum value of the rectifier output voltage, no more current could flow out of the rectifier. If, however, there were a high-resistance bleeder, and perhaps a load resistor of high value connected to the capacitor, it could discharge a little during each cycle and the load voltage would look like the curve shown in Fig. 5-14. The filter input current would flow in spurts, as indicated. If just enough inductance is present in the filter to prevent current cutoff, the waveform of the filter input current will be about as shown in Fig. 5-15. The d-c component is shown as the average value by the dashed line. Since the instantaneous

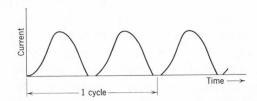

Fig. 5-13   Output current of full-wave rectifier with choke-input filter having insufficient inductance.

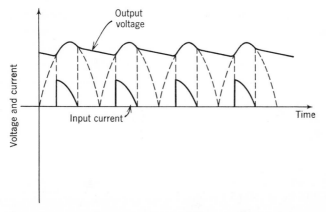

Fig. 5-14   Input current and output voltage of filter made of only a capacitor and a bleeder. The load is pure resistance.

total output current is zero once each cycle of fundamental ripple frequency, we may state that the smallest practical value of inductance is that which will make the amplitude of the a-c component of filter input current equal to the d-c value of the filter current. Note that the d-c component in the filter flows out of the filter to the load, since none can bypass the load through the capacitor. It is assumed that the components of current which have frequencies greater than fundamental frequency are negligible.

We may therefore say that the peak value of the a-c component of current in Fig. 5-15 is equal to $\sqrt{2}$ times the rms value of the fundamental component which we may call I_{r1}. The reactance of the capacitor C (Fig. 5-16) is very small compared with the reactance of L, at ripple frequency. We can neglect X_C and write

$$I_{r1} = \frac{E_{r1}}{2\pi f L} \tag{5-20}$$

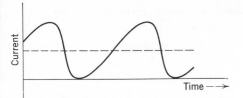

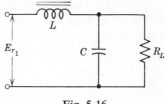

Fig. 5-15   Current input of a choke input filter that has just enough inductance to prevent current cut-off.

Fig. 5-16

Also, the direct current through the load is

$$I_{\text{d-c}} = \frac{E_{\text{d-c}}}{R_L} \tag{5-21}$$

and this is equal to $\sqrt{2}\,I_{r1}$, because it is the amplitude of I_{r1} (see Fig. 5-16), so that

$$\sqrt{2}\,I_{r1} = \frac{E_{\text{d-c}}}{R_L} \tag{5-22}$$

$$I_{r1} = \frac{E_{\text{d-c}}}{\sqrt{2}\,R_L} \tag{5-23}$$

Substituting Equation (5-23) into Equation (5-20),

$$\frac{E_{\text{d-c}}}{\sqrt{2}\,R_L} = \frac{E_{r1}}{2\pi f L} \tag{5-24}$$

$$L = \frac{\sqrt{2}\,R_L E_{r1}}{2\pi f E_{\text{d-c}}} \tag{5-25}$$

Notice that the ripple factor at the filter input is present here, i.e., $E_{r1}/E_{\text{d-c}} = \rho_1$. This ripple factor has a value 0.472 for a full-wave rectifier, and the ripple frequency (f) is 120 on a 60-hertz line. We then have

$$L \geq \frac{\sqrt{2}\,R_L \times 0.472}{2\pi \times 120} = \frac{R_L}{1,130} \tag{5-26}$$

From the foregoing we conclude that the inductance of a single-section choke input filter should be at least as large in henrys as the value obtained by dividing the sum of the load resistance and the choke resistance by 1,130. Since this is at best a close approximation, it is safe to make L at least as large as the value obtained by dividing the resistance equivalent of the bleeder in parallel with the maximum load resistance by 1,000. For a two-stage filter (two inductances, each followed by a capacitance), the first inductance should be as large as indicated above.

A relationship between the capacitance and inductance of a single-stage choke input filter in terms of the smoothing factor will now be formulated. Consider Fig. 5-16 and notice that E_{r1}, the rms value of the ripple voltage input to the filter, is shown. E_{r1}', the rms value of the ripple voltage output, would show up across the capacitor. It will be recalled that the smoothing factor is

$$S = \frac{E_{r1}}{E_{r1}'} \tag{5-27}$$

Considering the a-c component of current sent by E_{r1} to flow through L and C only (X_C is very much smaller than R_L), we may divide the applied voltage E_{r1} between L and C so that

$$E_{r1} = j2\pi f L I_{r1} + \frac{1}{j2\pi f C} I_{r1} \tag{5-28}$$

It is also true that the capacitor terminal voltage is

$$E_{r1}' = \frac{1}{j2\pi f C} I_{r1} \tag{5-29}$$

The smoothing factor is then obtained by substitution of Equation (5-28) and (5-29) into Equation (5-27):

$$S = \frac{[(j2\pi fL) + (1/j2\pi fC)]I_{r1}}{(1/j2\pi fC)I_{r1}} \tag{5-30}$$

Using absolute values,

$$S = 4\pi^2 f^2 LC - 1 \tag{5-31}$$

so that the L and C values must satisfy the relation

$$LC = \frac{S+1}{4\pi^2 f^2} \tag{5-32}$$

S is the ratio of the absolute values of E_{r1} and E_{r1}'. The j's in Eq. 5-30 were ignored. In these equations f is the frequency of the fundamental component of ripple voltage. For a full-wave rectifier, f is twice the line frequency and on a 60-hertz line the value of f would be 120. For a choke input filter having two similar sections, one following the other, the value of the smoothing factor becomes

$$S = (4\pi^2 f^2 LC - 1)^2$$

and for *each* of the sections

$$LC = \frac{\sqrt{S} + 1}{4\pi^2 f^2}$$

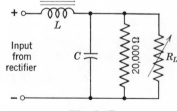

Fig. 5-17

EXAMPLE 5-2   A 60-hertz full-wave rectifier is to deliver between 20 and 100 mA at 200 V to a resistance load consisting of a 20,000-Ω bleeder and a maximum load resistance of 20,000 Ω. What minimum values of inductance and capacitance should be used in a choke input filter stage to provide a ripple factor of 0.01? (See Fig. 5-17.)

Solution.   The maximum total load resistance will be connected to the filter when R_L has its maximum value of 20,000 Ω. The equivalent resistance will then be 20,000/2 = 10,000 Ω.

$$L = \frac{R}{1,000} = \frac{10,000}{1,000} = 10 \text{ H}$$

The smoothing factor is equal to the ripple factor at the filter input divided by the ripple factor at the filter output:

$$S = \frac{\rho}{\rho'} = \frac{0.472}{0.01} = 47.2$$

$$LC = \frac{S + 1}{4\pi^2 f^2} = \frac{47.2 + 1}{4\pi^2 \times 120^2} = 84.8 \times 10^{-6}$$

$$C = \frac{84.8 \times 10^{-6}}{10} = 8.48 \times 10^{-6} \text{ F} = 8.48 \, \mu\text{F}$$

One of the standard sizes of commercial filter capacitors is 8 μF. The light-load current value is obtained by dividing the output voltage by the maximum load resistance:

$$I_L = \frac{200}{20,000} = 0.010 \text{ A} = 10 \text{ mA}$$

The bleeder will carry 10 mA also. The power dissipated in the bleeder is $P = I^2 R = (0.01)^2 \times 20,000 = 2$ W. The minimum load resistance will be such as to allow 100 mA to flow at 200 V.

$$\text{Minimum } R_L = \frac{200}{0.100} = 2,000 \, \Omega$$

The wire of the choke may have a resistance of about 75 Ω. Since the maximum choke current is 110 mA, the choke IR drop is $E = 0.110 \times 75 = 8.2$ V, which is quite small. There will be a drop of 15 to 60 V through the rectifier tube, depending on the type. Using a type 5Y3 tube, each half of the power supply transformer should produce an a-c voltage with a d-c component of $200 + 8 + 50 = 258$ V. From this, $E_m = 258/0.636 = 405$ V and $E_{rms} = 286$ V. The secondary should then be a 600-V winding with a center tap.

With a mercury-vapor tube, the tube drop would be about 15 V instead of about 50 V. The voltage available at the load would then be about 235 V.

5.12   Resistance-Capacitance Filter

When the current demand from a rectifier is very small, and also when the presence of more than the normal amount of ripple in the output can be tolerated, a filter consisting of resistors and capacitors (no inductances) may be used. The resistors are connected in the series line and the capacitors in parallel, of course.

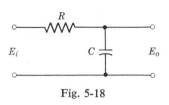

Fig. 5-18

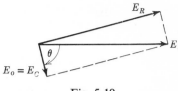

Fig. 5-19

Such a filter may be obtained by substituting resistors for the inductances in Fig. 5-8. In order to obtain a higher input voltage to the filter, and thus compensate for the increased d-c voltage drop in the series resistors, a capacitor-input type of filter is used. This also results in more effective filtering and smaller power loss because the capacitor will keep the higher-frequency components of current out of the resistors.

Consider the RC filter in Fig. 5-18. An a-c component in the input voltage, E_i, received from a rectifier must be reduced sufficiently so that it will not be a disturbance in the load circuit to which E_0 is applied.

By voltage divider action, the output voltage becomes a fraction of the input voltage with a shift in phase. The phasor diagram is shown in Fig. 5-19.

$$E_0 = \frac{X_C}{Z} E_i = \frac{1/j\omega C}{R + 1/j\omega C} E_i \tag{5-33}$$

$$E_0 = \frac{1}{1 + j\omega RC} E_i = \frac{1}{\sqrt{1 + (\omega RC)^2}} E_i \underline{/-\theta} \tag{5-34}$$

In practical applications it is found that $(\omega RC)^2 \gg 1$, so that

$$|E_0| \doteq \left| \frac{1}{\omega RC} E_i \right| \tag{5-35}$$

The angle of phase shift is unimportant in power supply filter applications, but in filters for higher-frequency operation the phase shift may be very important. Using the ratio of E_0 to E_i from Equation (5-34),

$$\frac{E_0}{E_i} = \frac{1}{1 + j\omega RC} = \frac{1 - j\omega RC}{1 + (\omega RC)^2} \qquad \text{(a complex number)}$$

For convenience, think of this ratio as a complex number $A + jB$, where

$$A = \frac{1}{1 + (\omega RC)^2}, \quad B = \frac{-\omega RC}{1 + (\omega RC)^2}$$

The phase-shift angle, θ, is the angle whose tangent is B/A:

$$\tan \theta = -\omega RC \tag{5-36}$$

Suppose $R = 10^5 \, \Omega$, $C = 1 \, \mu\text{F}$, $f = 120 \, \text{Hz}$.

$$\frac{E_0}{E_i} \doteq \frac{1}{\omega RC} \doteq \frac{1}{2\pi \times 120 \times 10^5 \times 10^{-6}} = 0.0133$$

This means, for example, that a 120-hertz component of 100 V amplitude in a rectifier output will be reduced to 1.33 V by this simple filter. Obviously an increase in R or C (preferably C) will reduce this further. The phase shift in this example is $\theta = -\tan^{-1} \omega RC = -\tan^{-1} 75.5 = -89.2°$.

Charts* are available for use in the design of filters. They facilitate the determination of RC and LC values required for the reduction of harmonic voltage ratios to desired low values. An understanding of the theory explained above makes them easy to use.

If two RC sections are used in tandem, the reduction ratio is the product of the ratios for the separate sections. A π-type CRC filter may be analyzed in the same manner as was done for the simple RC filter described above.

5.13 Voltage Stabilization

Stabilized d-c voltages may be obtained by using a voltage divider with one or more glow-discharge tubes. For example, it may be necessary to have 150 V direct current from a filtered power supply which is delivering 290 V to its main load. A 150-V glow-discharge tube connected in series with a resistance across the output of the filter will do the job.

Refer to the circuit of Fig. 5-20. Assume that 10 mA at 150 V, stabilized, are required at C. A type OD3 tube may be used. Select the value of R so

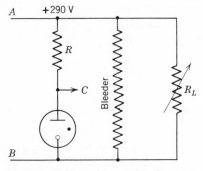

Fig. 5-20   Voltage divider with regulated output voltage.

* "Reference Data for Radio Engineers," International Telephone & Telegraph Corp.

that the tube will have 15 mA. Then

$$R = \frac{290 - 150}{0.010 + 0.015} = \frac{140}{0.025} = 5,600 \ \Omega$$

Assume that E_{AB} drops to 280 V for some reason. The voltage drop across R will then be $280 - 150 = 130$ V and the current through R is $130/5,600 = 0.0232$ A. Since the external resistance between C and B has not changed and the voltage is steady at 150 V, the current going out at C is still 10 mA. The tube current must then drop to

$$23.2 - 10 = 13.2 \ \text{mA}$$

There are more complicated voltage-regulator circuits in use with power supplies to prevent a drop or a rise in output voltage (E_{AB} here). They employ triode and pentode tubes, or transistors, which are described further on in the book.

5.14 Plate-Type Rectifiers

Copper oxide and selenium permit current flow in one direction much more readily than in the opposite direction. When they are used as covering layers on metal disks that are stacked and held firmly in intimate contact, they form efficient and rugged rectifiers of alternating current.

Copper Oxide Rectifier

The plates of a copper oxide rectifier are disks of copper that have been heat-treated in a special manner to cause the formation of a layer of copper oxide on one surface (see Fig. 5-21). A rectifier consists of a set of these plates stacked with the copper oxide faces all in the same direction, each disk separated from the next by a disk of lead or other metal that makes good

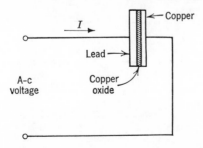

Fig. 5-21   Copper oxide rectifier unit.

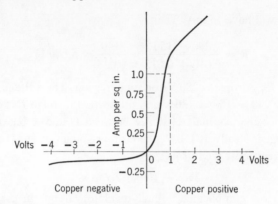

Fig. 5-22   Current-voltage characteristic of a copper oxide rectifier unit. The direction of current flow reverses when the polarity is reversed.

contact with the oxide layer. The disks are clamped together by a bolt made of insulating material.

When the copper oxide is made positive and the copper negative, electrons readily flow out of the copper, across the boundary, into the copper oxide layer, then on to the voltage source. That means the resistance to the flow of current is low. But, when the other polarity of voltage is applied, it is found that comparatively few electrons leave the oxide layer and pass through the boundary in the other direction to the copper. Thus the resistance is relatively high. Consequently, when an alternating voltage is applied to a stack of copper-copper oxide plates, practically all the current flow will be in only one direction. A "rectified" current is then obtained and a "rectified" voltage appears across the load supplied by the rectifier.

Figure 5-22 shows the current-voltage characteristic of a copper oxide rectifier unit consisting of a copper disk with a copper oxide layer in contact with another metal disk that makes good contact with the copper oxide. It is seen that, if an a-c voltage greater than 1 V at crest value is applied to the unit, a current of more than 1 A/sq in. will flow when the copper is positive

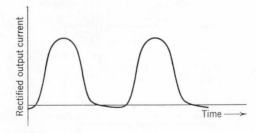

Fig. 5-23   Waveform of current delivered by a copper oxide rectifier unit.

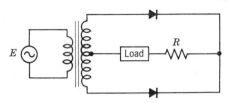

Fig. 5-24   Full-wave, dry-plate rectifier circuit.

with respect to the oxide, and a negligible current will flow when the copper is negative with respect to the oxide. This means that, if fifty units are bolted together and an alternating voltage of 100 V at crest is applied to them (or 2 V/unit), the forward current density will be about 1.25 A/sq in., while the reverse current will be practically negligible. It is readily seen that the current through a load resistor, or through one or more electron tubes supplied by the rectifier, will be flowing only in one direction for all practical purposes.

The wave of output current will be about as shown in Fig. 5-23, where the negative values are somewhat exaggerated. A typical circuit of a full-wave dry-plate rectifier is shown in Fig. 5-24. A rectifier is represented by an arrow and a line denoting a plate. Conduction takes place when the arrow is positive with respect to the plate. Electrons travel readily against the arrow but go with great difficulty in the direction in which the arrow points.

Selenium Rectifier

A disk of iron or steel, coated on one side with the element *selenium*, and specially heat-treated, forms a rectifying unit that responds to the application of an alternating voltage in much the same manner as a copper-copper oxide plate (see Fig. 5-25). A stack of selenium-coated disks performs rectification in a very satis-

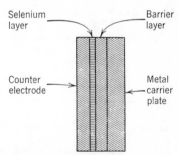

Fig. 5-25   Cross section of selenium dry-plate rectifier.

factory manner, although the forward current is smaller than the current allowed to flow by a copper oxide rectifier of the same size and number of plates. A photograph of a small selenium rectifier, having a capacity of about 100 mA, is shown in Fig. 5-26.

The characteristics of a typical selenium rectifier plate are shown in Fig. 5-27. It is to be observed that current begins to flow in the forward direction when the maximum value of the applied a-c voltage exceeds about $\frac{1}{4}$ V/plate If an a-c voltage having a maximum value of 1 V is applied, the maximum

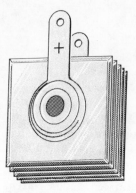

value of the forward current is about 1.25 A (curve *B* in Fig. 5-27) and the inverse current is practically zero. Figure 5-28 shows the approximate waveform of the current for that condition. The average value (d-c value) of the current would be somewhat less than the average value of a half-sine wave having a 1.25 A maximum. That is, the d-c current delivered to a filter and load by the selenium rectifier would be somewhat less than 0.318 × 1.25 (less than 0.398 A).

Selenium rectifiers are widely used instead of tube-type rectifiers in small power supplies, e.g., in portable television receivers. They are smaller, lighter in weight, more rugged, and much cheaper. They replace the tubes and sockets of the voltage doubler, and they require no filament current.

Fig .5-26 Type of selenium rectifier used in radio receivers. (Courtesy Radio Receptor Co., New York, N.Y.)

The initial charging current to the capacitor of a capacitor-input filter is quite high when a selenium rectifier is used. Relatively large currents flow for a minute or so after the rectifier has been turned on. It is advisable to use a series resistor between the rectifier and the filter, to limit the current.

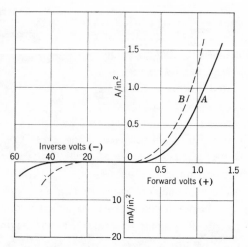

Fig. 5-27 Forward- and inverse-current characteristics of a typical selenium rectifier plate at ambient temperatures of (*A*) 25°C and (*B*) 60°C.

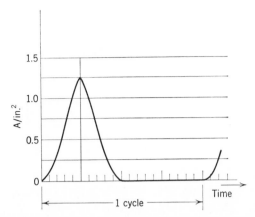

Fig. 5-28   Approximate waveform of current. Selenium rectifier unit.

In radio receiver selenium rectifier circuits, a 50-Ω resistor is used. The series resistor prevents damage to filter capacitors also.

Caution should be exercised in replacing tube-type rectifiers with selenium rectifiers. The low internal voltage drop of the selenium stack may raise the output voltage so much that the circuits supplied by the rectifier may not operate properly. The voltage ratings of the filter capacitors may be exceeded. The higher ripple content in the selenium rectifier output may necessitate changing the filter capacitors. Electrolytic capacitors cannot stand the inverse voltage of a ripple component of appreciable magnitude. One more precaution should be observed when changing from a tube rectifier to a dry-plate rectifier. Because no filament or heater voltage is required, it is necessary to install a resistor to take the place of the tube filaments that were formerly used (i.e., if the filaments of the rectifier tubes formerly used were part of a series circuit containing other tube filaments).

5.15   Bridge-Type Rectifier Circuit

Four diodes or rectifiers may be connected in a bridge circuit as shown in Fig. 5-29. This type of circuit is widely used because it does not require a center tap, although it produces full-wave rectification. The output must be filtered, of course. It is not well adapted to use with tube-type rectifiers, because the cathodes are not at the same potential and therefore may not be connected in parallel for a single filament-transformer supply. This circuit is widely used with dry-plate rectifiers in industrial electronic applications. It is extensively used with semiconductor rectifiers.

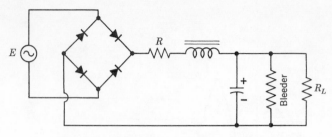

Fig. 5-29   Bridge-type dry-plate rectifier circuit.

Tubes are used, however, in bridge-type rectifiers on high-voltage x-ray machines for several good reasons. The entire transformer secondary winding is used at all times instead of each half being used alternately as in the full-wave rectifier. Although four tubes are used, there are always two in series; this means that, for a given output, the voltage ratings of the rectifier tubes need not be so high as for those in a full-wave two-element circuit. The peak-inverse voltage applied to each tube is equal to the load voltage plus the voltage drop across one tube. It must be understood that the *bridge-type rectifier provides full-wave rectification.* The equations that were developed in Section 5.5 apply to the operation of this circuit.

5.16   Full-Wave Voltage-Doubler Circuit

The average output voltage of a full-wave rectifier is 0.636 times the maximum instantaneous value of the a-c input voltage. For 115 V a-c input, this would be

$$0.636 \times 115\sqrt{2} = 103 \text{ V}$$

This is not enough voltage for the satisfactory operation of some tubes. It is possible to double the output of a full-wave rectifier by employing capacitors in what is called a *voltage-doubler circuit.* Such a circuit employing dry-plate rectifiers is shown in Fig. 5-30, and one employing a double-diode tube is shown in Fig. 5-31.

The action of the circuit in Fig. 5-30 is as follows: When the a-c input polarity is such that point 1 is positive and point 2 is negative, the upper capacitor charges through the upper rectifier. During this time the lower capacitor will not charge through the lower rectifier, because the rectifier polarity is wrong for conduction to take place. On the negative half-cycle of the a-c input the lower capacitor charges through the lower rectifier. Note that both capacitors are in series across the output terminals and the sum of

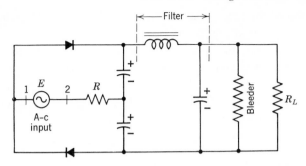

Fig. 5-30   Voltage-doubler circuit using dry-plate rectifiers.

heir voltages is applied to the load resistor, which, of course, could be
eplaced by the plate-to-cathode circuit of an electron tube. The series
esistor (R) is about 50 Ω. It protects the capacitors and rectifier elements
rom overload.

Capacitors of rather large size are used—as large as 40 or 50 μF each—so
hat they may hold up their voltage well while supplying load current.
The a-c input replenishes the charge on the capacitor that has delivered
:urrent to the load each half-cycle. While doing so, the input also supplies
ome load current through the other capacitor and the rectifier unit which
:onducts during the half-cycle.

The current delivered to the load is a pulsating direct current. It must be
iltered, but it does not require so much filtering as the output of a conven-
ional full-wave rectifier because the ripple voltage is much smaller. The
arge capacitors do not allow the output voltage to fall to zero, as happens
m the conventional rectifiers. The capacitors of the voltage doubler serve
ilso as smoothing filters.

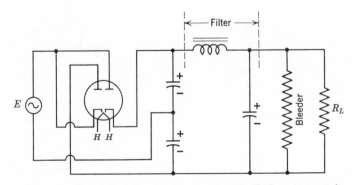

Fig. 5-31   Voltage-doubler circuit using double-diode vacuum tube.

The action of the tube circuit in Fig. 5-31 is exactly as described for th
plate-rectifier circuit in Fig. 5-30.

The tube types suitable for use in voltage-doubler circuits for small radi
receivers are the 25Z5 and 25Z6. Their heaters are usually supplied with th
required 0.3 A at 25 V from the 115-V a-c line through a suitable serie
resistor. The resistor must be heavy enough to dissipate $(115 - 25) \times 0.3 =$
27 W, unless other tube filaments are in series.

5.17   Half-Wave Voltage Doubler

In applications where only small currents are needed, the half-wav
voltage doubler is popular. The circuit is shown in Fig. 5-32. It may be fe
from the 115-V a-c line or through a step-up transformer. On the negativ
half-cycles when point 2 is positive, C_2 charges through T_2. On the positiv
half-cycles when point 1 is positive, the input current flows through C_2 an
T_1. C_2 can discharge only through T_1 because T_2 will not conduct downward
The capacitances are usually equal: $C_1 = C_2$.

Experimental data taken on this circuit reveal that for values of $2\pi f C_1 R$
greater than 200, the d-c voltage across R_L is nearly double the peak valu
of the input voltage for large values of R_L. R_L should be at least ten time
the sum of the source resistance and the internal resistance of one tube. Th
peak-inverse voltage of the series tube (T_1) is the output voltage plus the tub
drop across T_2. This is nearly double the peak supply voltage. The instan
taneous peak current is of the order of 100 times the d-c output current.

The a-c input voltage is shown in Fig. 5-33a, and the voltage across T
in series with C_1 and R_L is shown in Fig. 5-33b. The current in T_1 canno

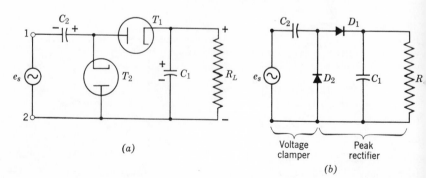

(a)

Voltage
clamper

Peak
rectifier

(b)

Fig. 5-32   Half-wave doubler. (a) Circuit using diode tubes. (b) Equivalent circuit

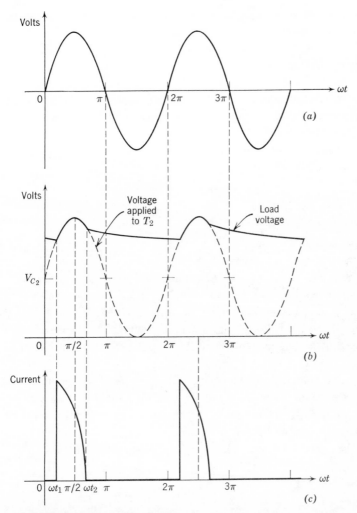

Fig. 5-33 Voltage and current waveforms in voltage doubler. (*a*) Input voltage waveform; (*b*) voltages applied to T_2 and across the load; (*c*) current pulses in T_1, waveform depends on $\omega R_L C_1$ as in half-wave rectifier with shunt-capacitor filter.

start flowing until the voltage across T_2 is greater than the voltage across C_1. Consequently, very short spurts of relatively large instantaneous values of current flow through T_1 and R_L. This explains why the instantaneous maximum peak current can be one hundred or more times the average (d-c output) current. It also indicates that the capacitances must be quite large in order to store enough energy to be able to supply continuous current to the load without severe drops in voltage between successive peaks of the charging half-cycles.

DRILL PROBLEMS

D5-6   Sketch the voltage waveform existing across capacitor C_2 in Fig. 5-32 for an input signal $e_s = 100 \sin \omega t$. Now suppose C_2 develops an internal short-circuit condition. Show the effect on load voltage.

D5-7   Sketch the waveform of current in $D_1(T_1)$ for $\omega R_L C_1 = 10;\ = 100$. Compare their maximum amplitudes.

5.18   Voltage Tripler

By connecting the output of a half-wave rectifier properly to the output of a voltage-doubler circuit, a voltage tripler is produced. The circuit is shown in Fig. 5-34. T_2 and T_3 conduct only on the *negative* half-cycles, charging C_2 and C_3. On the *positive* half-cycles T_1 conducts but not until the inverse voltage across T_2 exceeds the voltage across C_1, which is practically double the peak value of the a-c input voltage. Voltage tripling is thus achieved by adding this voltage to the output of the half-wave rectifier made up of T_3 and C_3.

Mention should be made of the fact that crystal diodes may be used in place of tube diodes in these circuits. Caution should be observed, however, in the matter of vulnerability of crystal diodes to excessive temperature rise due to peak-current values. Inverse-peak voltages may contribute to failure of the crystal diodes more readily than if diode tubes were used.

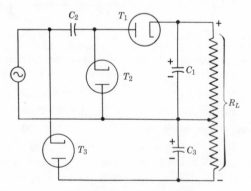

Fig. 5-34   Voltage tripler.

The drawing of a schematic diagram of a *voltage quadrupler*, using the outputs of two doublers, is left as an exercise for the student.

Calculations of resistances and power capacities of the sections of a voltage divider used at the output of a filter power supply are quite easily made. The student should be sure to solve Problem 5-14 for practice in the design of the multitapped voltage divider.

5.19   Semiconductor Stack Rectifiers

Silicon and germanium rectifiers are commercially available in pre-assembled stacks complete with heat sink and electrical interconnections for a great variety of typical applications of rectifier circuits. Standard units include half-wave, full-wave (center-tap), voltage-doubler, single-phase bridge, and arrangements for three-phase applications. Some of the stacks have semiconductor diodes connected in series and parallel for the greater voltage and current ratings needed in some applications.

Some typical silicon and germanium stack rectifiers are shown in Fig. 5-35. The larger units are designed with cooling fins as a part of the assembly. Although much more heat is generated by the increased power losses in the larger units, they are practical owing to the greatly increased heat-radiating capacity provided by the fins.

Semiconductor stack rectifiers are capable of operating at high voltages and currents. The General Electric type 4JA9013, having six fins each 7 in. × 7 in., has the following specifications:

Maximum current (free convection cooling at 55°C)	80 A/fin
Maximum current (cooling at 2,000 cfpm and 55°C)	207 A/fin
Maximum PIV	800 V/fin recurrent, and 1050 V/fin transient.

5.20   The Power Transformer

When more d-c power or higher voltages are needed than circuits such as the voltage doubler and tripler can handle, a *power transformer* is used. It is also convenient to reduce the a-c line voltage for supplying the filaments

Fig. 5-35   Typical silicon and germanium stacks. (Courtesy General Electric Co., N.Y.)

and heaters of electronic tubes used in a circuit. A properly designed power transformer can easily step up or step down a-c voltage to any desired practical value.

The transformer is composed of at least two windings of conducting wire, wound on an iron core and insulated from the core and from each other. When a voltage is applied to one of the windings, voltages are induced in the other windings. The windings usually have different numbers of turns of wire. The number of turns of wire in a given winding determines the magnitude of voltage that is induced in it.

In order to step up a voltage of 115 to 575 (5-to-1 ratio), it is necessary to use a transformer that has five times as many turns in one winding as in the other. The 115-V a-c supply is connected to the winding having the smaller number of turns, and 575 V alternating current is obtained at the ends of the other winding. The winding with the smaller number of turns is called the low-voltage winding, and the winding with the larger number of turns is called the high-voltage winding.

To use a power transformer in a full-wave rectifier circuit, it is necessary that the high-voltage winding be center-tapped, as shown in Fig. 5-5. Most power transformers have at least one winding, of relatively few turns of wire, that supplies filament (or heater) current for the cathode. The voltage of the filament winding may be 12.6, 6.3, 5, 2.5 V, or some other low value. The most commonly used values with small rectifier tubes are 6.3 and 5 V. These low-voltage windings are made of comparatively large-sized wire because filament currents are usually much greater than the plate currents supplied by the high-voltage winding. Transformers are built to operate at a definite voltage on each winding. In the case described above, the transformer could be used to "step-down" 575 V to 115 V.

Consider a transformer that has a rated voltage applied to one of its windings, the primary, while the other winding, the secondary, has nothing connected to it; i.e., the secondary is "open." An *exciting current*, being alternating in nature, flows in the primary and sets up an alternating magnetic field in the iron core. That is, the magnetic field increases to maximum strength in phase with the current, falls to zero, and then builds up strength in the reverse direction in step with the current.

This changing flux induces a voltage in the secondary winding of the transformer. The induced voltage will be proportional to the number of turns. Letting the subscript 1 denote the primary winding and 2 the secondary winding, and using the letter N to denote the number of turns, we may write

$$\frac{E_2}{E_1} = \frac{N_2}{N_1} \tag{5-37}$$

EXAMPLE 5-3. A transformer changes 115 V a-c to 575 V a-c. The low-voltage winding has 250 turns. How many turns are there on the high-voltage winding?

From Equation (5-37),

$$N_2 = \left(\frac{E_2}{E_1}\right) N_1$$

$$N_2 = \left(\frac{575}{115}\right) \times 250 = 1{,}250 \text{ turns}$$

EXAMPLE 5-4. How many turns are necessary on a 6.3-V filament winding of the transformer?

Solution.

$$\frac{E_3}{E_1} = \frac{N_3}{N_1}$$

$$N_3 = \frac{E_3}{E_1} \times N_1 = \frac{6.3}{115} \times 250 = 14 \text{ turns}$$

When the second winding of a transformer, usually called the secondary, is closed through an appropriate load, which may be a resistance or the plate circuit of one or more tubes or any other current-carrying device, current will be supplied to the load, and the current, of course, will flow in the secondary winding. This current produces a magnetic field that opposes the magnetic field produced by the current put into the first winding, which is called the primary winding. This opposing field builds up in strength as the current in the secondary winding builds up. The reduction in the net magnetic flux results in an increase in current flowing into the primary from the source of a-c supply.

We readily see why the input current to the primary of the transformer must increase when the secondary is connected to a load. The reason is that the power taken by the load must come from somewhere, and the only place is the source of supply serving the primary. In order to get more power into the primary, the current must increase since the voltage is constant. The flux produced by the secondary current opposes the flux produced by the primary current, thus reducing the counter EMF of self-induction of the primary winding. This results in an increase of primary current. The CEMF of self-induction determines the value of the current in the primary winding. The smaller the CEMF, the larger the primary current.

The currents in the windings are related to the turns in the following manner:

$$\frac{I_2}{I_1} = \frac{N_1}{N_2} \tag{5-38}$$

$$N_1 I_1 = N_2 I_2 \tag{5-39}$$

Equation (5-39) shows that the ampere turns (*NI*) of the primary are equal to the ampere turns of the secondary.

From Equation (5-37) we obtain

$$\frac{N_1}{N_2} = \frac{E_1}{E_2} \tag{5-40}$$

Substituting Equation (5-40) into Equation (5-38) we get

$$\frac{I_2}{I_1} = \frac{E_1}{E_2} \qquad (5\text{-}41)$$

$$E_1 I_1 = E_2 I_2 \qquad (5\text{-}42)$$

Equation (5-42) states that the primary voltamperes equal the secondary voltamperes. Equations (5-40) to (5-42) are not exact because they neglect such things as leakage flux, magnetizing component of current, and power losses in the core and in the windings. They are sufficiently exact for practical purposes, however.

SUGGESTED REFERENCES

1. Samuel Seely, *Electron-Tube Circuits*, 2nd ed., McGraw-Hill, New York, 1958.

2. Herbert J. Reich, *Theory and Applications of Electron Tubes*, 2nd ed., McGraw-Hill, New York, 1944.

3. Paul D. Ankrum, *Principles and Applications of Electron Devices*, International Textbook Co., Scranton, Pa., 1959.

4. Frederick E. Terman, *Radio Engineering*, 4th ed., McGraw-Hill, New York, 1955.

QUESTIONS

5-1 What purpose does a power supply serve in an electronic circuit?

The following six questions pertain to the vacuum-tube half-wave rectifier.

5-2 How do the d-c component, the fundamental component, and the double-frequency component of the rectifier output voltage compare in magnitude with the input voltage?

5-3 What is meant by peak-inverse voltage on the tube? How does it compare with the peak input voltage?

5-4 At a given voltage input, what determines the peak value of tube current?

5-5 How does the average value of tube current compare with the peak value? For what kind of load is this true—resistive, reactive, or a combination of the two kinds?

5-6 How is the rms value of the load current (resistive load) related to the d-c value?

5-7   Sketch the output-voltage waveform.

The following five questions pertain to the vacuum-tube full-wave rectifier.

5-8   What is the frequency of the largest a-c component of the output voltage of a full-wave rectifier?

5-9   How do the d-c component and the two lowest-frequency a-c components of the rectifier output voltage compare in magnitude with the input voltage?

5-10   How does the peak-inverse voltage compare with the instantaneous maximum input voltage?

5-11   How do the d-c and rms values of load current compare with the maximum instantaneous value?

5-12   Sketch the output voltage waveform.

5-13   Compare the performance of a full-wave rectifier with that of a half-wave rectifier in five important respects.

5-14   What is the purpose of a smoothing filter, and how does it accomplish its purpose? How is it that, if the desired effect is produced on the a-c component of lowest frequency, the other components will not cause trouble in the output?

5-15   A small amount of current, at a reduced voltage, is desired from a filtered rectifier output. Explain a simple circuit which will provide this and stabilize the voltage at the same time.

5-16   Draw from memory a full-wave rectifier circuit with a choke-input filter and a bleeder.

5-17   Define ripple factor; smoothing factor.

5-18   Why should a bleeder be used on a rectifier power supply?

5-19   Why is the input choke of a smoothing filter given a practical lower limit for its inductance value?

5-20   Explain how LC and RC filters like those studied in this chapter are able to reduce the magnitudes of a-c components of voltage so effectively. It is important to note that a-c components of *current* are simultaneously reduced since current equals voltage divided by impedance.

5-21   Write the approximate expression for the ratio of output to input a-c voltage of an RC filter. If $\omega RC = 5$, compute the percentage error caused by using the approximate equation.

5-22   What is the phase angle between the fundamental component of current in an unloaded RC filter and the fundamental component of the input voltage in terms of ω, R, and C?

5-23   What is the principle of the dry-plate rectifier?

5-24   What are the advantages of dry-plate rectifiers when compared with tube-type rectifiers?

5-25 Why should a series resistor be used with each selenium rectifier unit? Draw and explain the operation of a bridge-type rectifier circuit. What are its advantages? Does it have any disadvantages?

5-26 Draw from memory a full-wave voltage-doubler circuit.

5-27 Draw a half-wave voltage-doubler circuit.

5-28 Draw a voltage-tripler circuit.

5-29 Under what conditions are power transformers used instead of doubler and tripler circuits?

5-30 Write the important equations relating the following transformer quantities: (*a*) terminal voltages and turns: (*b*) terminal voltages and currents; (*c*) currents and turns.

PROBLEMS

5-1 A 1000-Ω resistor has applied to it a voltage $e = 50 + 50 \sin \omega t$ V. Determine the d-c current in the resistor and the required power rating for it to operate with this voltage.

5-2 A diode is rated for maximum current of 500 mA and average current of 200 mA. It is being considered for use in a full-wave rectifier circuit with a transformer supplying 220 V rms on each side of the center tap. (Neglect diode resistance.) (*a*) What value of load resistance may be used to obtain the greatest d-c power output without exceeding any diode rating? (*b*) Calculate the d-c load voltage and current for the resistance chosen in (*a*).

5-3 Determine the value of load resistance required for maximum d-c power output from a full-wave rectifier using two diodes rated at PIV $= 200$ V, $I_{max} = 2$ A (each diode), and $I_{av} = 1.2$ A (each diode).

5-4 A half-wave rectifier circuit is supplied from 120-V rms, 60-Hz a-c lines. The circuit uses a diode tube having an internal resistance of 200 Ω and supplies power to a load resistance of 5 kΩ. (*a*) What is the peak instantaneous value of diode current? Of load current? (*b*) What is the d-c power in the load? (*c*) Calculate the rms value of load current. (*d*) Calculate the power conversion efficiency. (*e*) What is the peak-inverse voltage on the diode?

5-5 Calculate the quantities asked for in Problem 5-4 for a full-wave rectifier circuit using the given diode and a center-tapped transformer that supplies a 120-V rms input to each tube. Compare the results with those for the half-wave circuit.

5-6 A half-wave rectifier is supplied with a 60-Hz sine-wave voltage of 120 V (rms) value. Calculate the d-c voltage and the rms values of the fundamental component and the next two largest a-c components in the output voltage.

5-7   A smoothing filter receives from a full-wave rectifier a pulsating voltage that has a fundamental component of 143 V peak. What is the value of the d-c component? What is the rms value of the voltage input to the rectifier?

5-8   The rectifier-filter of Problem 5-7 delivers current to a resistance load after a d-c voltage drop of 10 V in the series inductor of the filter. The rms ripple voltage across the 2125-Ω load is 2.5 V. Calculate the (a) ripple factor of the load voltage; (b) smoothing factor of the filter; (c) d-c voltage at the load; (d) power delivered to the load.

5-9   A single-section choke-input filter has an input voltage of 420 V peak, supplied by a full-wave rectifier from a 60-Hz line. The filter elements are $L = 10$ H, filter resistance $= 100 \ \Omega$, and $C = 20 \ \mu$F. The total load impedance is $750 \ \Omega$ pure resistance. Calculate the (a) d-c component of load current; (b) d-c load voltage; (c) smoothing factor of the filter; (d) fundamental component of ripple voltage across the load; (e) ripple factor at the load.

5-10   A 60-Hz full-wave rectifier is required to deliver a maximum of 150 mA and a minimum of 50 mA at 250 V to a resistance load consisting of a 5000-Ω bleeder in parallel with a variable resistance. An LC filter is to be used which will provide a load voltage with a ripple factor of 0.02. Determine the (a) minimum values of L and C; (b) minimum value of the variable load resistance and its maximum power; (c) power lost in the bleeder.

5-11   A one-section RC filter has $R = 10^4 \ \Omega$, $C = 8 \ \mu$F. What is its voltage reduction ratio at 60 Hz?

5-12   The RC filter of Problem 5-11 is used with a half-wave rectifier operating on 60 Hz. The maximum instantaneous value of the filter input voltage (E_m) is 170 V. (a) Calculate the ripple factor at the load, neglecting d-c voltage drop through the filter. (b) Calculate the ripple factor in the load again, allowing for a 25 per cent loss in voltage at the filter.

5-13   A power transformer suitable for a full-wave rectifier circuit will deliver 200 mA at 375 V to each half of a double-diode rectifier tube. At the same time it will deliver 5 A at 6.3 V rms to a filament circuit. The primary winding operates on 115 V. (a) What is the value of the total secondary voltamperes? of the primary voltamperes? (b) What is the full-load primary current? (c) Assuming 250 turns on the primary winding, calculate the number of turns on each of the secondary windings, assuming an ideal transformer.

5-14   A voltage divider is connected to the output terminals of a power supply filter which delivers direct current at 265 V. The various current demands and voltages are shown in Fig. P5-14. The current in R_3 is chosen to load the transformer properly. The potentials that are negative with respect to ground are needed for grid biases on tubes. No current is drawn in these cases. (a) Calculate the ohms values required for all the resistors and also the power dissipated in each. (b) If

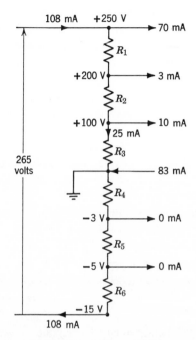

Fig. P5-14 Voltage divider.

the voltage divider were made of a continuous wire-wound spool with taps, the wire would have to be large enough to carry the maximum current safely. How much power would it then be capable of dissipating? (*c*) Compare the power *rating* found in (*b*) with the sum of the separate powers in (*a*). The arrangement in (*b*) would be lower in manufacturing cost. The power rating of the actual resistors used should be at least double the calculated value if the bleeder is to be confined in a poorly ventilated space.

5-15 In Fig. 5-5, a-c input = 120 V rms, and the transformer has a step-up ratio of 1:2. Assume no voltage drop across the diodes during conduction. (*a*) Calculate the value of resistance load to allow 200 mA average load current. (*b*) Suppose one of the diodes becomes inoperative. What average load current would then exist?

5-16 Explain the importance of the maximum plate current rating of a vacuum diode when it is used as a power rectifier with a π-section filter.

5-17 Why is "peak-inverse-voltage" rating of a vacuum diode of more significance: (*a*) in full-wave circuits than in half-wave circuits? (*b*) in filter networks having capacitance input than in those having inductance input?

5-18 Sketch and explain the output voltage waveform of a half-wave rectifier having a capacitor shunting the load resistance. Of what significance is the time constant of the R-C combination compared with one period of the input to the rectifier?

5-19 What must be the voltage rating of capacitors used in rectifier filter networks, compared with the rms value of the input voltage? Will this rating be the same for both full- and half-wave circuits? Explain.

5-20 Sketch a vacuum-diode full-wave rectifier circuit having a π-section filter and resistance load. Explain why plate current passes in the diode for less than a half-cycle of input voltage.

5-21 A voltage-regulated power supply with a smoothing filter delivers 100 mA d-c at 400 V. It is desired to draw a current which may vary between 10 mA and 25 mA but at a constant voltage of 105 V. Design a simple voltage-reduction circuit which will permit this.

5-22 A type CRC filter has equal capacitors. Assuming a load resistance R_L and an input fundamental component of voltage E_i rms at $\omega = 2\pi f$ rad/s, derive the equation of the output voltage, E_0, across R_L in terms of E_i and the other constants.

5-23 Develop the general form of Equation (5-1) by Fourier analysis.*

5-24 Verify that Equation (5-8) represents the waveform of the output of a full-wave rectifier.*

5-25 Explain the effect on circuit operation if the diode of Fig. 5-10a were connected backwards in the circuit. Would the circuit be useful?

5-26 The supply voltage for the circuit of Fig. 5-10a is $e_s = 100\sin 377t$. The load resistor and shunt capacitor are large enough so that there is no appreciable ripple in the load voltage. The load voltage is required to be 90 V and the diode is rated at 10 mA maximum recurrent current. Show the changes necessary in the circuit to meet these restrictions.

5-27 The periodic voltage shown in Fig. P5-27a is applied to the circuit containing a d-c microammeter (M) that has negligible resistance. The meter indicates the average current flowing in it. The diode is a semiconductor type and can be treated as an ideal element. (a) Determine the steady-state meter reading. (b) Sketch the waveform of output voltage, marking significant voltage levels and time intervals. (c) Repeat (a) and (b) for the case of an open-circuited capacitor.

5-28 A type of diode having negligible forward resistance is used in a half-wave rectifier whose load resistance is shunted by a large capacitance. Sketch and label the waveform of diode current for a total conduction angle of 90°. Sketch the waveform of the voltage across the diode during 1 cycle of input voltage.

5-29 Repeat Problem 5-28 for a full-wave rectifier circuit using the same type of diode. For a 90° conduction angle, what changes would have to be made in the filter and load? If the same R and C were used in the full-wave rectifier, what would be the total conduction angle of each diode?

* Solution requires the evaluation of trigonometric integrals.

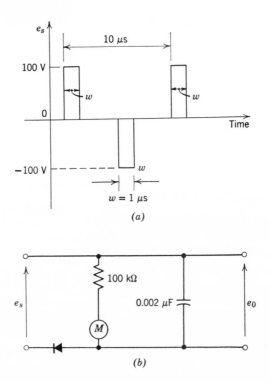

Fig. P5-27 Rectangular voltage wave in (*a*) is applied to the circuit in (*b*).

5-30 Sketch the waveform of diode current for a half-wave rectifier using the shunt-capacitor type of filter with $\omega RC = 10$.

5-31 It is desired to construct a half-wave rectifier and filter for a load resistance of 200 Ω. The supply voltage is 100 sin 377t. (*a*) Determine the ratings of a diode that could be used without filtering. (*b*) What value of capacitance is required to make $\omega RC = 10$? (*c*) If this value is used as a shunt-capacitor filter, will the diode chosen in (*a*) be adequate?

CHAPTER SIX

The Vacuum Triode, Tetrode and Pentode

The high-vacuum triode has many favorable characteristics and is capable of performing some important duties that cannot be accomplished by diode tubes. We know that, at a constant cathode temperature, the current passing through a high-vacuum diode may be varied only by changing the plate-to-cathode voltage. It is not possible to increase the plate-to-cathode voltage of a gas-type diode to any appreciable extent without driving excessive current through the tube. To change the current passing through a gas-type diode, we must either change the ohms value of the series resistor connected externally to the tube or change the voltage applied to the series circuit containing the tube.

The current flowing through a high-vacuum triode may be changed conveniently and effectively by changing the voltage of the grid with respect to the cathode. The grid is an electrode mounted between the plate and the cathode in the tube envelope.

6.1 Physical Characteristics of the Triode

The symbol that represents a high-vacuum triode tube, and a simplified vertical cross section of the electrode configuration, are shown in Fig. 6-1. The cathode structure is the same as that described for thermionic diodes in Chapter 4. It is represented here as a cylinder. The grid consists of a helix of fine wire coaxial with, and mounted close to, the cathode. The plate is a metallic cylinder also coaxial with the cathode but several times farther

216

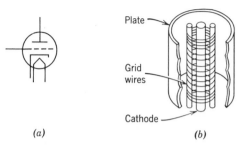

Plate

Grid
wires

Cathode

(a)   (b)

Fig. 6-1   (*a*) Symbol of a high-vacuum triode. (*b*) Sketch of electrode configuration
in a triode.

away from it than is the grid. In most triodes the cylindrical plate surrounds
the other elements of the tube, although in some tubes it is not a circular
cylinder.

6.2   Function of the Grid

The grid was introduced by Dr. Lee DeForest in the year 1906. It heralded
an era of rapid development in vacuum-tube applications and performance.
The ability to vary plate current by means of very small variations in grid-
to-cathode voltage made possible the strengthening of radio signals, which
was the first application of vacuum-tube amplification. Other kinds of tube
performance, including the generation of powerful high-frequency currents
by vacuum-tube oscillators, the variation (or modulation) of vacuum-tube
plate currents in accordance with frequencies and amplitudes of sound waves
of the human voice or musical instruments, and the separation of voice
frequencies from inaudible frequencies in very weak radio waves (detection),
soon were discovered. All these basic operations were accomplished with the
triode.

The grid controls the flow of electrons in the space between the cathode
and the anode. The potential on the grid wires determines in a large measure
the nature of the potential distribution within the tube. Figure 6-2 shows
how the potential within the tube is distributed when the grid is negative with
respect to the cathode and when the cathode is cold. The straight-line sections,
a, show the potential distribution along a straight line from cathode to plate
through a grid wire; the curved section, *b*, shows a potential distribution
along a straight line midway between the grid wires.

A three-dimensional representation (Fig. 6-3) of the potential distribution
between cathode and anode of a parallel-plane triode (i.e., a triode having

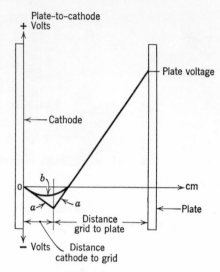

Fig. 6-2   Potential distribution in vacuum triode tube. No space charge.

flat electrodes lying in parallel planes) has been suggested. Imagine a rubber sheet stretched over two parallel rods mounted at different elevations. Then imagine an inverted picket fence to be lowered so that the points of the pickets depress the rubber sheet below the elevation of the lower rod. Consider the lower rod to have zero elevation (corresponding to the cathode at zero potential), and the elevation of the upper rod to be a positive value (corresponding to a positive potential on the plate). The elevations of all points of the rubber sheet under the tips of the pickets, and immediately between them

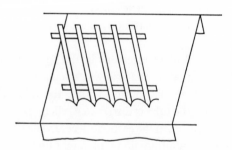

Fig. 6-3   Sheet of stretched rubber depressed by picket fence produces contour surface of potential distribution in high-vacuum triode not containing space charge.

in the vertical plane of the pickets, are negative and correspond to potentials below cathode potential. The force of the upper rod raises the rubber sheet in the regions between the picket tips, and so the elevations there are not so negative as the elevations immediately under the tips. This shows how the effect of the positive voltage on the plate in a triode can cause the *grid-plane potential* between the grid wires to be higher than the actual grid-wire potential and to reach a maximum midway between the wires.

When the cathode is heated to normal emitting temperature, an abundant supply of electrons appears immediately in front of the cathode. They are "boiled out" of the cathode, so to speak, and many of them leave it with sufficient velocity to carry them against the opposing forces of the electric field, so that they pass between the grid wires and on to the plate. If the plate potential is not high enough to produce a continually rising potential from cathode to plate so that all the electrons proceed immediately toward the plate, many of then reenter the cathode. When this situation prevails, there is a definite dip in the potential-distribution curve along the lines between grid wires, like the *b* segment of the line in Fig. 6-4. The amount of accumulated negative space charge (electrons) in front of the cathode determines how many *electrons per second* may pass on to the plate. The plate current that flows in this kind of situation is said to be *space-charge-limited*. It must be remembered that the potential of the grid wires is very influential in determining

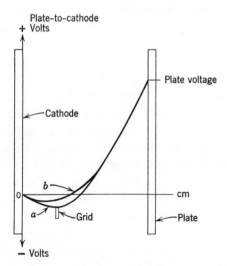

Fig. 6-4   Potential distribution in a triode (the current is space-charge-limited): (*a*) through a grid wire, (*b*) midway between two grid wires.

the nature of the potential distribution in the grid-cathode region. In normal operation the plate current of a triode is space-charge-limited.

The rate of flow of electrons to the plate is controlled by the potential of the grid. The electrons that leave the space-charge region and pass through the grid plane cause the potential of the space charge to rise, since negative charges are being removed. The result is that the repelling action of the space charge against electrons trying to enter the space-charge region from the cathode is decreased, and so more electrons enter to take the places of those that have left and passed on to the plate.

It may then be said that, although the plate current of a high-vacuum tube is space-charge-limited, the potential of the grid with respect to the cathode is the controlling factor in determining plate current when the plate voltage and cathode temperature are kept constant. The space charge limits the number of electrons per second that permanently leave the cathode. The potentials in the space between cathode and anode are not sufficiently high to produce an electric field strong enough to send all of the electrons immediately to the plate as fast as they leave the cathode.

The illustration using the rubber sheet and picket fence (Fig. 6-3) may be adapted to the practical case in which the cathode and anode are cylindrical surfaces and the grid is a helix coaxial with both. Imagine a funnel-shaped rubber membrane molded so that the larger end may, with a little stretching, be firmly fastened over a large metal ring. The smaller end may likewise be firmly fastened over a smaller ring having a diameter about one-tenth that of the larger. The rings are to be pulled apart, to remove wrinkles in the rubber membrane, and securely mounted coaxially in horizontal planes, the smaller ring at the bottom. Now imagine a circular picket fence being lowered into the rubber-membrane funnel, its axis coinciding with the axis of the funnel. The fence has a little larger diameter than the smaller ring, so that the pickets depress the membrane at points about one-fifth the radial distance between the rings. The lower ring represents the cathode, and its elevation (taken as zero) represents the cathode potential. The upper ring represents the plate, and its high elevation represents the highly positive plate potential. The elevation of the points of the rubber membrane touched by the tips of the pickets represents the grid-wire potential. In order to represent a grid potential below cathode potential, the picket tips must be forced below the zero elevation of the lower ring. When that is done, the entire surface of the rubber membrane will represent the potential distribution in a vacuum tube with cylindrical electrodes from cathode to plate, including the regions between individual turns of the grid wire. The potential distribution in the tube will have the same form as the elevation distribution of the rubber-membrane surface.

6.3 Types of Triode Tubes

High-vacuum triodes are classified generally as (*a*) voltage amplifiers, and (*b*) power amplifiers. Voltage-amplifier triodes carry small current, seldom more than 8 or 10 mA and usually less. Their function is to cause a current change in a load resistance which produces a relatively large voltage variation across the resistance. The current change is caused by very small changes in the voltage between the grid and cathode of the tube.

Twin triodes are used where two stages of amplification or two other separate triode functions are required. Figure 6-5 shows terminal connections for the 6F8G. The tube has double heaters in parallel connected to pins 2 and 7.

There are a number of new tube designs for very high-frequency work. The capacitance between electrodes inside the tube is a very disturbing matter. It limits the value of the highest frequency at which the tube will operate. If an attempt is made to separate the electrodes in order to reduce the capacitance, the time required for electrons to pass from cathode to

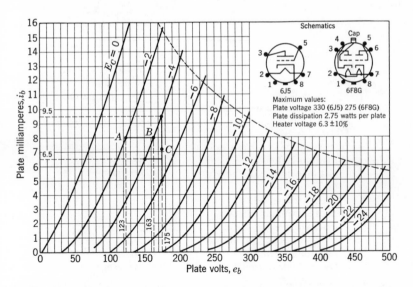

Fig. 6-5 Average plate characteristics of 6J5, 12J5, 6F8G vacuum triodes.

anode (called the *transit time*) is increased. We can imagine the effect on tube operation if transit time becomes an appreciable fraction of the time of a positive half-cycle of voltage on the plate of the tube.

The GE type 7077 ceramic triode and the RCA type 7586 Nuvistor are recent types of very high-frequency triodes. The GE 7077 ceramic triode has a ceramic body measuring 0.335 in. in diameter, although the grid-terminal ring near the middle is 0.48 in. in diameter. The overall length is 0.438 in. It operates as a grounded-grid amplifier at 450 MHz.

The RCA 7586 is a general-purpose industrial triode, measuring 0.8 in. in length and 0.435 in. in diameter. It is designed for critical industrial applications such as communications equipment, control and instrumentation

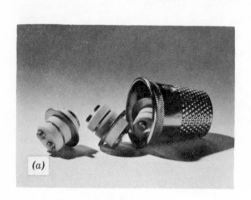

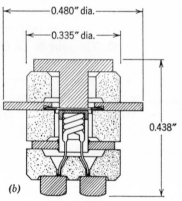

Specifications:

Frequency limit	above 6,000 MHz
Grid-wire diameter	0.0004 in.
Grid-wire density	800 turns per in.
Rated maximum temperature	250°C
Heater voltage, a-c or d-c	6.3 ± 5% volts
Heater current	0.24 A
Interelectrode capacitances:	
Plate to cathode and heater	0.01 pF
Cathode and heater to grid	1.7 pF
Plate to grid	1.0 pF
Heater to cathode	1.1 pF
Design maximum values:	
Plate voltage	250 volts
Limits of peak grid voltage	0 to −50 volts
Plate dissipation	1.0 watt
D-c cathode current	10.0 mA

Fig. 6-6   Type 7077 VHF-UHF ceramic triode. (*a*) Illustrating physical size; (*b*) internal structure. (Courtesy General Electric Co.)

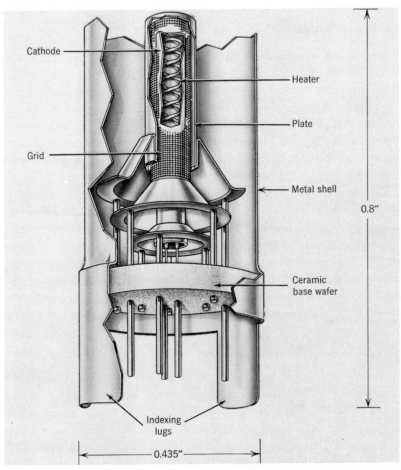

Fig. 6-7   RCA Nuvistor very-high-frequency triode. (Courtesy Radio Corporation of America.)

apparatus, test and measurement instruments, television cameras, and medical electronic equipment.

These two types are shown in Figs. 6-6 and 6-7.

Power amplifiers carry much larger currents than do voltage amplifiers; even small ones handle plate currents of 100 mA. The power-amplifier tube controls relatively large plate-circuit currents through grid action. The a-c power represented by the a-c component of plate current flowing through the load impedance may be utilized in a number of ways. It may drive a loud speaker, operate a relay that controls more power, run a motor, or drive the tube of a larger power amplifier.

6.4   Plate-Characteristic Curves of a Triode

Inasmuch as plate current in a high-vacuum triode depends on the potential of the grid with respect to the cathode (hereafter called either grid potential or grid voltage), it is natural to inquire what happens when the grid potential is kept constant and the plate potential is varied. A "family" of curves that give this information is shown in Fig. 6-5. Such a group of curves is required for the study of tube operation because all three quantities— plate voltage, plate current, and grid voltage—change in normal operation of the tube.

Consider a definite condition of operation such that the grid potential is −4 V and the plate potential is 163 V. The plate current for these potentials is 8 mA. If the grid is held at 4 V negative with respect to the cathode while a positive potential of 163 V is applied to the plate, the plate current will remain at 8 mA. This situation is called a static condition because no potentials or currents are changing. No current flows in the *grid circuit* of a high-vacuum tube if the grid is negative with respect to the cathode.

The static resistance of this tube for the static condition just described is

$$R_{\text{d-c}} = \frac{\text{plate volts}}{\text{plate amperes}} = \frac{163}{0.008} = 20{,}375 \ \Omega$$

It is readily understood that there will be a static resistance corresponding to every point on the plate-characteristic curves. Any point not located on one of the characteristic curves given in Fig. 6-5 will lie on a curve representing the proper grid potential that corresponds to the plate current and voltage for that point. That is, we can readily see that curves may be determined for many grid-voltage values between those represented by any two adjacent curves on the chart.

In Fig. 6-5, the grid voltage curve for $E_c = 0$ V is very similar to the forward voltage-current curve for a diode tube. Of course it should be, because the grid has no effect when its voltage with respect to the cathode is zero and the triode operates as a diode. The curves for increasing negative grid voltage also resemble the diode characteristic but are shifted toward positive plate voltages.

6.5   Letter Symbols for Vacuum Tubes and Circuits

Many voltages and currents in a circuit containing a vacuum tube are important in its operation. In order to distinguish easily a particular voltage or current for discussion or to represent it in an equation, letter symbols are used according to the following legends for various parameters that we shall use in the remainder of the book.

CONTROL GRID

e_c Instantaneous total voltage
i_c Instantaneous total current
E_c Average or quiescent voltage
I_c Average or quiescent current
e_g Instantaneous value of alternating-voltage component
i_g Instantaneous value of alternating-current component
E_g Effective value of alternating-voltage component
I_g Effective value of alternating-current component
E_{cc} Supply voltage
C_g Grid capacitance
e_{c1} Instantaneous total control-grid voltage
e_{c2} Instantaneous total screen-grid voltage, and so forth
     (Subscripts 1, 2, 3, etc., are used to specify the grids.)

CATHODE

E_f Voltage impressed
I_f Current
E_{ff} Voltage of supply
I_s Saturation current
C_k Capacitance (bypass)
R_k Resistor

PLATE

e_b Instantaneous total voltage
i_b Instantaneous total current
E_b Average or quiescent voltage
I_b Average or quiescent current
e_p Instantaneous value of alternating-voltage component
i_p Instantaneous value of alternating-current component
E_p Effective value of alternating-voltage component
I_p Effective value of alternating-current component
E_{bb} Supply voltage
P_p Plate dissipation
r_p Dynamic plate resistance

MUTUAL

C_{gp} Grid-plate capacitance
C_{gk} Cathode-grid capacitance
C_{pk} Plate-cathode capacitance

C_{gh} Grid-heater capacitance
C_{ph} Plate-heater capacitance
g_m Transconductance
μ Amplification factor
$E_{gm}, I_{gm}, E_{pm}, I_{pm}$ Maximum instantaneous values of the alternating components
R_b Direct-current resistance of plate load
r_b Alternating-current resistance of plate load
R_g Direct-current resistance of grid load
C Blocking capacitor
L_g Grid-inductance coil
L_p Plate-inductance coil

6.6   Three Dynamic Factors

The analysis of vacuum-tube operation requires that certain tube factors, generally called *parameters*, be known. These parameters tell how two of the three important quantities—plate voltage, grid voltage, and plate current—are related while the third is kept constant. They have specific names and their values may be determined from the plate-characteristic curves.

The *amplification factor* of a vacuum tube may be defined as *the ratio of a small change in plate voltage to the change in grid voltage required to restore the plate current to the value it had before the plate voltage was changed.* If the grid potential is changed slightly and if it is possible to change the plate voltage at the same time at such a rate that the plate current is prevented from changing, the rate of change of plate voltage with respect to grid voltage would be another way of defining the amplification factor of the tube.

Actually, the amplification factor is the *ratio* of the *rate of change of plate voltage with respect to time* to the required *rate of change of grid voltage with respect to time*, such that the *plate current of the tube is constant*. Assuming that small changes in the two voltages occur in the same short time interval, we say that the rate of change of plate voltage with respect to time is $\Delta e_b/\Delta t$ and the rate of change of grid voltage with respect to time is $\Delta e_c/\Delta t$. Δe_b represents a *very small* change in plate voltage that takes place in a *very small* time interval, Δt. The amplification factor (μ) is, then,

$$\mu = \frac{\Delta e_b}{\Delta t} \div \frac{\Delta e_c}{\Delta t} \qquad (i_b \text{ constant})$$

$$\mu = \frac{\Delta e_b}{\Delta e_c} \qquad (i_b \text{ constant}) \tag{6-1}$$

There should actually be a minus sign on one of the voltage terms here because the voltages change in opposite directions. That is, if Δe_b is positive, Δe_c must be negative in order to restore the plate current to its original value, and vice-versa. The amplification factor μ is a positive number, so it is the *absolute value* of the ratio given in Equation (6-1).

The amplification factor tells *how many times more effective the grid voltage is than the plate voltage in controlling the plate current.* For example, the amplification factor of the type 6J5 tube, as given in the tube manual, is 20. This means that a change of 1 V in grid potential will require a change of 20 V in plate potential in the opposite direction to keep the plate current from changing. This does not mean that when this tube is used in an amplifier circuit it will multiply the voltage twenty times, but it does give some information about how much voltage amplification may be expected. Tubes with high amplification factors are chosen for voltage-amplifier work. It will be seen later that pentode tubes, which have three grids instead of one, possess more advantages than triodes for voltage-amplification service, and higher amplification factor is one of these advantages.

The amplification factor of a tube may be determined from measurements made on the plate characteristics. Its value depends on where the "operating point" is located. There are certain regions, however, over which the amplification factor is nearly constant. Observe the graphical construction at point A in Fig. 6-5. At a current of 8 mA, a change in grid voltage from -2 to -4 V requires a change in plate voltage from 123 to 163 V to keep the current constant.

$$\mu = -\frac{\Delta e_b}{\Delta e_c} = -\frac{163 - 123}{(-4) - (-2)} = -\frac{40}{-2} = 20 \qquad (6\text{-}2)$$

The *dynamic plate resistance,* which is often called the a-c plate resistance, is a tube parameter defined as the *rate of change of plate voltage with respect to plate current, grid voltage being kept constant.* That is, with constant potential on the grid, a small change in plate voltage will cause a small change in plate current, and the ratio of those two changes is the dynamic plate resistance.

The graphical construction at point B in Fig. 6-5 shows how to determine graphically the dynamic plate resistance of the tube. At point B, the characteristic curve has a certain slope. The slope is defined as the slope of a straight line drawn tangent to the curve at the point. The slope of the tangent line is the ratio of the altitude to the base of a right triangle that has the tangent line for its hypotenuse.

The slope of the curve is given by $\Delta i_b/\Delta e_b$. The dynamic plate resistance is the reciprocal of the slope of the curve at the point; thus

$$r_p = \frac{\Delta e_b}{\Delta i_b} \qquad (e_c \text{ remaining constant}) \qquad (6\text{-}3)$$

The construction at point B shows that

$$r_p = \frac{175 - 150}{0.0095 - 0.0065} = \frac{25}{0.003} = 8,333\ \Omega$$

The dynamic plate resistance is a very important tube factor. It is used, as will be shown later, in setting up an equivalent circuit (representing the tube and other circuit parts) that is useful in performance calculations. The dynamic plate resistance is present only while the plate current is changing. Electronic and radio circuits accomplish nothing unless tube currents and voltage change.

It is useful to know how effective the grid of a high-vacuum tube is in *controlling plate current*. This effectiveness is expressed by the third important tube factor, the *transconductance*, which is represented by the symbol g_m. Conductance is defined as the reciprocal of resistance, and, since resistance may be calculated by dividing voltage by current (Ohm's law), conductance, is given by current divided by voltage. The "trans" part of the word trans-conductance denotes a transition through the tube from one side (the input, or grid side) to the other (the output, or plate side). In short, the trans-conductance of a tube is the *ratio of a slight change in plate current to the slight change in grid voltage that caused it, the plate voltage being held constant while the changes take place.*

That is,

$$g_m = \frac{\Delta i_b}{\Delta e_c} \qquad (\text{plate voltage remaining constant})$$

The transconductance (often called *mutual conductance*, indicating that it is mutual to the grid and plate circuits) is determined graphically in Fig. 6-5 at point C. Here Δe_c was chosen to be -2 V, and the corresponding i_b turned out to be 4.5 mA.

$$g_m = \frac{\Delta i_b}{\Delta e_c} = \frac{0.0095 - 0.0050}{-4 - (-6)} = \frac{0.0045}{2} = 0.00225 \text{ mho} = 2{,}250\ \mu\text{mho}$$

It may be seen algebraically that the amplification factor is equal to the transconductance times the dynamic plate resistance.

$$g_m r_p = \frac{\Delta i_b}{\Delta e_c} \times \frac{\Delta e_b}{\Delta i_b} = \frac{\Delta e_b}{\Delta e_c} = \mu$$

That is,

$$\mu = g_m r_p \tag{6-4}$$

The numerical values obtained above for these parameters do not check exactly, principally because they must all be determined at the same point on the plate-characteristic curves instead of at three separate points A, B, and C, as was done here. Three separate points were chosen in the explanation above in order to prevent complications on the diagram. Another source of error is the accuracy to which the curves can be read. The values determined agree fairly well with the values given in the tube manual. It must be kept in mind that the location of the point on the chart influences all three numerical values. Usually the dynamic plate resistance is called simply plate resistance. It is so listed in tube manuals.

The effect of the grid voltage on plate current can be expressed by an equation that includes the amplification factor

$$i_b = k(e_b + \mu e_c)^n$$

where n is about $\frac{3}{2}$, and k is an empirical constant. Note that this is similar to the expression of Child's law for a diode.

DRILL PROBLEMS

D6-1 Determine the three dynamic parameters of a 12AU7A triode at the operating point $e_b = 200$ V, $e_c = -6$ V; at $e_b = 100$ V, $e_c = -4$ V.

D6-2 Assuming the modified Child's law for a certain triode in which the plate current is 10 mA when $e_b = 200$ V and $e_c = -6$ V, calculate the approximate plate current at $e_b = 100$ V and $e_c = -2$ V. Assume that the amplification factor remains constant and equal to 12 over this range.

6.7   Grid-Plate Transfer Characteristics of a Triode

A useful set of curves is shown in Fig. 6-8. As indicated, the plate voltage is constant for each curve. The points that determine a curve may be obtained from the curves of Fig. 6-5 by drawing a vertical line at each desired plate-voltage value (this is equivalent to holding the voltage constant) and obtaining a plate-current reading at each value of grid voltage desired. A plot of a set of data, for each plate-voltage value chosen, gives the curves of Fig. 6-8. A set of curves of this type will be used later to help describe what happens in the important process called amplification.

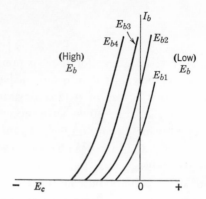

Fig. 6-8   Static-transfer characteristics of a triode.

DRILL PROBLEM

D6-3   Determine the three dynamic parameters of a 6AU6 pentode at the operating point $e_b = 200$ V, $e_c = -3$ V; at $e_b = 50$ V, $e_c = -1$ V. (*Hint:* The transfer curve may be helpful.)

6.8   Grid Bias

A triode tube in a simple electronic circuit is shown in Fig. 6-9. Although plate current is usually supplied by a rectifier and filter, a battery is shown here for convenience. Imagine first that the terminals 1 and 2 in the grid circuit are wired together instead of being supplied with an alternating voltage (E_1). Then the steady value of plate current may be set at any desired value from zero to the maximum the tube can carry, by merely adjusting the sliding contact on the voltage-dropping resistor across the bias-supply battery that has a fixed voltage (E_{cc}). The d-c voltage (E_c) in the grid circuit of an electron tube is called the *bias voltage*. Often it is referred to merely as the *bias*. E_1, when replaced, will be in series with E_c.

6.9   Grid Bias Arrangements

Ordinarily a battery is used to supply grid bias only in laboratory exercises or in test work. Bias batteries are not used in practice for a number of reasons. They are expensive, require too much space, and wear out.

It was shown in Chapter 5 how bias voltages are obtainable from taps on a power supply bleeder resistance. This is a convenient and commonly used method.

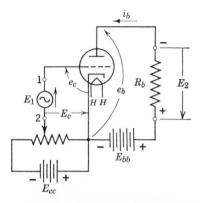

Fig. 6-9   Simple triode circuit.

A resistor in series with the cathode of a tube will provide a negative voltage for use as grid bias. The circuit of Fig. 6-10 shows the arrangement called *self-bias*. This circuit works well in applications where the plate current never decreases to zero. Its operation is easy to understand.

The cathode capacitor C_k serves to bypass the a-c components of plate current around R_k so that its voltage drop is a steady d-c voltage. It will be explained later that, although the tube current flows *downward only* in this circuit, it continually changes in *instantaneous value*. This means that it has a-c components superimposed upon a steady d-c component. Only the d-c component flows through R_k, making the cathode *positive* in polarity with respect to ground. It is important to note here, then, that the voltage drop across R_k makes the grid negative with respect to the cathode. The input signal voltage, e_g, is *generally* kept small enough so that its instantaneous

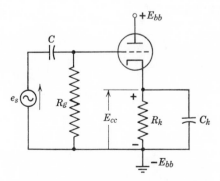

Fig. 6-10   Self-bias by cathode resistor.

maximum value *never makes the grid positive with respect to the cathode.* However, in certain applications the grid does go positive with respect to the cathode. This situation will be discussed later.

DRILL PROBLEMS

D6-4   A triode type 12AU7A is used in the circuit of Fig. 6-9. Determine the value of R_b if $E_1 = 0$ V, $E_c = 4$ V, and $e_b = 150$ V. What magnitude of supply voltage E_{bb} can be used?

D6-5   The cathode resistor R_k in Fig. 6-10 has the value 1 kΩ. If the capacitive reactance of C_k is to be always less than or equal to one-tenth of R_k, what is the lowest frequency permissible in the plate-current variations?

6.10   Grid-Leak Bias

It is possible to provide negative grid-bias voltage by connecting a resistor to the grid terminal of the tube and bypassing the a-c signal around it with a capacitor as in Fig. 6-11. This is done in high-frequency amplifiers, oscillators, television circuits, and radio receivers.

Some electrons in a vacuum tube land on the grid when it is not made negative with respect to the cathode by connecting it to a negative point, as has already been described.

The very small electron current entering the grid circuit will produce a d-c voltage across the series *grid-leak* resistor connected to the grid terminal of the tube. This electron flow is interpreted as a *positive current* flowing toward the grid which makes the grid end of the resistor negative with respect to its other end. The parallel capacitor is thus charged to the same polarity, and it will discharge slightly during negative half-cycles of the signal voltage on the grid, thus maintaining the bias voltage across the resistor at a practically constant level. Electrons not used in charging the capacitor back up during

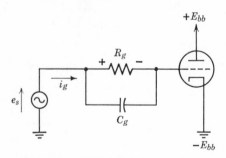

Fig. 6-11   Grid-leak bias circuit.

positive half-cycles of signal voltage, pass down through the input circuit to the cathode of the tube, and are neutralized by positive charges. The cathode is usually grounded when grid-leak bias is used. Grid-leak resistance values are of the order of $\frac{1}{2}$ MΩ or more.

6.11   The Triode in a Circuit

A plate-circuit resistor (R_b) is connected in series with the plate of the tube in Fig. 6-9. This is an important circuit element because the *a-c component* of the voltage E_2 across this resistor is the *output voltage* of this simple amplifier stage. To make this more meaningful, let us consider that an a-c generator or a transformer winding is applying a small alternating voltage E_1 to terminals 1 and 2. It will be seen later, in detail, how the tube causes a variation of current in R_b to such an extent that the *variations* in the voltage E_2, across R_b, will be *an enlarged reproduction of the input voltage E_1*. It will furthermore be shown that the variations in E_2 will be displaced 180 degrees in phase with respect to those of E_1.

The vast majority of vacuum electronic tubes operate with their grids always negative with respect to their cathodes. It will be recalled that the plate-characteristic curves of the 6J5 tube (Fig. 6-5) did not include a condition of positive grid voltage. In Fig. 6-9, however, the input voltage E_1 is alternating in nature and so it has positive and negative half-cycles. Therefore, in order to prevent the grid from going positive at any time, the negative bias voltage (E_c) should be sufficiently large. A graph of the voltages in the grid circuit is shown in Fig. 6-12a. Hereafter in this text, the alternating component of grid-cathode voltage will be denoted by the symbol e_g or E_g, which is standard notation, instead of by E_1 as is done in Fig. 6-9.

It is a very important matter to learn that *the alternating component of plate current (i_p) is in phase with the alternating component of grid voltage* in any high-vacuum tube with resistance load in which only the control-grid voltage is forced to vary. Figure 6-12b shows the alternating component of plate current (i_p) superimposed upon the direct component of plate current (I_b). The direct component of plate current flows in the tube even without the input voltage E_g applied. The plate current flowing when there is no varying component in the input voltage is denoted by the symbol I_{bo}. Note that the instantaneous total plate current is represented by the symbol i_b. Figure 6-12c shows how the voltage between the plate and cathode of the tube varies with time while the performance just described above is going on. The voltage E_b is the average value (d-c component) of plate-to-cathode voltage; e_b

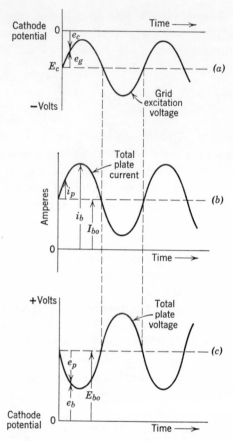

Fig. 6-12 (*a*) Voltages in the grid circuit of Fig. 6-9. (*b*) Currents in the plate circuit of Fig. 6-9. (*c*) Voltages in the plate circuit of Fig. 6-9.

is the instantaneous total value of plate-to-cathode voltage; and e_p is the instantaneous value of the varying component of plate-to-cathode voltage. It will be seen later that e_p shows up entirely across the load resistor (R_b) because the alternating component of plate current does not encounter enough resistance in going through the battery to cause the battery-terminal voltage to have a varying component.

6.12 Simple Circuit Equations

An examination of the circuit of Fig. 6-9 and a familiarity with the symbols for current and voltage components will permit the setting up of mathematical relationships among those quantities.

The instantaneous total plate current will be the sum of the steady d-c component and the instantaneous value of the varying component:

$$i_b = I_b + i_p \qquad (6\text{-}5)$$

Using Kirchhoff's voltage law, we see that, in the plate circuit, E_{bb} is a voltage rise above cathode potential, E_2 is a voltage fall below the positive battery potential, and e_b (plate to cathode) is a voltage fall back to cathode potential. Hence

$$E_{bb} - E_2 - e_b = 0$$

That is, the output voltage is

$$E_2 = E_{bb} - e_b \qquad (6\text{-}6)$$

but E_2 is also equal to $i_b R_b$, so that

$$E_{bb} - e_b = i_b R_b \qquad (6\text{-}7)$$

In the grid circuit we also start at the cathode, go down the voltage drop (E_c), up the voltage rise E_1, now called e_g, and down the amount e_c to get back to cathode level. Hence

$$-E_c + e_g - e_c = 0$$

or, the instantaneous total grid-to-cathode voltage is

$$e_c = -E_c + e_g \qquad (6\text{-}8)$$

Although $i_p R_b$ is the instantaneous a-c component of voltage drop across the plate-circuit resistor (R_b), there is a constant d-c voltage drop across it equal to the product of the d-c component of current and the resistance. Letting E_{bo} represent the d-c component of plate-to-cathode voltage, we have

$$E_{bb} = E_{bo} + I_{bo} R_b \qquad (6\text{-}9)$$

The amount of negative bias voltage provided by the cathode-bias arrangement is easily calculated. The d-c component of tube current and the ohms value of R_k obey Ohm's law. In a triode tube the cathode resistor current is equal to the d-c plate current. In a tube with a screen grid and a control grid the cathode current is the sum of the plate current and the screen-grid current. The latter is usually much smaller than the plate current, but both flow through R_k.

Assume that a grid bias voltage $E_c = -4$ V is needed in a particular application where the d-c component of plate current, I_b, is 5 mA, and the screen current is 1 mA. The required value of cathode-bias resistance is

$$R_k = \frac{4}{0.005 + 0.001} = 667\ \Omega$$

It is desirable that the bypass capacitor have a reactance not greater than one-tenth the resistance of R_k at the *lowest* frequency at which the tube is to operate. Assume this to be 50 Hz. The capacitance of C_k should then be such that

$$\frac{1}{\omega C_k} = \frac{667}{10}$$

$$C_k = \frac{10}{2\pi \times 50 \times 667} = 47.7 \times 10^{-6}\ \text{F} = 47.7\ \mu\text{F}$$

A standard-size electrolytic capacitor of 50 μF would be chosen. The voltage drop across it would be evaluated as the product of the current in R_k and the resistance:

$$E_c = 0.006 \times 667 = 4\ \text{V}$$

yielding a bias voltage of -4 V.

6.13   The Screen-Grid Tetrode and the Pentode

The screen-grid tube was developed through research in the tube design directed toward the elimination of difficulties that arise when triodes are used at high frequencies. The capacitance between the control grid and the plate of a triode allows a small part of the plate-circuit energy to be fed back, within the tube, to the grid circuit. This starts surges of current, known as oscillations, in both plate and grid circuits and puts an end to the amplification process.

The presence of a mesh of fine wire between the control grid and the plate appreciably reduces the grid-plate capacitance. This mesh is the screen grid. It usually is connected to the cathode through a capacitor of appropriate size to prevent a-c current from flowing through the power supply. The control-grid-plate is further reduced by bringing the control-grid connecting wire to a metal cap at the top of the tube envelope.

The screen-grid tetrode was found to have other desirable characteristics not possessed by the triode. The electrostatic field between the screen grid and the cathode, especially in the control-grid region, is affected only slightly by variations in plate voltage in the tube's normal range of operation. On he other hand, the control grid is just as effective as it is in the triode tube n determining the potential distribution of the electrostatic field. This means hat the effectiveness of the grid voltage as compared to that of the plate •oltage in determining plate current is much greater in the tetrode than in the riode, i.e., the amplification factor of the screen-grid tetrode is much larger.

The plate resistance is much higher in a tetrode than in a triode of compar-.ble size, because Δe_b must be substantially arger to produce a given Δi_b at constant E_c han in the case of the triode. The transconduc-.ance may be made high also by appropriate lesign and construction.

There is a tendency for electrons, that are knocked out of the plate of a screen-grid tet-·ode by the electrons forming the plate cur-·ent, to pass to the screen. This is objectionable when the a-c component of voltage in the plate ·ircuit (superimposed on the d-c plate-supply voltage, of course) is high. Under these con-litions the plate voltage dips to low values and .he secondary-emission electrons from the plate pass to the screen in large numbers. Then distortion becomes excessive and oscillations are likely to start.

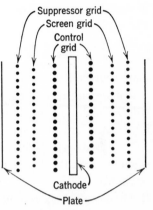

Fig. 6-13 Elements of a pentode tube.

To overcome this effect, a third grid, called the *suppressor grid*, was installed just in front of the plate, and the tube became a pentode (Fig. 5-13). The suppressor is usually connected to the cathode of the tube and is thus at full negative potential with respect to the plate. It forces secondary electrons back into the plate while offering practically no obstruction to the passage of the main stream of electrons from the cathode to the plate.

The pentode tube has a higher amplification factor and plate resistance than the tetrode. The suppressor grid isolates the plate still more from the control grid and the cathode. As a result, the capacitance between the control grid and the plate is even less than in the tetrode. An example of the reduction in this capacitance is seen when the grid-plate capacitance of the 6J5 triode, which is 3.4 pF, is compared to that of the 6SJ7 pentode, which is 0.005 pF, a reduction ratio of 680 to 1.

The suppressor grid is able to cause the return of secondary electrons to the plate, even when the plate voltage is low. This is because those electrons are emitted with comparatively low velocities and the opposing electric field between the plate and suppressor is sufficiently strong to prevent those electrons from passing back through the suppressor-grid mesh.

The greatly reduced capacitance between plate and control grid in pentodes makes possible their use in amplifiers at high frequencies where triodes would not be practical. Oscillations produced by feedback of energy through the tube from the plate circuit to the grid circuit would cause the triode to quit amplifying. Tetrodes are seldom used now except in special applications. A set of plate characteristics of a voltage-amplifier pentode is shown in Fig. 6-14.

The plate characteristics are shown for a screen-grid voltage of 100 V, which is a typical operating potential for the screen. Superimposed on the plate characteristics is a curve showing the variation of screen-grid current with increasing plate voltage, when the control grid is at 0 volts. Screen current will be smaller for higher (more negative) bias values. Although it is small it must be taken into account if the pentode is self-biased, as explained earlier.

The screen-grid potential may be obtained from the plate supply (E_{bb}) when the screen voltage is to be equal to or smaller than E_{bb}. The screen may be connected directly to E_{bb} when the design requires operation at this voltage. If the screen voltage is to be less than E_{bb}, a voltage-dropping resistor such as R_s in Fig. 6-15 is required.

The screen current can be used to reduce the screen-to-cathode voltage to the design value. For example, if $E_{bb} = 250$ V and the screen voltage

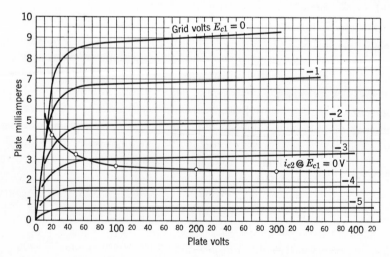

Fig. 6-14   Average plate characteristics of type 6SJ7 pentode. $E_{c2} = 100$ V, $E_{c3} = 0$ V.

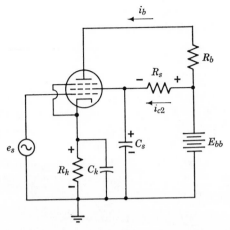

Fig. 6-15   Pentode circuit supplying screen-grid voltage from plate supply through bias network $R_s - C_s$.

is to be 100 V, then R_s must drop 150 V across it. Then the value of R_s can be calculated,

$$R_s = \frac{150 \text{ V}}{\text{screen current, A}}$$

Suppose the circuit is operating at a bias voltage of -4 V and that the screen current is 1.5 mA. The value of R_s will be 100 kΩ.

In an operating circuit the control-grid potential will be varying, and the screen current will vary in the same manner. In order to avoid variations in the screen voltage, a large capacitor (C_s) can be connected between the screen grid and the negative terminal of E_{bb}, as shown in Fig. 6-15. The reactance of C_s is usually chosen to be one-tenth or less of R_c at the *lowest* frequency of current variations. The capacitor C_s provides a low-reactance path for a-c around R_s, so that the screen voltage will remain practically constant at 100 V as required.

DRILL PROBLEM

D6-6   A voltage amplifier is to be designed using a 6SJ7 pentode in the circuit of Fig. 6-15. A power supply of 250 V is available. Based on certain requirements for the circuit, the plate load will be 50 kΩ and cathode bias will be provided by 1 kΩ shunted by 20 μF. Determine the required values for the screen-bias network to hold $E_{c2} = 100$ V. (*Hint:* Determine the operating point, and consult a tube manual for the magnitude of screen current flowing at that point.)

6.14   The Compactron Vacuum Tube

The Tube Department of General Electric Company has developed a vacuum tube, called a *compactron*, which either (*a*) combines the functions of several conventional tubes into a single envelope, or (*b*) puts a single function into a significantly smaller unit. Its 12-pin base construction is not much larger than the 9-pin conventional base, 0.750 in. diameter vs. 0.468 in. Mounting heights vary from 1.5 to 4.5 in., depending on the purpose of the tube. Larger sizes are needed, of course, for power applications where heat must be dissipated from a larger surface area. Figure 6-16 compares some conventional tubes with comparable compactrons.

A compactron tube may contain a single pentode design for high-frequency operation and three voltage-amplifier triodes, or it may contain a combination of two diodes and two triodes. As an example, the type 6U10 compactron tube contains three triodes, one similar to the 12AX7 and two sections similar to the 12AU7A. This combination makes it extremely suitable for television receiver circuits, while requiring only one socket and other mounting hardware. The type 6AF11 is a twin-triode pentode similar to

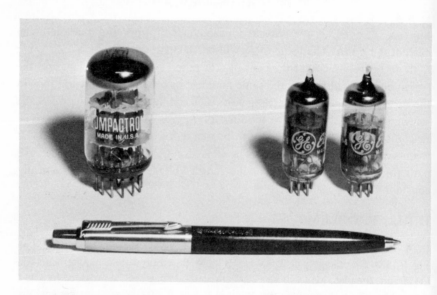

Fig. 6-16   Comparison of conventional tubes and compactron. Tube at the left (twin pentode) replaces two pentodes on right. (Courtesy General Electric Company, Tube Department, Owensboro, Ky.)

a video pentode 6GN8, a high-mu triode, and a triode similar to the 6CX8.

The use of compactrons in circuit applications reduces the number of sockets, amount of wiring, and other components, and also reduces weight and space requirements. These tubes appear to have an increasing role to play in circuits where initial costs are competitive, such as home television receivers and commercial sound systems.

6.15   Variable-Mu Tubes

It is often necessary to amplify large signals, especially those in the radio-frequency range. This requires a wide range of grid-voltage variation in the tube. An ordinary pentode has a transfer curve as shown in Fig. 6-17*a* (uniform grid). Plate current will go to zero rather abruptly as the bias voltage is increased, leading to the name *sharp-cutoff pentode* for this type of tube. In order to have a wider control over plate current with a variable bias voltage, the *variable-mu*, or *remote-cutoff* pentode was developed. It is called variable-mu because the amplification factor (μ) of the tube varies with plate current. The transfer curve (variable-mu grid) of a remote-cutoff tube may be compared to the curve for a sharp-cutoff tube.

Figure 6-17*b* illustrates the specially wound grid structure used in a remote-cutoff pentode. The grid wires are spaced close together at the top and bottom but are farther apart in the middle section of the grid. The effect of this

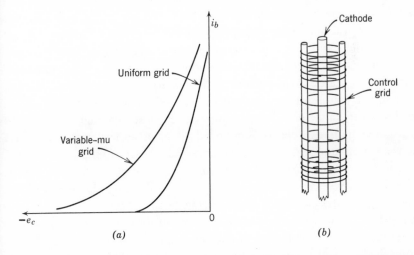

Fig. 6-17   Variable-mu pentode transfer curve and grid structure. (*a*) Transfer curves. (*b*) Grid structure.

variable spacing may be explained as follows. When the plate current is small, few electrons are getting through the grid because of its high negative bias. As the bias voltage is decreased, however, the effect of the widely spaced wires on the current is reduced. At very low bias voltages, the larger plate current is practically limited only by the grid wires at the top and bottom of the tube.

A 6BA6 pentode is a remote-cutoff tube that is often used in radio-frequency amplifiers. It would be instructive for the student to construct transfer curves for pentodes identified in a tube manual as remote-cutoff and sharp-cutoff types. For example, compare the 6AU6A with the 6BA6.

An evaluation of the parameters of the type 6BA6 pentode shows that its transconductance varies from 4,400 μmho at -1.5 V bias to about 40 μmho at -20 V bias. Its amplification factor changes correspondingly over the range of bias voltages. The transconductance of the 6AU6A varies from about 4,400 μmho at -0.5 V to a small value at only -4 V bias. Hence, 6AU6A is a sharp-cutoff pentode, and the 6BA6 is a remote-cutoff pentode.

6.16   Beam-Power Tube

The pentode performs exceedingly well as an amplifier of small signal voltages, but when large signals are used it has several minor faults. The suppressor grid, which operates at cathode potential, is supposed to prevent *secondary electrons* from reaching the screen grid. These electrons are knocked out of the plate by primary electrons that originate at the cathode. The suppressor grid must have openings to allow the primary electrons to pass through, and it is through these openings that some secondary electrons pass to reach the screen. The resulting additions to screen current and unwanted variations in plate current cause distortion in large-signal operation. Another undesirable feature of the conventional pentode is that the positive screen grid wires are in the paths of the primary electron stream, and they receive a substantial number of electrons intended for the plate. The result is excessive screen current and power loss in the screen and its series resistor, if present.

The *beam-power tube*, or simply *beam tube*, is a special form of tetrode designed to reduce these shortcomings. The screen-grid wires and the control-grid wires are wound with the same pitch and lined up so the screen-grid wires are in the electron-beam shadows of the control-grid wires. This materially reduces the number of electrons captured by the screen-grid wires from the beams of the primary electrons as they speed toward the plate.

Beam-forming metal plates, connected to the cathode and therefore at cathode potential, condense the electron beams and cause them to hit the plate in two well-defined areas diametrically opposite each other in the tube. The high density of electrons in these beams just before they reach the plate forms a negative-potential region that performs the function of a suppressor grid. Secondary electrons have a difficult time trying to pass through this region to reach the screen grid. At large signal voltages on the grid, the grid bias voltage is large and the screen current is very small. Beam-power tubes with concentric electrodes do not have beam-forming plates, but they operate in a similar manner.

Beam tubes are used in sound systems as drivers of loud speakers, in power amplifiers in automatic control systems, and in other power-controlled systems.

SUGGESTED REFERENCES

1. W. G. Dow, *Fundamentals of Engineering Electronics*, 2nd ed., Wiley, New York, 1952.

2. E. Milton Boone, *Circuit Theory of Electron Devices*, Wiley, New York, 1953.

3. G. E. Happell and W. M. Hesselberth, *Engineering Electronics*, McGraw-Hill, New York, 1953.

4. Paul D. Ankrum, *Principles and Applications of Electron Devices*, International Textbook Co., Scranton, Pa., 1959.

QUESTIONS

6-1 Describe the physical form of the electrodes in a conventional vacuum triode. What are the advantages of cylindrical construction?

6-2 Explain how the grid controls the flow of electrons in a vacuum triode.

6-3 What is meant by the potential distribution in a vacuum tube? How is the potential distribution affected by making the grid more negative with respect to the plate?

6-4 Consider a vacuum triode with proper d-c voltages on its electrodes. What conditions inside it result in a dip in the potential-distribution curve just off the cathode? What will be the effect of increasing only the plate potential? Of making the grid still more negative?

6-5 Plate current flows in a vacuum triode with normal d-c voltages on its electrodes. Why is it called space-charge-limited current? Why are some electrons that are emitted from the cathode unable to reach the plate? Where do they go?

6-6   What are the meanings of the symbols E_c, e_g, i_g, I_g, R_k, E_b, e_p, i_p, I_b, r_p, R_b?

6-7   Draw from memory a plate-characteristic curve of a triode, labeling all important lines. Then draw a few more, including one that goes through the origin. What quantity is constant on each of the curves?

6-8   Give the definition, in words, of each of the following tube parameters: (*a*) amplification factor, (*b*) dynamic plate resistance, (*c*) transconductance.

6-9   Write the symbolic equations that define each of the three tube parameters. How is g_m related to r_p and μ? Check the units.

6-10   What is a grid-plate transfer characteristic? What quantity is constant when a static transfer characteristic is drawn?

6-11   How does a power amplifier tube differ from a voltage amplifier tube in respect of currents?

6-12   Name three ways of obtaining grid bias for a triode.

6-13   Why is a capacitor used with cathode bias? With screen bias?

6-14   How is the ohms value of a cathode-biasing resistor determined? Of a screen resistor?

6-15   How is the capacitance of a cathode bypass capacitor calculated? Why is its value determined for the *lowest* operating frequency?

6-16   Refer to the simple triode circuit of Fig. 6-9. The *a-c component* of grid-to-cathode voltage is properly denoted by e_g. (*a*) At what instant is the plate current a maximum? Answer in terms of possible value of e_g. What is true about e_c at that instant? (*b*) At the instant of maximum plate current, what is known about i_b, i_p, e_b, e_p? (*c*) Since the instantaneous value of total plate current is a maximum when E_g is at its *positive* maximum, why is e_b at its *minimum* at that instant? (*d*) Consider that E_p may be measured with a very high-resistance voltmeter across R_b. Explain why E_p is 180° out of phase with respect to E_g. The a-c voltage across E_{bb} is entirely negligible because its impedance is essentially zero at moderate frequencies.

PROBLEMS

6-1   A vacuum triode operating under space-charge-limited current conditions has a 3-V dip in front of its cathode. Calculate the normal component of velocity with which electrons must leave the cathode in order to reach the lowest point of the dip.

6-2   An electron that was just able to pass the lowest point of the dip in Problem 6-1 goes on to the plate, which is at +100 V with respect to the cathode. (*a*) With how many electronvolts of energy will it strike the plate? (*b*) With what velocity, in meters per second, will it strike the plate?

6-3   In a simple voltage-amplifier circuit like Fig. 6-9, the maximum and minimum instantaneous values of e_b are 150 V and 90 V, respectively. (*a*) What is the rms value of e_p? (*b*) What is the average (d-c) value of plate-to-cathode potential? (*c*) If $R_b = 5,000 \, \Omega$ what is the rms value of i_p? (*d*) If $E_{bb} = 250$ V calculate the d-c component of plate current, using Equation (6-9). Should this current be obtainable from $E_b \div R_b$? Why? Check it.

6-4 On a set of plate characteristics for a 12AU7A tube, determine graphically values of μ, g_m, and r_p in the neighborhood of $E_b = 170$ V, $E_c = 0$ V. Check $\mu = g_m r_p$.

6-5 Draw several static transfer characteristics for the 12AU7A tube, using the average plate characteristics.

6-6 Plot a set of curves showing plate voltage on the ordinate and grid voltage on the abscissa for several values of constant plate current in the 12AU7A tube. Measure $\Delta e_b / \Delta e_c$. What parameter is represented by the slope of these curves?

6-7 What is the transconductance of a tube that has an amplification factor of 12 and a dynamic plate resistance of 5,000 Ω?

6-8 A tube in a circuit like that of Fig. 6-9 is biased at -10 V. What is the largest rms value of a-c input voltage that may be used without driving the grid positive?

6-9 Cathode bias is produced by a resistor $R_k = 800 \, \Omega$ carrying 6 mA of current. What is the bias voltage? Calculate the required size of the bypass capacitor for a minimum frequency of 159 Hz.

6-10 What should be the resistance of a bleeder section on a power supply so that it will provide -5 V of bias when the current is 60 mA before the bias connection is made to the tube?

6-11 Using the characteristic of the 2A3 triode, determine the amplification factor, plate resistance, and transconductance for a plate supply voltage of 200 V and a grid voltage of -35 V.

6-12 The characteristics of a certain triode may be approximated by the expression $i_b = 10^{-5} (155e_c + 8e_b)$. Determine the values of μ, r_p, and g_m.

6-13 Draw the dynamic transfer characteristic for a 12AU7A tube for $E_{bb} = 250$ V and plate load $R_L = 2,500 \, \Omega$.

6-14 The plate current in a triode is 8 mA when $E_{bb} = 225$ V and $E_{cc} = -14$ V. Assuming μ as approximately constant and equal to 10, what should be the approximate plate current if $E_{bb} = 300$ V and $E_{cc} = -25$ V?

6-15 The plate resistance of a triode is 8,000 Ω and the transconductance is 3,000 μmho. If the plate voltage is increased by 40 V, what is the increase in plate current, assuming the grid voltage is maintained constant?

6-16 A certain triode has the following tube coefficients:

$$\mu = 40, \quad g_m = 2,000 \; \mu\text{mho}, \quad r_p = 20,000$$

If the plate current is to go up 0.2 mA when the grid voltage is increased 1 V in the negative direction, in what direction, and by how much, must the plate voltage be changed?

6-17   The plate current of a certain pentode connected as a triode can be expressed approximately by the equation

$$i_b = 41(e_b + 10e_c)^{1.41} \times 10^{-6} \text{ A}$$

What is the amplification factor?

6-18   A 6J5 triode operates into a 20-kΩ plate load and is powered by a 200-V d-c source in the plate circuit. A 6-V battery is used to provide fixed bias in the cathode circuit. (*a*) Draw from memory the schematic diagram of the circuit. (*b*) Determine the values of plate current, plate voltage, and grid bias when the grid circuit has no a-c signal applied; i.e., the *operating point*.

6-19   In the circuit shown in Fig. P6-19, tube 2 is to operate at $i_b = 1$ mA, $e_c = -6$ V. Determine the values of i_b, e_b, and e_c for tube 1 and the necessary value for R_k. (*Hint*: The plate current of both tubes flows through R_k and produces the −6 V bias for both of them.)

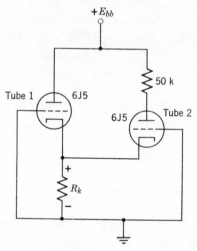

Fig. P6-19

6-20   The grid circuit of a 6J5 triode uses a fixed bias of −4 V. The plate circuit contains a 25 kΩ load and is supplied from a 250-V supply. Calculate the range of plate supply voltage required to hold plate current constant for plate loads varying from 10 kΩ to 100 kΩ. (*Hint*: The plate supply voltage equals the sum of load voltage and plate voltage, e_b, at all times.)

6-21   For a signal $e_s = 20 \sin(\omega t + \phi)$ in the plate circuit as shown in Fig. P6-21, determine the peak-to-peak variation of the plate current. (*Hint*: The grid bias will remain constant.)

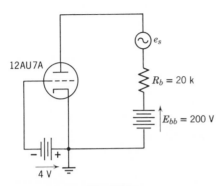

Fig. P6-21

-22 One section of a 12AU7A triode is used in the circuit of Fig. 6-9. Supply voltage E_{bb} is 200 V and $R_b = 2$ kΩ. What value of grid bias is required to make the quiescent plate voltage $= 100$ V?

5-23 A 12AU7A triode is used in the circuit of Fig. 6-9. Determine the necessary value of R_b for $E_1 = 0$ V, $E_c = 0$ V, $e_b = 200$ V, $E_{bb} = 300$ V.

6-24 The cathode resistor R_k in Fig. 6-10 has the value 1.2 kΩ and $C_k = 10$ μF. If the reactance of C_k is to be always less than or equal to $\frac{1}{10}$ of R_k, what is the lowest permissible frequency of variation in plate current?

6-25 A triode has $g_m = 4{,}000$ μmho and $r_p = 12$ kΩ. (*a*) What change in plate current will occur for a grid voltage change from -2 to -6 V? (Assume e_b remains at 150 V.) (*b*) How much change in plate voltage is required to return plate current to its original value, with e_c remaining at -6 V? (State the magnitude and direction of the change in plate voltage.)

6-26 Using the plate characteristics for the 6SN7GT triode, (*a*) plot the static transfer characteristic for $E_b = 200$ V, and (*b*) plot the dynamic transfer curve for $E_{bb} = 200$ V and $R_b = 20$ kΩ.

6-27 State the purpose of each of these circuit elements in Fig. 6-15: R_b, R_k, R_s, C_k, C_s, E_{bb}.

6-28 A 6SJ7 pentode is operated with $E_{bb} = 300$ V and $R_b = 30$ kΩ. The screen voltage is set at 100 V and the suppressor grid is connected to the cathode. Fixed bias holds the quiescent control grid potential at -2 V. For a signal $e_s = 2 \sin \omega t$ in the grid circuit, determine the variation of plate current and the variation of plate voltage. Make graphical plots of the time variations of e_s, i_b, and e_b. Label the graph with numerical values at significant points, in a manner similar to the symbolic markings in Fig. 6-12.

6-29 In the circuit of Fig. P6-29, if the grid is made 1 V more positive than normal relative to the cathode, the voltage across R_k is 15 V larger than normal. What value of e_s is required to cause this to happen?

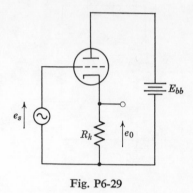

Fig. P6-29

6-30   In the circuit shown in Fig. P6-30, it is known that plate current is related to grid voltage by the expression $i_b = (30 + 2e_c)$, where i_b is in milliamperes and e_c is in volts. How much plate current flows in this circuit?

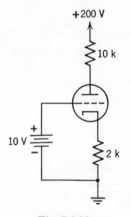

Fig. P6-30

6-31   Determine numerical values for each of these quantities in Fig. P6-31: I_b, E_b, E_c, e_p, i_p, e_g.

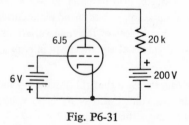

Fig. P6-31

CHAPTER SEVEN

Transistors

Since the discovery of transistor action in semiconductor crystals in 1948, there has been a rapidly expanding technology to utilize the unique properties and characteristics of many types of semiconductor devices. Applications of electronics in missiles, earth and lunar satellites, and digital computers (in all of which space and weight are at a premium) have made critical the need for smaller electronic components. Vacuum tubes have been miniaturized, printed circuits have replaced wiring on a chassis, and crystal triodes and tetrodes called *transistors* are performing many jobs formerly done by bulky vacuum and gas tubes. Some circuit functions are performed better by transistors than by tubes.

Transistors offer several advantages over tubes. Among them are: (1) no heater or filament is required; in fact, transistors behave better at room temperatures than at the elevated temperatures required for tubes; (2) very low operating voltages can be used; (3) they consume low power, resulting in better efficiency and less heat loss; (4) they have a long life with essentially no aging effects; (5) they resist damage from shock and vibration; and (6) extreme flexibility in circuit design is possible. Since no heat is generated from heater filaments as in vacuum tubes, transistors allow a substantial reduction in cooling capacity in air-conditioning equipment for giant electronic computers.

The material in this chapter describes the physical structure, electrical characteristics, and theory of operation of transistors. The reader should be familiar with the discussion of *P-N* junction operation given in Chapter 2.

7.1 Historical Background

Between the year 1900 and the end of World War II, the vacuum tube was developed into many complicated and specialized types. Some even perform three duties in one envelope such as oscillation, rectification, and amplification. But during this period the crystal units used in radio and electronic apparatus remained in diode form.

In 1948, Dr. John Bardeen and Dr. Walter H. Brattain achieved success in their experiments with semiconductors at Bell Telephone Laboratories. They made the first successful *crystal triode* and named it the *transistor*, by combining the words *transfer* and *varistor*. Varistor is the name given to the germanium diode by the Bell Laboratories, and the name transistor signifies the transfer of signals through a varistor.

7.2 Point-Contact Transistor

Space will not be devoted in this book to detailed explanations of manufacturing techniques by which transistor materials are produced and transistor construction is accomplished. There are many references on these subjects in technical libraries which may be consulted. The nature of *N*- and *P*-type semiconductors was explained in the discussion of crystal diodes in Chapters 2 and 3.

The first commercially produced transistor was developed by Dr. Bardeen and Dr. Brattain as the Bell Laboratories type A. It had two fine wires with their sharp points touching the surface of a crystal of *N*-type germanium. The points of contact were only a few thousandths of an inch apart. The germanium crystal was a flat plate about $\frac{1}{16}$ in. sq. and about 0.02 in. thick. An enlarged sketch of this kind of transistor, which is the *point-contact* type, is shown in Fig. 7-1.

It was stated in the discussion of the point-contact diode that a *P*-type region was formed in the crystal around the point where the fine wire was welded to the germanium surface by a surge of electric current. In the manufacture of the point-contact transistor, two *P*-type regions are formed, one at the contact point of each fine wire. The result is a *P-N-P* type of transistor. One of the wires is called the *emitter* and the other is called the *collector*. The block of germanium is called the *base*. A wire is fastened to the base for use in making the necessary connection to a circuit.

The point-contact transistor can be expected to have the same principles

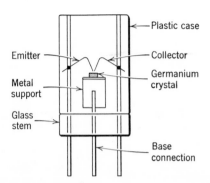

Fig. 7-1   Elements of point-contact type of transistor.

of operation as the junction-type transistor, which will be described below.

The point-contact transistor is not widely used because it has certain undesirable characteristics and limitations when compared with junction transistors. One of these is an interaction between collector and emitter circuits, due mostly to the requirement that emitter and collector contact points must be very close together. As a result, this type of transistor is unstable in a circuit. Another objection is the low limiting value of current that can be used. Local heat is developed at the contact points, and the temperature rises to excessive values with currents that a junction transistor of the same physical size can handle easily at safe temperatures.

7.3   Junction Transistor

A junction transistor is made so that its physical configuration is two *P-N* junctions back-to-back with a thin *P*-type or *N*-type region between them. The sequence of the layers of *P*-type and *N*-type materials classifies the transistor as either a *PNP* or an *NPN* type, depending on which is the center region. The junctions are either grown or diffused, rather than being mechanical joints of the materials, as explained in Chapter 3. The structures of two typical junction transistors are shown in Fig. 7-2. These are called *PNP* transistors because the thin *N*-type region separates the two *P*-type junctions. The active *base* region separating the two junctions is the most significant feature of this type of transistor.

Several junction transistors, useful as small-signal amplifiers, are shown in Fig. 7-3. The ball-point pen, the standard 6J5, and the miniature 6C4 tubes are used to indicate the actual size of the transistor. The outside dimensions

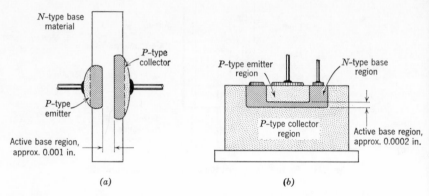

Fig. 7-2 Structures of typical junction transistors. (a) *PNP* alloy-junction type. (b) *PNP* diffused-junction type.

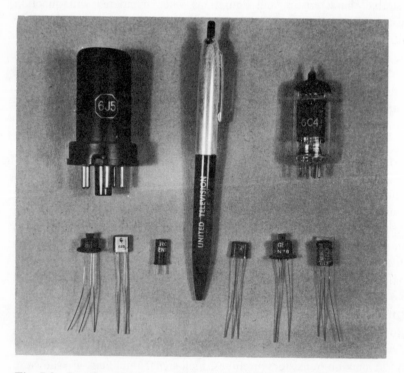

Fig. 7-3 Junction transistors. (Courtesy United Electronics Laboratories, Louisville, Ky.)

of the transistor are of the order of $\frac{3}{8}$ in. long, $\frac{1}{4}$ in. thick, and $\frac{1}{4}$ in. wide, excluding leads. Some are cylindrical in shape, about $\frac{1}{3}$ in. in diameter and $\frac{1}{4}$ in. long.

DRILL PROBLEMS

D7-1 It will be instructive for the student to obtain a transistor manual or manu-facturers' data sheets, and determine for these transistors the following information: (1) *NPN* or *PNP*; (2) silicon or germanium; (3) point-contact, junction, or other method of fabrication; (4) voltage and current ratings; (5) type of case and lead configuration; (6) typical forward current-transfer ratio; (7) maximum power dissipation.

2N 107	2N 450	2N 1015A
2N 109	2N 508	2N 1510
2N 334	2N 524	2N 1671
2N 388	2N 635	2N 1830
2N 404	2N 656	2N 1833

D7-2 These JEDEC case numbers represent different sizes and shapes. (Refer to last paragraph, Section 7.10.) Determine how they differ:

TO-5, TO-12, TO-18, TO-33, TO-39

also which lead is collector, emitter, and base.

7.4 Symbols and Bias Polarities for Transistors

The two usual symbols for transistors in circuit diagrams are given in Fig. 7-4, together with the vacuum triode symbol for comparison. The only difference between the symbols for transistors is the direction of the arrow on the emitter lead. The arrow points in the direction of conventional forward

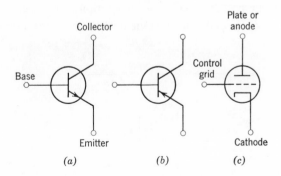

Fig. 7-4 Conventional symbols for transistors and vacuum triode. (*a*) *NPN*. (*b*) *PNP*. (*c*) Triode.

current in the base-emitter diode, the direction for a positively charged hole current. The similarity of a vacuum triode and a transistor is in the control of a current by a voltage. In the triode, the current from the plate to the cathode is controlled by the voltage between the grid and cathode. In a transistor, a similar action occurs: the current between the collector and emitter is controlled by the voltage between the base and emitter. A simple mnemonic device for remembering the direction of the emitter arrow is this: the arrow is *Not Pointing iN* on the *NPN* transistor.

The transistor is an electrical valve that works on the principle of controlling a current by a voltage. The controlling voltage is applied between the base and the emitter and controls the current between the collector and the emitter by means of the current from the base to the emitter. In the normal operation of a transistor as an amplifier, the base-emitter *P-N* junction diode is forward biased and the collector-base *P-N* junction diode is reverse biased.

The circuit symbol and voltage and current variables used in transistor circuit analysis are shown in Fig. 7-5. For this *NPN* transistor, $-V_{BE}$ produces forward bias on the emitter-base junction, and V_{CB} places reverse bias on the collector-base junction. Correct bias connections for normal operation are easily remembered, because forward bias requires the *Positive* battery connected to a *P-type* material or the *Negative* battery connected to an *N-type* terminal. Reverse bias is obtained by using the opposite battery connections. The collector is always biased in the reverse mode, and the emitter is always biased in the forward mode. This is easily seen in the figure.

The interaction of the internal currents may be explained as follows. An electron entering the emitter from its source, V_{BB}, will be conducted into the base region because of the forward bias of the base-emitter junction. An electron that enters the base may have one of two fates: (1) It may combine with a hole in the base, triggering the generation of an electron-hole pair at the external base terminal, which will maintain the charge equilibrium in the

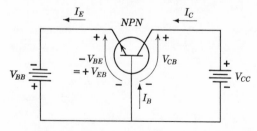

Fig. 7-5   Voltage and current symbols for *NPN* transistor. Note that V_{BE} is negative in the chosen direction; also I_E is usually taken as negative in the opposite direction.

base region. The generated electron will then pass into the external base connection and contribute to base current, I_B. (2) It may pass right on through the base and enter the collector region. Since the concentration of holes in the base can be small in the narrow region between the emitter and the collector regions, the second fate is the more likely of the two. There will be few holes in the base region to permit electron-hole recombination there. When the electron enters the collector region, the internal equilibrium is upset, and one electron must leave the external collector terminal and contribute to the collector current.

Many electrons may enter the emitter from its source. By avoiding electron-hole recombinations in the base, they will cause a continuous current in the collector-emitter circuit. The base current will be very small compared with the collector current, typically $\frac{1}{10}$ to $\frac{1}{200}$ of the collector current. The emitter current is the sum of the base and collector currents, from Kirchhoff's law, so that the collector current will be almost equal to the emitter current.

The strength of the emitter current, and, consequently, the collector current, is determined by the forward bias on the emitter-base diode. Thus, the junction voltage, V_{BE}, controls the collector current by controlling the base current. For this reason transistors are often called *current amplifiers* and are said to be current-operated devices, although the junction voltage determines base current.

The operation of a *PNP* transistor is similar to that described for the *NPN* type. The only difference is the opposite connection of external sources to provide the correct bias voltages.

7.5　Common-Terminal Connections

There are three possible arrangements in which transistors can be connected in useful circuits. These are shown in Fig. 7-6. These have *common-emitter*, *common-base*, and *common-collector* connections, respectively, depending on which transistor terminal is connected to both bias supplies. Each of these circuit connections has specific advantages in some applications, but the common-emitter connection is by far the most widely used, partly because it is capable of more power gain than the others.

Correct bias connections for normal operation are easily remembered because polarity of the emitter, with respect to the base, *is determined first:* negative if the emitter semiconductor is *N*-type, and positive if it is *P*-type. The base is connected to the other side of the bias supply, either directly or through a signal source. Also, it should be easy to remember that the collector

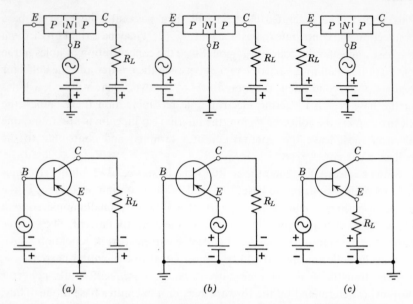

Fig. 7-6 Physical connections and schematic diagrams of *PNP* transistors. (*a*) Common-emitter; input signal applied in base circuit. (*b*) Common-base; input signal applied in emitter circuit. (*c*) Common-collector; input signal applied in base circuit.

bias is always opposite in polarity to the emitter bias, because one of the internal *P-N* junctions must be forward biased, and the other reverse biased.

The emitter-base junction is forward biased so that it offers *low impedance* to input signal currents which may then be sufficiently large for transistor operation. Because of the relationship of the junction voltage and current in a forward-biased diode, the emitter-base voltage will change very little for a wide variation in the junction current. Thus, the *current* in the emitter-base junction *effectively* is in control of the collector current. Note that this kind of performance is similar to grid-control action in a vacuum tube. In a transistor, however, *currents* and *current changes* in the emitter-base circuit will vary the current in the collector circuit, whereas in a vacuum tube *grid-voltage changes* vary the plate current. This means that a transistor is essentially a *current-operated device*. Many respected authors prefer to call a transistor a voltage-controlled device, and others prefer to think of it as a current-operated device. Both are correct descriptive terms, of course. It should be clear that current cannot operate a transistor unless a controlling voltage is properly applied.

We shall find that the *amplification factor* of a transistor is expressed in terms of a *current ratio*, instead of a voltage ratio as in vacuum-tube amplifiers. Voltage amplification may also occur from input to output in these circuits.

Referring to Fig. 7-6, let us first note that a *P-N-P* transistor is shown. An *N-P-N* transistor would serve just as well for illustration, but in that case the bias-voltage polarities would be reversed. Both types have wide use in applications, serving equally well in some cases and one or the other having some small advantage in others.

The emitter is always given forward bias, for reasons explained above, so it is *positive* when the emitter is made of *P*-type material. The collector is then biased negatively. The load resistor R_L is, of course, in the output circuit. Its voltage drop, its current, or its power is the output quantity desired. The a-c component of either the voltage across R_L or the current through it is expected to be an enlarged reproduction of the input-signal voltage or current.

Note the arrow on the emitter electrode in each of the transistor symbols. Note also that the d-c component of emitter current will flow from E toward B, i.e., from emitter into base. The arrow is drawn in the conventional direction of current flow (not electron flow) from plus to minus and consequently the symbols in the three figures represent a *P-N-P* transistor. Without the arrow there would be no way of telling from the symbol which type of transistor is represented, unless d-c bias polarities were given. The d-c bias polarities are not readily determined in the circuit diagrams of many transistor applications. We should surmise then (and we would be correct) that the arrow on the emitter electrode *points away from the base* in the symbol for an *N-P-N* transistor. These facts concerning transistor bias polarities and symbols should be committed to memory at once. A convenient way to identify the type of transistor represented by a symbol is to look at the emitter terminal. If the arrow points *toward the base*, it means that bias current is flowing from the outside *into* the emitter and from there toward the base and collector. This is what happens when the emitter is biased *positively*, and the transistor must therefore be the *P-N-P* type.

If the arrow in the transistor symbol points *toward the emitter terminal* (i.e., away from the base) the emitter bias must be made *negative*. This is the polarity for an *N-P-N* transistor.

7.6 Transistor Operation

The transistor symbol in schematic circuit diagrams is usually oriented as shown in Fig. 7-6, with the line representing the base vertical and its connection to the left, and with the emitter in the lower right region. Oriented in this way the transistor operation can be described in the same way as the operation of a vacuum triode. The emitter terminal is like the cathode of a tube; both are emitters of current carriers. The base current is similar in action to the grid voltage of the tube, and the collector behaves in much the same way as the plate of a tube. The input circuit usually connects to the base as the input circuit connects to the grid of a vacuum triode, and the other side of the input circuit goes to the emitter in the same manner as it goes to the cathode of the tube. The load is connected between the collector and emitter in the common-emitter circuit, as it is between the plate and cathode of a tube. This connection is sometimes referred to as a *grounded emitter*.

An important fact concerning the relative amounts of impurity in the emitter and base materials should be presented. The impurity concentration should be much larger in the emitter than in the base. To understand why, consider the *P-N-P* transistor (Fig. 7-6). It is desired that the collector, which is biased negatively, receive as many *holes* as possible. The emitter, which is strongly *P*-type with an abundance of free holes, will supply them and send them on their way toward the base. If the base has a much lower impurity concentration than the emitter, only relatively few of the holes will be neutralized, because of the *relative* scarcity of free electrons in the *N* material of the base. Thus the vast majority of the holes will go down the potential hill to the collector.

Another good reason for having a much smaller impurity concentration in the base than in the emitter is that electron flow *from base to emitter* can be held small. That such *reverse current* is undesirable may be gathered by observing what should happen in the circuit of Fig. 7-6a. Without a signal present, the current from *E* to *B* is due entirely to *holes* leaving the emitter. We need to remind ourselves that in a *P-N-P* transistor the emitter emits *holes* not *electrons*. Now, a signal voltage applied in the base-emitter circuit should alter the rate at which holes pass on through the *N* material, separating the emitter and collector, and then to the collector. For faithful reproduction of the wave form of the signal voltage, the collector current which passes through R_L should not be altered by the neutralization of holes by electrons of reverse current flowing from the base to the emitter. The base should not have so high a concentration of impurity that excess electrons are present to neutralize holes that should pass from the emitter to the collector. It will be seen that there is some loss in the number of carriers between emission and collection in both kinds of transistors, the *P-N-P* and *N-P-N*.

Figure 7-6b shows the common-base connection and Fig. 7-6c the common-collector connection. Notice that in all cases the *common* electrode is connected to the junction between the two bias sources and grounded.

7.7   Equalization of Fermi Levels in a *P-N* Junction

In the discussion of *N*-type and *P*-type semiconductors in Chapter 2, the concept of a *Fermi level* of energies was introduced. In any kind of semiconductor at room temperature or above, whether pure or doped, *N*-type or *P*-type, the electrons have a great variety of energy values and thus occupy many energy levels. In a chosen short-time interval, some energy levels are occupied more frequently than others, as electrons gain or lose energy levels in maintaining an equilibrium condition. There is an energy value, called the Fermi level, for which there is a 50 *per cent probability of occupancy*. The Fermi level is the *most probable energy value* that will be found among the electron energies.

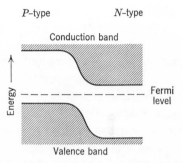

Fig. 7-7   Energy levels in a *P-N* junction in equilibrium (unbiased.)

Donor impurities in a semiconductor increase the number of electron energy levels in or very near the conduction band. This raises the level at which there is a 50 per cent probability of occupancy, i.e., the Fermi level is raised. *Acceptor* impurities, on the other hand, reduce the Fermi level in the semiconductor.

When *N*-type and *P*-type semiconductors are brought together as a *P-N* junction, the two different energy values of their separate Fermi levels will equalize at one value. This situation is illustrated in Fig. 7-7. The common Fermi level of the two types is in equilibrium across the *P-N* junction, i.e., no bias is applied to the junction. The net effect on the separate Fermi levels is to raise the energy levels of the *P*-type relative to those of the *N*-type.

7.8   Effects of Bias on Energy "Hills" at a *P-N* Junction

Consider first the equilibrium (no bias) state of the *P-N* junction. The *equalization action* resulting in a common Fermi level involves the addition of permissible higher-energy levels in the conduction band of the *P* region and the removal of some permissible lower-energy levels in that band. (Refer to the discussion of the *P-N* junction barrier region in Chapter 2.) In the *N* region, some high-energy levels have been lost and some low-energy levels added in its conduction band. Since the *N* region has an excess of electrons, *some of these electrons pass over to the P region*; likewise *holes pass from the P to the N region*. Equilibrium is established when the Fermi level has reached a common value for both regions, but the *boundary face* of the *P* region has a layer of negative charge (acceptor holes have been neutralized) and the boundary face of the *N* region has a layer of positive charge (donated electrons have left impurity ions there). Thus the *potential barrier* is formed and a *potential hill exists at the junction*.

Now consider the application of *forward bias* (Fig. 7-8a). Since the *P* side of the barrier has a *net negative charge* in the no-bias case, the application of an external positive potential to the *P* region will lower the *Fermi level* on that side of the junction. The Fermi levels try to become aligned, and this action *lowers the potential hill* across the junction with the result that electrons move more readily across the barrier from the *N* to the *P* region. It is important to note here that although the potential hill at the junction is favorable to electron passage from the *N* to the *P* region, there are *many fewer* electrons available for passage than in a conductor. A semiconductor current with forward-conduction bias is much greater than with reverse bias, for in the latter case it will be found that in order for electrons to pass through the junction they must go down a large potential hill. In doing so they lose energy.

Figure 7-8c shows the reverse-potential hill of the reverse-bias case. The effect of reverse bias is to increase the energy difference between the valence and conduction bands. Notice in Fig. 7-8c that current flow requires electrons to go down the high potential hill in crossing the junction barrier from left to

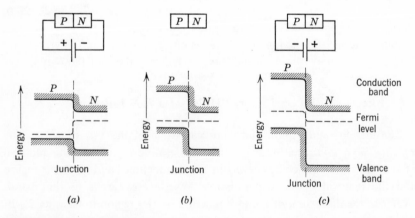

Fig. 7-8   Energy diagrams of a *P-N* junction. (*a*) Forward bias, majority-carrier conduction; (*b*) no bias, equilibrium; (*c*) reverse bias, minority-carrier conduction.

right. This means going against the electric field produced by the potential difference across the barrier, so comparatively few electrons have enough energy to make it and the current is extremely small. It should be remembered that the *electric field force* on an *electron* urges it to go *up a potential hill.* If we could release an electron at rest (no kinetic energy) in an electric field, it would start up the hill and gain velocity as a result of the force exerted on it by the field.

When reverse bias is applied, the majority carriers of both regions move away from the junction. Electrons in the *N* material move toward the positive terminal—to the right in Fig. 7-8*c*—and holes to the negative terminal. This causes a *depletion of majority carriers* on both sides of the junction, with the result that current is due to passage of *minority carriers* across the junction. Remembering that *forward bias* results in conduction by *majority carriers* (electrons from *N* type to *P* type and holes from *P* type to *N* type) and *reverse bias* results in *minority-carrier* conduction (electrons from *P* to *N* and holes from *N* to *P*) should help in understanding the mechanism of conduction across a *P-N* junction.

7.9   Characteristic Curves and Parameters

Before studying the relations among the parameters of a transistor (resistances of emitter, base, collector; current gain, etc.), we should learn the symbols by which they, and the various currents and voltages, are represented. A few are introduced now, and others will be added as needed. A statement follows each symbol to explain what it represents.

i_c Instantaneous value of a-c component of collector current

i_C Instantaneous total collector current

I_c Rms value of a-c component of collector current

I_C Average value of collector current (the d-c component)

v_{cb} Instantaneous value of a-c component of collector-to-base voltage

v_{CB} Instantaneous total collector-to-base voltage

V_{cb} Rms value of a-c component of collector-to-base voltage

V_{CB} Average value of collector-to-base voltage (the d-c component)

V_{CC} Collector-circuit supply voltage (d-c)

I_{CEO} Reverse collector-to-emitter current when base is open circuited (not connected)

Note that the lower-case *letters* are used to denote instantaneous values; upper-case *letters* denote *rms* and average values. Lower-case *subscripts* denote a-c components; upper-case *subscripts* indicate average values. Obviously, when the letters e, E, b, B are used as first-letter subscripts, the currents and voltages of emitter or base are represented. V, rather than E, is used for voltage with transistors.

Of the three different ways to connect a transistor in a circuit, the common-emitter and the common-base connections are the most frequently used. There are various groups of characteristic curves showing relationships between two quantities, for example, collector current and collector-to-base voltage (I_C vs. V_{CB}) for various constant values of a third quantity (I_E, emitter current), just as there are characteristic curves for a triode vacuum tube. The tube, of course, almost always has the input signal applied to the grid-cathode terminals. It is seen that the transistor input signal may be applied between base and emitter terminals (through a bias supply, of course) with either terminal grounded, and also between base and collector with the collector grounded. Observe, also, that the load resistor separates the third electrode from the grounded terminal in every case.

7.10   Some Important Characteristics and Maximum Ratings

Transistor manufacturers supply data sheets containing operating characteristics and maximum ratings that apply to a particular transistor. The information will usually include a description of the device, followed by sections on environmental tests, mechanical data (type of case, size, shape, etc.), absolute ratings, electrical characteristics, and typical characteristic curves. Application data and parameter information are usually included.

Before inserting a transistor in a circuit design, it is necessary to determine its ability to meet the specifications as to gain, power levels, etc., and to avoid using it where its absolute maximum ratings may be exceeded. These ratings, beyond which degradation or destruction of a transistor may occur, are established by the manufacturer. They are based on the semiconductor material, the manufacturing process, and internal physical construction. Since they represent the extreme capabilities of a transistor, they are not recommended for design conditions. A transistor will not necessarily withstand all maximum ratings simultaneously, so that care is necessary in the design of transistor circuits. Transistors in a design that exceeds a maximum rating are more susceptible to destruction than are vacuum tubes.

The section of a data sheet that lists electrical characteristics will be referred to most often. The limits to the electrical parameters that are important in a circuit design will be found there. Reverse-breakdown voltage, reverse-current values, internal capacitances, and amplification factors are among the parameters. Several of these are not important in vacuum-tube operation, but because of the diode construction of transistors, several diode parameters become important in their application.

As an example of the information supplied by a transistor manufacturer, the following is given for a Texas Instruments type 2N1302 Alloy-Junction Germanium Transistor:

Absolute Maximum Ratings (25°C)

Collector-base voltage	25 V
Emitter-base voltage	25 V
Collector current	300 mA
Total dissipation*	150 mW

Typical Electrical Characteristics (25°C)

I_{CBO}	Collector reverse current $(V_{CB} = 25$ V, $I_E = 0)$	3 μA
BV_{CBO}	Collector-base breakdown voltage $(I_C = 100\ \mu$A)	25 V
h_{FE}	DC Forward current ratio (in common-emitter circuit; ratio of collector and base currents)	50

This list is not exhaustive but is given as an example of the information supplied for a transistor. Neither is this type of transistor available only from Texas Instruments. The type number 2N1302 indicates that this transistor has been registered with the Joint Electron Device Engineering Council (JEDEC). The purpose of registration is to facilitate the purchase and distribution of semiconductor devices and to standardize electronic devices. Registration procedures are designed to ensure that devices which differ from one another in their characteristics and performance are identified by different type numbers. Type numbers are assigned in numerical sequence as they are requested and approved. The numbers 1Nxxxx usually denote diodes or rectifiers, 2Nxxxx denote triode devices, and 3Nxxxx denote four-terminal (tetrode) construction. Any manufacturer that makes transistors that have the same characteristics as this one will number his product 2N1302.

---

* Derate 2.5 mW/°C above 25°C ambient.

7.11 Characteristic Data and Curves; Alpha and Beta Factors

The behavior of a transistor may be studied by looking at the characteristic data and curves supplied by the manufacturer. Figure 7-9 shows a set of static curves for the 2N334 transistor, a grown-junction silicon *NPN* type. These curves describe the operation of the transistor in a common-base connection. These *output characteristics* show how the collector current varies with the output voltage between the collector and base terminals, for various values of emitter current. Again we are reminded that the transistor is a current-operated device.

For a constant emitter current, the collector current is practically unaffected by a substantial change in collector-to-base voltage. However, the *ratio* of a change in I_C to a change in I_E (at a constant V_{CB}) is an important factor. When this ratio is expressed as a *positive number*, it is called the *forward current gain* of the common-base connection and is designated by the Greek letter *alpha* (α).

$$\alpha = -\frac{\text{change in } I_C}{\text{change in } I_E \; V_{CB} \text{ constant}} = -\frac{\Delta I_C}{\Delta I_E \; V_{CB} \text{ constant}} \tag{7-1}$$

We may get an idea of the magnitude of this factor by calculating its value

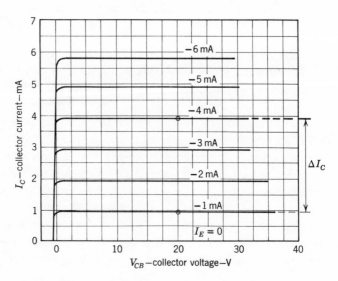

Fig. 7-9 Common-base output characteristics of 2N334 transistor. (Courtesy Texas Instruments, Inc., Dallas, Tex.)

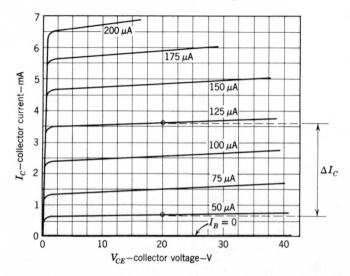

Fig. 7-10   Common-emitter output characteristics for 2N334 transistor. (Courtesy Texas Instruments, Inc., Dallas, Tex.)

from the curves. At a constant collector-base voltage of 20 V, a change in I_E from -1 mA to -4 mA ($\Delta I_E = 3$ mA) causes a change in the collector current from 1 mA to 3.9 mA ($\Delta I_C = -2.9$ mA). The value of *alpha* is $-(-2.9)/3$, or 0.97. It is important to note that *alpha* is always less than unity, and it usually is greater than 0.9. The difference between the value of *alpha* and unity represents the magnitude of base current in a transistor, because the emitter current is the sum of base current and collector current. In general, the closer *alpha* is to unity, the better the transistor quality.

It should also be noted that the factor *alpha* applies only to the common-base connection of the transistor. The current gain for other common-terminal connections is defined in a different way.

The common-emitter operation of a transistor may be described by a set of output characteristic curves as in Fig. 7-10. Notice that the collector current changes very little for substantial changes in collector-to-emitter voltage. The *ratio* of changes in collector current to changes in base current is the current-gain factor for the common-emitter connection. It is denoted by the Greek letter *beta* (β) and is defined as

$$\beta = \frac{\Delta I_C}{\Delta I_B} \quad V_{CE} \text{ constant} \tag{7-2}$$

The base current is much smaller than the emitter current or collector current, because almost all of the emitter current passes to the collector, i.e.,

$$I_E - I_C = I_B \tag{7-3}$$

Connected in the common-emitter circuit, the base is on the input side, and the input signal will vary the base current. A corresponding change will take place in the collector current. The curves may be used to calculate a value for β:

$$\beta = \frac{(3.2 - 0.7)\,\text{mA}}{(125 - 50)\,\mu\text{A}} = \frac{2.5}{0.075} = 33.3$$

It can be seen that this current-gain factor is closely related to the spacing between constant base-current lines. The farther apart they are, the larger will be the value of β.

The two factors alpha and beta are closely related. One may be calculated from the other. From Equation (7-3) we obtain

$$\Delta I_E - \Delta I_C = \Delta I_B$$

so that

$$\beta = \frac{\Delta I_C}{\Delta I_E - \Delta I_C} \tag{7-4}$$

The factor β can be expressed in terms of α by recognizing that the numerator and denominator may be divided by ΔI_E,

$$\beta = \frac{\Delta I_C/\Delta I_E}{\Delta I_E/\Delta I_E - \Delta I_C/\Delta I_E}$$

$$\beta = \frac{\alpha}{1 - \alpha} \tag{7-5}$$

Similarly, α may be expressed in terms of β by solving Equation (7-5) for α:

$$\alpha = \frac{\beta}{1 + \beta} \tag{7-6}$$

7.12   Circuit Equations

The input and output characteristic curves for operation of a type 2N334 silicon transistor in a common-emitter circuit are given in the Appendix. This *NPN* transistor is typical of general-purpose types. Its characteritic

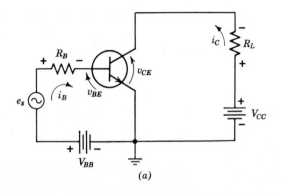

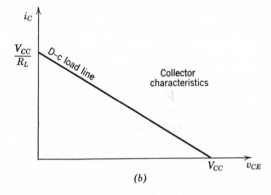

Fig. 7-11 Amplifier circuit using *NPN* transistor and graph of load line on collector characteristics. (*a*) Circuit. (*b*) Load line location.

curves will be used to set up mathematical relations between the voltage and current quantities in the circuit of Fig. 7-11.

In order to analyze the circuit operation, we shall assume that R_L is 6 kΩ and the voltage sources, V_{CC} and V_{BB}, are 30 V and 5 V, respectively. These are typical values in this type of circuit. An examination of the output characteristic curves ($V_{CE} - i_C$) shows that changes in base current will cause simultaneous changes in collector voltage and current. Since the changes in collector current exist in the load (R_L) the voltage developed across the load will change.

When collector current is zero, no voltage will be developed across R_L and the collector voltage will equal the source voltage V_{CC}. When collector current is flowing, conditions in the collector circuit may be expressed as

$$V_{CC} - v_{CE} = i_C R_L$$

This will be recognized as a statement of Kirchhoff's voltage law for the collector circuit. The expression may be rearranged in the form of a straight-line equation in terms of the two variables, i_C and v_{CE}

$$i_C = v_{CE}\left(-\frac{1}{R_L}\right) + \frac{V_{CC}}{R_L} \qquad (7\text{-}7)$$

Equation (7-7) is called the "D-C load line equation" for this circuit. It may be drawn on the output characteristic curves between $v_{CE} = V_{CC}$ (when $i_C = 0$) and $i_C = V_{CC}/R_L$ (when $v_{CE} = 0$). The intercepts on the axes in this case are 30 V and 5 mA. This load line represents all possible combinations of collector voltage and current that may exist in this particular circuit. Circuits containing another type of transistor or other bias sources and load resistors will have their own unique d-c load lines. The use of this concept in circuit analysis will be explained in much more detail in subsequent chapters.

From the input base-emitter curve for the 2N334 transistor given in the Appendix it can be seen that the 5-V V_{BB} source is much too large for proper operation of the circuit. It can be seen from the curve that the base voltage V_{BE}, should be in the range from about 0.5 to 0.7 V. This requires a voltage-dropping resistor, R_B, in the base circuit to limit the base-emitter voltage to this range. What should be the value of R_B? Let us assume that the circuit is to operate about a nominal base current of 100 μA. The resistance value may be calculated

$$R_B = \frac{V_{BB} - V_{BE}}{I_B}$$

For this circuit the value will be, taking the base-emitter voltage at the center of its range,

$$R_B = \frac{(5.0 - 0.6)\ \text{V}}{100 \times 10^{-6}\ \text{A}} = 44\ \text{k}\Omega$$

It will be noted that an *increase* in base current will cause a *decrease* in collector voltage, resulting in the characteristic 180° phase shift of the common-emitter circuit.

DRILL PROBLEMS

D7-3 Suppose it is desired to use a single bias source for both the collector and base circuits. The source V_{CC} can be used provided the base dropping resistor is properly chosen. Sketch a circuit to show the connections that are made to use V_{CC} for biasing the base circuit, and calculate the value of R_B necessary to hold static base current at 100 μA.

D7-4 Determine the d-c load lines for the circuit of Fig. 7-11 using a 2N334 transistor and resistor loads of 2 kΩ and 5 kΩ. Assume V_{CC} remains constant at 30 V.

D7-5 In D7-4, assume the static base current is 100 μA. For each of the load resistors, calculate or show the maximum amplitude of base current variations that may occur without distorting the output voltage. Assume the base current varies sinusoidally.

7.13 Amplifying Action in a Transistor

The performance of a transistor in a simple amplifier circuit is similar to that of a triode vacuum tube. A small varying potential in the input circuit alters the current entering the transistor, and the variations in current are reproduced as larger current variations in the output. The output current variations may produce voltage changes in a load resistor. This means that *voltage* amplification can occur, as well as *current* amplification.

The circuit of Fig. 7-11 will be used to describe the amplification in a common-emitter amplifier. The circuit is assumed to operate under these conditions:

Collector supply voltage = 30 V
Load resistance = 6 kΩ
Base current variation = 150 to 50 μA

The corresponding limits of collector current, obtained from a load line, will be 4.7 mA and 0.7 mA, with an amplitude of $(4.7 - 0.7)/2 = 2.0$ mA. The base current amplitude is $(150 - 50)/2 = 50$ μA.

The current gain is $2.0/0.05 = 40$. The voltage gain is calculated by dividing the amplitude of the load voltage by the input signal amplitude. The load voltage is $-0.002 \times 6,000 = 12$ V. The input signal voltage may be calculated in terms of base current and the other parameters of the base circuit:

$$e_s = v_{BE} + i_B R_B = 0.27 + 50 \times 10^{-6} \times 44 \times 10^3 = 2.47 \text{ V}$$

The voltage amplification is

$$\frac{12.00}{2.47} = 4.85$$

Power amplification, or power gain, is then $40.00 \times 4.85 = 194$.

There is a voltage phase shift through this amplifier. When the signal voltage increases, it increases the base current and hence decreases collector current. The output load voltage goes into its negative half-cycle when input signal is in its positive half-cycle. Conversely, decreases in signal voltage will increase the output, generating an output that is 180° out of phase with the input.

The concepts of amplification and phase shift in other circuits are presented in detail in Chapters 8 and 9.

7.14   Junction Field-Effect Transistor

The *unipolar field-effect transistor* (*FET*) is a relatively new addition to the many semiconductor devices manufactured for use in electronic circuits. Two types have been constructed to date (1) the junction *FET*, and (2) the insulated-gate *FET*.* The former controls the conductivity of its "channel" by means of the reversed bias of its *P-N* junctions. The latter controls the conductivity of the channel by means of an electric field impressed through an insulated metal electrode. Only the former will be discussed here. The *FET* was first proposed as a device by W. Shockley in 1952† but has only recently become a part of the commercial market.

The block of *P*-type semiconductor with two *N*-type regions introduced into opposite sides of it (shown in Fig. 7-12) is representative of a field-effect transistor. The output current of the *FET* flows between the terminals marked *source* and *drain*. The conductance of the current path is controlled by applying an electric field perpendicular to the current path, between the *source* and *gate* terminals. The *FET* is called "unipolar" because current flow is due to only one type of charge. The usual transistor action in junction

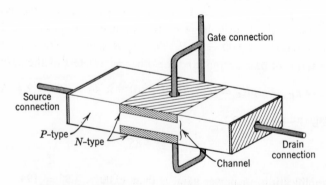

Fig. 7-12   Basic structure of field effect transistor.

* Also called the MOS or metal-oxide semiconductor version of the *FET*.
† W. Shockley, "A Unipolar Field-Effect Transistor," *Proc. I.R.E.*, **40,** 1365–76 (1952).

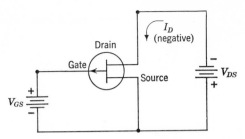

Fig. 7-13 Symbol and bias polarities for *P*-channel *FET*.

types is bipolar, depending on both holes and electrons as charge carriers. The terminals labeled drain, gate, and source, respectively, are analogous to the collector, base, and emitter of a junction transistor and to the plate, grid, and cathode, respectively, of a vacuum triode.

The part of the semiconductor bar between the two *P-N* junctions (at the gate) is called the *channel*. It is the active part of the *FET*, because its height controls the amount of current passing through the device. The space-charge layer that builds up in the channel, because of a potential on the gate terminal, is the controlling factor in *FET* operation.

The schematic symbol for the *FET* is given in Fig. 7-13. It will be noted that the drain and source terminals are interchangeable. However, in a circuit application the gate is biased to whichever is chosen as the source. The correct polarities for external biasing are shown for a *P-channel FET*. All voltages and currents, and the arrow on the gate, are reversed for an *N-channel FET*.

7.15 Static Characteristic of *FET*

The *FET*, like other three-terminal devices, can be operated with any one of its terminals common to the biasing sources. For the purpose of illustrating the characteristics of a *P*-channel *FET*, the characteristic curves for a common-source connection are shown in Fig. 7-14. This is analogous to common-emitter and common-cathode connections of junction transistors and vacuum triodes, respectively.

These characteristics for the *FET* closely resemble those for the vacuum pentode. The polarities of voltages and currents are just the opposite of those of the vacuum-tube "look-alike." In fact, the *P*-channel *FET* has been facetiously called "*PNP* vacuum tube" because of this resemblance. The polarities are reversed for an *N*-channel *FET*, of course.

Three regions of the curves are of interest. The normal operating region is

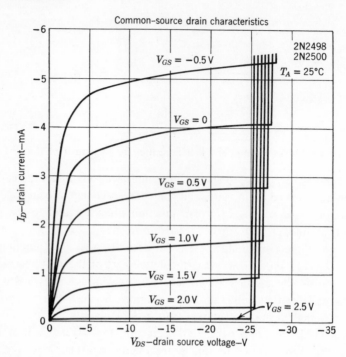

Fig. 7-14   Common-source characteristics for *P*-channel *FET*. (Courtesy Texas Instruments, Inc., Dallas, Tex.)

in the center where the curves are essentially flat, representing the performance of a nearly constant-current source between the drain and source. The region of breakdown at the extreme right should be avoided in normal circuit operation of the *FET*. Similarly, the region of nonlinearity at the left is not generally used.

The shape of the curves may be explained. A certain value of gate-source voltage, V_{GS}, will set up an electric field in the channel and thereby permit a certain current, I_D, to get through. This is shown by the nearly constant-current portions of the curves. When drain-source voltages (V_{DS}) are less than gate-source voltages (V_{GS}) the device behaves much as a triode structure. At the other extreme, within the breakdown region, the reverse-voltage breakdown of the gate-to-drain diode shows up where drain current increases rapidly.

Electrical parameters describe a particular junction *FET*. The following are the important parameters of the 2N2498, whose characteristics are given in Fig. 7-14:

Absolute maximum ratings (25°C free-air temperature)

Gate current   10 mA
Total device dissipation
    25°C free-air temperature   0.5 W*
    25°C case temperature   1.5 W**

Electrical characteristics (25°C free-air temperature)

BV_{DGO} Drain-gate breakdown voltage (source open) -20 V
I_{GSS} Gate cutoff current ($V_{GS} = 10$ V, $V_{DS} = 0$)   0.01 μA
$I_{D(on)}$ Zero-gate-voltage drain current -2 to -6 mA
    ($V_{DS} = -10$ V, $V_{GS} = 0$)
$I_{D(off)}$ Pinch-off drain current -10 μA
    ($V_{DS} = -15$ V, $V_{GS} = 6$ V)

7.16   Basic Scheme for Biasing the *FET*

Figure 7-15 is the basic common-source circuit for an amplifier using an
FET. The characteristics of an *FET* vary over a wide range with temperature
variations and from one device to another of the same type. For this reason,
and to eliminate the cost of an extra power supply, some form of *self-bias* is
usually employed in practical circuits. The necessary reverse voltage between
the gate and source may be obtained from the drain current passing through
the $R_S - C_S$ network in Fig. 7-15. This circuit operates in the same way
as the cathode-bias network described in vacuum-tube circuits (Chapter 6).

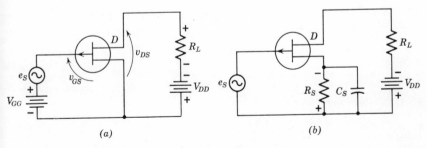

(a)   (b)

Fig. 7-15 Basic circuit and self-bias arrangement in common-source *FET*
amplifier. (*a*) Basic circuit. (*b*) Self-bias.

  * Derate linearly to 175°C by 3.3 mW/°C
 ** Derate linearly to 175°C by 10 mW/°C

DRILL PROBLEMS

D7-6 Using the 2N2498 *FET* whose characteristics are given in Fig. 7-14, it is desired to operate a circuit at the point $V_{DS} = -15\,\text{V}$, $I_D = -1.5\,\text{mA}$. What should be the values of R_L, V_{DD}, V_{GS}?

D7-7 What magnitude of e_s will essentially cut off drain current?

D7-8 For e_s having maximum amplitude of 0.5 V, what will be the variation in V_{DS}?

D7-9 To use self-bias in the circuit when the lowest frequency of variation is 100 Hz, what values of R_S, C_S will be needed?

SUGGESTED REFERENCES

1. Marlin P. Ristenbatt and Robert L. Riddle, *Transistor Physics and Circuits* 2nd ed., Prentice-Hall, Englewood Cliffs, N.J., 1966.

2. Paul D. Ankrum, *Principles and Applications of Electron Devices*, International Textbook Co., Scranton, Pa., 1959.

3. Douglas M. Warschauer, *Semiconductors and Transistors*, McGraw-Hill, New York, 1959.

4. A. Coblenz and H. L. Owens, *Transistors*, McGraw-Hill, New York, 1955.

5. Leonce J. Sevin, Jr., *Field-Effect Transistors*, McGraw-Hill, New York, 1965.

QUESTIONS

7-1 What is the origin of the word "transistor"?

7-2 Describe the structure of the point-contact transistor. In what respects is it inferior to the junction transistor? What are the reasons for these shortcomings?

7-3 What are the basic elements of a junction transistor? What are the approximate dimensions?

7-4 Draw from memory the symbol of the *PNP* and *NPN* transistors as they are represented in schematic diagrams.

7-5 Draw from memory an *NPN* transistor in a common-base circuit. Show bias voltages and label all parts.

7-6 Draw from memory a *PNP* transistor in a common-emitter circuit. Show bias voltages and label all parts.

7-7 With what polarity is the emitter normally biased? Why?

7-8 Why is a transistor classified as a *current-operated* device?

7-9 Compare the control action in a junction transistor with the control action in a vacuum triode.

7-10 Why is the impurity concentration made much greater in the emitter than in the base?

7-11 What is the advantage of having a very low impurity concentration in the base?

7-12 What is emitted by the emitter of a *PNP* transistor? Answer also for *NPN* transistor.

7-13 In what units would you expect a Fermi level to be expressed?

7-14 What is meant by the statement that the Fermi level is the "most probable energy value" that will be found among the energies of electrons in a crystal?

7-15 What is meant by the statement that the Fermi level has a "fifty per cent probability of occupancy"?

7-16 What is the effect on the Fermi level when a *donor* impurity is added to a semiconductor?

7-17 What is the effect on the Fermi level when an *acceptor* impurity is added to a semiconductor?

7-18 What is an energy hill at a *P-N* junction?

7-19 Does forward bias across a *P-N* junction produce conduction by majority carriers or by minority carriers? Which way do they go across the junction?

7-20 Answer the preceding question for reverse-bias conditions.

7-21 In what direction do majority carriers in both regions move when reverse bias is applied?

7-22 Referring to the partial list of symbols in Section 7.9, tell what the following represent: v_{ce}, V_{CE}, I_e, i_E, v_{CE}, I_B.

7-23 Define current gain in a common-base transistor amplifier. What is the symbol?

7-24 Why can α never be greater than unity in a junction transistor? In fact, why can it not quite reach unity?

7-25 Define current gain in a common-emitter transistor amplifier.

7-26 Why does β have a large value compared with α?

7-27 Explain in a general way how amplification takes place through a transistor.

7-28 What can you say about the impedances of emitter and collector circuits of transistor amplifiers?

7-29 Explain the phase relationship between the output and input currents and voltages of the common-base amplifier. Do the same for the common-emitter amplifier.

7-30   Which form of transistor amplifier circuit has the largest power gain? Why?

7-31   Why is the field-effect transistor called a "unipolar" device? How is its behavior different from that of a junction transistor?

7-32   What is the significance of the direction of the gate arrow on the *FET* symbol?

7-33   Why is normal operation of the *FET* limited to the center region of its characteristics?

PROBLEMS

7-1   The Fermi level of intrinsic (pure) germanium has an energy value of 0.36 eV above the valence-band energy. Compute the relative linear velocity of an electron at this level, using $KE = \frac{1}{2}mv^2$.

7-2   *N*-type germanium with an impurity density of 10^{16} arsenic atoms per cubic centimeter has a Fermi level of 0.256 J below the conduction band. How many volts of energy must an electron at the Fermi level receive to be elevated to the conduction band? (Volts × coulombs = joules.)

7-3   A potential hill of -3 V lies in the path of an electron moving toward it at a velocity of 10^6 m/s. Can it go all the way down the hill without first receiving more energy? If not, how much more energy would it need, and how many electronvolts of energy does this represent?

7-4   Choose a constant collector-to-base voltage of 5 V on the common-base characteristic curves for the 2N334 transistor (see Appendix) and determine the value of α. Is it constant for (*a*) all collector current, (*b*) all collector-base voltages?

7-5   On the common-emitter curves for the 2N334 transistor (see Appendix), choose a collector-to-emitter voltage of 5 V and determine β. Is it constant for(*a*) all collector current, (*b*) all collector-emitter voltages?

7-6   Show that α and β are related by Equations (7-5) and (7-6). Calculate α in terms of β starting with Equation (7-1).

7-7   The output characteristics of a transistor that is used in a common-emitter amplifier can be approximated by horizontal lines with $i_C = 50i_B$. Sketch and dimension a set of these curves for i_B in increments of 50 μA from 0 to 250 μA. (*a*) The circuit uses a supply voltage $V_{CC} = 20$ V and $R_L = 2$ kΩ. Construct a load line for the circuit on the characteristic curves. (*b*) Determine the static collector current and collector-emitter voltage necessary to operate the circuit at the mid-point of its load line. Estimate the value of base dropping resistor necessary to operate at this point.

7-8   The maximum permissible collector dissipation for the transistor of Problem

7-7 is given as 50 mW. Is this value exceeded in the circuit of Problem 7-7? (The collector dissipation is the product of collector current and collector voltage.)

7-9 It is of interest to know how the circuit of Problem 7-7 will behave if its supply voltage should decrease to 10 V. For this purpose, determine the effect of decreasing V_{CC} to 10 V, with other components in the circuit remaining the same.

7-10 An *NPN* silicon transistor is assumed to have linear output characteristics and a current amplification factor $\beta = 80$. It is used in the common-emitter connection and the circuit uses only one bias supply. The supply voltage $V_{CC} = 20$ V, $R_L = 3.3$ kΩ, and $R_B = 330$ kΩ. (*a*) Determine the load line and static operating point. (*b*) If the input signal is a sinusoidal current, determine the approximate amplitude of collector current that is permissible without serious waveform distortion in R_L. Is the limit set by collector current cutoff or saturation of the collector?

7-11 For the common-collector configuration, the short-circuit current gain may be approximated by β/α. Find an expression for this ratio in terms of (*a*) α only, (*b*) β only.

7-12 Using Fig. P7-12, sketch curves showing phase relations among e_s, i_B, i_C, and v_{CE}, for $e_s = E_m \sin \omega t$. What is the relative phase shift between e_s and v_{CE}?

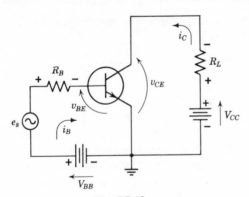

Fig. P7-12

7-13 A common-emitter amplifier using a 2N334 transistor has $V_{CC} = 30$ V and collector load resistor $R_L = 6$ kΩ. Plot the collector current i_C vs. base current i_B. Over what range can the curve be considered linear? What value of bias current in the base circuit would be most desirable?

7-14 (*a*) In the circuit of Fig. 7-11 discussed in Section 7.13, for no signal input, what power is being dissipated by the collector? (*b*) What power is supplied by the bias source? (*c*) With the signal applied, what is the ratio of a-c power in R_L and the d-c power input from V_{CC}? This is the collector circuit efficiency.

7-15   The accompanying circuit in Fig. P7-15 uses a 2N334 transistor with $V_{CC} = 30$ V and $R_L = 6,000$ Ω. Find R_f, assuming that static operating point is approximately at the center of the load line.

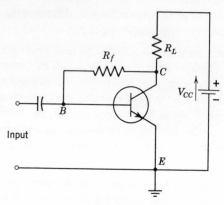

Fig. P7-15

7-16   If the effective a-c resistance of a junction is 150 Ω and its capacitance is 2.5 pF at a certain frequency, what is the time constant of the junction?

7-17   The internal potential barrier that results from the alignment of the *P*-side and *N*-side Fermi level is

$$V_0 = \frac{kT}{q_e} \ln \frac{n_e n_h}{n_i^2}$$

Compute V_0 if

$$n_e = 1 \times 10^{16} = N\text{-side doping}$$
$$n_h = 1 \times 10^{14} = P\text{-side doping}$$
$$n_i^2 = 2.25 \times 10^{26} = \text{intrinsic carrier density}$$
$$T = 300°K = \text{room temperature}$$
$$k = 1.38 \times 10^{-23} \text{ J/°K} = \text{Boltzmann's constant}$$
$$q_e = 1.602 \times 10^{-19} \text{ C} = \text{electron charge}$$

7-18   An *NPN* silicon transistor is assumed to have linear output characteristics and a current amplification factor $\beta = 100$. It is used in a common-emitter connection and uses a single bias supply for both collector and base circuits, $V_{CC} = 20$ V. The circuit is to be designed for a Q-point at $I_C = 1$ mA and $V_{CE} = 10$ V. Determine the required values for R_B and R_L to operate at this Q-point.

7-19   A *PNP* germanium transistor has a beta of 100 and a maximum permissible collector dissipation of 150 mW. Assuming it has a linear characteristic, design a common-emitter circuit that will develop a sinusoidal voltage of 5 V peak value across a load $R_L = 2$ kΩ. Use the smallest possible supply voltage V_{CC}. Determine the required value of R_B. (A sketch of the load line on i_C–v_{CE} coordinates may be helpful.)

7-20 An *NPN* silicon transistor having linear characteristics may have a value of beta that lies between 30 and 120, depending on its method of manufacture, with a typical value being 80. (*a*) With a supply voltage $V_{CC} = 20$ V and $\beta = 80$, find the values of R_B and R_L that will place the Q-point at $I_C = 2$ mA and $V_{CE} = 4$ V. (*b*) Using the resistance values of part (*a*), determine the locations of Q-point when $\beta = 30$ and $\beta = 120$. What effect on circuit operation is noted?

7-21 A 2N334 transistor is used in a common-emitter connection with $V_{CC} = 30$ V and $R_L = 6$ kΩ. The base current varies sinusoidally between 50 and 150 μA. (*a*) What is the variation of collector current? (*b*) Calculate the current gain of the circuit.

7-22 The 2N2498 *FET* whose characteristic is given in Fig. 7-14 is used in a circuit that operates at a Q-point where $V_{DS} = -10$ V, $I_D = -2.5$ mA. What should be the values of R_L, V_{DD}, V_{GS}?

7-23 Using the 2N2498 *FET* in a common-source circuit, as in Fig. 7-15*b*, determine the values of R_L, R_S, and C_S necessary to operate at the Q-point defined by $V_{DS} = -15$ V, $I_D = -1.5$ mA, when $V_{DD} = -25$ V. The lowest frequency of e_s is 159 Hz.

7-24 A 2N404 *PNP* transistor is being used in a common-emitter circuit that has $V_{CC} = -10$ V and operates at a Q-point determined by $I_B = -0.15$ mA. Determine the smallest value of R_L that may be used without exceeding the maximum dissipation rating of 150 mW.

7-25 If the circuit described in Problem 7-24 uses the smallest permissible value of R_L and a single source for biasing the collector and base circuits, calculate the required value of R_B.

7-26 Using the curves in the Appendix, calculate the value of h_{fe} at the point $V_{CE} = 2$ V, $I_B = 0.15$ mA, for each of the transistors 2N1302, 2N1304, 2N1306, and 2N1308. Compare these values with those listed for these types and discuss the differences, if any.

7-27 Repeat Problem 7-26 for the complementary types 2N1303 through 2N1309.

7-28 A 2N404 *PNP* transistor is to be used in free air at a temperature of 75°C. If it is used in a common-emitter circuit with $V_{CC} = -10$ V, what is the smallest permissible value of R_L that may be used without exceeding the dissipation rating for the device?

7-29 On a graph of common-emitter characteristics for the 2N334 transistor, plot maximum device dissipation curves for operation at 25°C, 100°C, and 150°C. Does the 2N334 have a *linear* derating factor for temperature?

7-30 The maximum ratings for a 2N334 transistor include $V_{CB} = 45$ V, $V_{BE} = 1$ V, and $I_C = 25$ mA. If this transistor is used in a common-emitter circuit in which $V_{CC} = 40$ V and $R_L = 2$ kΩ, will the device operate in excess of any of its maximum ratings? Assume operation at 25°C.

CHAPTER EIGHT

Triodes in Action—Voltage Amplification

An electronic circuit does not usually perform a useful task unless changes occur in one or more voltages or currents in the system. Varying voltages on one or more electrodes of an electronic tube or transistor can produce a control voltage or current of sufficient magnitude to perform a variety of tasks at the output of the system. The control voltage or current may be needed to operate a switching circuit, stabilize another voltage or current, produce sound or music, produce an image on a television screen, run a motor with precise control, vary the illumination in a room, or do many other tasks.

In this chapter we shall analyze the performance of the triode vacuum tube and transistor in simple amplifier circuits and become familiar with circuit terminology. Voltage amplification is explained in detail through the use of load lines and equivalent circuits.

8.1 Actual and Equivalent Circuits of Vacuum-Tube Amplifier

The circuit of a simple amplifier is shown in Fig. 8-1a. The heater circuit for the cathode has been omitted, since it serves merely to provide thermionic emission at the cathode and does not enter otherwise into the circuit operation.

The grid-circuit resistor, R_g, usually has a high ohmic value, perhaps 500,000 Ω, and serves to accept the signal voltage and to connect the negative-bias voltage to the grid. The varying input voltage, E_1, is placed on R_g and so is "superimposed" upon the steady d-c grid-bias voltage supply

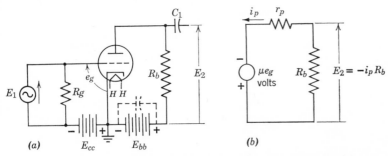

Fig. 8-1   (*a*) Simple amplifier circuit. (*b*) Its equivalent circuit.

E_{cc}. Thus E_1 causes the grid to be, alternately, first less, and then more negative with respect to the cathode than the d-c bias-voltage value. It is very important that no grid current should flow, and this will be the case if the grid is not allowed to go positive with respect to the cathode.

If grid currents were to flow in a small-signal voltage-amplifier circuit, several detrimental effects would result. The grid of a vacuum-tube voltage amplifier is not designed to carry more than an extremely small current. Such grids are easily damaged. When grid current flows, the a-c component is supplied by the input circuit, which usually cannot function properly when "loaded" in this way. Distortion of the input-voltage waveform results; this means that the output voltage of the amplifier will be distorted.

Although the a-c component of the grid-to-cathode voltage, e_g, is across R_g and equal to E_1, there is no a-c voltage loss in the bias battery, so it is correct to consider that e_g exists at the grid-cathode terminals of the tube.

The plate-load resistor, R_b carries the plate current at all times. Since the plate current will vary because of variations in grid-to-cathode voltage, E_2 will be a varying voltage which will have a larger amplitude than E_1.

Figure 8-1*b* shows how the actual circuit may be depicted by a simple series circuit in which the tube is represented by a generator having an induced voltage equal to μe_g. It is important to observe that *no d-c voltages appear in the equivalent circuit*. Only *varying voltages* are used in working with the equivalent circuit.

The *coupling capacitor* C_1 is large enough to provide negligible reactance at the operating frequencies. Furthermore, the a-c voltage drop across the power supply battery is also negligible. A bypass capacitor across E_{bb} could be used to ensure zero a-c voltage drop, if necessary. This means that the a-c output voltage across R_b is available where E_2 is shown. It has the same value between the right-hand plate of C_1 and ground.

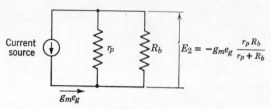

Fig. 8-2   Circuit equivalent to Fig. 8-1*a*; constant-current source.

The equivalent circuit, Fig. 8-1*b*, is derived by applying Thévenin's theorem. The open-circuit a-c voltage (R_b disconnected) is the plate-cathode voltage e_b which is equal to $-\mu e_g$. The impedance looking back toward the tube, with R_b disconnected, is the internal a-c resistance of the tube, denoted by r_p. The polarity of the equivalent generator is as indicated, because i_p will be in its *positive* half-cycle and flowing *into the plate* while e_g is in its *positive* half-cycle since the two are in phase. It is worth mentioning here that, when e_g is in its *negative* half-cycle, i_p will also be in its negative half-cycle *even though current is still flowing into the plate* because i_b *is always positive.* The plate current has smaller instantaneous values during the negative half-cycle of i_p than it has during the positive half-cycles. Reexamine the curves of Fig. 6-12.

Another type of equivalent circuit for Fig. 8-1*a* may also be set up with the help of Norton's theorem. It employs a *constant-current* source. The current is given by $g_m e_g$.

A short circuit across R_b will carry a-c current limited only by r_p:

$$i_p = \frac{\mu e_g}{r_p} = g_m e_g$$

Looking back toward the tube and determining the impedance to be put in parallel with the load, as required by Norton's theorem, we see only r_p. The equivalent circuit, using a constant-current source shown in Fig. 8-2, is more often used in connection with pentode tubes than with triodes, because its plate resistance is usually much larger than any practical R_b used with it. The pentode, it will be remembered, acts almost as a constant-current source.

8.2 The Load Line

Consider the actual circuit, Fig. 8-1*a*, and assume that the voltage E_1 is applied. This naturally results in variations in plate current. The

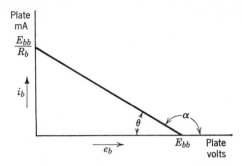

Fig. 8-3   Load line of circuit of Fig. 8-1.

instantaneous value of total plate current is i_b, and of plate-to-cathode voltage is e_b. It is readily seen that

$$E_{bb} = i_b R_b + e_b \tag{8-1}$$

The variables in this equation are i_b and e_b, and it is an equation of a straight line, since it contains no powers higher than the first.

Figure 8-3 shows the straight line represented by Equation (8-1). Note that, when $i_b = 0$, $e_b = E_{bb}$; and when $e_b = 0$, $i_b = E_{bb}/R_b$. The tangent of the smaller angle (θ) that the line makes with the voltage axis is

$$\tan \theta = \frac{E_{bb}}{R_b} \div E_{bb} = \frac{1}{R_b} \tag{8-2}$$

The tangent of the larger angle (α) is $-1/R_b$, and this is precisely the slope of the straight line represented by Equation (8-1). In the analysis of vacuum-tube circuits, the load line is an important instrument. It may readily be drawn on the plate-characteristic-curve chart. In this case only the plate-load resistance, R_b, and plate-supply voltage need be known. That is, a straight line is drawn on the chart through the E_{bb} voltage point on the horizontal axis such that $\tan \theta = 1/R_b$, or $\tan \alpha = -1/R_b$. The load line may be drawn simply by locating one point on each axis through which the line will pass. This is illustrated in the example in Section 8.3.

8.3   Graphical Analysis of Triode-Tube Performance

Let us first examine only the right side of Fig. 8-4. A load resistor ($R_b = 40,000 \ \Omega$) is in the plate circuit, and so the slope of the load line is determined by $-1/R_b = -1/40,000$.

The load line for this circuit passes through the 320-V point on the voltage

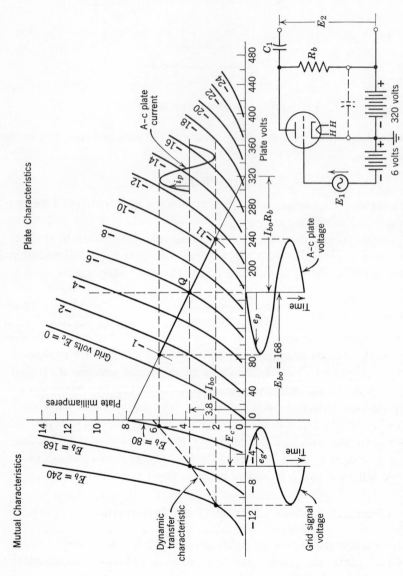

Fig. 8-4 Showing dynamic operation of triode and graphical determination of voltage amplification.

axis, and the point $e_b = 0$, $i_b = 0.008$ A   (8 mA) on the current axis, obtained from Equation (8-1).

The grid bias is $E_c = -6$ V. Observe that, if no other grid voltage than this existed, the plate current would be $I_{bo} = 3.8$ mA, and the plate-to-cathode voltage would be $E_{bo} = 168$ V. This point on the chart where the load line intersects the grid-bias voltage curve is called the *quiescent point* (*Q*). The quiescent plate current (I_{bo}), the quiescent plate-to-cathode voltage (E_{bo}), and the grid-bias (quiescent grid-to-cathode voltage) are given by the coordinates of this point. The voltage drop across the resistor is

$$I_{bo}R_b = E_{bb} - E_b = 320 - 168 = 152 \text{ V}$$

as obtained from the chart. This checks with the product of R_b and the plate current as read from the chart; thus

$$I_{bo}R_b = 0.0038 \times 40{,}000 = 152 \text{ V}$$

Now let us examine the upper left-hand portion of Fig. 8-4. The static-transfer characteristics are shown for convenient values of constant plate voltage. Referring to the curve for $E_b = 80$ V, it is easy to see how three points were obtained from the plate characteristics to use in plotting the dynamic-transfer characteristic curve. A vertical line at 80 volts would show that, when $E_c = 0$, $i_b = 8$ mA; when $E_c = -2$, $i_b = 3.3$ mA; and, when $E_c = -4$, $i_b = 0.4$ mA. That is how the static-transfer characteristics may be plotted when only the plate characteristics are available. However, data for plotting the static-transfer characteristics may be obtained directly by placing the tube in a test circuit arranged so that grid and plate voltages may be controlled. Readings are taken of plate current and grid bias at fixed values of plate voltage.

The static-transfer characteristics are useful in locating points that determine the dynamic-transfer characteristic. We know that, when the grid voltage of the tube is varied, the variation in plate current is accompanied by a variation in plate-to-cathode voltage. The dynamic-transfer characteristic relates the instantaneous values of grid voltage and plate current.

The value of the load resistance plays an important part in determining the location and shape of the dynamic-transfer characteristic, since it fixes the slope of the load line. For every instantaneous value of the plate current, there is a corresponding value of instantaneous plate voltage, and these pairs of current and voltage values are coordinates of points that lie on the load line. Since each of these points corresponds to an instantaneous grid-voltage value, it can be seen that there are points in the static-transfer characteristic chart that correspond to points on the load line. Corresponding points are connected in Fig. 8-4 by dash lines, to show how to locate the dynamic-transfer characteristic.

Locations of points on the dynamic-transfer characteristic may also be determined by calculation. After the static-transfer characteristics (I_b against E_c at constant values of plate voltage) have been plotted and the quiescent point (Q) determined, one may use the equation

$$e_b = E_{bb} - i_b R_b$$

For example, at the instant the plate voltage is 80 V, i_b will be

$$(320 - 80)/40,000 = 0.006 \text{ A}$$

Again, when $e_b = 240$ V, i_b is equal to

$$(320 - 240)/40,000 = 0.002 \text{ A}$$

and so on.

Figure 8-5 shows how the shape of the dynamic-transfer characteristic determines the wave shape of the plate current flowing in the tube and the load, and therefore the faithfulness of the amplification effected by the circuit. The grid-bias value and the amplitude of the signal voltage are also very influential in determining plate current wave form.

The grid-bias voltage, E_c, is -6 V, and the a-c input voltage E_g, has been chosen to be a sine-wave voltage of 5 V maximum value. This is shown superimposed upon the grid bias in the lower left-hand region of Fig. 8-4. As the a-c voltage increases in the positive direction, the plate current increases and the *instantaneous operating point moves up the load line.* At the

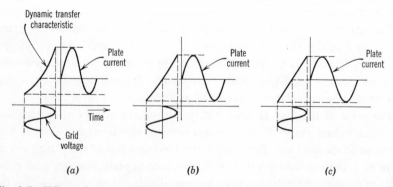

Fig. 8-5   Effect of shape of dynamic-transfer characteristic on waveshape of plate current in a vacuum tube. (*a*) Much curvature, much distortion; (*b*) little curvature, little distortion; (*c*) no curvature (straight line), no distortion.

highest position of the point the instantaneous grid-to-cathode voltage, e_c, is -1 V, the instantaneous plate current is 5.8 mA, and the instantaneous plate voltage is 87 V.

The a-c voltage then decreases, and, while this is going on, the *instantaneous operating point* moves *down* the load line through Q and on down to its lowest position. The lowest position is reached when the grid is most negative (-11 V); this occurs, obviously, at the negative peak of the a-c input voltage. At this position we have the maximum instantaneous plate voltage, which is 238 V. The a-c input voltage continues and comes up to zero from its negative value of -5 V, and so the plate current rises with the total instantaneous grid voltage, e_c, until the quiescent value, 3.8 mA, is again reached. All the variations just described occur during every cycle of the applied voltage.

Note the a-c component of the plate voltage superimposed upon the d-c, or average, value $(E_{bo}) = 168$ V. Its variation is such that its instantaneous value (e_p) goes from zero (corresponding to 168 V) up 70 V in the positive direction ($238 - 168$), and down 82 V in the negative direction. Because the dynamic-transfer characteristic *is not a straight line*, the output voltage is not a perfect reproduction of the sine-wave form of the input voltage. It is very important to observe that the a-c component of plate voltage is 180 *degrees out of phase* with the a-c component of grid voltage. This is a natural result, because, when the a-c component of grid voltage is at its positive maximum, the instantaneous plate current is maximum and the $i_b R_b$ drop through the plate-circuit resistor (R_b) is maximum. Thus at that instant the plate-to-cathode voltage is a minimum, and the a-c component of the plate voltage is at its negative maximum. When two a-c voltages of the same frequency are so timed that one has its negative maximum value at the same instant the other has its positive maximum value, they are 180 degrees out of phase.

At the instant the a-c component of grid voltage is at its greatest negative value, the plate current is a minimum and the $i_b R_b$ drop is a minimum. At that instant the plate voltage on the tube is a maximum; this means that the a-c component of the plate voltage is at its positive maximum.

A few words must be said here about the output voltage (E_2) being the a-c component (e_p) of the plate-to-cathode voltage of the tube. Referring to the circuit diagram in Fig. 8-4, we see that the load resistor (R_b) is in series with the plate-supply battery and that this series circuit is connected between the plate and cathode of the tube. Properly, e_p should be divided so that part of it shows up across R_b as E_2 and the remainder across the battery. Actually the battery has such low resistance to the flow of plate current that the part of e_p that exists across the battery is negligible. Hence all of e_p shows up across R_b because it is considered to be connected between plate and cathode *when only a-c voltage is involved.*

We may find a capacitor connected across a plate battery, and usually there is one in parallel in a power supply filter, which bypasses the a-c components of plate current, offering very low impedance to them and thus preventing a-c variations in the plate-supply voltage (E_{bb}).

Thus we see that the tube receives a voltage of 5 V maximum value and puts out a voltage of about 76 V maximum value across its load resistor. This is a *voltage amplification* of 76/5 = 15.2.

8.4   Simple Amplifier Equations

The equivalent circuit of Fig. 8-1*b* is reproduced here for convenience. It represents the amplifier stage for the purpose of mathematical analysis. It is convenient to assume that the voltages and currents are of sine-wave form and to use effective values in the analysis. The symbols for the effective values of the a-c components of grid voltage, plate voltage, and plate current are E_g, E_p, and I_p, respectively.

It is seen in Fig. 8-6 that

$$I_p = \frac{\mu E_g}{r_p + R_b} \tag{8-3}$$

Considering E_2 to be the effective value of the voltage across R_b,

$$E_2 = -I_p R_b = \frac{-R_b}{r_p + R_b} \times \mu E_g \tag{8-4}$$

In the original circuit of Fig. 8-1*a* we call E_1 the effective value of the a-c component of grid-to-cathode voltage. Then $E_1 = E_g$, and we find that the

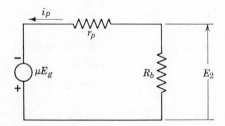

Fig. 8-6   Equivalent circuit of Fig. 8-1*a*.

ratio of the output voltage to the input voltage is, from Equation (8-4),

$$\frac{E_2}{E_1} = \frac{-\mu R_b}{r_p + R_b} = A_v \tag{8-5}$$

This is called the *voltage amplification* of the amplifier, and it is sometimes denoted by A_v. The voltage amplification is, then, the ratio of the effective value of the alternating output voltage to the effective value of the alternating input voltage. The negative sign indicates the $180°$ phase shift between E_g and E_2.

The tube does not actually enlarge the input voltage. It does, however, make possible the creation of a larger voltage that has the same frequency and practically the same waveform as the input voltage. With proper load resistance, grid bias, and amplitude of input voltage (often called signal voltage), the amplifier stage will produce an output voltage that is practically an exact reproduction (although larger) of the input voltage, even though its waveform differs very much from sine-wave form. The complicated waveforms produced in a microphone circuit by musical tones and speech may be reproduced faithfully (with high fidelity) by voltage amplifiers.

Although the purpose of a voltage amplifier is to produce a large voltage output, the plate-circuit resistor, commonly called the load resistor, must dissipate power. The equation for the a-c power in the load resistor, called power output, and denoted by P_o, is

$$P_o = I_p{}^2 R_b = \frac{\mu^2 E_1{}^2 R_b}{(r_p + R_b)^2} \tag{8-6}$$

It is seen that the *output power involves only the a-c component of plate current*. The d-c component of plate current also flows through the load resistor in this circuit and contributes to the power that the resistor must dissipate, but not to the useful power output.

A more detailed discussion of power amplifier analysis will be found in Chapter 10.

8.5 Alternating-Current Load Line

The load line shown in Fig. 8-3 for the circuit of Fig. 8-1 is usually called the *d-c load line*, because the load resistance, R_b, carries the d-c current of the tube. It usually happens that the a-c output voltage E_2 is applied to a circuit element, such as a resistor, which must be taken into account in determining the effective *a-c load* on the amplifier. Figure 8-7 shows such an

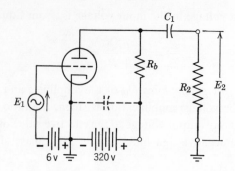

Fig. 8-7   Amplifier with a-c load resistance (R_2) capacitively coupled.

arrangement. C_1 is large enough so that its reactance is negligible compared to the resistance of R_2, which could well be R_g for a second stage of amplification.

In the discussion of the a-c load line we shall use the circuit values of Fig. 8-4 and add $R_2 = 40,000\ \Omega$, as shown. The load line is now the *d-c load line*, and we shall need to construct another (a-c) load line to represent the modified a-c load conditions. It will pass through the point Q because the tube current under static conditions is still 3.8 mA. The addition of R_2 by *capacitive coupling* did not change those conditions. However, the a-c component of tube current will be larger because some will flow through C_1 and R_2.

Neglecting the very small reactance of C_1, the a-c load resistance, R_L, will be the equivalent resistance of R_b and R_2 in parallel. This is their product divided by their sum, and it turns out to be 20,000 Ω in this case because they are 40,000 Ω each. The slope of the a-c load line $(-1/R_L)$ is therefore $-1/20,000$. It is drawn in Fig. 8-8 on the same plate characteristics used in Fig. 8-4. It is convenient to use the intercepts on the axes and draw the line as shown in dash form, and then draw the actual line parallel to it through the quiescent point Q. To do this, we may assume any convenient value of plate-supply voltage—for example, 80 V here—and use it to calculate the current-axis intercept of the line of proper slope:

$$I_b = 80 \div 20,000 = 0.004\ \text{A}$$

The intercepts are thus 4 mA on the current axis and 80 V on the voltage axis.

After the actual a-c load line is drawn through Q, parallel to the line of proper slope, it is extended to the grid-volts curves representing the extremes of grid-cathode potential encountered with full voltage, e_g, applied between grid and cathode. Waves of a-c plate current and output voltage are then easily drawn.

The smaller the ohms value of R_b, the less will be the d-c power loss. R_b may be replaced by an inductance with low d-c resistance but high reactance to reduce the alternating current through it as much as possible; this amounts to establishing as high an a-c voltage buildup across it as possible. If such an inductance is used, the *d-c load line* will be almost directly vertical, and the a-c load resistance will be practically equal to R_2, the capacitively coupled resistance.

An illustration involving a d-c load line with a very steep slope is given in the study of a power amplifier with a transformer-coupled load in Drill Problem D8-2. The resistance of the transformer primary winding is usually very small compared to the a-c resistance "coupled in" from the load.

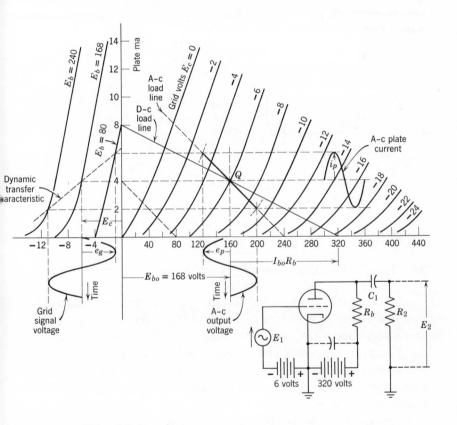

Fig. 8-8   Graphical analysis for circuit shown; a-c load line differs from d-c load line because of $C_1 - R_2$.

DRILL PROBLEMS

D8-1   In the circuit of Fig. 8-7 (*a*) determine the end points of the d-c load line when $R_b = 20\text{ k}\Omega$; (*b*) what is the slope of the load line on the tube characteristic curves? (*c*) what will be its slope if the plate-supply voltage is reduced to 200 V? (*d*) Determine the a-c load line for $R_2 = 1\text{ M}\Omega$ and $R_b = 20\text{ k}\Omega$, $E_{bb} = 320$ V.

D8-2   In the circuit shown (Fig. D8-2), the plate is loaded with an *output transformer* that supplies power to the 8-Ω load in its secondary winding. (This is typical of a circuit used to drive a loudspeaker.) The primary winding has a resistance of 100 Ω. The turns ratio of the transformer is such that the 8-Ω load is *coupled* into the primary winding as an effective a-c resistance of 2,500 Ω. On a set of characteristic curves for the tube (see Appendix), choose a value for E_{bb} and locate a Q-point so that a 30-V peak-to-peak a-c grid-voltage variation may be used without appreciable distortion of the output waveform. (*a*) Locate a suitable Q-point on the characteristics. (*b*) Sketch the d-c and a-c load lines. (*c*) What will be the expected variation of plate voltage?

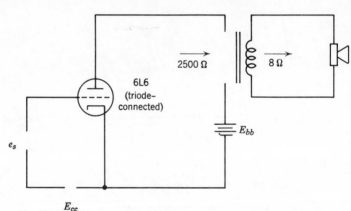

Fig. D8-2

8.6   Effect of Cathode-Biasing Resistor on Q-Point

When a resistor R_k, bypassed for alternating current by a capacitor, is used in the cathode branch to provide grid-bias voltage, the *d-c load line* is drawn with slope $-1/R_b + R_k$. The quiescent point Q will be located on the *bias line* at its intersection with the d-c load line. The quiescent grid voltage will, in most cases, have a value such that the Q-point will not lie on one of the constant-grid-voltage curves (E_c) of the static characteristics, as published by a tube manufacturer.

A self-biased triode and its average-plate-characteristic curves are shown in Fig. 8-9. The d-c load line is drawn for the 20,000-Ω load in the conventional manner. The bias line is drawn through points P_1 and P_2, which are

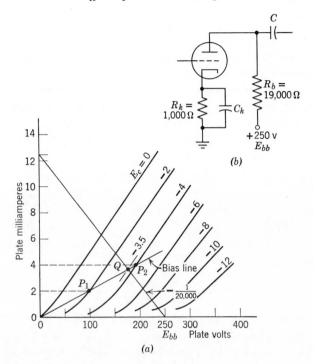

Fig. 8-9 Location of Q-point for cathode-bias operation. (a) Bias line construction; (b) plate circuit; cathode-bias operation.

located in the following manner: Without a signal on the grid, a 2-V self-bias requires a direct current of $2/1,000 = 2$ mA; a 4-V self-bias requires $4/1,000 = 4$ mA, and so on. Of course, zero self-bias requires zero current. The bias line is drawn through the origin and points P_1, P_2. The Q-point must lie on the d-c load line as well as on the bias line; therefore their intersection locates the Q-point for cathode-bias operation. In this example, the Q-point falls on the static line at $E_c = -3.5$ V.

Because the bypass capacitor has the effect of taking R_k out of the circuit insofar as alternating current is concerned, R_k is not included in the calculation for the location of the a-c load line.

DRILL PROBLEMS

D8-3 In the circuit associated with Fig. 8-9, show the effect of the value of R_k on the location of the Q-point, when it is varied from 500 Ω to 5,000 Ω.

D8-4 Can cathode bias cut off plate current in the circuit? Why?

8.7 Graphical Analysis of a One-Stage Transistor Amplifier

The graphical method of analyzing the performance of a transistor amplifier to determine the gains will now be demonstrated. Refer to Fig. 8-10.

We shall design the circuit to have an a-c load resistance of 2,000 Ω. This means that R_2 must be larger than 2,000 Ω since it is in parallel with R_3. Choosing $R_2 = 3,000\ \Omega$ and $R_3 = 6,000\ \Omega$ would make the load resistance 2,000 Ω. The a-c load line would fall as shown in Fig. 8-11 after the d-c load line has been drawn for 3,000 Ω, allowing 100 Ω for R_1. The determination of the locations of the load lines will now be explained.

The quiescent point is selected at 20 μA base current. The d-c load line passes through the quiescent point and has a slope of $-1/3,100$. It may be drawn as a straight line having this slope and passing through the quiescent point. It will intercept the collector-current axis and collector-emitter voltage axis according to the equation for the load line:

$$ i_C = v_{CE}\left(-\frac{1}{R_{DC}}\right) + \frac{V_{CC}}{R_{DC}} $$

where V_{CC} is the voltage intercept at $i_C = 0$, and i_C is the current intercept at $v_{CE} = 0$. Current and voltage values at Q are: $i_C = -0.8$ mA, $v_{CE} = -2.25$ V. Using these values in the equation we may obtain V_{CC}, the voltage intercept. This value is -4.75 V. The current intercept is $V_{CC}/R_{DC} = -1.525$ mA.

The a-c load line also passes through Q, and is a straight line having a slope of $-1/2,000$. Using the point-slope form of the equation for the straight line

$$ -\frac{1}{R_{AC}} = \frac{(i_C - i_Q)}{(0 - v_Q)} $$

the current intercept will be -1.925 mA.

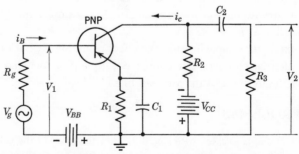

Fig. 8-10   Transistor amplifier to be analyzed graphically. $R_1 = 1\ \mathrm{k}\Omega; R_2 = 30\ \mathrm{k}\Omega;$ $R_3 = 60\ \mathrm{k}\Omega.$

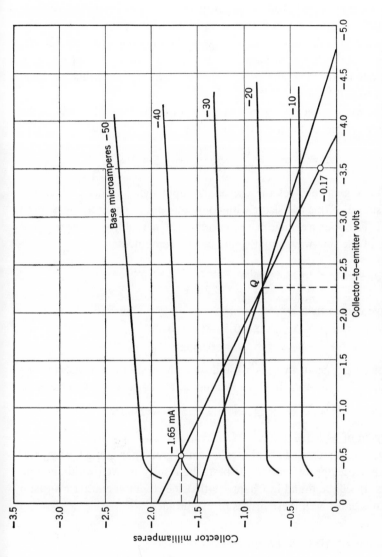

Fig. 8-11 Common-emitter characteristic curves of a transistor, Q at -2.25 V, -0.8 mA.

295

Changes in collector-emitter voltage will result from changes in base current along the a-c load line. It can be seen that the base current changes caused by an input signal produce large changes of collector-emitter voltage. Assume that the input signal causes a sinusoidal variation in base current corresponding to collector-emitter voltage variation from −0.5 to −3.5 V. The corresponding change in base current is from −40 μA to −4 μA, as a close approximation. The change in collector current, meanwhile, is from −1.65 mA to −0.17 mA.

The gains will now be computed from the graphical data. The ratio of the a-c collector current to the a-c base current is

$$A_i = \frac{i_C}{i_B} = \frac{-1.65 - (-0.17)}{[-40 - (-4)] \times 10^{-3}} = 41.2$$

The voltage gain involves the use of the input-signal voltage value. The input resistance of this transistor with common-emitter connections is given as 1,200 Ω. To obtain a change in base current of 36 μA, (40 − 4), the required amplitude of the sine-wave input signal can be computed.

$$\Delta V_i = R_i \times \Delta i_B = 1200 \times 36 \times 10^{-6}$$
$$\Delta V_i = 43.2 \text{ mV}$$

The voltage gain is then

$$A_v = \frac{\Delta v_{CE}}{\Delta V_i} = \frac{-0.5 - (-3.5)}{0 - 0.0432} = -69.5$$

The negative sign means there is a 180-degree phase shift in voltage through the amplifier. The power gain is the product of the current and voltage gains

$$A_p = A_i \times A_v = 41.2 \times 69.5 = 2860$$

This is a gain of 34.6 dB.

Although the curves used are plotted on small graph paper and the readings can be only close approximations at best, the results are well within acceptable limits. It will be found in Chapter 9 that the common-emitter transistor amplifier has gains with these orders of magnitude.

8.8 Theory of Hybrid Parameters

The hybrid parameters, or *h-parameters* as they are commonly called, are gaining favor with manufacturers for use in specifications for transistors. It is helpful in learning about them to describe their use in setting up equivalent circuits for general, linear, four-terminal networks. Then they may be extended for use in describing the behavior of transistors in their small-signal mode of operation.

We have learned that a linear four-terminal network of any configuration between a generator and a load can be represented by a new, simple, series circuit derived by applying Thévenin's theorem or by a new, simple, parallel circuit obtained by applying Norton's theorem.

The *hybrid* equivalent circuit is the most widely used in the analysis of transistor circuits. It is termed *hybrid* because it contains both *admittance* and *impedance* parameters. The ease of measurement of the these parameters, which are the *h-parameters* of the circuit, has contributed to the widespread use of the equivalent circuit.

A set of *h*-parameters can be derived for any *black box* network having linear elements and two input and two output terminals. Consider the situation in Fig. 8-12, where the external currents and voltages at the terminals of the black box network are measurable quantities. In general, the output current is a function of the output voltage and input current; the input voltage can be written as a function of input current and output voltage:

$$v_1 = f(i_1, v_2) \tag{8-7}$$

$$i_2 = g(i_1, v_2) \tag{8-8}$$

This is merely another way of saying that any two of the input and output quantities may be expressed in terms of the other two quantities.

Expressing these two equations in total differential form yields:

$$dv_1 = \frac{dv_1}{di_1}\bigg|_{v_2\,\text{constant}} di_1 + \frac{dv_1}{dv_2}\bigg|_{i_1\,\text{constant}} dv_2 \tag{8-9}$$

$$di_2 = \frac{di_2}{di_1}\bigg|_{v_2\,\text{constant}} di_1 + \frac{di_2}{dv_2}\bigg|_{i_1\,\text{constant}} dv_2 \tag{8-10}$$

Now if the differential changes in voltages and currents are taken as single-frequency sinusoidal variations, these equations may be written in terms of

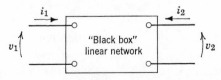

Fig. 8-12   Four-terminal "black box" network.

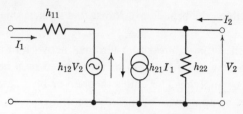

Fig. 8-13   Equivalent circuit for "black box" network, showing placement of *h*-parameters.

phasors to represent the varying quantities:

$$V_1 = \frac{dv_1}{di_1}\bigg|_{v_2 \text{ constant}} I_1 + \frac{dv_1}{dv_2}\bigg|_{i_1 \text{ constant}} V_2 \tag{8-11}$$

$$I_2 = \frac{di_2}{di_1}\bigg|_{v_2 \text{ constant}} I_1 + \frac{di_2}{dv_2}\bigg|_{i_1 \text{ constant}} V_2 \tag{8-12}$$

The coefficients of I_1 and V_2 are the *h*-parameters of the network. They are usually defined as follows:

$$V_1 = h_{11}I_1 + h_{12}V_2 \tag{8-13}$$

$$I_2 = h_{21}I_1 + h_{22}V_2 \tag{8-14}$$

Since Equations (8-13) and (8-14) apply to the black box network, an equivalent circuit may be drawn that fits these equations. The equivalent circuit is given in Fig. 8-13. It can be seen that the input circuit (Equation 8-13) is represented by Thévenin's equivalent circuit, and the output circuit is a Norton equivalent of Equation (8-14). The input circuit equation is the sum of voltages existing there, whereas the output circuit equation is a nodal sum of currents in the output side of the network.

Each of the *h*-parameters may be determined from measurements in the network, when the input and output voltages and currents are sinusoidal, according to this tabulation derived from Equations (8-13) and (8-14):

$$h_{11} = \frac{V_1}{I_1}\bigg|_{V_2=0} \ \Omega \tag{8-15}$$

$$h_{12} = \frac{V_1}{V_2}\bigg|_{I_1=0} \ \text{dimensionless} \tag{8-16}$$

$$h_{21} = \frac{I_2}{I_1}\bigg|_{V_2=0} \ \text{dimensionless} \tag{8-17}$$

$$h_{22} = \frac{I_2}{V_2}\bigg|_{I_1=0} \ \text{mho} \tag{8-18}$$

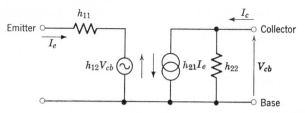

Fig. 8-14   *h*-parameter equivalent circuit for transistor in common-base circuit.

Figure 8-14 is the circuit for a transistor operating in the common-base configuration. Note that the input and output voltages and currents can be related to those of the black box network by this set of equations:

$$V_{eb} = h_{11}I_e + h_{12}V_{cb}$$
$$I_c = h_{21}I_e + h_{22}V_{cb}$$

(8-19)

The values of the common-base *h*-parameters may be measured for the transistor by using the definitions and restrictions of Equations (8-15) through (8-18). Since they are defined for sinusoidal quantities, the measurements will be made using the small-signal operation of the transistor, for which linearity may be assumed in the network.

8.9   Transistor Equivalent Circuit Using *h*-Parameters

We are now equipped to draw the equivalent circuit of a transistor for dynamic operation employing *h*-parameters. The grounded-base circuit is discussed first.

Figure 8-15 shows the actual and equivalent circuits. As in the case of

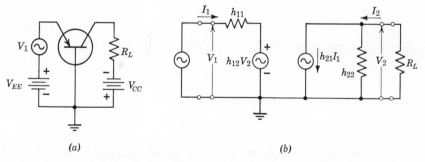

(a)   (b)

Fig. 8-15   Common-base amplifier. (*a*) Actual circuit; (*b*) equivalent circuit.

an equivalent circuit for a vacuum-tube amplifier, d-c voltages do not appear. The impedance h_{11} is put in place of its equivalent in the circuit of the "black box." By definition, h_{22} is a *conductance*. The element labeled h_{22} is a resistance whose value is $1/h_{22}$, so that when V_2 is multiplied by the h_{22} value a *current* is obtained since this multiplication actually amounts to *dividing* V_2 by $1/h_{22}$.

The constant-current generator in the output part of the equivalent circuit for the black box can be replaced by $h_{21}I_1$. Thus the equivalent-circuit nomenclature is established.

We are now prepared to write the operating equations for the equivalent circuit of the *grounded-base transistor* amplifier.

8.10   Mathematical Analysis of Transistor Amplifier

Refer to the equivalent circuit of the common-base transistor in Fig. 8-15. The input section lends itself to the writing of a loop-voltage, or *mesh*, equation. The output section lends itself to the writing of a current-node or nodal equation.

$$V_1 = h_{11}I_1 + h_{12}V_2 \qquad (8\text{-}20)$$

$$I_2 = h_{21}I_1 + h_{22}V_2 \qquad (8\text{-}21)$$

where, as explained in Section 8.8, h_{11} is a resistance and h_{22} is a conductance.

Consider the quantities which are known in these equations. The *h*-parameters are given for the transistor by the manufacturer. Typical values for a specific transistor are as follows:

Input resistance, output circuit shorted:

$$h_{11} = 40 \ \Omega$$

Reverse voltage transfer ratio, input circuit open:

$$h_{12} = 3.23 \times 10^{-4}$$

Forward current transfer ratio, output circuit shorted:

$$h_{21} = -0.95$$

Output conductance, input circuit open:

$$h_{22} = 1.4 \ \mu\text{mho}$$

The remaining known quantities are the applied voltage, V_1, and the load current, I_2. Evidently, the simultaneous equations may be solved for the unknowns, I_1 and V_2. This is a simple task, and the results are:

$$V_2 = \frac{h_{11}I_2 - h_{21}V_1}{h_{11}h_{22} - h_{12}h_{21}} \quad \text{V} \tag{8-22}$$

$$I_1 = \frac{h_{22}V_1 - h_{12}I_2}{h_{11}h_{22} - h_{12}h_{21}} \quad \text{A} \tag{8-23}$$

The voltage gain, A_v, is the ratio of V_2 to V_1. We can get an expression for V_2 in terms of only the h-parameters, V_1, and the load resistance.

$$I_2 = -V_2/R_L \tag{8-24}$$

Substituting this in Equation (8-22),

$$V_2 = \frac{-h_{11}\dfrac{V_2}{R_L} - h_{21}V_1}{h_{11}h_{22} - h_{12}h_{21}}$$

Calling this denominator a constant D to save needless writing,

$$V_2 D = -h_{11}\frac{V_2}{R_L} - h_{21}V_1$$

$$V_2\left(D + \frac{h_{11}}{R_L}\right) = -h_{21}V_1 \tag{8-25}$$

$$V_2 = \frac{-h_{21}V_1}{D + \dfrac{h_{11}}{R_L}} = -\frac{h_{21}V_1}{h_{11}h_{22} - h_{12}h_{21} + \dfrac{h_{11}}{R_L}}$$

The voltage gain, $V_2 \div V_1$, is then

$$A_v = \frac{-h_{21}}{h_{11}\left(h_{22} + \dfrac{1}{R_L}\right) - h_{12}h_{21}} \tag{8-26}$$

Checking the dimensions, A_v should be dimensionless (volts $\div$ volts).

The numerator h_{21} is the forward-current transfer ratio, therefore dimensionless. The denominator terms will be written in dimensional form:

$$\text{ohms}\left(\frac{1}{\text{ohms}} + \frac{1}{\text{ohms}}\right) - \frac{\text{volts}}{\text{volts}} \times \frac{\text{amperes}}{\text{amperes}}$$

Thus the denominator is also dimensionless.

The h-parameters supplied by the manufacturer of the transistor hold for small-signal operation *with a specified Q-point*.

This analysis is intended to present the general procedure for setting up the equivalent circuit and writing equations for finding the important quantities involved in transistor amplifier behavior. In the next chapter we shall present a more detailed analysis of the three types of transistor amplifier—common base, common emitter, and a common collector—using the equivalent circuit for each case.

DRILL PROBLEM

D8-5 Calculate the voltage gain of a transistor whose common-base h-parameter values are $h_{11} = 50 \ \Omega$, $h_{22} = 0.5 \ \mu$mho, $h_{12} = 200 \times 10^{-6}$, $h_{21} = -0.98$, and which is driving a load resistance of $1,000 \ \Omega$.

8.11 Manufacture Designation of h-Parameters

The subscripts on the h-parameters used in the preceding sections are not generally used by transistor manufacturers. Instead, they use a system of letter subscripts to designate which terminal of the transistor is common in a circuit. For example, h_{11} is designated as h_{ib} in the common-base circuit connection and as h_{ie} in the common-emitter circuit. The first letter refers to the *input* parameter and the second letter designates the common terminal. The h-parameters are labeled as follows, depending on the common terminal of the circuit in which the transistor is being used:

$$h_{11}: \quad h_{ib}, h_{ie}, h_{ic} \qquad h_{12}: \quad h_{rb}, h_{re}, h_{rc}$$

$$h_{21}: \quad h_{fb}, h_{fe}, h_{fc} \qquad h_{22}: \quad h_{ob}, h_{oe}, h_{oc}$$

The first letter of the double-subscript system has a special meaning. The i refers to the *input* impedance; r designates the *reverse* voltage feedback ratio; f refers to the *forward* current ratio; o is used to designate the *output* admittance.

8.12 Calculation of h-Parameters from Manufacturer's Data

The data supplied by a transistor manufacturer may include values of the h-parameters for a specific Q-point. A circuit in which the transistor is to operate may require a different Q-point. The values of h-parameters may be quite different at this operating point. Before making an analysis of the circuit behavior by the methods of the next chapter, it will be necessary to calculate the values of h-parameters that apply to the circuit operating point.

As an example of the necessary calculations, we will use the following data supplied for the 2N334 transistor:

Common-Base Design Characteristics, at 25°C

h_{ib}	Input impedance	$(V_{CB} = 5 \text{ V}, I_E = -1 \text{ mA})$	56 Ω
h_{ob}	Output admittance	$(V_{CB} = 5 \text{ V}, I_E = -1 \text{ mA})$	0.5 μmho
h_{rb}	Feedback voltage ratio	$(V_{CB} = 5 \text{ V}, I_E = -1 \text{ mA})$	350×10^{-6}
h_{fb}	Current transfer ratio	$(V_{CB} = 5 \text{ V}, I_E = -1 \text{ mA})$	-0.97

In addition to these numerical values of the common-base characteristics, charts are provided to permit their calculations for other operating points. The given values apply only to the condition $V_{CB} = 5 \text{ V}$, $I_E = -1 \text{ mA}$. For other emitter-current conditions, the chart of Fig. 8-16 may be used.

It will be noted that the quantity $1 + h_{fb}$ is plotted instead of h_{fb}. The

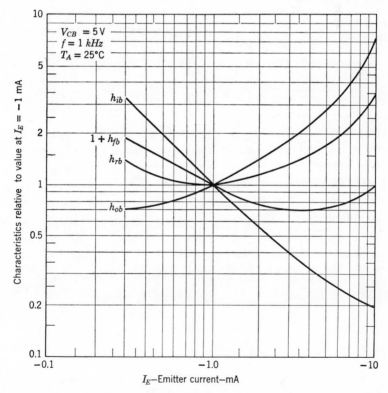

Fig. 8-16   Common-base characteristics vs. emitter current–2N334 transistor h-parameter variations with changes in I_E.

reason for this is clear. Since the parameter h_{fb} is nearly unity and changes very little for different emitter currents, the small changes would not be apparent in the chart of Fig. 8-16. However, the quantity $1 + h_{fb}$ may change greatly with only a small change in h_{fb}. For example, suppose h_{fb} varies from -0.97 to -0.94. The change is only 0.03; compared with the value of h_{fb} it is small. Now look at the change in $1 + h_{fb}$—from 0.03 to 0.06, a ratio of 2 to 1.

An important result of plotting the variations of the h-parameters is their wide range of possible values with emitter currents. Over the range of currents from -0.3 to -10 mA; they may change their values by an order of magnitude or more. Therefore, it becomes important to determine the h-parameters for the particular Q-point being used in a circuit.

8.13   Formulas for Calculating h-Parameters

The manufacturer data sheets usually list the common-base characteristics and h-parameters. However, a transistor is more often used in the common-emitter mode and occasionally will be connected in the common-collector configuration. In order to analyze circuit behavior using the h-parameter equivalent circuit for the transistor, it is necessary to calculate the values of the h-parameters for these circuit modes, in addition to any variations because of the Q-point conditions. The formulas of Table 8-1 are useful in making the necessary calculations.

An example will show how to use these formulas. Suppose it is desired to determine the values of h-parameters for a common-emitter circuit that uses a 2N334 transistor. The common-base parameters are listed above and will be used in the equations of Table 8-1 to calculate the values of the common-emitter parameters. For the purpose of this example, it is assumed that the Q-point of the circuit will be fixed at the same conditions in the two circuits.

The values of the common-emitter parameters may be calculated from their common-base values by using the equations listed under common-base in Table 8-1. The value of h_{ie} is calculated

$$h_{ie} = \frac{h_{ib}}{1 + h_{fb}} = \frac{56 \ \Omega}{1 - 0.97} = \frac{56}{0.03} \ \Omega$$
$$= 1,870 \ \Omega$$

Similarly, the value of h_{fe} is calculated

$$h_{fe} = -\frac{h_{fb}}{1 + h_{fb}} = -\frac{-0.97}{1 - 0.97} = \frac{0.97}{0.03} = 32.3$$

The other parameters are calculated in a similar manner. Here again we see the factor $1 + h_{fb}$, which is plotted in Fig. 8-16.

TABLE 8-1 Conversion Equations for h Parameters (Numerical Values Are Typical at 25°C)

Symbol	Common-Base	Common-Emitter	Common-Collector
h_{ib}	30–80 Ω	$\dfrac{h_{ie}}{1 + h_{fe}}$	$-\dfrac{h_{ic}}{h_{fc}}$
h_{rb}	5–200 × 10⁻⁶	$\dfrac{h_{ie}h_{oe}}{1 + h_{fe}} - h_{re}$	$h_{rc} - 1 - \dfrac{h_{ic}h_{oc}}{h_{fc}}$
h_{fb}	−0.90 to −0.99	$-\dfrac{h_{fe}}{1 + h_{fe}}$	$-\dfrac{1 + h_{fc}}{h_{fc}}$
h_{ob}	0.5 to 1.5 μmho	$\dfrac{h_{oe}}{1 + h_{fe}}$	$-\dfrac{h_{oc}}{h_{fc}}$
h_{ie}	$\dfrac{h_{ib}}{1 + h_{fb}}$	600–2000 Ω	h_{ic}
h_{re}	$\dfrac{h_{ib}h_{ob}}{1 + h_{fb}} - h_{rb}$	3 × 10⁻⁴	$1 - h_{rc}$
h_{fe}	$-\dfrac{h_{fb}}{1 + h_{fb}}$	10 to 100	$-(1 + h_{fc})$
h_{oe}	$\dfrac{h_{ob}}{1 + h_{fb}}$	20 μmho	h_{oc}
h_{ic}	$\dfrac{h_{ib}}{1 + h_{fb}}$	h_{ie}	600–2000 Ω
h_{rc}	1	$1 - h_{re} \approx 1$	1
h_{fc}	$-\dfrac{1}{1 + h_{fb}}$	$-(1 + h_{fe})$	−10 to −100
h_{oc}	$\dfrac{h_{ob}}{1 + h_{fb}}$	h_{oe}	20 μmho

DRILL PROBLEM

D8-6 Using the equations of Table 8-1 and Fig. 8-16, determine the values of common-emitter h-parameters for emitter current $= -5$ mA, $V_{CB} = 5$ V, for the 2N334 transistor.

8.14 Effects of Temperature on Q-Point

Several variable factors associated with the temperature of a transistor affect the location of the Q-point. Interaction between these variables may shift the Q-point to an undesirable location on the characteristics of the transistor. For the best reproduction of an input signal without appreciable distortion, the Q-point should be located at approximately the center of the load line. Also, it should remain below the maximum dissipation level, to prevent self-destruction of the transistor because of internal heating. The location of the Q-point, and possible variations of it because of temperature effects, should remain in a region where cumulative self-heating is not possible. Self-heating that may result in the self-destruction of a transistor is referred to as *thermal runaway*.

The characteristics of a transistor that are primarily affected by temperature variations are the forward current ratio (h_f), I_{CEO} (reverse collector current),

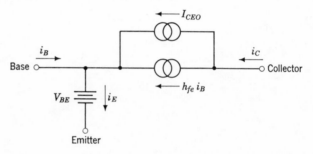

Fig. 8-17   Model of common-emitter operation of a transistor.

and the base-emitter quiescent voltage (V_{BE}). When it is operated in the common-emitter mode, a transistor may be represented by the linear model of Fig. 8-17. In this figure, the battery represents the quiescent value of the base-emitter voltage, and the collector current includes the reverse collector current due to reverse bias of the collector-base diode and the current component due to the forward current ratio.

The value of h_{fe} can vary through a wide range of values from one transistor to another, even for those of the same type. For example, the 2N334 has a guaranteed h_{fe} of 18 to 90, a 5 to 1 ratio. For a constant value of base current, the value of collector current at the Q-point can vary over a range of values in the ratio of 5 to 1 for transistors of the 2N334 type. The variation of h_{fe} can become quite a problem when it is necessary to replace a given transistor, unless some provision is made in the circuit to allow stabilization of the Q-point. This topic is discussed in the next section. It is a greater problem in transistor circuits than in comparable vacuum-tube circuits, because of the possible parameter variations from one transistor to another.

The curves of Fig. 8-18 show how the Q-point may be shifted when base current is held fixed, and a transistor having a low value of h_{f_e} is replaced with one having a much higher h_{f_e}. The value of h_{f_e} is also dependent on the operating temperature of the transistor, so that the Q-point may be shifted because of temperature effects, as well.

Increasing the operating temperature will increase the reverse collector current in a transistor. This component of the collector current is generated primarily by thermal ionization, although surface effects contribute a small

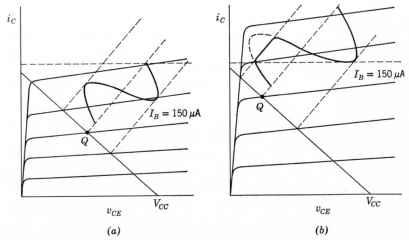

Fig. 8-18   Shifted Q-point resulting from transistor replacement by high h_{f_e} device, causes distortion of signals. (*a*) Low h_{f_e} transistor. (*b*) High h_{f_e} transistor.

fraction of the total. The component that occurs because of thermal ionization within the semiconductor crystal is usually predominant in shifting the Q-point because of temperature variations. The value of I_{CEO} approximately doubles for each 10°C increase in temperature.

In silicon transistors the change in I_{CEO} with a change in temperature is usually not troublesome except at higher temperatures, because its value is small compared to the total collector current. In germanium transistors, however, thermal ionization may generate 1,000 times as many current carriers as the same temperature would produce in silicon. The reverse collector current component due to thermal ionization may become large enough so that the total collector current is increased an appreciable amount. For a fixed base current, the Q-point will be shifted upward as in Fig. 8-18.

The value of the quiescent base-emitter voltage *decreases* about 2.5 mV/°C with *increases* in temperature. Since the base current is determined by this voltage in a fixed-base-current biasing arrangement

$$i_B = \frac{V_{BB} - V_{BE}}{R_B}$$

decreasing V_{BE} will increase i_B, and the Q-point will shift upward to higher collector current regions. In germanium transistors, V_{BE} is smaller than in comparable silicon units but can be important in either, depending on the base-bias source V_{BB}.

All of the factors considered, h_{fe}, I_{CEO}, and V_{BE}, cause the Q-point to shift upward toward increasing collector current when the temperature of the transistor is increased. The operating temperature of a transistor is determined by two factors—(1) ambient conditions, and (2) internal self-heating. Ambient conditions may be controlled externally, but the self-heating effects must be controlled internally, by controlling the variations of these three factors.

Self-heating occurs internally because of the power that is dissipated in the collector junction. It results in an increased temperature of the junction above the ambient temperature. The rate of dissipation of the heat will be determined by the rate of heating and the rate of cooling permitted by the physical conditions at the junction. The increase in junction temperature with increased power in a transistor results in the derating factors for transistors. For example, the 2N334 is derated at 1 mW/°C. Since its maximum collector dissipation rating is 150 mW, it cannot safely dissipate any power at 150°C and only 50 mW at 100°C. The derating necessary is dependent on the ability of the transistor to dissipate heat. Some high-power transistors are mounted on large metal bases of copper and aluminum, called *heat sinks*, and their outside surfaces may be formed as fins to radiate the heat more readily. In any event, self-heating must not be permitted at a rate that will be self-destructive.

The curves of Fig. 8-19 show the effect of a shift of the Q-point on the collector dissipation. If the initial location of the Q-point is below the "knees" of the dissipation curves, then increases in operating temperature may shift the Q-point to higher collector currents and thus increase the operating temperature even more. If the rate of self-heating is greater than the cooling rate of the assembly, the effect will be cumulative and may lead to destruction of the transistor. However, if the initial location of the Q-point is above this region, increases in collector current because of the upward shifting of the Q-point will result in a lower dissipation.

We have looked at the possible variations that may occur in the operation

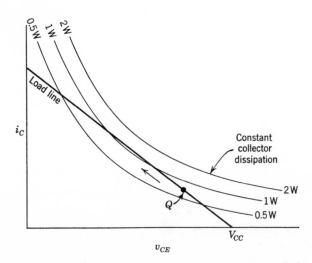

Fig. 8-19   Effect of shifting Q-point location on collector dissipation.

of a transistor because of temperature variations. If any practical use is to be made of transistors in reliable circuits, some means must be provided to counteract internal parameter changes. This requires that external circuit components be used and connected in such a way that their actions will reduce the effects of parameter variations from one transistor to another and counteract the effects of temperature variations.

8.15   *Q*-Point Stabilization

There is some objection to using a single biasing resistor in the base circuit. The high external resistance of R_B tends to free the transistor for rather wide shifts in the Q-point location resulting from the effects of temperature on the internal parameters. In order to stabilize the operation of a transistor circuit so that it occurs in some desirable region of the characteristics, the arrangement shown in Fig. 8-20 is often used.

This scheme for base biasing overcomes the difficulties of temperature effects on internal parameters and stabilizes the operating point against variations between individual transistors that may be used as replacements in the circuit.

The use of two resistors, R_1 and R_2, in the bias circuit permits the total resistance in the base circuit to be adjusted to a desirable value and at the same time permits the quiescent base current to be adjusted to the bias value.

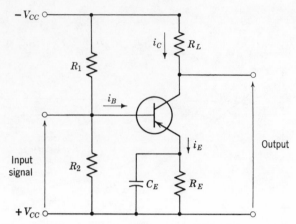

Fig. 8-20   Bias stability network used in common-emitter circuit to stabilize Q-point location.

Stability in this circuit results primarily from the presence of R_E in the emitter lead. Changes in emitter current, which is largely collector current, will change the emitter-base voltage. An increase in I_{CEO} or $h_{fe}I_B$ in the collector circuit due to an increase in operating temperature will decrease the voltage V_{BE}. A smaller voltage reduces the quiescent base current and, in turn, the collector current. Thus, the operation tends to maintain the Q-point at its original location on the characteristics.

The capacitor shunting R_E bypasses a-c signal currents around R_E. Its action is similar to that of C_k in self-biased tube circuits. If there were no path for a-c currents around R_E, the voltage developed across R_E would tend to reduce any incoming a-c signal at the base and a lower amplification would result in the circuit. This action is called *degenerative feedback* and is usually avoided in amplifying circuits. There are special circumstances in which it is purposely introduced for its advantages, but these are reserved for discussion in a subsequent chapter.

SUGGESTED REFERENCES

1. Samuel Seely, *Electron Tube Circuits*, McGraw-Hill, New York, 1958.

2. W. G. Dow, *Fundamentals of Engineering Electronics*, 2nd ed., Wiley, New York, 1952.

3. K. R. Spangenburg, *Vacuum Tubes*, McGraw-Hill, New York, 1948.

4. Paul D. Ankrum, *Principles and Applications of Electron Devices*, International Textbook Co., Scranton, Pa., 1959.

5. G. E. Happell and W. M. Hesselberth, *Engineering Electronics*, McGraw-Hill, New York, 1953.

6. C. L. Searle et al., *Elementary Circuit Properties of Transistors*, Wiley, New York, 1965.

QUESTIONS

8-1 The grid circuit resistor, or coupling resistor as it is usually called, serves two purposes in a vacuum-tube amplifier stage. What are they?

8-2 Why should there be no grid current in a small-signal voltage amplifier?

8-3 Refer to the circuit of a vacuum-tube amplifier and describe in words how amplification takes place.

8-4 Draw from memory the Thévenin equivalent circuit of a vacuum-tube amplifier and label all elements. Why are d-c voltages not included in the circuit?

8-5 Explain why i_b is always positive in the actual circuit.

8-6 Draw from memory the Norton equivalent circuit of a vacuum-tube amplifier. What kind of generator is in this circuit? What kind of generator is in the Thévenin equivalent circuit?

8-7 Explain how to draw the d-c load line on a set of average plate characteristics. When is the d-c load line the same as the a-c load line?

8-8 What is meant by quiescent point and what determines its location?

8-9 What is the dynamic-transfer characteristic curve and how is it located?

8-10 Explain the effect of the shape of the dynamic-transfer characteristic on the wave shape of the a-c plate-current wave and on the shape of the output-voltage wave.

8-11 What would be the effect on the amount of voltage amplification if R_b were increased? If r_p were larger? Can you deduce what things put a practical upper limit on values of R_b?

8-12 When is the a-c load line different from the d-c load line? When does the a-c load line have a greater slope than the d-c line, and when does it have a lesser slope?

8-13 Explain how the a-c load line for a voltage amplifier is drawn.

8-14 What is the effect of a cathode-biasing resistor on both load lines? Justify your answer for the a-c load line.

8-15 What is grid-leak bias? Explain how this method of bias can be practical when grid current is usually not tolerated in amplifiers. How much resistance is commonly used in a grid-leak-bias arrangement? What is the purpose of the bypass capacitor?

8-16 What is understood to be a "black box" in network-analysis terminology?

8-17 Recite the word definition of the hybrid parameter h_{11}. Write its definition in mathematical form. What are its units?

8-18   Do as directed in the preceding question, but for h_{12}.

8-19   Do as directed in question 8-17, but for h_{21} and h_{22}.

8-20   Which h parameter is the input resistance when the output circuit is shorted? Determine this by making a mental picture of what ratio will give this resistance.

8-21   Which h parameter is the forward-current transfer ratio, output circuit-shorted?

8-22   Which h parameter is the reverse-voltage transfer ratio, input circuit open?

8-23   Which h parameter is the output conductance, input circuit open?

8-24   Why are I_1 and V_2 the unknowns in the general equations for the transistor equivalent circuit, rather than V_1 and I_2? Are these a-c or d-c quantities?

PROBLEMS

8-1   A voltage amplifier with the same circuit as that of Fig. 8-7 has the following data: Tube 6SF5, high-mu triode; plate-supply volts, 250; fixed bias, -1.0 V; $R_b = 100,000 \ \Omega$; $C_1 = 0.1 \ \mu$F; $R_2 = 100,000 \ \Omega$. (a) Draw the two load lines on the plate characteristics. (b) Assume an input-signal voltage of 0.707 V rms at 10 kHz. Determine graphically the maximum and minimum instantaneous values of plate-to-cathode voltage. (c) Determine the rms output voltage and the voltage amplification. Express the voltage gain in decibels. (d) Compute the reactance of C_1 and see if it is negligible when compared with the a-c load resistance.

8-2   Plot the static and dynamic transfer characteristic curves for the 6SF5 tube with the a-c load given in Problem 8-1. Is there a range of reasonable linearity? Between what values of E_b does it extend?

8-3   A 6J5 tube is used in a voltage-amplifier circuit similar to that of Fig. 8-7 except that a well-bypassed cathode-bias resistor of 1,430 Ω is used instead of fixed bias. The plate-supply voltage is 250, R_b and R_2 are 20,000 Ω each. $C_1 = 0.1 \ \mu$F. Do what is requested in Problem 8-1 for this amplifier with the same input-signal voltage.

8-4   Draw the equivalent circuit for the voltage amplifier of Problem 8-1. Using $\mu = 100$ for the 6SF5 triode, calculate the rms plate current (I_p), the output voltage, the voltage amplification, and the a-c power in R_b and in R_1.

8-5   Using the average-collector characteristics for the 2N334 transistor in the Appendix, determine α and β for that transistor.

8-6   The 2N334 transistor is used in a common-base amplifier circuit. The input resistance $R_i = 52 \ \Omega$. $V_{CB} = +25$ V and the load resistance is 3,000 Ω. The bias current is -3 mA. The rms a-c component of emitter current is $3/\sqrt{2}$ mA. Determine (a) the a-c output voltage, (b) the current gain, (c) the power gain.

8-7 When the 2N334 transistor is used as a common-emitter amplifier it has $V_{CC} = +12$ V and a load resistance of 600 Ω. I_B is $+100$ μA. The base current is made to vary sinusoidally from $+50$ to $+150$ μA. Determine (*a*) the current gain, (*b*) the voltage gain, assuming v_{be} to be 0.1 V in *amplitude,* (*c*) the power gain.

8-8 Calculate the theoretical voltage amplification of a transistor having the following parameters in common-emitter circuit:

$$h_{11} = 2{,}720 \ \Omega \qquad\qquad h_{21} = 55$$
$$h_{12} = 3.12 \times 10^{-4} \qquad\quad h_{22} = 14 \ \mu\text{mho}$$
$$R_L = 0.5 \ \text{M}\Omega$$

How many decibels gain is this?

8-9 For the circuit shown in Fig. P8-9, $\mu = 15$, $r_p = 10$ kΩ, $R_L = 6{,}000$ Ω, $R_k = 1{,}500$ Ω. Draw Thévenin's equivalent circuit and solve for I_p, the rms a-c plate current, if $E_i = 1 \underline{/0°}$ V.

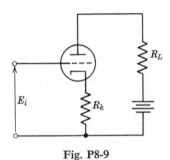

Fig. P8-9

8-10 In the circuit shown in Fig. P8-10, the tube parameters are $\mu = 20$, $r_p = 7{,}000$ Ω. The input signal E_i is impressed between the grid and ground. Determine the ratio E_o/E_i.

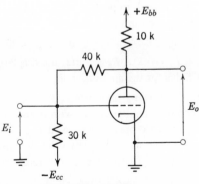

Fig. P8-10

8-11   Find an expression for the voltage gain of the circuit shown in Fig. P8-11. Assume that d-c path exists from cathode to grid.

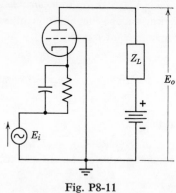

Fig. P8-11

8-12   For the circuit given in Fig. P8-12, draw the d-c load line, the bias line, and the a-c load line.

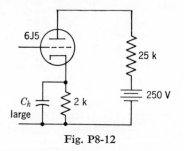

Fig. P8-12

8-13   A 2-kHz signal of 2 V peak-to-peak is applied to the input of the circuit in Problem 8-12. What is the a-c output?

8-14   In the circuit of Fig. P8-14, $E_g = 1 \big/ 0° $ V, $\mu = 20$, and $r_p = 7,000 \, \Omega$. The plate load R_b is 5 kΩ, $R_1 = 1$ kΩ, $R_2 = 1$ kΩ, and C is large enough so that R_1 is adequately bypassed. Solve for the a-c output, E_o.

8-15   A transistor has the following parameters: $\beta = 60$, $I_{CEO} = 4 \, \mu\text{A}$, $V_{CC} = 16$ V, $I_C = 2$ mA. The common-emitter circuit with fixed bias is used. Determine (*a*) circuit resistances required, (*b*) operating point if I_{CEO} increases to 10 μA, (*c*) operation point if I_{CEO} remains at its original value but β increases to 80, (*d*) operation point if I_{CEO} increases to 10 μA and β increases to 80 simultaneously.

8-16   For each of the circuits in Fig. P8-16, $\mu = 20$, $r_p = 10$ kΩ, $E_s = 2 \big/ 0°$ V,

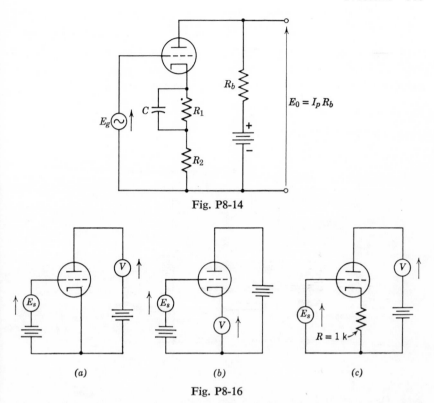

Fig. P8-14

Fig. P8-16

$V = 1\underline{/0°}$ V, and operation may be considered linear. (*a*) Draw and label Thévenin's equivalent circuit. (*b*) Calculate I_p. (*c*) Determine the impedance presented to the source V. (Consider that $E_s = 0$ for this calculation.)

8-17 The circuit of Fig. P8-17 may be used to measure the value of μ in terms of

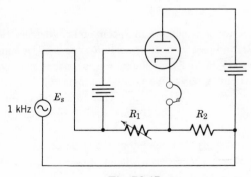

Fig. P8-17

the resistors R_1 and R_2 only. The value of μ is a ratio of the resistance values when no tone is heard in the earphones. Determine the ratio that is a measure of μ.

8-18   For the circuit of Fig. P8-18, calculate an expression for the voltage gain $A_v = E_o/E_s$ in terms of tube coefficients and circuit impedances only. Assume identical tubes and small capacitive reactance compared with R_g.

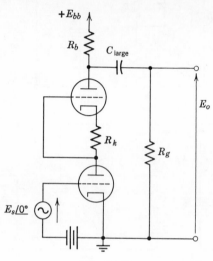

Fig. P8-18

8-19   Determine the common-base h-parameters for a 2N1308 transistor when $V_{CB} = 5$ V and $I_E = 10$ mA, using the curves in the Appendix.

8-20   Calculate the values of common-emitter parameters for the 2N1308 transistor at the operating point specified in Problem 8-19.

8-21   (*a*) Sketch the h-parameter equivalent circuit for a transistor operating in the common-base mode. (*b*) Sketch the equivalent circuit for common-emitter mode.

8-22   A *PNP* transistor is connected in a common-emitter circuit with $V_{CC} = -10$ V, $R_L = 2$ kΩ, and base bias set by some R_B and V_{CC}. A sinusoidal voltage source drives the base of the transistor and the output of the circuit is taken between the collector and circuit ground (the positive terminal of V_{CC}). (*a*) Sketch the circuit, showing the schematic of all components. (*b*) Sketch characteristics of the transistor, assuming it to have $h_{fe} = 100$ and linear properties. (*c*) What value of base current will permit the input signal to have the largest excursion without distortion in the output? (*d*) What should be the value of R_B to set the operating point for largest input signal? (*e*) What should be the collector dissipation rating for the transistor to operate at this point?

8-23 A 2N334 transistor has $V_{CC} = 30$ V and $R_L = 5$ kΩ in a common-emitter circuit. Using V_{CC} to provide the base bias, what value of base resistor R_B is needed to set $V_{CE} = 20$ V? 10 V?

8-24 A 2N404 transistor is to be used in a common-emitter circuit having $V_{CC} = -10$ V. Choose values for R_L that will permit a maximum collector current variation for sinusoidal variations in base current, without exceeding any maximum ratings (*a*) when operating at 25°C; (*b*) when operating at 75°C.

CHAPTER NINE

Analysis of Amplifier Performance

This chapter presents principles of operation of tube and transistor amplifiers and their special features. The discussion is confined, for the most part, to the audio-frequency range, although some attention is given to high-frequency amplifiers.

9.1 Two-Stage Voltage Amplifier

The elementary amplifier circuits studied in Chapter 8 produced an output voltage across a load resistor in the plate circuit of the tube or the collector circuit of the transistor. It often happens that this voltage is not sufficiently large to perform the task desired, and so further amplification is required.

Figure 9-1 shows two-stage voltage amplifiers, each of which has resistance-capacitance coupling between its stages. C_1 and R_{g2} form the coupling network for the tube amplifier and C_1, R_{b2} for the transistor amplifier. In each case, C_1 serves to block the d-c potential of the first stage from reaching the input circuit of the next stage while allowing the amplified a-c voltage to "drive" the second stage. C_1 is made large enough in capacitance that the a-c voltage loss across it is negligible in the range of working frequencies of the amplifier.

9.2 Operating Conditions; Classes of Amplifiers

It is possible to select the grid bias and the input-voltage amplitude so that plate current may be caused to flow in the tube during the complete cycle of input voltage or during any fraction of it. The letters A, B, and C are used to designate operation under different bias conditions.

Class A. Figure 9-2 shows the nature of operation called Class A_1. "Class A" means that plate current flows during the complete grid-voltage cycle

318

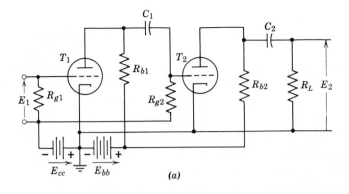

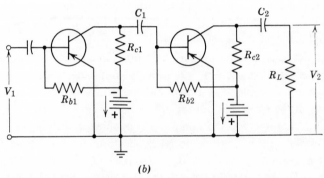

(b)

Fig. 9-1 Two-stage resistance-capacitance coupled voltage amplifiers. (a) Vacuum-tube amplifier, (b) common-emitter transistor amplifier.

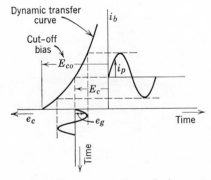

Fig. 9-2 Class A_1 operation. Plate current flows during complete cycle. Grid does not go positive.

319

and the subscript 1 means that grid current does not flow. The grid is not driven positive.

If the bias (E_c) were not so large, the grid would go positive, grid current would flow, and the operation would be Class A_2. This type of operation is not desirable, because the flow of grid current usually prevents proper performance of the source supplying the grid-input voltage, and the advantage gained from larger plate-current amplitude is offset by larger losses within the tube itself.

Class AB. When the grid bias has a value that will permit current cut-off for less than a half-cycle, the operation is Class AB. If the grid does not go positive, the operation is Class AB_1; if it does, grid current flows and the operation is Class AB_2.

It is seen in Fig. 9-3 that the plate-current wave is far from sinusoidal in shape. This is called distortion. It prohibits the use of Class AB operation in an amplifier stage employing a single tube. It will be found later in a push-pull amplifier, a type which employs two tubes in a special circuit in a single stage, Class AB operation is practical and desirable. Indeed, satisfactory performance is possible with a push-pull stage in which plate current flows alternately in the two tubes but for only a half-cycle at a time.

Class B. Although Class B operation is confined mostly to the push-pull type of amplifier, it is used in special applications with one tube. One-tube amplifier operation is often called single-ended operation to distinguish it from push-pull operation. In Class B the grid bias is set at the current cutoff point. The grid-input voltage is necessarily quite large and usually is made to drive the grid a few volts positive, producing Class B_2 operation (Fig. 9-4). Class B is mostly confined to power amplification in push-pull stages.

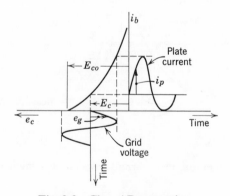

Fig. 9-3　Class AB_1 operation.

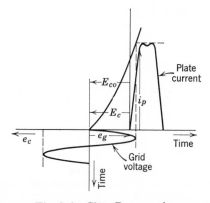

Fig. 9-4   Class B$_2$ operation.

Class C. When power amplification at only one frequency or over a narrow band of frequencies is desired, we may use Class C in either a single-ended or a push-pull stage. In Fig. 9-5 it is seen that plate current flows during appreciably less than a half-cycle. The bias of a Class C amplifier is set anywhere between approximately $1\frac{1}{2}$ and 2 or more times cutoff bias. The plate current flows only in "spurts," and therefore the distortion is great. Class C amplifiers usually are driven with sinusoidal grid voltages. The use of a parallel-resonant circuit in the plate circuit results in the production, across its terminals, of a practically pure sine wave of voltage having large amplitude. The result is that there is no distortion problem when the grid voltage is of only one frequency. Class C amplifiers are usually confined to high-frequency power work. They have greater efficiency than the other classes.

Comparisons of some general properties of amplifiers operated as classified are shown in Table 9-1.

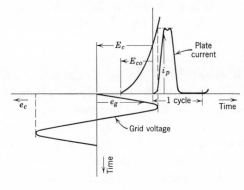

Fig. 9-5   Class C operation.

TABLE 9-1   General Properties of Triode Amplifiers

Class	Bias (Approx.)	Plate-Current Flow	Approx. Relative Efficiency of Triode Power Amplification	Distortion
A_1	$\frac{1}{2}$ cutoff	Full cycle	Low (15–20%)	Minimum
A_2	$\frac{1}{2}$ cutoff	Full cycle	Low—seldom used	—
AB_1	$\frac{2}{3}$ cutoff	Less than full cycle	Appreciably better than A_1—still relatively low (25–30%)	Low
AB_2	$\frac{2}{3}$ cutoff	Less than full cycle	Better than A_1 (25–35%)	Low
B_1	Cutoff	One-half cycle	Not often used, but in PP less than with Class B_2	—
B_2	Cutoff	One-half cycle	Good (40–60% in PP)	Low in PP, high in single tube
C	$1\frac{1}{2}$–$2\frac{1}{2}$ times cutoff	$120°–160°$	High (60–85%)	Low at single frequency

9.3   Frequency Response of Amplifiers

The performance of voltage amplifiers throughout the range of audible frequencies, from about 40 to 15,000 Hz, is an important and interesting topic. If a microphone is placed against a rotating machine, it will develop a weak voltage due to the machine's vibration. The frequency of this voltage may be low or high, depending on how fast the machine rotates or vibrates. It is usually desirable to amplify this voltage so that an indicating instrument may be activated or some other registration effected. An ordinary motor may run at 1,750 r/min. This is approximately 29 r/s. The signal picked up with the microphone would in all probability have a frequency very near to the 29 Hz. A voltage amplifier supplied with this signal should be capable of amplifying it without much distortion.

Speech is easily understood—even when coming from a loud speaker driven by amplifiers incapable of good reproduction above 2,500 or 3,000 Hz —and music has a good sound if a frequency band from 100 to 4,000 Hz is employed. However, the handling of overtones that enrich musical sounds requires that the frequency range of the amplifier be much greater. The amplification should be constant from approximately 40 to 15,000 Hz, for audible response by the average human ear.

Frequency distortion is present in an amplifier when it does not amplify signal voltages of all frequencies by the same amount. Consider, for example, a musical sound that produces in a microphone a voltage wave composed of only two frequencies. One of the components (the fundamental) has a frequency of 500 Hz and a peak value of 1 mV; the other has a frequency of 1,000 Hz and a peak value of $\frac{1}{2}$ mV. If the fundamental is amplified ten-fold to 10 mV, the 1,000-hertz component should also be multiplied tenfold to 5 mV. If, instead, the components are multiplied by different amounts, the amplifier has frequency distortion.

Phase distortion is another undesirable characteristic of an amplifier. Consider a voltage made up of a 500-hertz component and a 1,500-hertz component that happens to be displaced, say, 10 degrees lagging, from the fundamental. At the output of the amplifier the 1,500-hertz component should still be lagging 10 degrees behind the 500-hertz component. If it does not, the amplifier has phase distortion.

Unless phase distortion is severe, it is not very noticeable in the amplification of sound. In high-frequency work it is very noticeable, and in some applications, such as television, it cannot be tolerated.

Nonlinear distortion exists when an amplifier puts out frequencies that are not present in the input signal. Curvature of the dynamic-transfer characteristic results in nonlinear distortion. The generation of a second harmonic by the tube, due to curvature of the dynamic-transfer characteristics, results from large signal excursions. This is discussed in more detail in Chapter 10.

Proper choice of operating voltages and load impedances will help to minimize this type of distortion. Larger load impedance tends to straighten the dynamic-transfer characteristic, reducing its curvature. Operating voltages may be chosen so that the amplifier operates over the most linear part of the characteristic.

DRILL PROBLEMS

D9-1 The signal input to an amplifier is expressed as $e_{in} = 10 \sin \phi - 3 \cos 3\phi$. Its output signal may be expressed as $e_{out} = 100 \sin \phi - 20 \cos 3\phi$. Is the signal distorted in passing through the amplifier? If so, what type of distortion is it?

D9-2 Show an example of a signal that is *phase distorted* in passing through an amplifier.

D9-3 The input signal to an amplifier is $e_{in} = 10 \sin \phi - 5 \cos 2\phi$. The output signal is $e_{out} = 100 \sin \phi - 30 \cos (2\phi - 30°)$. What type(s) of distortion is introduced by this amplifier?

D9-4 The output signal from an amplifier may be expressed in terms of its input signal as $e_{out} = Ke_{in}^2$. For an input signal of $e_{in} = a \sin \phi - b \sin \theta$, calculate the output and determine whether distortion has been introduced and, if so, what type.

9.4 Analysis of Resistance-Capacitance Coupled Vacuum-Tube Amplifier

The ratio of output voltage to input voltage (voltage amplification, A_v) of a resistance-capacitance coupled amplifier is shown plotted against frequency in Fig. 9-6. The region on the frequency scale over which the curve is flat is called the *midband range*. It is seen that the amplification does not change with frequency in this range; hence a mathematical expression for use in calculating the amplification in this range should not contain a term representing frequency. Such an equation will be derived (Equation 9-4). It seems advisable, then, to analyze the performance of the amplifier in its three important ranges of performance, namely, the midband range, the low-frequency range, and the high-frequency range.

Midband Range

The single-stage RC coupled voltage amplifier circuit to be analyzed is shown in Fig. 9-7. Note that the second resistor (R_2), coupled by the capacitor (C) may be the grid-circuit resistor (R_{g2}) of a second stage or it may represent some other kind of load. The imaginary capacitor (C_2) is not actually connected in the circuit, but it represents a real capacitive effect caused by wires being close to each other and to the metal chassis on which

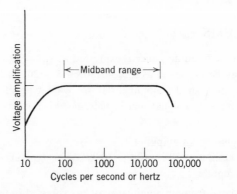

Fig. 9-6 Frequency-response curve of a resistance-capacitance coupled amplifier, designed to amplify audio signals.

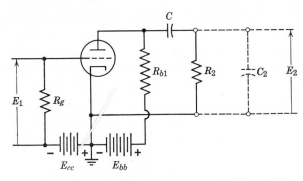

Fig. 9-7   One-stage resistance-capacitance coupled amplifier.

the amplifier is built. It also represents the capacitive effect of the second tube if a second stage is present.

The drop in amplification in the high-frequency range is due to the effect of the output capacitance represented by C_2. At high frequencies the reactance of the coupling capacitor (C) is negligible because its capacitance is relatively large. On the other hand, C_2 is of the order of 10 or 20 pF, and its reactance becomes low enough so that it acts more and more like a short circuit across the output as the frequency of the input and output voltages increases above the midband values.

The drop in amplification in the low-frequency range is largely due to the action of C. It should offer as little opposition as possible to current flow through R_2. At high and midband frequencies its reactance is nearly zero, and so for the purpose of analysis its effect may be ignored. But at low frequencies its reactance becomes great enough to cause it to claim a portion of the amplified voltage developed across the plate resistor (R_{b1}), and so there is less voltage at the output terminals, i.e., across R_2.

The equivalent circuit for analysis of midband operation is shown in Fig. 9-8. Since C_2 and C have negligible reactances at midband frequencies,

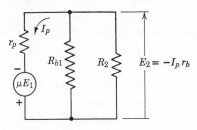

Fig. 9-8   Equivalent circuit of *R-C* coupled amplifier in the midband range.

they are omitted. We shall now obtain a mathematical expression by means of which the voltage amplification may be calculated.

Let r_b represent the equivalent resistance of R_{b1} and R_2 in parallel. This (r_b) may be called the a-c load resistance, since a-c components of plate current flow through both R_2 and R_{b1}. Incidentally, in this type of amplifier the d-c component of plate current flows through only R_{b1}. From earlier work on parallel resistances we see that

$$\frac{1}{r_b} = \frac{1}{R_{b1}} + \frac{1}{R_2} \tag{9-1}$$

Using r_b in place of R_{b1} and R_2, we find that the voltage amplification is

$$A_v = \frac{E_2}{E_1} = \frac{-I_p r_b}{E_1} = -\frac{\mu E_1}{r_b + r_p} \times \frac{r_b}{E_1} = -\frac{\mu r_b}{r_b + r_p} \tag{9-2}$$

I_p is expressed as a negative quantity because E_2 is negative whenever E_1 is positive. Their ratio must be negative. Dividing numerator and denominator by r_b,

$$A_v = -\frac{\mu}{1 + r_p/r_b}$$

Dividing now by r_p, and using Equation (9-1),

$$A_v = -\frac{\mu/r_p}{1/r_p + 1/r_b} = -\frac{\mu/r_p}{1/r_p + 1/R_{b1} + 1/R_2} \tag{9-3}$$

That is, the voltage amplification in the midband range, which we shall denote by A_{vm}, is given by

$$A_{vm} = -\frac{g_m}{1/r_p + 1/R_{b1} + 1/R_2} = -g_m R_e \tag{9-4}$$

in which R_e is the equivalent resistance of r_p, R_{b1}, and R_2 in parallel. None of the quantities in Equation (9-4) depends on frequency. Therefore the amplification in the midband range does not change with frequency. When the frequency is low enough to make X_c no longer negligible, or high enough to make X_{c2} no longer so high that its current drain may be neglected, the limit of the midband range of frequencies has been passed.

Let us examine Equation (9-4) to see what contributes to high-voltage amplification. Evidently the tube should have high transconductance and a high tube-amplification factor. Increasing either R_2 or R_{b1} (or both) helps to decrease the denominator of Equation (9-4), and thus increases the voltage amplification.

There are practical limits to how high these resistance values may be made. Tube design and manufacture produce g_m and μ as high as practicable. R_{b1} is limited by the $I_b R_{b1}$ drop which causes the plate-to-cathode voltage to be small or the required plate-supply voltage (E_{bb}) to be too high. If R_2 represents R_{g2}, the grid-circuit resistor of the following circuit may not be larger than a certain value recommended for the tube by the tube manufacturer, because electrons tend to become "trapped on the grid" instead of "leaking off to the cathode," and then they adversely affect tube operation, by altering the bias value.

EXAMPLE 9-1. Calculate the midband voltage amplification of the amplifier of Fig. 9-7 in which a 6SF5 tube is used having $g_m = 1,500$ μmho; $r_p = 66,000$ Ω; and in which $R_{b1} = 250,000$ Ω and $R_2 = 500,000$ Ω. $C_c = 0.01$ μF.

Solution.   From Equation (9-4)

$$A_{vm} = -\frac{1,500 \times 10^{-6}}{1/250,000 + 1/500,000 + 1/66,000}$$

First multiply the numerator and the denominator by 10^6:

$$A_{vm} = -\frac{1,500}{4 + 2 + 15.2} = -\frac{1,500}{21.2} = -70.7$$

This is a theoretical value and is probably a little higher than that which can actually be obtained. The minus sign is the result of the division of E_2 by E_1 and of E_2 being opposite in phase to E_1.

DRILL PROBLEM

D9-5   A 6SF5 triode is used in the circuit of Fig. 9-7 with $E_{bb} = 300$ V and $E_{cc} = -1.5$ V. The plate load R_{b1} is 150 kΩ and R_2 is 470 kΩ. (*a*) Determine d-c and a-c load lines, assuming operation in the midband frequency range. (*b*) Estimate from the a-c load line the voltage gain for an input signal having a 2-V swing peak-to-peak. (*c*) Calculate the voltage gain using Equation (9-4) and compare with the result of (*b*). (Use $\mu = 100$, $g_m = 1000$ μmho, $r_p = 66,000$ Ω.)

Low-Frequency Range

In analyzing the low-frequency performance of an amplifier stage that is coupled to the next stage by means of a capacitor, the reactance of the coupling capacitor is a matter of primary importance. As the frequency decreases, $X_C = 1/2\pi f C$ becomes larger. The a-c component of voltage across the load resistor, being applied to the series combination of C and R_{g2} (Fig. 9-1), is then divided between these two circuit elements. That is,

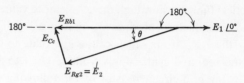

Fig. 9-9 Phasor diagram for R-C coupled voltage amplifier, with resistive load, operated at frequency below midband range. The output voltage $E_{R_g}2$ leads its midband frequency position by an angle θ.

part of the a-c voltage available for output appears across C and cannot be utilized as output voltage. Furthermore, the actual output voltage across R_{g2} is no longer 180 degrees out of phase with the input voltage (E_1). It leads the 180-degree phase position by an angle θ, as shown in Fig. 9-9. For a given amplifier, the value of the phase-shift angle θ depends on the frequency of operation. Figure 9-10 shows how it varies with the product of frequency, the capacitance of the coupling capacitor, and the equivalent resistance R_{el} defined in terms of the resistances in the equivalent circuit for *low-frequency operation*. The voltage amplification at any particular frequency below the midband range will be a certain fraction of the midband value, depending on the frequency. This fraction is shown by the voltage-amplification curve in Fig. 9-10.

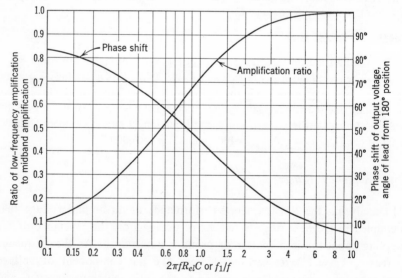

Fig. 9-10 Curves showing amplification ratio and phase shift in low-frequency range. Resistance-capacitance coupled amplifier.

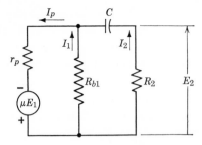

Fig. 9-11   Equivalent-circuit diagram of a resistance-capacitance coupled amplifier operating at frequencies below the midband range.

The equivalent circuit for low-frequency operation is shown in Fig. 9-11. Note that C_2 does not appear. As stated before, it is very small and its reactance is extremely high except at high frequencies. Because of this, its shunting (bypassing) effect is negligible in the midband and low-frequency ranges. The symbol R_2 is used instead of R_{g2} for the sake of consistency with the midband-analysis equation. R_{g2} symbolizes the grid-current resistors of a second amplifier stage, which would not be part of the circuit of a single-stage amplifier. R_2, however, may be used in general to represent any resistive load fed by the amplifier. The symbols for current and voltage represent effective values of a-c components. It should be reemphasized that the a-c component of grid-to-cathode voltage is used in the equivalent circuit. The symbol for the effective value of the a-c component of grid-to-cathode voltage is E_g, but we are using E_1 here because $E_1 = E_g$ since there is no other a-c voltage in the grid circuit that could make E_g different from E_1.

In Fig. 9-11 we see that, starting at the top of the equivalent generator and going counterclockwise, the summation of voltages around the left-hand loop gives

$$+\mu E_1 - I_1 R_{b1} - I_p r_p = 0 \tag{9-5}$$

Using the other loop we obtain, going clockwise,

$$-I_1 R_{b1} + I_2 R_2 + I_2 X_C = 0 \tag{9-6}$$

But

$$I_p = I_1 + I_2 \quad \text{and} \quad X_C = \frac{1}{j\omega C} = -\frac{j}{\omega C}$$

where $\omega = 2\pi f$. Substituting these relations into Equations (9-5) and (9-6), and rearranging terms:

$$I_1 R_{b1} + (I_1 + I_2) r_p = \mu E_1 \tag{9-7}$$

$$I_1 R_{b1} - I_2 (R_2 - j/\omega C) = 0 \tag{9-8}$$

$$I_1 (R_{b1} + r_p) + I_2 r_p = \mu E_1 \tag{9-9}$$

The output voltage (E_2) is $-I_2R_2$. The minus sign is used because, at the instant the currents flow as indicated, the top of R_2 is negative with respect to the bottom (a-c ground connection of actual circuit), and at that instant E_1 is positive with respect to ground. Hence E_2 is in its range of negative values with respect to E_1. If X_C were zero, E_2 would be directed exactly opposite to E_1 in the phasor diagram.

Using $E_2 = -I_2R_2$, we obtain the expression for the voltage amplification in the low-frequency range:

$$A_{vl} = \frac{E_2}{E_1} = -\frac{I_2R_2}{E_1} \tag{9-10}$$

The minus sign indicates 180-degree phase shift through the tube. Let us now solve Equations (9-8) and (9-9) for I_2. From Equation (9-7),

$$I_1 = \frac{\mu E_1 - I_2 r_p}{R_{b1} + r_p} \tag{9-11}$$

Substituting this into Equation (9-8),

$$\frac{\mu E_1 R_{b1} - I_2 r_p R_{b1}}{R_{b1} + r_p} - I_2\left(R_2 - \frac{j}{\omega C}\right) = 0 \tag{9-12}$$

Multiplying both sides by $(R_{b1} + r_p)$,

$$\mu E_1 R_{b1} - I_2 r_p R_{b1} - I_2(R_{b1} + r_p)\left(R_2 - \frac{j}{\omega C}\right) = 0$$
$$\mu E_1 R_{b1} - I_2 r_p R_{b1} - I_2\left(R_{b1}R_2 + r_p R_2 - j\frac{R_{b1} + r_p}{\omega C}\right) = 0 \tag{9-13}$$

$$I_2 = \frac{\mu R_{b1} E_1}{r_p R_{b1} + R_{b1}R_2 + r_p R_2 - \dfrac{j(R_{b1} + r_p)}{\omega C}} \tag{9-14}$$

Using Equation (9-10), the voltage amplification in the low-frequency range is

$$A_{vl} = \frac{-\mu R_{b1} R_2}{r_p R_{b1} + R_{b1}R_2 + r_p R_2 - \dfrac{j(R_{b1} + r_p)}{\omega C}} \tag{9-15}$$

Equation (9-15) may be used to compute the voltage amplification at any value of frequency in the low-frequency range. Inasmuch as the midband amplification is a constant and rather easy to calculate, the amplification in the low- and high-frequency ranges may be expressed in terms of A_{vm}.

Since $\mu = g_m r_p$,

$$A_{vl} = \frac{-g_m r_p R_{b1} R_2}{r_p R_{b1} + R_{b1} R_2 + r_p R_2 - \dfrac{j(R_{b1} + r_p)}{\omega C}} \tag{9-16}$$

Dividing numerator and denominator by $(r_p R_{b1} + R_{b1} R_2 + r_p R_2)$,

$$A_{vl} = -\frac{g_m \dfrac{r_p R_{b1} R_2}{r_p R_{b1} + R_{b1} R_2 + r_p R_2}}{1 - \dfrac{j(R_{b1} + r_p)}{\omega C(r_p R_{b1} + R_{b1} R_2 + r_p R_2)}} \tag{9-17}$$

Referring to Equation (9-4) and converting it to an equivalent form,

$$A_{vm} = -g_m \frac{1}{\dfrac{1}{r_p} + \dfrac{1}{R_{b1}} + \dfrac{1}{R_2}}$$

$$= -g_m \frac{r_p R_{b1} R_2}{r_p R_{b1} + R_{b1} R_2 + r_p R_2} = -g_m R_e \tag{9-18}$$

in which R_e represents the parallel resistance combination shown in two forms. Equation (9-17) may then be written

$$A_{vl} = \frac{-g_m R_e}{1 - \dfrac{j}{\omega C R_{el}}} = \frac{A_{vm}}{1 - \dfrac{j}{\omega C R_{el}}} \tag{9-19}$$

in which R_{el} is an equivalent resistance for the low-range computation:

$$R_{el} = \frac{r_p R_{b1} + R_{b1} R_2 + r_p R_2}{R_{b1} + r_p} \tag{9-20}$$

Since every quantity on the right-hand side of Equation (9-19) is constant except ω, it is only necessary to substitute the value of the frequency and to solve for the amplification *after A_{vm} has been determined.* Note that, if the frequency is large enough to make the j term of the denominator very small compared to unity, the amplification becomes the midband value, as it should. The ratio of A_{vl} to A_{vm} is therefore

$$\frac{A_{vl}}{A_{vm}} = \frac{1}{1 - \dfrac{j}{\omega C R_{el}}} \tag{9-21}$$

The frequency at which the denominator of Equation (9-21) becomes $(1 - j1)$ will be useful as a reference to which all other frequencies may be compared. It is seen that this frequency can be determined from

$$1 = \frac{1}{\omega CR_{el}} = \frac{1}{2\pi f_1 CR_{el}}$$

where f_1 is the frequency that makes the expression equal to unity. Solving for the frequency f_1 yields

$$f_1 = \frac{1}{2\pi CR_{el}}$$

Substituting this value into Equation (9-21) yields

$$\frac{A_{vl}}{A_{vm}} = \frac{1}{1 - j(f_1/f)} \qquad (9\text{-}21a)$$

The frequency f_1 is often referred to as the *low-frequency half power frequency*. At this frequency the output voltage of an amplifier is reduced from its midband amplitude by an amount that represents only half the power in the load. Let us see that this is so. Since the amplification is a measure of the output voltage and the output voltage squared is proportional to the load power, then we may use Equation (9-21a) to calculate load power at different frequencies. We will calculate the frequency f at which the power ratio is one-half in the following manner:

$$\frac{1}{2} = \frac{1^2}{[1 - j(f_1/f)]^2} = \frac{A_{vl}^2}{A_{vm}^2}$$

or

$$2 = 1^2 + \frac{f_1^2}{f^2}$$

This expression is true when $f = f_1$.

It is convenient to use a curve giving the value of this ratio for all values of ωCR_{el} or f_1/f. Reich[*] has shown a set of amplification-ratio and phase-shift curves that are convenient for determining quite accurately the voltage amplification in the low-frequency and high-frequency ranges. Figure 9-10

---

[*] H. J. Reich, *Theory and Application of Electron Tubes*, McGraw-Hill, New York, 1944.

shows the curves for the low-frequency range. The constant R_{el} and the frequency in the low range at which the amplification is desired being known, it is necessary only to compute the product $2\pi f R_{el} C$ and find its value on the horizontal scale. The point directly above, on the amplification curve, has an ordinate on the relative-amplification scale denoting the ratio of the amplification in question to the midband amplification, which is known from Equation (9-4).

EXAMPLE 9-2.   An example of the determination of amplification in the low range follows. The circuit of Example 9-1 will be used again. Assume that the voltage amplification at 60 Hz is desired. The equivalent resistance for low-range determinations is obtained by using Equation (9-20):

$$R_{el} = \frac{66,000 \times 250,000 + 250,000 \times 500,000 + 66,000 \times 500,000}{250,000 + 66,000}$$

$$R_{el} = \frac{10^6(16,500 + 125,000 + 33,000)}{10^6(0.25 + 0.066)}$$

$$R_{el} = \frac{174,500}{0.316} = 552,200$$

$$\omega R_{el} C = 2\pi \times 60 \times 552,200 \times 0.01 \times 10^{-6} = 2.08 = f_1/f$$

The curve of Fig. 9-10 shows the ratio A_{vl}/A_{vm} to be 0.9. Since $A_{vm} = 70.7$, from Example 9-1 above,

$$A_{vl} = 0.9 \times 70.7 = 63.6.$$

The phase shift, as read from the other curve of Fig. 9-10, is found to be $26°$ beyond the $180°$ position, making $206°$ between E_1 and E_2. To determine A_{vl} directly, use Equation (9-15) ($\mu = 100$ as given in the tube manual):

$$A_{vl} = -\frac{100 \times 0.25 \times 0.5 \times 10^{12}}{10^{12}(0.066 \times 0.25 + 0.25 \times 0.5 + 0.066 \times 0.5) - \dfrac{j10^6(0.25 + 0.066)}{2\pi \times 60 \times 0.01 \times 10^{-6}}}$$

$$= -\frac{12.5}{0.0165 + 0.125 + 0.033 - j0.316/3.77}$$

$$= -\frac{12.5}{0.1745 - j0.0839}$$

$$A_{vl} = -\frac{12.5}{0.1745 - j0.0839} \times \frac{0.1745 + j0.0839}{0.1745 + j0.0839}$$

$$= -\frac{2.18 + j1.05}{0.0305 + 0.00706} = -57 - j28 = 63.6\underline{/206.2°}$$

The angle of lead, θ, with reference to the 180-degree phase position is seen to be $26.2°$ (Fig. 9-12). The relative positions of the phasors verify the general phasor diagram shown earlier in Fig. 9-9.

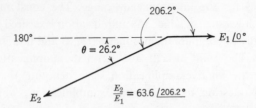

Fig. 9-12 Phasor diagram showing phase shift at low frequency of output voltage (E_2) with respect to its midband 180° position. Shift is in the leading direction.

High-Frequency Range

Attention has been called to the presence of capacitance at the output terminals of an amplifier. Plate-to-cathode capacitance of the tube, capacitance between plate-circuit wiring and the input capacitance of the following stage or other connected load all contribute to the bypassing of high-frequency energy. The combined capacitance of these effects is represented by C_2 in the equivalent-circuit diagram shown in Fig. 9-13. The current I_2 flows through an equivalent impedance (Z_2) which is given by the equation

$$\frac{1}{Z_2} = \frac{1}{R_2} + \frac{1}{X_{C_2}} \tag{9-22}$$

Since

$$\frac{1}{X_{C_2}} = \frac{1}{1/j\omega C_2}$$

$$\frac{1}{Z_2} = \frac{1}{R_2} + j\omega C_2 \tag{9-23}$$

The current I_p flows through a total equivalent impedance made up of R_{b1}

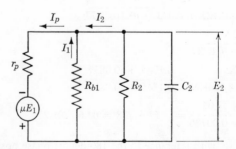

Fig. 9-13 Equivalent-circuit diagram of a resistance-capacitance coupled amplifier operating at frequencies above the midband range.

in parallel with Z_2. Calling this total equivalent impedance Z_e, we have

$$\frac{1}{Z_e} = \frac{1}{R_{b1}} + \frac{1}{R_2} + j\omega C_2 \tag{9-24}$$

The plate current (I_p) is then given by

$$I_p = \frac{\mu E_1}{r_p + Z_e} \tag{9-25}$$

in which Z_e is a complex number. The output voltage (E_2) is equal to I_p multiplied by the total equivalent inpedance (Z_e) through which it flows.

$$E_2 = -I_p Z_e = \frac{-\mu E_1 Z_e}{r_p + Z_e} \tag{9-26}$$

and the voltage amplification in the high-frequency range, A_{vh}, is

$$A_{vh} = \frac{E_2}{E_1} = \frac{-\mu Z_e}{r_p + Z_e} \tag{9-27}$$

Now insert the expression for Z_e in order to make this equation useful. Let us first divide the numerator and denominator by $r_p Z_e$:

$$A_{vh} = \frac{-\mu/r_p}{\dfrac{1}{Z_e} + \dfrac{1}{r_p}} \tag{9-28}$$

Using Equation (9-24),

$$A_{vh} = \frac{-\mu/r_p}{\dfrac{1}{R_{b1}} + \dfrac{1}{R_2} + \dfrac{1}{r_p} + j\omega C_2} = \frac{-g_m}{\dfrac{1}{R_{b1}} + \dfrac{1}{R_2} + \dfrac{1}{r_p} + j\omega C_2} \tag{9-29}$$

Observe that this equation is very similar to Equation (9-4), which is for the midband amplification. Indeed, if the frequency is so low that the value of $j\omega C_2$ is negligible, it becomes identical with Equation (9-4), as it should.

Equation (9-4) allows us to write Equation (9-29) in the form

$$A_{vh} = \frac{-g_m}{(1/R_e) + j\omega C_2} = \frac{-g_m R_e}{1 + j\omega C_2 R_e} \tag{9-30}$$

And, from Equation (9-18),

$$A_{vh} = \frac{A_{vm}}{1 + j\omega C_2 R_e} \tag{9-31}$$

$$\frac{A_{vh}}{A_{vm}} = \frac{1}{1 + j\omega C_2 R_e} \tag{9-32}$$

We may calculate a frequency in the high-frequency operating region at which the load power is one-half its value in the midband region. This frequency is often referred to as the *upper half power frequency*. We will label it f_2, and calculate its value in the manner used for the lower half power frequency:

$$1 = 2\pi f_2 C_2 R_e$$

or

$$f_2 = \frac{1}{2\pi C_2 R_e}$$

Substituting this value into Equation (9-32) yields

$$\frac{A_{vh}}{A_{vm}} = \frac{1}{1 + j(f/f_2)} \tag{9-32a}$$

To show that the load power is one-half its midband value at $f = f_2$, square both sides of Equation (9-32a) and set equal to $\frac{1}{2}$:

$$\frac{1}{2} = \frac{1^2}{[1 + j(f/f_2)]^2} = \frac{A_{vh}{}^2}{A_{vm}{}^2}$$

or

$$2 = 1^2 + \frac{f^2}{f_2{}^2}$$

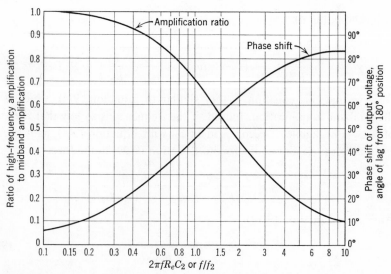

Fig. 9-14   Curves showing amplification ratio and phase shift in high-frequency range. Resistance-capacitance coupled amplifier.

This expression will be true only when $f = f_2$. Therefore the load power is one-half of its midband value when the driving frequency is f_2.

The range of frequencies between f_1 and f_2 is called the *bandwidth* of the amplifier. This concept is presented in some detail in Chapter 1. It can be used as a *figure of merit* when comparing the performance of one circuit design with another. It is a measure of the range of driving frequencies that may be amplified by the circuit in order to maintain load power between its full value and one-half that amount. Further, it will be seen that within the bandwidth the phase shift of the amplifier will remain within $\pm 45°$ of the $180°$ phase shift at midband frequencies.

Equation (9-32) suggests that a curve of A_{vh}/A_{vm} plotted against $\omega C_2 R_e$ or f/f_2 would be very useful in determining A_{vh} after A_{vm} is known. Such a curve, and one for phase shift, is shown in Fig. 9-14.

EXAMPLE 9-3. Now determine the voltage amplification of the circuit in Example 9-1 at a frequency in the high range. Assume that the total output capacitance C_2 is 50 pF and that the frequency in question is 50,000 Hz.

Solution. Assuming that we do not know the amplification in the midband range, let us use Equation (9-29).

$$A_{vh} = \frac{-g_m}{\dfrac{1}{R_{b1}} + \dfrac{1}{R_2} + \dfrac{1}{r_p} + j\omega C_2}$$

$$= -\frac{1,500 \times 10^{-6}}{10^{-6}\left(\dfrac{1}{0.25} + \dfrac{1}{0.5} + \dfrac{1}{0.066}\right) + j2\pi \times 50,000 \times 50 \times 10^{-12}}$$

$$= -\frac{1,500}{4 + 2 + 15.3 + j15.71} = -\frac{1,500}{21.3 + j15.71}$$

$$= -\frac{1,500}{21.3 + j15.71} \times \frac{21.3 - j15.71}{21.3 - j15.71} = -\frac{31,800 - j23,565}{449.44 + 246.80}$$

$$A_{vh} = -45.8 + j33.8 = 56.9\,\underline{/143.6°}$$

We shall verify this result by the shorter method of solution that employs the universal curves of Fig. 9-14. The mid-frequency-band amplification was found to be 70.7. Solving for the value to be used on the horizontal scale, we obtain

$$\omega C_2 R_e = 2\pi \times 50,000 \times 50 \times 10^{-12} \times \frac{10^6}{21.3} = 0.742 = \frac{f}{f_2}$$

$R_e = 10^6/21.3$ was determined in the solution for A_{vm} in Example 9-1 of this section. At this value (0.742) on the horizontal axis of Fig. 9-14, we find the amplification ratio to be 0.80, so that $A_{vh} = 0.80 \times A_{vm} = 0.80 \times 70.7 = 56.5$. The phase shift is seen to be $36°$ lagging the midband $180°$ position or an advance in phase of $144°$.

DRILL PROBLEMS

D9-6   Using the curves of Figs. 9-10 and 9-14, sketch the relative voltage gain of an amplifier that has a bandwidth extending from 100 Hz to 100 kHz.

D9-7   Show that the gain of an *R-C* coupled amplifier is "down 3 dB" from its midband gain at f_1 and f_2. (*Hint:* Start with Equations (9-21) and (9-32) and calculate the absolute magnitudes; then express in terms of decibels.)

9.5   General Conditions in Transistor Amplifier Analysis

Having analyzed the single-stage vacuum-tube amplifier with a mathematical approach, we find it logical to move on to a mathematical analysis of the single-stage transistor amplifier. As with the tube-type amplifiers, the analysis of single-stage performance may be applied to a multistage amplifier if correct values of source voltage and impedances of the source and the load for each stage are known.

We shall need to analyze the performance of the transistor in each of the three possible circuit arrangements, namely, common-base, common-emitter, and common-collector. Input and output resistances must be determined. The *h*-parameter symbols will be used.

In addition to voltage gain, current gain, and power gain, input impedance and output impedance are of prime importance. Only pure resistance loading will be studied; this means that these impedances will turn out to be pure resistances in the frequency range of operation. Reactive loads on tube and transistor amplifiers cause the load line to become an ellipse and introduce such extensive complications that mathematical analysis is far beyond the scope of this book. At very high frequencies, inductance and capacitance values show up in tubes and transistors to add further complications.

Input resistance of a transistor amplifier is simply defined as the input voltage divided by the input current *when the load is connected. Output resistance* is defined as the output voltage divided by the output current *when the supply source is connected.* It is the resistance presented to the output terminals *by the amplifier circuit when the source is connected.* Of course, the load resistance is not included. However, the load resistance should be selected to match the output resistance of the stage for maximum power to the load. *The transistor amplifier is essentially a power amplifier.*

It is, as previously emphasized, a current-actuated device. This means that its control (input) circuit takes power, and the predominant objective in the circuit design is to obtain maximum power amplification consistent with other desirable features of performance such as minimum distortion, long lift, and stability.

Power gain is, as in all cases, the ratio of power output to power input. Because only the a-c power is useful in an amplifier, P_o and P_i will be understood to signify a-c power.

$$\text{Power gain} = A_p = P_o/P_i \tag{9-33}$$

$$\text{Current gain} = A_i = \text{load current/input current} = I_L/I_i \tag{9-34}$$

with I_L = a-c load current and I_i = a-c input current.

$$\text{Voltage gain} = A_v = \text{load voltage/input voltage} = V_L/V_i \tag{9-35}$$

where V_L = a-c load voltage and V_i = a-c input voltage.

9.6 Common-Base Amplifier Analysis

The equivalent circuit of the common-base transistor amplifier is used for the purpose of mathematical analysis and it is presented again (Fig. 9-15) for convenience. In order to obtain the expression for input resistance, the input end must be opened at points 1 and 2. In order to derive the formula for output resistance, the load R_L must be disconnected.

Input Resistance

Using Kirchhoff's voltage law on the input loop to the right of points 1, 2 gives

$$V_1 - I_1 h_{ib} - h_{rb} V_2 = 0 \tag{9-36}$$

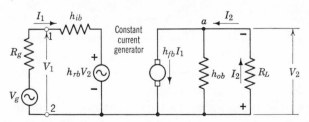

Fig. 9-15 Equivalent circuit of common-base transistor amplifier.

In the output circuit, the currents at node *a* are added. Remember that h_{ob} is an *admittance* ($1/Z_{o2}$), so its current is given by $h_{ob}V_2$.

$$V_2/R_L + h_{ob}V_2 + h_{fb}I_1 = 0 \tag{9-37}$$

The *known quantities* are V_1 and I_2 (in addition to the *h*'s), but the *unknowns* are I_1 and V_2. These should be kept together on the same side of the equality sign, while the known quantities are put on the opposite side.

$$h_{ib}I_1 + h_{rb}V_2 = V_1 \tag{9-38}$$

$$h_{fb}I_1 + (1/R_L + h_{ob}V_2) = 0 \tag{9-39}$$

Solving for V_2 in Equation (9-39) and substituting into Equation (9-38) gives

$$I_1 = \frac{(1/R_L + h_{ob})V_1}{(1/R_L + h_{ob})h_{ib} - h_{fb}h_{rb}} \tag{9-40}$$

By means of this equation the a-c input current could be computed when the applied signal voltage and the load resistance are known. The *h*-parameters are obtainable for the particular transistor from the manufacturer, or they can be measured in the laboratory.

To obtain the *expression for input resistance*, we simply divide V_1 by I_1, and this inverts the form of the right-hand side of Equation (9-40).

$$\frac{V_1}{I_1} = R_i = \frac{(1/R_L + h_{ob})h_{ib} - h_{fb}h_{rb}}{1/R_L + h_{ob}} \, \Omega$$

This equation may be simplified somewhat by multiplying numerator and denominator by R_L and rearranging terms.

Input resistance common-base

$$R_{ib} = h_{ib} - \frac{h_{fb}h_{rb}R_L}{1 + h_{ob}R_L} \, \Omega \tag{9-41}$$

Checking the dimensions by substituting units for each symbol,

$$\text{Ohms} - \frac{\dfrac{\text{amperes}}{\text{amperes}} \times \dfrac{\text{volts}}{\text{volts}} \times \text{ohms}}{\text{mho} \times \text{ohms}} = \text{ohms} - \text{ohms} = \text{ohms}$$

The numerator of the second term turns out to be ohms and the denominator dimensionless, giving ohms as the result.

Equation (9-41) shows that the input resistance of a transistor amplifier depends very much on the value of R_L. Input-circuit values are influenced by changes in output-circuit values. This is very different from vacuum-tube operation, where the tube quite effectively isolated the grid circuit from the load circuit.

Output Resistance

To set up the equations needed for determining the expression for output resistance, we refer again to the general equivalent circuit (Fig. 9-15). The generator is left connected, but R_L must be disconnected.

Using node a again, we write a current equation,

$$I_2 - h_{ob}V_2 - h_{fb}I_1 = 0 \tag{9-42}$$

and, in the input circuit,

$$h_{rb}V_2 + I_1(h_{ib} + R_g) = 0 \tag{9-43}$$

It will be recalled that V_1 and I_2 are known quantities, but V_1 does not appear here. Rearranging the terms of Equations (9-42) and (9-43),

$$h_{ob}V_2 + h_{fb}I_1 = I_2 \tag{9-44}$$

$$h_{ib}V_2 + (h_{ib} + R_g)I_1 = 0 \tag{9-45}$$

Solving for V_2 gives

$$V_2 = \frac{(R_g + h_{ib})I_2}{(R_g + h_{ib})h_{ob} - h_{rb}h_{fb}} \text{ V} \tag{9-46}$$

A check of dimensions is left to the student. It will be beneficial to review the parameter definitions again.

The output resistance is the ratio V_2/I_2, and its expression is obtained by dividing Equation (9-46) by I_2.

Output resistance common-base

$$\frac{V_2}{I_2} = R_{ob} = \frac{R_g + h_{ib}}{(R_g + h_{ib})h_{ob} - h_{rb}h_{fb}} \tag{9-47}$$

Observe that the resistance of the generator, be it a signal source or the impedance looking back into an amplifier stage driving this one, is significant in determining the value of the output resistance of the amplifier. This is not true of the vacuum-tube amplifier.

Gains

The expressions for the various gains of the common-base transistor amplifier will now be found. In order to obtain the one for *voltage gain*, an expression for V_2 must be derived. Furthermore, it must come from the equations that involve V_1 so that when we divide by V_1 it will cancel out and leave an expression involving only the circuit constants. Therefore we use Equations (9-38) and (9-39), solving for V_2, with the result

$$V_2 = \frac{-h_{fb}R_LV_1}{(h_{ob}R_L + 1)h_{ib} - h_{rb}h_{fb}R_L}$$ (9-48)

The ratio of V_2 to V_1 is now apparent.

Voltage gain, common-base

$$A_{vb} = \frac{V_2}{V_1} = \frac{-h_{fb}R_L}{(h_{ob}R_L + 1)h_{ib} - h_{rb}h_{fb}R_L}$$ (9-49)

Current gain is the ratio of I_2 to I_1. If we can get an equation containing both currents, it will be a simple matter to solve for *their ratio*. In the output circuit of Fig. 9-15, I_2 is shown flowing *upward* through R_L so that the top point is negative and the bottom positive. But V_2 is designated as the voltage *from bottom to top*, which is therefore negative. Accordingly, we must write

$$V_2 = -I_2R_L$$ (9-50)

The loop equation on the input side is given above as Equation (9-37). Putting in $-I_2R_L$ for V_2,

$$\frac{-I_2R_L}{R_L} - I_2h_{ob}R_L + h_{fb}I_1 = 0$$

$$I_2(1 + h_{ob}R_L) = h_{fb}I_1$$

Current gain, common-base

$$A_{ib} = \frac{I_2}{I_1} = \frac{h_{fb}}{h_{ob}R_L + 1}$$ (9-51)

The fact that this ratio is positive indicates that in this case *the currents I_1 and I_2 are in phase*.

Power gain is the ratio of output power to input power. When no reactances are present, the powers are obtained by straight multiplication of voltage and current.

$$A_p = \frac{V_2I_2}{V_1I_1} = \frac{V_2}{V_1} \times \frac{I_2}{I_1}$$ (9-52)

Evidently, power gain is expressible as voltage gain times current gain in transistors. Multiplying Equation (10-49) by Equation (10-51), we get

$$A_{pb} = \frac{h_{fb}{}^2 R_L}{(h_{ob}R_L + 1)^2 h_{ib} - (h_{ob}R_L + 1)h_{rb}h_{fb}R_L}$$

(9-53)

It would be a waste of time to memorize these gain equations. It is sufficient to understand their derivations and to memorize the meanings of the symbols used.

The forward-current ratio h_{fb} is, by definition, I_2/I_1 when $V_2 = 0$ (output shorted). For the common-base transistor $\alpha_b = -\Delta I_C/\Delta I_E$ with V_{CB} constant. The only way V_{CB} can remain constant while I_C has even a small change is for the collector to be shorted to the base. It follows, then, that *for the common-base connection* $h_{fb} = -\alpha_b$.

9.7 Common-Emitter Amplifier Analysis

The actual and equivalent circuits for the common-emitter amplifier are shown in Fig. 9-16.

At the node a, the current relations are

$$h_{fe}I_1 + h_{oe}V_2 + V_2/R_L = 0$$

(9-54)

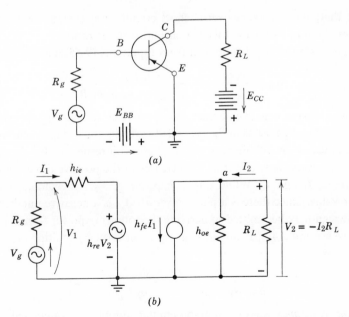

(a)

(b)

Fig. 9-16 Circuits of common-emitter transistor amplifier. (*a*) Actual circuit; (*b*) equivalent circuit.

from which we obtain

$$V_2 = \frac{-h_{fe}R_L I_1}{1 + h_{oe}R_L} \tag{9-55}$$

The output current is $-V_2/R_L$,

$$I_2 = \frac{-h_{fe}I_1}{1 + h_{oe}R_L} \tag{9-56}$$

The input voltage is

$$V_1 = h_{ie}I_1 + h_{re}V_2 \tag{9-57}$$

Using Equation (9-55),

$$V_1 = h_{ie}I_1 - \frac{h_{re}h_{fe}R_L I_1}{1 + h_{oe}R_L} \tag{9-58}$$

$$V_1 = \frac{I_1[h_{ie}(1 + h_{oe}R_L) - h_{re}h_{fe}R_L]}{1 + h_{oe}R_L} \tag{9-59}$$

The *current gain*, A_{ie}, as before, is I_2/I_1, which is obtained from Equation (9-56):

$$A_{ie} = \frac{I_2}{I_1} = \frac{-h_{fe}}{1 + h_{oe}R_L} \tag{9-60}$$

Notice that, if R_L becomes zero (short circuit), the current amplification becomes equal to $-h_{fe}$, the forward-current transfer ratio.

The *voltage gain*, A_{ve}, is obtained from Equations (9-55) and (9-59):

$$A_{ve} = \frac{V_2}{V_1} = \frac{-h_{fe}R_L}{h_{ie}(1 + h_{oe}R_L) - h_{re}h_{fe}R_L} \tag{9-61}$$

Since this ratio is a negative number, this means that the same phase relations between V_2 and V_1 exist as in the grounded-cathode single-stage vacuum-tube amplifier, i.e., V_2 is 180 degrees out of phase with respect to V_1. To explain it in another way, notice that as V_1 increases in the positive direction so do the currents I_1 and I_2. While this is happening, V_2 is becoming a larger *negative* value. Since there is a phase reversal, A_{ve} is a negative number. The negative sign may be omitted in deriving the following power formula.

The *power gain*, A_{pe}, is obtained in equation form by multiplying A_{ve} by A_{ie}:

$$A_{pe} = \frac{(h_{fe})^2 R_L}{h_{ie}(1 + h_{oe}R_L)^2 - h_{re}h_{fe}R_L(1 + h_{oe}R_L)} \tag{9-62}$$

The *input resistance* for the common-emitter amplifier is calculated in the same manner as for the common-base amplifier, except that *h*-parameters for the common-emitter connection must be used.

$$R_{ie} = h_{ie} - \frac{h_{re}h_{fe}R_L}{1 + h_{oe}R_L} = h_{ie} - \frac{h_{re}h_{fe}}{h_{oe} + \dfrac{1}{R_L}} \tag{9-63}$$

Similarly, the *output resistance* is:

$$R_{oe} = \frac{R_g + h_{ie}}{(R_g + h_{ie})h_{oe} - h_{re}h_{fe}} = \frac{1}{h_{oe} - \dfrac{h_{re}h_{fe}}{h_{ie} + R_g}}$$

The *common-collector* amplifier may be analyzed in a similar manner. The analysis will lead to the same form of expressions for gains and resistances, except that the common-collector *h*-parameters are used.

9.8   Comparison of the Three Connections

The expressions for gains and resistances seem rather cumbersome, but they are easy to evaluate. The expressions are similar for each of the three common connections and only require that the correct *h*-parameters be used. Table 9-2 lists the expressions that are used to evaluate the various gains and resistances. The factor D represents $(h_{i-}h_{o-} - h_{r-}h_{f-})$, which may be calculated separately and used as a multiplier in several of the expressions.

As an example of the calculations necessary to evaluate these properties of transistor amplifiers operating in the three basic connections, the *h*-parameters for the 2N334 transistor shown in Table 9-3 will be used. The common-base *h*-parameters are furnished by the manufacturer, and the remaining *h*-parameters have been calculated using the expressions of Table 8-1.

TABLE 9-2   Equations for Calculating Gains and Resistances

Current gain	$A_{i-} = \dfrac{h_{f-}}{h_{o-}R_L + 1}$
Voltage gain	$A_{v-} = \dfrac{-h_{f-}R_L}{h_{i-} + DR_L}$
Power gain	$A_{p-} = A_{v-}A_{i-}$
Input resistance	$R_{i-} = \dfrac{h_{i-} + DR_L}{h_{o-}R_L + 1}$
Output resistance	$R_{o-} = \dfrac{R_g + h_{i-}}{h_{o-}R_g + D}$

TABLE 9-3 *h*-Parameters of 2N334 *NPN* Transistor

	Common-Base *h*-Parameters		
$h_{ib} = 56\,\Omega,$	$h_{ob} = 0.5\,\mu\text{mho},$	$h_{rb} = 350 \times 10^{-6},$	$h_{fb} = -0.97$
	Common-Emitter *h*-Parameters		
$h_{ie} = 1870\,\Omega,$	$h_{o} = 16.7\,\mu\text{mho},$	$h_{re} = 9.2 \times 10^{-4},$	$h_{fe} = 32.3$
	Common-Collector *h*-Parameters		
$h_{ic} = 1870\,\Omega,$	$h_{oc} = 16.7\,\mu\text{mho},$	$h_{rc} = 1,$	$h_{fc} = -33.3$

EXAMPLE 9-4 For the purposes of these calculations, we will assume that the source and load resistors are equal and have a value of 1,000 Ω. We will calculate and compare the numerical values of voltage gain, current gain, power gain, input resistance, and output resistance for each of the three basic connections of a transistor amplifier. The equivalent circuit of Fig. 9-16 is used for the analysis.

First, we calculate the factor *D* given in Table 9-2.

Common-Base

$$D_b = h_{ib}h_{ob} - h_{rb}h_{fb} = (56)(0.5 \times 10^{-6}) - (350 \times 10^{-6})(-0.97)$$

$$= (28 + 339.5) \times 10^{-6}$$

$$= 367.5 \times 10^{-6}$$

Common-Emitter

$$D_e = (1870)(16.7 \times 10^{-6}) - (920 \times 10^{-6})(32.3)$$

$$= (3120 - 2970) \times 10^{-6}$$

$$= 150 \times 10^{-6}$$

Common-Collector

$$D_c = (1870)(16.7 \times 10^{-6}) - (1)(-33.3)$$

$$= 33.303$$

$$\doteq 33.3$$

Next, we will evaluate the *current gain* for each of the three connections.

Common-Base

$$A_{ib} = \frac{h_{fb}}{h_{ob}R_L + 1} = \frac{-0.97}{(0.5 \times 10^{-6})(10^3) + 1}$$

$$\doteq -0.97$$

Common-Emitter

$$A_{ie} = \frac{32.3}{(16.7 \times 10^{-6})(10^3) + 1}$$

$$\doteq 32.3$$

Common-Collector

$$A_{ic} = \frac{-33.3}{(16.7 \times 10^{-6})(10^3) + 1}$$
$$\doteq -33.3$$

The *voltage gain* may be calculated using the equation of Table 9-2.

Common-Base

$$A_{vb} = \frac{-h_{fb}R_L}{h_{ib} + DR_L} = \frac{-(-0.97)(10^3)}{56 + (377.5 \times 10^{-6})(10^3)}$$
$$= 17.2$$

Common-Emitter

$$A_{ve} = \frac{-(32.3)(10^3)}{1870 + (150 \times 10^{-6})(10^3)}$$
$$= -17.3$$

Common-Collector

$$A_{vc} = \frac{-(-33.3)(10^3)}{1870 + (33.3)(10^3)}$$
$$= 0.95$$

The *power gain* for each basic connection may be calculated.

Common-Base

$$A_{pb} = A_{vb}A_{ib} = 17.2 \times 0.97$$
$$= 16.68$$

Common-Emitter

$$A_{pe} = 17.3 \times 32.3$$
$$= 558.$$

Common-Collector

$$A_{pc} = 0.95 \times 33.3$$
$$= 31.6$$

The *input resistance* for each of the basic connections is found in a similar manner, using the equation of Table 9-2.

Common-Base

$$R_{ib} = \frac{56 + (377.5 \times 10^{-6})(10^3)}{(0.5 \times 10^{-6})(10^3) + 1} = 56.3 \ \Omega$$

Common-Emitter

$$R_{ie} = \frac{1870 + (150 \times 10^{-6})(10^3)}{(16.7 \times 10^{-6})(10^3) + 1} = 1870 \ \Omega$$

Common-Collector

$$R_{ic} = \frac{1870 + (33.3)(10^3)}{(16.7 \times 10^{-6})(10^3) + 1} \doteq 34.6 \ \text{k}\Omega$$

TABLE 9-4 Comparison of Transistor Performance in the Three Basic Common-Terminal Connections

Type of Connection	R_i	R_o	A_v	A_i	A_p
Common-base	56.3 Ω	1.2 MΩ	17.2	−0.97	16.68
Common-emitter	1870 Ω	170 kΩ	−17.3	32.3	558.
Common-collector	33.5 kΩ	86.5 Ω	0.95	−33.3	31.6

The values of *output resistance* may be evaluated in the same way. The resulting numerical values are listed in Table 9-4 for comparison of the three common connections.

In general, transistor amplifiers (except those using the *FET*) have very low input resistance when compared with the input resistance of low-frequency vacuum-tube amplifiers. The output impedance, on the other hand, is usually high, with the exception of the common-collector circuit. This connection has a high input resistance and a very low output resistance.

The values obtained for the voltage gains show that the common-emitter amplifier is the only one through which a 180-degree phase reversal occurs. This is shown by the minus sign associated with the calculated value of voltage gain. This circuit is the one most frequently used in transistor amplifiers, because it has the largest power gain and its voltage and current gains are quite favorable. The values of input and output resistances are also favorable for practical operation of the circuit. The same transistor was used in the example solely for the purpose of comparing the performances of the three circuit connections. It should be clear that another type would yield entirely different numerical values. However, the methods used here for evaluating the circuit performance would be valid for any type of transistor and for any source and load resistors.

In practical circuits it is often necessary to match the impedances of a circuit and its load, and to match the circuit to its source generator. As indicated in Table 9-4, the common-base amplifier could match a low-resistance driver to a high-resistance load, while the common-collector amplifier could match a high-resistance driver to a low-resistance load. The latter connection is often used for the primary purpose of obtaining its input and output resistance-matching properties. It will be seen in later discussions that the cathode-follower vacuum-tube amplifier has the same property.

Another important property of the common-emitter amplifier circuit has a practical value. Only one supply battery is needed to drive the collector and base circuits, as was discussed in connection with biasing methods. The

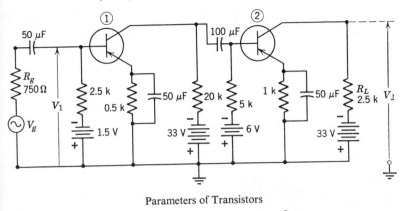

Parameters of Transistors

	1	2
h_{ie}	800 ohms	2,400 ohms
h_{fe}	24	50
h_{re}	8×10^{-6}	25×10^{-4}
h_{oe}	20×10^{-6} mho	45×10^{-6} mho

Fig. 9-17   Two-stage small-signal transistor amplifier.

other two basic connections need two separate supplies, a distinct disadvantage in practical designs where costs of components are significant.

9.9   Analysis of Small-Signal Two-Stage Transistor Amplifier

Two common-emitter amplifiers connected in cascade are shown in Fig.
9-17. The parameters of the transistors are listed below the circuit diagram.

Analysis of the performance will be done by first drawing the equivalent
circuit and then applying the equations developed in this chapter to one
stage at a time. Although there are other kinds of equivalent circuits such
as the T or π type, we shall again use the loop and nodal forms with which
we are now familiar.

The complete equivalent circuit is shown in Fig. 9-18. It is advantageous
to think of the parts inside the circles as active coupling networks having

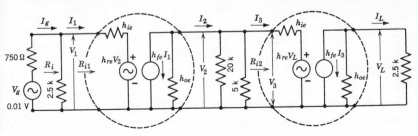

Fig. 9-18   Equivalent circuit of amplifier in Fig. 9-17.

obtainable input and output impedances. We have determined such quantitie
before and they will be useful in this analysis. We desire to determine curren
gain and power gain.

Since the load resistance is known, we should begin there and wor
toward the input end. The equations already derived will be used.

$$A_{i2} = \frac{I_L}{I_3} = \frac{-h_{fe}}{1 + h_{oe}R_L}$$

$$= \frac{50}{1 + (45 \times 10^{-6})(2.5 \times 10^3)} = 44.4 \qquad (9\text{-}60)$$

$$R_{i2} = h_{ie} - \frac{h_{re}h_{fe}R_L}{1 + h_{oe}R_L}$$

$$= 2{,}400 - \frac{(2.5 \times 10^{-4})(-50)(25 \times 10^3)}{1 + 0.1125} = 2{,}680 \ \Omega \quad (9\text{-}63)$$

Observe that the current I_2 entering the second amplifier circuit is divided
among 20 kΩ, 5 kΩ, and R_{i2} in parallel. The equivalent resistance of these
three is R_e kΩ, such that

$$\frac{1}{R_e} = \frac{1}{20} + \frac{1}{5} + \frac{1}{2.68} = 0.623$$

$$R_e = 1.6 \text{ k}\Omega$$

The voltage drop across each parallel branch is then $1{,}600I_2$, and the current
input to the second stage is I_3, where

$$I_3 = \frac{1{,}600I_2}{2{,}680} = 0.597I_2$$

so that

$$I_3/I_2 = 0.597$$

Going now to the first stage, we see that $R_e = 1{,}600 \ \Omega$, through which I_2
flows, is the load (R_{L1}) on this stage. The current gain through the first stage
and its input resistance are now found as they were above for the second stage.

$$A_{i1} = \frac{I_2}{I_1} = \frac{-h_{fe}}{1 + h_{oe}R_L} = \frac{24}{1 + (20 \times 10^{-6})(1.6 \times 10^3)} = 23.2$$

$$R_{i1} = h_{ie} - \frac{h_{re}h_{fe}R_{L1}}{1 + h_{oe}R_{L1}} = 800 - \frac{8 \times 10^{-4}(-24)(1.6 \times 10^3)}{1 + (20 \times 10^{-6})(1.6 \times 10^3)} = 829.8 \ \Omega$$

The 2,500-Ω bias resistor at the input of the first stage is paralleled by R_{i1}.
Consequently not all of the generator current enters the transistor. This cuts
down the current gain of the whole amplifier insofar as the ratio of input
current to output current is concerned. This two-branch parallel circuit will

vide I_g, so that

$$I_1 = \frac{2{,}500 I_g}{2{,}500 + 829.8} = 0.752 I_g$$

he current gain ahead of the first stage is then

$$I_1/I_g = 0.752$$

he *input resistance to the amplifier*, as *presented to the generator terminals*, is

$$R_i = \frac{2{,}500 R_{i1}}{2{,}500 + R_{i1}} = \frac{2{,}500 \times 829.8}{2{,}500 + 829.8} = 623 \ \Omega$$

he *overall current gain*, A_{iT}, is now determined by multiplying all of the urrent gains together:

$$A_{iT} = \frac{I_L}{I_g} = \frac{I_L}{I_3} \times \frac{I_3}{I_2} \times \frac{I_2}{I_1} \times \frac{I_1}{I_g} = 44.4 \times 0.597 \times 23.2 \times 0.752 = 463$$

he *overall power gain*, A_{pT}, can be expressed in terms of the power gain nd the terminal resistances. Since power gain in pure resistance circuits is he ratio of output volt-amperes to input volt-amperes, it is also equal to the ratio of output $I^2 R$ o input $I^2 R$:

$$A_p = \frac{V_2 I_2}{V_1 I_1} = A_v A_i = \frac{I_2{}^2 R_2}{I_1{}^2 R_1} = (A_i)^2 \frac{R_2}{R_1} \quad (9\text{-}64)$$

The total power gain through the amplifier is then

$$A_{pT} = (A_{iT})^2 \frac{R_L}{R_i} = 463^2 \times \frac{2{,}500}{623} = 860{,}000$$

The equivalent decibel gain is 59.3 dB.

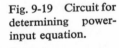

Fig. 9-19   Circuit for determining power-input equation.

Power output is obtained from the power input by multiplying it by the overall power gain. The generator circuit is represented in Fig. 9-19.

The power input, P_i, to the amplifier is $V_1 I_g$ and $I_g = V_1/R_i$. Therefore

$$P_i = V_1{}^2/R_i$$

By voltage-divider action,

$$V_1 = \frac{R_i}{R_g + R_i} V_g$$

so that

$$P_i = \frac{R_i V_g{}^2}{(R_g + R_i)^2} \quad (9\text{-}65)$$

Using values already determined, including $V_g = 0.01$ V,

$$P_i = \frac{623 \times (0.01)^2}{(750 + 623)^2} = 3.31 \times 10^{-8} \text{ W}$$

The power output, P_o, is then

$$P_o = 3.31 \times 10^{-8} \times 8.60 \times 10^5 = 28.42 \times 10^{-3} \text{ W} = 28.42 \text{ mW}$$

The foregoing analysis illustrates the application of node and loo equations to the problem of evaluating the performance characteristics • a transistor amplifier. Because the stages are not isolated from each othe the load on one influences the load on the preceding one. Starting the analys at the final output end makes it possible to determine the load impedance presented to preceding stages in easy, logical steps.

9.10 Analysis of Transistor-Amplifier Operation Outside Midband

Transistor-amplifier operation in the low-frequency range, where th coupling capacitor (C_1 in Fig. 9-1b) no longer has negligible reactance is analyzed in the same manner as for vacuum tubes. However, the couplin network between stages has a slightly more complicated form because th electrode currents must pass through "internal resistances" in the tw adjacent stages.

For low-frequency operation the equivalent circuit of the interstag network is shown in Fig. 9-20. R_{o1} is the output resistance of the first stage and R_{i2} is the input resistance of the second stage. The output voltage of th

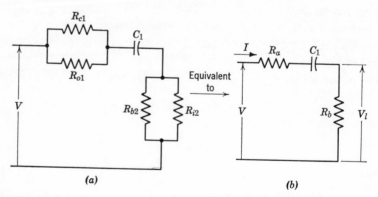

(a)   (b)

Fig. 9-20 Equivalent interstage coupling network of transistor amplifier in Fig. 9-1b

first stage must be V_2 in the analysis previously presented, so here V must be of proper value to produce V_2 across R_{b2} and R_{i2} in the midband *when C_1 is not present*. Therefore this current will be

$$I_m = V_2/R_b$$

To determine the voltage V,

$$V = I_m(R_a + R_b)$$

and, combining these two equations,

$$V_2 = V \frac{R_b}{R_a + R_b}, \qquad V = \frac{R_a + R_b}{R_b} V_2$$

Now consider Fig. 9-20b, where V is applied:

$$I = \frac{V}{R_a + R_b - j/\omega C_1}$$

The low-frequency output voltage is

$$V_l = \frac{V R_b}{R_a + R_b - j/\omega C_1}$$

and

$$\frac{V_l}{V_2} = \frac{R_a + R_b}{R_a + R_b - \dfrac{j}{\omega C_1}} = \frac{1}{1 - \dfrac{j}{\omega C_1(R_a + R_b)}}$$

This is the same form as Equation (9-21) for the vacuum-tube amplifier and has the same meaning, since $A_{vl} = V_l/V_i$ and $A_{vm} = V_2/V_i$; where V_i is the input-signal voltage.

In the high-frequency range, where C_1 is neglected but the shunting capacitance C_s is appreciable, there are four resistances in parallel: R_{o1}, R_{C1}, R_{b2}, and R_{i2}. The total shunt capacitance is made up, as for the tube, of the output capacitance of the transistor and the capacitance of the circuit wiring. The output capacitance is supplied by the transistor manufacturer. Using the equivalent circuit in a procedure similar to that for the low-frequency range, the ratio of output voltage (or gain) in the high range to output voltage (or gain) in the middle range turns out to be

$$\frac{V_h}{V_2} = \frac{1}{1 + j\omega C_T R_{eq}}$$

where C_T is the total shunting capacitance and R_{eq} is the parallel equivalent of the four resistors listed above. This is the same form as Equation (9-32) for operation of a vacuum-tube amplifier in its high-frequency region. It has the same meaning, and the transistor amplifier performance may be determined from Figs. 9-10 and 9-14 in the same way.

9.11   Interelectrode Capacitances of a Triode

The electrodes in a triode tube are insulated conductors with small distances between them. This means that they act together as small capacitances called *interelectrode capacitances*. Alternating current will flow through these small capacitors when the frequency of operation is high enough to lower their reactances to the point where they can no longer be ignored. In this frequency range the input (grid) circuit is no longer isolated by the tube from the output (plate) circuit, but rather *coupled to it by the* interelectrode capacitances. Figure 9-21 shows the situation in actual and equivalent circuit form.

The a-c voltage across C_{gp} is substantial because as the grid increases in potential the plate potential decreases (180-degree phase shift in voltages) and there can be appreciable current flow which must come from the input-signal source. This current may be expressed in terms of known quantities. We shall use *admittances* and voltages, rather than impedances and voltages, to represent the currents.

$$I_1 = E_1 Y_1$$

$$I_2 = E_{gp} Y_2 = (E_1 - E_2) Y_2$$

E_2 is equal to the voltage gain times E_1, so that

$$E_2 = A_v E_1$$

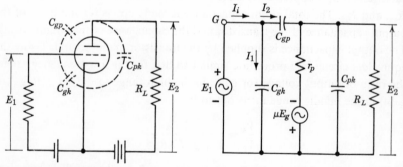

Fig. 9-21   Actual and equivalent circuits of a triode amplifier, including interelectrode capacitances.

The input current is then

$$I_i = I_1 + I_2 = E_1 Y_1 + (E_1 - A_v E_1) Y_2$$
$$I_i = E_1 [Y_1 + (1 - A_v) Y_2]$$

The input admittance is I_i/E_1,

$$Y_i = Y_1 + (1 - A_v) Y_2 \qquad (9\text{-}66)$$

The expression for A_v is taken from previous analysis (Equation 8-5), and it must have the negative sign here to indicate the phase relationship between E_2 and E_1.

$$Y_i = Y_1 + \left(1 + \frac{\mu R_L}{r_p + R_L}\right) Y_2$$

$$Y_i = j\omega C_{gk} + j\omega \left(1 + \frac{\mu R_L}{r_p + R_L}\right) C_{gp}$$

$$Y_i = j\omega \left[C_{gk} + \left(1 + \frac{\mu R_L}{r_p + R_L}\right) C_{gp}\right] \qquad (9\text{-}67)$$

This means that the *equivalent capacitor* accounting for input admittance has a capacitance value given by the total capacitance represented in the brackets. This is known as the *Miller effect*. This equivalent capacitance is between grid and cathode and will be represented by C_i:

$$C_i = C_{gk} + \left(1 + \frac{\mu R_L}{r_p + R_L}\right) C_{gp} \qquad (9\text{-}68)$$

Note that the original grid-plate capacitance in the tube has an amount added to it which is A_v times as large. Using values for the 6J5 tube, the maximum input capacitance (at midband frequencies) is

$$C_i = 3.4 + (1 + 15)3.4 = 57.8 \text{ pF}$$

This is enough capacitance to seriously affect the operation of the amplifier at high frequencies. The *pentode tube* was developed to reduce very materially the grid-plate interelectrode capacitance, with the result that much higher frequencies may be used. The 6SJ7, a popular pentode, has a control-grid-to-plate capacitance of only 0.005 pF. The total input capacitance of this tube is then

$$C_i = 6 + (1 + 200)0.005 \doteq 7.0 \text{ pF}$$

The value used for C_{gk} is seen to be larger than that for the triode. The reason is that the capacitances to cathode of the *screen grid* and the *suppressor grid* of the pentode are included.

9.12   Output Impedance

The *output impedance* of a voltage generator, such as a vacuum-tube amplifier, is defined as the *rate of change of output-terminal voltage with respect to output current*. That is,

$$Z_o = \Delta E_o / \Delta I_o$$

A change in load will cause the output current to change by the amount ΔI_o, and this will be accompanied by a corresponding change in output voltage ΔE_o *due to impedances in the circuit ahead of the load impedance.*

In the Thévenin equivalent circuit, Fig. 9-22, we can write

$$-E' + E_o = I_o Z_{oc}$$

It will be recalled that in this form of equivalent circuit E' is the open-circuit voltage when the load is removed from the original circuit and Z' is the *open-circuit impedance* (Z_{oc}) looking back toward the generator after it has been replaced by its internal impedance. Both E' and Z_{oc} are constant. A small change in load impedance will then cause a small change in output current, ΔI_o, which is accompanied by a small change in output voltage ΔE_o. Accordingly, we can write an equation showing the *increments of change.* Since E' is constant, $\Delta E' = 0$ and

$$\Delta E_o = \Delta I_o Z_{oc}$$
$$\Delta E_o / \Delta I_o = Z_{oc} \tag{9-69}$$

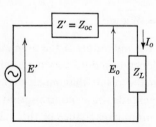

Fig. 9-22 Thévenin equivalent circuit of generator, coupling network and load.

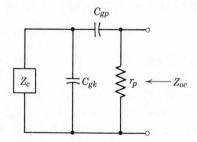

Fig. 9-23 Circuit for determining output impedance at high frequencies.

That is, the output impedance is what we have called the open-circuit impedance looking back from the output terminals toward the generator.

When the operating frequency is sufficiently high that interelectrode capacitances cannot be neglected, the output impedance may be determined from the equivalent circuit in Fig. 9-23. The grid side of the tube is no longer isolated from the plate side, owing to the presence of the interelectrode capacitances. The impedance Z_c coupling the generator to the input is a known quantity. Z_{oc} can be determined by using simple series-parallel circuit relations.

9.13   Improving Amplifier Response at Low Frequencies

The gain of an amplifier falls off as the operating frequency is reduced in the low-frequency range. Increase in the capacitance of the coupling capacitor to the upper limit imposed by space, cost, or other conditions is the first step in improving low-frequency response.

Another effective way to maintain the voltage gain at, or nearer to, the midfrequency value while the frequency is decreasing in the low range is to use the "decoupling" arrangement of C_d and R_d in parallel shown in Fig. 9-24. At the low end of the *midband range* of frequencies, the effect of R_d as an a-c load resistor is canceled by the low reactance of C_d. But, as the frequency *decreases*, the impedance of the parallel circuit *increases* and the a-c voltage drop across the load impedance of the amplifier (this is the output voltage) holds up very well. The effect is as if the a-c load line were tilted to become more nearly horizontal, thus increasing the a-c voltage swing, E, and offsetting the growing voltage drop across C so far as the output voltage E_o or transistor current I_2 is concerned.

A pentode tube is shown because it will be used instead of a triode if the objective is constant voltage gain over as wide a frequency range as possible. Pentodes perform the same functions as triodes, and in some respects they do them better.

The a-c plate resistance of the tube and the internal-collector resistance of the transistor are not shown because they are large compared with $(R_1 + R_d)$. It is desired that the output voltage $I_2 R_2$ in the tube circuit and I_2 in the transistor circuit be independent of frequency. This is asking a lot of the transistor circuit because the input resistance of a second stage is so low. In order to make the reactance of C very small compared to R_{ie}, a capacitance of many microfarads must be used.

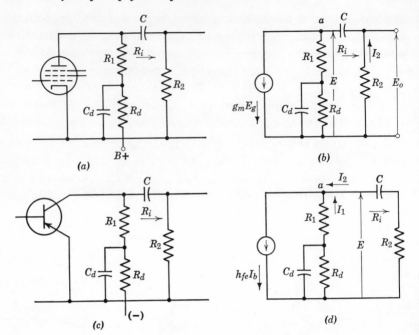

Fig. 9-24 Low-frequency compensation arrangement for vacuum-tube and transistor amplifiers.

The current equation at node a in the transistor circuit is

$$h_{fe}I_b = -\frac{E}{R_i - \dfrac{j}{\omega C}} - \frac{E}{R_1 + \dfrac{R_d(-j/\omega C_d)}{R_d - j/\omega C_d}} \tag{9-70}$$

and

$$E = -I_2\left(R_i - \frac{j}{\omega C}\right) \tag{9-71}$$

From these two equations I_b, which is the base current of the transistor stage, can be expressed in terms of I_2, the input current to the next stage or load. The algebra is cumbersome, although the assumption that $R_d = 1/\omega C_d$ helps. Eliminating E gives the result

$$h_{fe}I_b = \left[\frac{\omega^2(R_i + R_1)CC_d - j\omega(C + C_d)}{\omega^2 R_1 CC_d - j\omega C}\right]I_2 \tag{9-72}$$

By means of a procedure in calculus, an expression which is the *rate of change* of I_2 with respect to frequency can be obtained. Since it is desirable that I_2

does not change with frequency, this *rate of change* is ideally *zero*. Setting the rate-of-change expression equal to zero leads algebraically to the following circuit requirement:

$$R_1(C + C_d) = (R_1 + R_i)C$$

Canceling R_1C on both sides gives

$$R_1C_d = R_iC \qquad (9\text{-}73)$$

This says that the time constants of the compensating circuit and the coupling circuit should be equal. Note that, although R_i is shown as a coupling resistor in the transistor circuit, it really includes the input resistance to the next stage. For the tube, the input impedance to the next stage will be practically infinite at such low frequencies unless the grid goes positive.

The analysis for the pentode amplifier will be the same as for the transistor, as can be seen from the fact that their equivalent circuits are identical. The current I_2 flows through the grid-leak resistor of the next stage or through an adequately large load resistance. Another beneficial effect is derived from this compensation circuit, in that making the time constants equal (Equation 9-73) results in zero phase shift in the coupling circuit at the output of this stage in the low-frequency region.

9.14   Improving Amplifier Response at High Frequencies

It has been seen that, as the operating frequency of a tube amplifier is increased above the midband range, the gain drops off because of plate-to-cathode capacitance, input capacitance to the next stage, and stray capacitances in the wiring of the output circuit. Neutralization of these capacitance effects, insofar as possible, is done by using an inductance in the output circuit. We shall again use pentodes because of their greatly reduced plate-to-cathode capacitance.

Two methods of compensation for decrease in voltage gain at high frequencies are shown in Fig. 9-25. On the amplifier-response curve at the high-frequency end, the half-power point is of practical interest and it is desired to move this point as far as possible on the graph in the direction of increasing frequency.

The Norton equivalent circuit for a vacuum-tube amplifier operating in the high-frequency range shows a constant-current generator supplying a parallel set of impedances made up of r_p, R_L, and the capacitances across the output. The plate-to-cathode capacitance of the operating tube is included because it draws alternating current from the plate circuit.

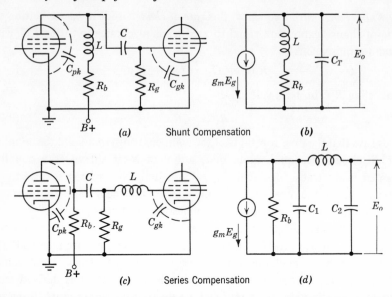

Fig. 9-25 High-frequency compensation arrangement for vacuum tube amplifiers.

The a-c plate resistance of pentodes is around 0.8 MΩ, and the grid-leak resistor of the following stage is approximately 1 MΩ. These are so much larger than R_b, the resistance carrying d-c plate current to the tube, that they may be neglected, leaving R_b as the only resistance in the equivalent circuit.

The total shunting capacitance, C_T, is

$$C_T = C_{pk} + C_w + C_{in}$$

where C_w = stray capacitance due to wiring and C_{in} = input capacitance to the next stage (Equation 9-68). At the upper half-power point, where f_2 is the frequency,

$$R_b = 1/2\pi f_2 C_T \tag{9-74}$$

and

$$f_2 = 1/2\pi C_T R_b \tag{9-75}$$

Let us first consider the *uncompensated* case, which would be represented in Fig. 9-25b *with the inductance L short-circuited*. The voltage across the output terminals would be

$$E_o = -g_m E_g \left[\frac{R_b(+1/j\omega C_T)}{R_b + 1/j\omega C_T} \right] \tag{9-76}$$

and the gain would be

$$A_{vh} = -g_m \left[\frac{R_b(-j/\omega C_T)}{R_b - j/\omega C_T} \right] = -g_m R_b \left[\frac{-j}{\omega C_T R_b - j} \right] \tag{9-77}$$

The absolute value of the gain is*

$$|A_{vh}| = g_m R_b \left(\frac{1}{1 + \omega^2 C_T{}^2 R_b{}^2}\right)^{\frac{1}{2}} \tag{9-78}$$

Expressing the frequency term within the parentheses as $f^2 (2\pi C_T R_b)^2$, let us compare it with $1/f_2{}^2$ obtained from Equation (9-75):

$$1/f_2{}^2 = (2\pi C_T R_b)^2$$

Evidently the term in Equation (9-78) is f^2 times $1/f_2{}^2$, so we can write it as $(f/f_2)^2$.

$$|A_{vh}| = g_m R_b \left(\frac{1}{1 + (f/f_2)^2}\right)^{\frac{1}{2}} \tag{9-79}$$

From this, the phase angle θ involved in the gain is given by

$$\tan \theta = -\omega C_T R_b = -f/f_2$$

Using *shunt compensation* changes the equivalent circuit to that of Fig. 9-25b. For our purpose we use the same equivalent circuit as before and simply remove the short circuit imposed upon the inductance L for the discussion above. The expression for E_o is exactly like that of Equation (9-76) except that $j\omega L$ is added in both numerator and denominator. However, when the expression for the absolute value of the gain is derived, it is a very complicated affair.** From it, a relation between L and the product $C_T R_b{}^2$ is obtained which would keep the gain *most nearly at the midband value* in the extended high-frequency range. The relation is

$$\frac{L}{R_b{}^2 C_T} = 0.414$$

The curves in Fig. 9-26 show how the high-frequency gain is boosted from the uncompensated values when L is not present ($L = 0$) to conditions of extreme overcompensation in which the above ratio is 10.

Series compensation to improve the high-frequency response of amplifiers is illustrated in Fig. 9-25c with the equivalent circuit shown in Fig. 9-25d. The expression for gain, derived from analysis of the equivalent circuit, is much too complex for purposeful presentation in a text at this level.

---

* The absolute value of a complex quotient is the absolute value of the numerator divided by the absolute value of the denominator.

** J. D. Ryder, *Electronic Fundamentals and Applications*, 3rd ed., Prentice-Hall, Englewood Cliffs, N.J., 1964.

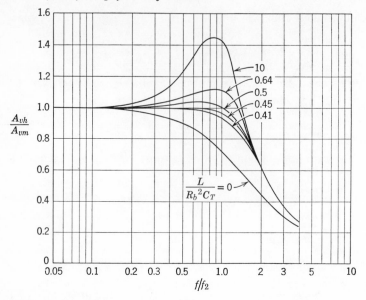

Fig. 9-26   High-frequency response curves of a shunt-compensated amplifier. The gain without compensation is 3 dB down at $f = f_2$.

The result is two equations relating L, C, and R:

$$\left(\frac{C_2}{C_1 + C_2}\right)\frac{\omega_2 L}{R_b} = \frac{1}{2} \tag{9-80}$$

$$\frac{\omega_2 L}{R_b} = \frac{2C_1}{C_2} = q \tag{9-81}$$

We shall call this ratio a constant q because we shall refer to it and determine possible numerical values it may have.

From Equations (9-80) and (9-81) is obtained a condition for optimum response:

$$\frac{C_1}{C_1 + C_2} = \frac{1}{4} = m \tag{9-82}$$

$$4C_1 = C_1 + C_2$$

$$3C_1 = C_2$$

This means that

$$q = \frac{\omega_2 L}{R_b} = \frac{2}{3} \tag{9-83}$$

For the uncompensated amplifier, for which the equivalent circuit of Fig. 9-25b holds if we let $L = 0$, the "3 dB down frequency," f_2 is given by

$$f_2 = \frac{1}{2\pi(C_1 + C_2)R_b}$$

$$\omega_2 = \frac{1}{(C_1 + C_2)R_b}$$

Using this for ω_2 in Equation (9-83) rearranged,

$$L = \frac{qR_b}{\omega_2} = q(C_1 + C_2)R_b^2 \qquad (9\text{-}84)$$

The required inductance can be computed with this equation. It has been found that, using $m = 0.2$ in Equation (9-82) and $q = 0.56$, the high-frequency compensation is about as good as can be achieved by this simple method.

It will be recalled that in *series* resonance the voltage across the capacitor, and also across the inductance, may rise to an unusually high value. This phenomenon is one of the contributing factors in the success of series compensation for improving gain in the high-frequency range. This method produces even better results than the shunt-compensation method. Parallel-resonance effects enter into the performance with shunt compensation.

Starting with as small a value of C_T as possible, the frequency f_2 at which the gain of the uncompensated amplifier is 3 dB down may be made fairly high even though R_b is quite small. The addition of L in series compensation

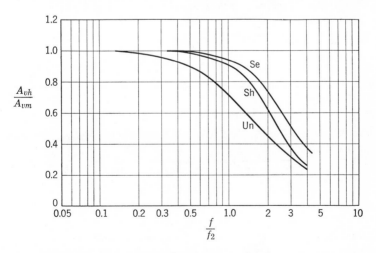

Fig. 9-27 High-frequency response curves of an amplifier: uncompensated (Un), shunt-compensated (Sh), series-compensated (Se).

will then extend f_2 much farther into the high range. In practical amplifiers the new f_2 frequency has been made 2 to 3 times the value for the uncompensated amplifier. A comparison of the high-frequency responses of an amplifier with and without compensation is shown in Fig. 9-27.

SUGGESTED REFERENCES

1. Herbert J. Reich, *Theory and Applications of Electron Tubes*, McGraw-Hill, New York, 1944.

2. Frederick E. Terman, *Electronic and Radio Engineering*, 4th ed., McGraw-Hill, New York, 1955.

3. Marlin P. Ristenbatt and Robert L. Riddle, *Transistor Physics and Circuits*, Prentice-Hall, Englewood Cliffs, N.J., 1958.

4. Paul D. Ankrum, *Principles and Applications of Electron Devices*, International Textbook Co., Scranton, Pa., 1959.

5. J. D. Ryder, *Electronic Fundamentals and Applications*, 3rd ed., Prentice-Hall, Englewood Cliffs, N.J., 1964.

QUESTIONS

9-1 What are the functions of the coupling resistor and capacitor in an *R-C* coupled amplifier? Must they have large or small values, and why?

9-2 Compare the four classes of amplifiers described in regard to current flow and amount of grid bias.

9-3 Name and identify the types of distortion that may be present in an amplifier.

9-4 Draw from memory a curve showing the variation of voltage amplification with frequency of an *R-C* coupled amplifier. What are some of the causes of drop in amplification in the low- and high-frequency ranges?

9-5 Explain why there is a 180° phase shift in voltage through a vacuum-tube amplifier stage.

9-6 Draw the equivalent circuit that represents a voltage amplifier stage in the midband range. Using it, develop the equation for output voltage in terms of input voltage.

9-7 Explain how to use the "universal" curves (Fig. 9-10) showing voltage gain and phase shift as a function of frequency and circuit constants.

9-8 What difference, if any, is there in the methods of using the "universal" curves for amplifier operation in the high-frequency and low-frequency ranges?

9-9 Why is a transistor amplifier regarded as a *current-controlled* device? Compare its most useful amplifying function with that of the vacuum-tube amplifiers studied thus far.

9-10 Define the following for a transistor amplifier stage: (*a*) input resistance, (*b*) output resistance, (*c*) voltage gain, (*d*) current gain, (*e*) power gain.

9-11 It is desired to compare the various gains and impedances of transistor amplifiers classified as common-base, common-emitter, and common-collector. Make a table and designate the gains and input and output impedances as high, medium, or low. Which are negative? In which is there phase shift?

9-12 Name the principal advantages of the common-emitter connection of transistors in amplifiers.

9-13 Show that voltage gain times current gain must equal power gain. Would this be true at power factors other than unity?

9-14 What is interelectrode capacitance? What is its effect in high-frequency operation of amplifiers?

9-15 Describe the Miller effect. Which interelectrode capacitance is responsible for this effect?

9-16 Describe the pentode tube and tell how it is superior in performance to the triode, and why.

9-17 Define output impedance of an amplifier stage.

9-18 Describe a means of improving the low-frequency response of an amplifier by making an addition of elements to the plate circuit. Explain how the system works.

9-19 Describe two methods of improving the high-frequency response of an amplifier. What phenomenon comes into action in each case?

9-20 Refer to the curves of Fig. 9-26. Which should be used in the design of the compensating circuit if the gain must not go *above* the midband value? Which should be used if a slight increase in gain is allowable?

9-21 In Fig. 9-26, once a curve is selected how is the value of L determined?

9-22 How would you determine the value of L to be used to provide series compensation in an amplifier?

PROBLEMS

9-1 An *R-C* coupled amplifier stage and its triode tube have the following constants: $R_{b1} = 0.25$ MΩ, $R_2 = 0.5$ MΩ, $C = 0.005$ μF, $r_p = 7,700$ Ω, $g_m = 2,600$ μmho. Determine the voltage amplification and the angle by which the output voltage leads the input voltage (*a*) in the midband range, (*b*) at 100 Hz. Do this by direct computation and check your results by using the curve of Fig. 9-10.

9-2 The amplifier of Problem 9-1 has a total shunting capacitance of 1,200 pF. Calculate the voltage gain at 15,920 Hz. Check your result by using the curve of Fig. 9-14.

9-3 Calculate the frequencies at the half-power points for the amplifier of Problems 9-1 and 9-2. For the low range $R_{el} = X_C$ and for the high range $R_e = X_{C2}$.

9-4 A 6SF5 triode is used in a preamplifier with cathode bias. $E_{bb} = 300$ V, $R_b = 0.147$ MΩ, $R_g = 0.1$ MΩ (next stage), $r_p = 66,000$ Ω, $g_m = 1,500 \times 10^{-6}$ mho, $C = 0.02$ μF. (a) Draw the d-c and a-c load lines, choosing a Q-point that will give linear operation. Then calculate the bias resistance. (b) Using a signal input of 0.5 V peak, determine the peak value of output voltage and the voltage gain in the midband range.

9-5 A manufacturer lists the following data for using a high-mu triode in a small-signal voltage amplifier.

E_{bb} (volts)	R_b (megohms)	R_g (megohms)	R_k (ohms)	C (μF)	g_m (μmho)	r_p (ohms)
180	0.25	0.5	2,150	0.006	1,325	53,000
300	0.50	1.0	2,980	0.003	1,325	53,000

(a) Calculate the midband voltage gain in each type of operation. (b) Calculate the rms signal-voltage input for 39 V output when E_{bb} is 180 V, and for 48 V output when E_{bb} is 300 V.

9-6 Assume that the total output capacitance of the amplifier of Problem 9-4 is 75 pF. (a) Determine the voltage gain at 40,000 Hz with and without C_2. (b) At what frequency above midband is the gain down 3 dB?

9-7 Compute the gain of the amplifier of Problem 9-4 at 80 Hz. What is the low-frequency half-power point?

9-8 An amplifier stage using a 6SJ7 tube operates with $E_{bb} = 250$ V, bias -3 V, $R_b = 30$ kΩ, $r_p = 1.25$ MΩ, $R_2 = 30$ kΩ. (a) Draw the d-c and a-c load lines. (b) Find the maximum and minimum instantaneous values of plate-to-cathode voltage for a grid signal with 1-V peak. (c) Find the amplitude of the plate-current swing and the rms value of plate current, assuming sine-wave form. Calculate the power that must be dissipated by the plate: $P_p = E_{bo}I_{bo}$.

9-9 A pentode tube is used in a small-signal amplifier circuit with constants given below. (a) Calculate the voltage gain in the midband. (b) What are the upper and lower half-power frequencies? (c) How much signal voltage is needed to produce 100 V rms across R_{g2}?

$$r_p = 1.0 \text{ MΩ}, g_m = 4 \times 10^{-3} \text{ mho}, R_b = 0.25 \text{ MΩ},$$
$$R_{g2} = 0.25 \text{ MΩ}, C = 0.0032 \text{ μF}, C_T = 159 \text{ pF}$$

9-10 A two-stage amplifier uses a 6SN7 twin triode with circuit elements as listed and identified with reference to Fig. 9-1. $R_{g1} = 0.10$ MΩ, $R_{b1} = 0.1$ MΩ, $C_1 = 0.014$ μF, $C_{w1} = 500$ pF (due to wiring), $R_{g2} = 0.22$ MΩ, $R_{b2} = 0.22$ MΩ, $C_2 = 0.0065$ μF, $C_{w2} = 500$ pF, $R_L = 0.47$ MΩ. Tube constants are: $g_m = 2,600$ μmho, $r_p = 7,700$ Ω, $C_{gp} = 4$ pF, $C_{gk} = 2.4$ pF, $C_{pk} = 0.7$ pF. (a) What is the overall midband gain? (b) How much input signal is needed for an output voltage of 100 V rms from the second stage? (c) At what frequency above the midband range is the output voltage of the first stage only 0.707 of the midfrequency output? (*Note:* Take into account the Miller effect.) (d) Calculate, for an input signal of 0.1 V rms, the output of the second stage at the frequency determined for part (c).

9-11 After working Problem 9-10, determine the frequency above the midband range at which the gain of the *whole amplifier* is down 3 dB below the midband gain of the unit as a whole.

9-12 A transistor has the following parameters for common-emitter, base-input connection, when $V_{CE} = -1.3$ V and the d-c collector-current is -0.3 mA: $h_{ie} = 4,800$ Ω, $h_{re} = 9.1 \times 10^{-4}$, $h_{fe} = -45$, $h_{oe} = 12.4 \times 10^{-6}$ mho. Calculate the input resistance R_i, and the output resistance R_o. Assume the driving generator resistance $= 500$ Ω. Calculate the current gain and power gain with $R_L = 4,700$ Ω.

9-13 A two-stage transistor amplifier is shown in Fig. P9-13. The parameters are listed below it. (a) Calculate the stage gain at the midband frequencies. (b) Calculate the input and output resistances of the transistors. (c) Calculate V_l/V_2, the stage gain at 15.92 Hz. (d) Calculate V_h/V_2, the stage gain at 15,920 Hz (assume a total shunting capacitance of 0.003 μF), C_{CE} + shunt output capacitance $= 100 \times 10^{-12}$ F (or 100 pF).

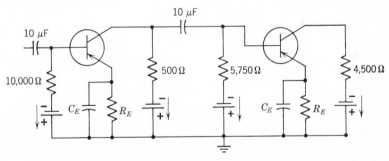

Fig. P9-13 $h_{ie} = 2,000$ Ω; $h_{fe} = 42$; $h_{re} = 4.5 \times 10^{-4}$; $h_{oe} = 28 \times 10^{-6}$ mho; $C_w = 15$ pF = wiring capacitance, $C_{CE} = 42$ pF = output capacitance. Assume that C_E has zero reactance.

9-14 Solve Problem 9-13 after replacing the transistors with two having the following parameters: $h_{ie} = 1,667$ Ω, $h_{fe} = -44$, $h_{re} = 4.95 \times 10^{-4}$, $h_{oe} = 22.8 \times 10^{-6}$ mho, $C_{CE} = 40$ pF. Shunt output capacitance is 60 pF.

9-15 Calculate the output impedance of the network of Fig. 9-23, using the constants $Z_c = 100\ \text{k}\Omega$, $C_{gp} = 3.8\ \text{pF}$, $C_{gk} = 2.6\ \text{pF}$, $r_p = 7,700\ \Omega$, $f = 2\ \text{MHz}$.

9-16 (*a*) In the amplifier circuit of Fig. P9-16, what is the voltage gain if the effect of C is neglected? Assume $\mu = 20$, $r_p = 10,000\ \Omega$. (*b*) What is the lower half-power frequency?

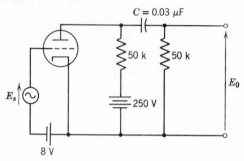

Fig. P9-16

9-17 An *R-C* coupled amplifier consists of a single stage using a 6J5 tube. For the circuit, the following values apply: $E_{bb} = 250\ \text{V}$, $R_b = 50,000\ \Omega$, $R_g = 500,000\ \Omega$, $R_k = 3,900\ \Omega$ and is well bypassed. The coupling capacitor is $0.01\ \mu\text{F}$; stray capacitance $= 12\ \text{pF}$. Assume $r_p = 1,400\ \Omega$ and $\mu = 20$. Determine the input admittance at the upper half-power frequency.

9-18 A broad-band amplifier uses four stages. The shunt capacitance of each stage is 20 pF and $g_m = 3,000\ \mu\text{mho}$. The overall bandwidth of the compensated amplifier, using $L/R_b{}^2C_T = 0.41$, is 3.5 MHz. What is the overall gain of the amplifier?

9-19 A shunt-compensated broad-band amplifier has a bandwidth of 3.5 MHz with $R_L = 1.8\ \text{k}\Omega$. What value of L is required for the flattest gain? Use Fig. 9-26.

9-20 A wide band amplifier is to have a flat gain characteristic and a phase-shift proportional to frequency up to 4 MHz. What value of the upper half-power frequency should be used for the design of the uncompensated amplifier?

9-21 The circuit of an amplifier is shown in Fig. P9-21, and d-c supply voltages have been omitted for convenience. Calculate the voltage gain E_o/E_s.

9-22 Sketch the small-signal equivalent for the circuit shown in Fig. P9-22 using *h*-parameters for the transistor.

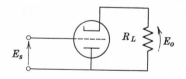

Fig. P9-21

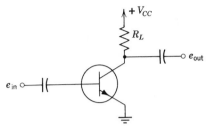

Fig. P9-22

9-23 For a 2N334 transistor connected in common-emitter mode, determine the maximum V_{CC} that may be used with a 6-kΩ collector load without exceeding the maximum collector dissipation of 150 mW at 25°C.

9-24 An amplifier produces an output voltage that is related to an input voltage by the expression $e_{out} = ke_{in}^2$. Does the amplifier produce distortion of the input signal? What type of distortion will it produce for $e_{in} = 10 \sin \omega t$?

9-25 An R-C coupled, single-stage voltage amplifier has $f_1 = 100$ Hz and $f_2 = 100$ kHz. Its response is flat in the middle-frequency region of operation. Sketch and label curves showing its relative amplification and phase shift throughout the frequency spectrum, to include f_1 and f_2. Express the relative amplification as a ratio compared to the midband gain, and in dB, at f_1 and f_2.

9-26 Using the *h*-parameters listed in the Appendix for the 2N404 transistor, calculate and construct a table of values similar to Table 9-4, using $R_L = 1$ kΩ and $R_g = 1$ kΩ.

9-27 By making calculations of input resistance, determine the effect of R_L in a common-emitter circuit. Sketch your results (R_i vs. R_L) on semilogarithmic graph paper, for a 2N404 transistor operating at $V_{CE} = -6$ V, $I_C = -1$ mA.

9-28 Calculate the effect of R_g on input resistance for a 2N404 operating at $V_{CE} = -6$ V, $I_C = -1$ mA. Sketch the results (R_o vs. R_g) on semilogarithmic graph paper.

9-29 A common-emitter amplifier uses the 2N334 at an operating point $V_{CE} = 12$ V, $I_C = 2.5$ mA. (a) Determine the values of *h*-parameters, assuming V_{CB} is constant at 5 V. (b) Calculate voltage and current gains, assuming $V_{CC} = 25$ V.

9-30 For the circuit conditions of Problem 9-29, calculate the effect of R_L on voltage gain and on current gain, as its value varies from zero to infinity (from short-circuit to open-circuit conditions).

9-31 Using the values of *h*-parameters found in part (a) of Problem 9-29, calculate the required value of C so that the common-base amplifier of Fig. P9-31 will have a lower half power frequency of 100 Hz.

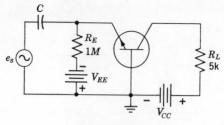

Fig. P9-31

9-32 The transistor of Fig. P9-32 has these parameters: $h_{ie} = 2\,k\Omega$, $h_{fe} = 300$, $h_{re} = 2 \times 10^{-4}$, $h_{oe} = 30 \times 10^{-6}$ mho. Calculate the value of coupling capacitor C needed to reduce the gain by 3 dB at 100 Hz.

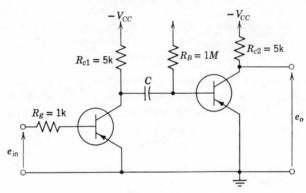

Fig. P9-32

CHAPTER TEN

Power Amplifiers

The amplification and control of power at low and high frequencies involve many of the principles already studied. The location of operating regions along load lines plays a major role in the analysis of performance. The signals are large and extend over the audio-frequency range in amplifiers of speech, music, and many kinds of control signal. At higher frequencies used in radio and TV transmitters the frequency range of operation of power amplifiers is very narrow; in fact, it would be ideal in some cases if operation were at a single frequency rather than in a narrow band. This can be appreciated when it is learned that a parallel-resonant circuit is often used to obtain high selectivity of frequencies, resistive impedance, and maximum signal between stages and at the output of the amplifier.

Power tubes are built to have large current capacities, low values of plate resistance, and good power-handling capabilities. Triodes, pentodes, and beam-power tubes are used. *Power transistors* are built to have good heat dissipation capabilities and are often required to be attached to special shapes and sizes of metal called *heat sinks*.

We shall discuss several concepts that become important in the operation of power amplifiers but which are usually negligible in voltage amplifiers. The large signals introduce some effects of nonlinearities in tube and transistor characteristics, such as distortion and variations of parameters that affect amplification. We shall learn that some classes of operation are more suited to particular load applications than others.

10.1 Single-Stage Power Amplifier

The simple one-tube and one-transistor amplifier circuits we have studied in previous chapters may be used as power amplifiers, provided the active

371

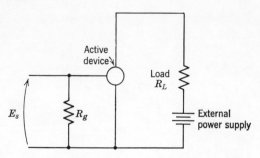

Fig. 10-1 Basic power amplifier; active device may be tube or transistor in electronic applications.

device is capable of handling the voltages and currents necessary and the load is properly chosen to receive power that may be generated. The circuit shown in Fig. 10-1 is the basic configuration of a single-stage power amplifier.

The component labeled *load* in the diagram may be a single resistor in some cases, but usually the loads required are relatively small resistances that require an impedance-matching transformer between the active device and the load. Three general types of loading may be encountered in power amplifiers: (*a*) direct connection, (*b*) transformer-coupled connection, and (*c*) shunt, or *L-C* coupling. These are shown in Fig. 10-2.

Power-amplifier operation usually requires a large driving voltage or current, which may be the output of a tube or transistor voltage amplifier. If the operation of the circuit in Fig. 10-1 is limited to Class A_1, then grid current does not flow, so the only current taken from the driving source is the very small amount taken by R_g. Thus, the power delivered by the driving source may be very small. However, the power delivered to the load may be

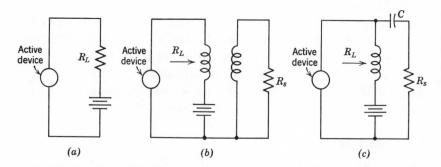

Fig. 10-2 Types of load coupling in power amplifiers. (*a*) Direct connection. (*b*) Transformer coupling. (*c*) Shunt *L-C* coupling.

many times larger, because the load current is being controlled by the active device.   The load power is derived from the external power supply, rather than from the driving source. Similarly, for a transistor as the active device, base current may be very small but may control a much larger collector current in the load. In this way, power amplification occurs in the circuit; it is a *power amplifier*.

10.2   A-C Load of a Power Amplifier

A power amplifier primarily controls power. It causes sufficiently large current variations in the load circuit so that substantial a-c power is present in the load. This output power is controlled by the signal coming from the driving source, and the power delivered from this source may be very small. The output power is generated in the load, designated R_L in the circuit of Fig. 10-1. If R_L were merely a resistor, the output power would be dissipated as heat. If, on the other hand, R_L represented the effective a-c resistance of a control circuit that required a-c power for its operation, or of a loudspeaker, the output power would do something useful in addition to disappearing as heat. The important idea to be gained is that R_L *represents an a-c load.*

We are interested in the a-c component of current in R_L, because only that component of the total current is useful in the production of output power. It is desirable, then, to minimize the resistance path for d-c components of current, because the power developed in the d-c paths is lost as far as useful power in the load is concerned. This may be done by employing the shunt-feed circuit (Fig. 10-1c) or by using a transformer to couple the load device to the output circuit (Fig. 10-1b).

A triode tube performs most efficiently as a power amplifier when its output-circuit impedance is a pure resistance ($Z_L = R_L$) that is roughly one-half to two times the value of the dynamic internal resistance. When the load is chosen to be within this range of values, maximum power is developed in the load and reasonably low distortion (about 5 per cent) results. The load will often be a device that has a low value of a-c resistance, for example, a loudspeaker that has an 8-Ω voice coil. The 8-Ω a-c load cannot be connected directly into the plate circuit for maximum power, because the dynamic plate resistance of the triode will be many times this value. The actual load must be coupled into the plate circuit in such a way that an *effective a-c load* is present in the plate circuit. Either shunt coupling or transformer coupling may be used to make the tube plate load "look like" a much larger value than the actual load.

We will now study these two concepts separately. We are interested in knowing that the dynamic plate resistance must be matched reasonably well to an effective load and how transformer coupling achieves this result.

It will be recalled that Class A operation of a single-stage amplifier yields a-c plate current that may be expressed as

$$I_p = \frac{\mu E_g}{r_p + R_L}$$

Then the power developed in R_L is

$$P = I_p{}^2 R_L = \frac{\mu^2 E_g{}^2 R_L}{(r_p + R_L)^2}$$

Now let us define a *power sensitivity* as the ratio of output power and the square of input voltage

$$\text{Power sensitivity} = \frac{P}{E_g{}^2}$$

Then the sensitivity may be written as

$$\text{Power sensitivity} = \frac{\mu^2 R_L}{(r_p + R_L)^2}$$

$$= \frac{(\mu^2/r_p{}^2) R_L}{(1 + R_L/r_p)^2} = \frac{\mu g_m (R_L/r_p)}{(1 + R_L/r_p)^2} \qquad (10\text{-}1)$$

Equation (10-1) may be used to determine the ratio of the load resistance to plate resistance that will permit the maximum output power to be developed. The factor in Equation (10-1) that contains this ratio is plotted in Fig. 10-3 as a function of the ratio. It will be noted that the output power remains essentially constant (between 100 and 88 per cent) for the load varying within the range $\frac{1}{2} r_p$ to $2 r_p$.

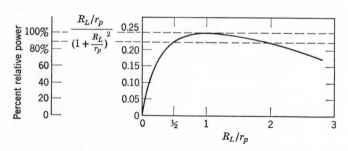

Fig. 10-3   Power sensitivity related to R_L/r_p ratio.

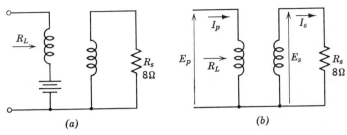

Fig. 10-4   Analysis of transformer coupling. (*a*) Output circuit. (*b*) Equivalent circuit.

Suppose it is desired to drive an 8-Ω loudspeaker with a tube whose plate resistance is 2 kΩ and to operate the power amplifier so that maximum power can be delivered to the 8-Ω load. What should be the properties of the transformer to be used as the impedance-matching device? Consider the output circuit in Fig. 10-4. The 8-Ω load is designated R_s and the plate load, R_L. If the transformer can be considered almost ideal, i.e., the a-c power supplied to the primary winding is transformed to the secondary without significant losses, then we may write

$$E_p I_p = E_s I_s$$

For an ideal transformer, the primary and secondary voltage and current values will be related by the number of turns in the primary and secondary windings as

$$\frac{n_p}{n_s} = \frac{E_p}{E_s} = \frac{I_s}{I_p}$$

The secondary load may be expressed as the ratio of secondary voltage to secondary current, and the primary load may be written as the ratio of primary voltage to primary current:

$$R_L = \frac{E_p}{I_p}$$

and

$$R_s = \frac{E_s}{I_s}$$

Then the tube plate load, R_L, may be written in terms of the transformer turns ratio and the secondary load

$$R_L = \left(\frac{n_p}{n_s}\right)^2 R_s \qquad (10\text{-}2)$$

In order to make the 8-Ω loudspeaker appear as 2 kΩ to the plate circuit of the tube, the transformer turns ratio must be

$$\frac{n_p}{n_s} = \left(\frac{2,000}{8}\right)^{\frac{1}{2}} = 15.8$$

Therefore a transformer having a turns ratio of 15.8 would be used.

Now we see the reason for using a so-called *output transformer* in the plate circuit of a vacuum tube power amplifier. For similar reasons, a transistor power amplifier may use a transformer to couple a small load into its output circuit. Coupling transformers are also used between amplifier stages operating at radio frequencies, and the primary and secondary windings are parts of *tuned circuits* to permit efficient operation over a narrow range of frequencies. Tuned-circuit coupling will be discussed later when Class C operation is introduced. The fundamental reason for using transformer coupling is to match two dissimilar impedances.

10.3   Power Calculations

To determine the equation for power output (P_o) when there is no distortion, we make use of the fact that the sine-wave forms of output voltage and current have maximum values which are one-half the difference between the instantaneous maximum and minimum values; thus

$$I_{p \text{ max}} = \tfrac{1}{2}(i_{b \text{ max}} - i_{b \text{ min}}) \tag{10-3}$$

$$E_{p \text{ max}} = \tfrac{1}{2}(e_{b \text{ max}} - e_{b \text{ min}}) \tag{10-4}$$

Since the maximum value of a sine wave is $\sqrt{2}$ times the effective value, we have for the effective values of output current and voltage

$$I_p = \frac{i_{b \text{ max}} - i_{b \text{ min}}}{2\sqrt{2}} \tag{10-5}$$

$$E_p = \frac{e_{b \text{ max}} - e_{b \text{ min}}}{2\sqrt{2}} \tag{10-6}$$

The power output is the product of these two effective values:

$$P_o = \frac{(e_{b \text{ max}} - e_{b \text{ min}})(i_{b \text{ max}} - i_{b \text{ min}})}{8} \tag{10-7}$$

The power output may also be computed by using the a-c load resistance (R_L):

$$P_o = I_p^2 R_L \tag{10-8}$$

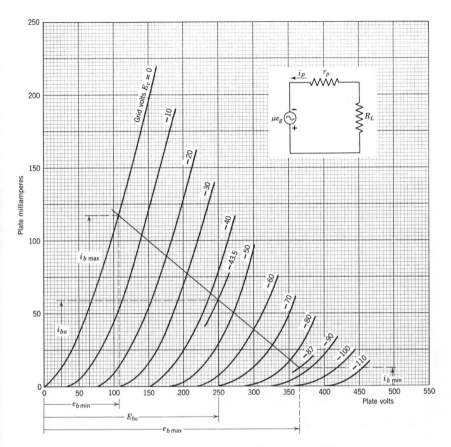

Fig. 10-5   Graphical analysis of power amplifier.

The characteristics in Fig. 10-5 will be used to calculate the output power of an amplifier whose a-c load line is constructed on the graph. In this example, the Q-point is chosen at $E_c = -43.5$ V to allow a large input signal swing and to prevent operation in the low current, high distortion region. In the example being studied, we have

$$P_o = \frac{(365 - 107)(0.117 - 0.0125)}{8} = 3.38 \text{ W}$$

$$I_p = \frac{0.117 - 0.0125}{2\sqrt{2}} = 0.0369 \text{ A}$$

$$E_p = \frac{365 - 107}{2\sqrt{2}} = 91.3 \text{ V}$$

As a check, the output power may be computed by $I_p^2 R_L$:

$$P_o = 0.0369^2 \times 2500 = 3.41 \text{ W}$$

and also

$$P_o = E_p I_p = 91.3 \times 0.0369 = 3.37 \text{ W}$$

There is actually some distortion present. The percentage of second-harmonic distortion is given by the following equation, which is developed in Section 10-4.

$$\text{Per cent second-harmonic distortion} = \frac{\dfrac{i_{b\,max} + i_{b\,min}}{2} - I_{bo}}{i_{b\,max} - i_{b\,min}} \times 100\%$$

and this becomes

$$\frac{(0.117 + 0.0125)/2 - 0.060}{0.117 - 0.0125} \times 100 = 4.6\%.$$

The power input (P_i) to the tube is

$$P_i = E_{ba} I_{ba}$$

We take the E_{bo} and I_{bo} values for this computation. I_{ba} is a little larger than I_{bo}, while E_{ba} may be slightly smaller than E_{bo} because of the slight distortion.

It will be seen that, when distortion is present, an additional d-c component of plate current comes into existence.

$$P_i = E_{bo} I_{bo} = 250 \times 0.060 = 15 \text{ W}$$

The plate dissipation (P_p) is the tube-input power minus the power output.

$$P_p = P_i - P_o = 15 - 3.4 = 11.6 \text{ W}$$

The maximum rating for this tube must be greater than 15 W. The plate efficiency is

$$\frac{P_o}{P_i} = \frac{3.4}{15} = 0.226 = 22.6\%$$

The foregoing example illustrates that the power output of an amplifier is the power associated with the fundamental components of plate current and voltage. The amplitudes of these fundamental components are calculated from readings on the plate-characteristic chart at different points on the load line. The product of their effective values $(E_p I_p)$ is the output power. The output power may also be computed by $I_p^2 R_L$, where I_p is the effective value of the fundamental component of current and R_L is the resistance to the flow of *alternating current* in the plate circuit.

When the output transformer is used to couple a load to the power amplifier, the resistance of the primary winding consumes some of the a-c power. This should be taken into account when calculating the power being supplied by the plate source and to the transformer primary.

DRILL PROBLEM

D10-1 A transistor Class A amplifier supplies power to a transformer-coupled load. The load requires 1 W of power. The primary winding resistance is $\frac{1}{10}$ as large as the a-c load resistance in the collector circuit. (*a*) How much a-c power must be developed by the transistor? (*b*) How much power is lost in the winding resistance?

10.4 Maximum Power Output Considerations

Graphical analysis may be used to determine optimum conditions for the largest power output commensurate with allowable distortion in tube and transistor amplifiers. The trial load lines drawn on the plate characteristics in Figs. 10-6 and 10-7 show that the a-c plate current decreases and the a-c plate voltage increases as the load resistance increases. Since output power is the product of a-c voltage and current, it will be a maximum for a particular value of a-c load resistance. This turns out to be about 2 or 3 times the dynamic plate resistance of the tube, rather than equal to r_p as we might expect from the maximum power transfer theorem. This is the result of the entrance of amplitude distortion of the a-c plate current into the picture, and this distortion must not be allowed to become excessive.

As the load resistance of the triode is increased from 1,000 Ω toward 20,000 Ω, the lower ends of the load lines leave the nonlinear regions of the characteristic curves and go into the more linear areas. Operation with 10,000 Ω with a signal voltage swing of ±30 V would give the best performance of the four trial values shown. A load line for a still larger resistance would perhaps give such a small current amplitude that the output power would decrease.

In the case of the pentode tube, operation with a load resistance of 2,500 Ω would result in a flattening of the negative half-cycles of the a-c plate-current wave, while with 5,000 Ω the positive half-cycles would be flattened as can be seen by assuming a 12.5-V grid swing. The *Q*-point is at $E_c = -12.5$ V. Incidentally, this tube is capable of operation in push-pull amplifiers, in which the grid may go moderately positive as in Class AB_2 operation. Evidently a load resistance value between 2,500 and 5,000 Ω would be most

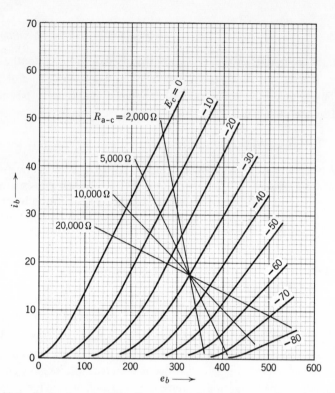

Fig. 10-6 Plate characteristics for power triode with several a-c load lines.

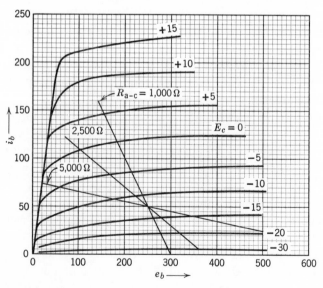

Fig. 10-7 Plate characteristics of 6AQ5A beam-power pentode with several a-c load lines.

desirable. There is no simple relationship between the optimum load resistance for the pentode and its dynamic plate resistance.

The average collector characteristic curves for an *NPN* silicon power transistor with common-base connection are very similar in shape to the curves for the 6AQ5-A power pentode tube in Fig. 10-7, although the curves gradually become closer to each other as the collector current rises toward its highest allowable value. The graphical analysis for optimum loading is carried out exactly as that described for tubes.

For the common-emitter connection, the curves for the *NPN* power transistor are bunched together in the high base-current region with less uniformity in spacing than for the common-collector connection. The procedure of graphical analysis is again the same as described above.

10.5 Amplitude Distortion in Amplifiers

The following analysis applies to both voltage and power amplifiers. This analysis will present only second-harmonic distortion and leave the matter of distortion due to higher harmonics for reference reading by the interested student.

When an amplifier stage employing a single triode is driven hard (relatively large input-voltage swing), the effect of curvature in the dynamic transfer characteristic becomes manifest in the production of harmonic components of plate-current. The second harmonic is the most pronounced. Figure 10-8 shows the existence of a second harmonic only, together with an added d-c component when the dynamic transfer characteristic is shaped like part of a parabola.

Taking the point Q as the origin of a pair of axes on which instantaneous values i_p and e_g are plotted, the equation of the parabola is

$$i_p = Ae_g + Be_g{}^2 \tag{10-9}$$

Let E_{gm} = the maximum value of the grid-excitation voltage, of which e_g is the instantaneous value. Then

$$e_g = E_{gm} \sin \omega t \tag{10-10}$$

Substituting Equation (10-10) into (10-9),

$$i_p = AE_{gm} \sin \omega t + BE_{gm}{}^2 \sin^2 \omega t \tag{10-11}$$

Since $\sin^2 \omega t = \frac{1}{2} - \frac{1}{2} \cos 2\omega t$,

$$i_p = AE_{gm} \sin \omega t + \frac{B}{2} E_{gm}{}^2 - \frac{B}{2} E_{gm}{}^2 \cos 2\omega t \tag{10-12}$$

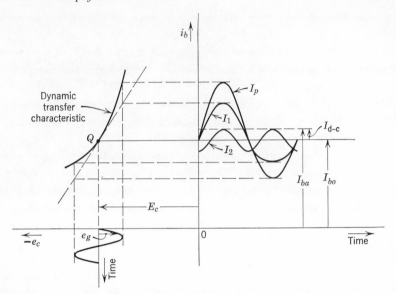

Fig. 10-8   Parabolic dynamic transfer characteristic—tube generates a second-harmonic component of plate current plus d-c component of the same amplitude.

Note that there is a fundamental component of current having a magnitude of AE_{gm}, a d-c component $(B/2)E_{gm}^2$ which is added to the I_{bo} current, and a second-harmonic component having the same magnitude as the d-c component. Calling these three new components I_1, $I_{\text{d-c}}$, and I_2, respectively, we may write Equation (10-12) in the following symbolic form.

$$I_p = I_1 + I_{\text{d-c}} + I_2$$

These components of the a-c plate current are labeled in Fig. 10-8. Note that I_{pm}, the maximum value of the total a-c component of current, is equal to $I_{1\ \max} + I_{\text{d-c}} + I_{2\ \max}$. It is also true that the amplitude A of the *negative half-cycle* of the a-c component of plate current is equal to $I_{1\ \max} - I_{\text{d-c}} - I_{2\ \max}$. These two relationships give the following equations for the maximum and minimum values of instantaneous total plate current.

$$i_{b\ \max} = I_{bo} + I_{1\ \max} + I_{\text{d-c}} + I_{2\ \max} \qquad (10\text{-}13)$$

$$i_{b\ \min} = I_{bo} - I_{1\ \max} + I_{\text{d-c}} + I_{2\ \max} \qquad (10\text{-}14)$$

The maximum value of the fundamental component is obtained in terms of e instantaneous total values by subtraction of these equations.

$$I_{1\,max} = \tfrac{1}{2}(i_{b\,max} - i_{b\,min}) \tag{10-15}$$

dding them gives the maximum value of the second-harmonic component.

$$2I_{2\,max} = I_{b\,max} + i_{b\,min} - 2I_{bo} - 2I_{d-c}$$

d, since $I_{d-c} = I_{2\,max}$,

$$4I_{2\,max} = i_{b\,max} + i_{b\,min} - 2I_{bo}$$

$$I_{2\,max} = \tfrac{1}{4}(i_{b\,max} + i_{b\,min} - 2I_{bo}) \tag{10-16}$$

he *second-harmonic distortion is defined as the ratio of the amplitude of the
cond harmonic to the amplitude of the fundamental.* Multiplying this ratio,
$_{max}/I_{1\,max}$, by 100 gives the percentage of second-harmonic distortion.

ercentage of second-harmonic distortion $= \dfrac{i_{b\,max} + i_{b\,min} - 2I_{bo}}{2(i_{b\,max} - i_{b\,min})} \times 100\%$

his assumes that all harmonics above the second are negligible.

Distortion in single-ended triode amplifiers is due mostly to the second-
armonic component. Pentode tubes are used extensively in voltage ampli-
ers. Third and fifth harmonics of plate current are more pronounced in
entodes than in triodes. The second harmonic is not so prominent.

The foregoing type of analysis is applicable to pentode amplifiers. To
clude higher harmonics, Equation (10-9) for i_p must be written with more
rms, as a power series in e_g rather than for a parabola.

The direct measurement of second-harmonic distortion in the output of
n amplifier may be done by means of a wave analyzer, or a harmonic
nalyzer as it is called by some manufacturers. This analyzer measures the
mplitudes of the fundamental component and harmonic components of a
oltage or current wave, up to as many as fifteen or more. Knowing the
mplitudes of the fundamental and the second harmonic, we simply divide
he latter by the former to get the second-harmonic distortion. The same
rocedure will give the distortion caused by any other harmonic if its ampli-
ude is known.

Distortion analyzers are also available which measure total distortion
lue to all harmonics of the fundamental. The output voltage of the amplifier
s applied to a filter circuit that takes out only the fundamental component.
The instrument measures the ratio of the amplitude of the remaining com-
onents, taken together, to that of the fundamental. That ratio times 100
s the total distortion in per cent.

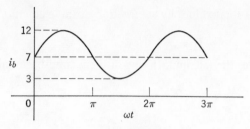

Fig. DP10-2

DRILL PROBLEM

D10-2 The plate current in a certain power amplifier has the waveshape shown i Fig. DP10-2. Assuming that second-harmonic distortion is the most significan estimate the per cent·of second harmonic distortion.

10.6 Transistor Power Amplifiers

The power that may be delivered to a load in a transistor power amplifie depends on several factors. These include: (1) maximum current rating (2) maximum breakdown voltage ratings, (3) permissible distortion, an (4) maximum permissible dissipation. The last factor is influenced by tem perature and often requires special care in designing the amplifier circuit t avoid exceeding the maximum dissipation rating when temperature effect may become important.

The maintenance of bias current as nearly constant as possible is of grea importance. A shift of the Q-point may cause the collector dissipation t rise, thereby further raising the temperature and resulting in thermal runawa and eventual destruction of the transistor. The saturation current I_{C_0} which flows when emitter current is zero is greatly affected by temperatur and can cause a shift in the Q-point if some provision is not made to counter act its effect on the dissipation.

There are various methods of stabilizing collector current against effect of temperature variations. One of them is illustrated in Fig. 10-9, where th R_E-C_E combination in the emitter circuit reduces the base voltage and bas current for increases in total collector current. The effect is to maintain th Q-point below dissipation values that may destroy the transistor. Thi circuit was described in some detail in the chapter on transistors.

The design of a power amplifier is concerned primarily with obtaining th required power in the load, using the smallest size and weight possible while maintaining a suitably low level of distortion. This requires the hig

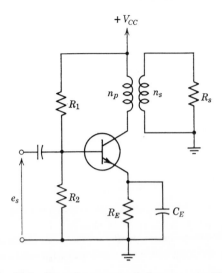

Fig. 10-9   Transformer-coupled power amplifier.

fficiency of Class B or C, but because of its high distortion Class C is
ractically limited to tuned loads, where its inherent filtering provides
reedom from distortion. Classes A and B are generally used for audio and
ow r-f applications. Class B is usually employed as a *push-pull* arrangement,
equiring two transistors and a center-tapped transformer for coupling the
oad. Even in Class A operation, an output transformer is generally required
o match the load to the transistor.

The procedures for analysis of amplifier operation and distortion developed
or tube amplifiers are fully applicable to transistor amplifiers. It is more
mportant in transistor amplifiers, however, to assure that maximum ratings
re not exceeded in their operation. Transistors are more prone to destruction
han are tubes. The additional circuit components needed in transistor
ircuits will be described in the following sections, as the various classes of
peration are presented.

EXAMPLE. The circuit of Fig. 10-9 uses a transistor that has a maximum
$_{CE}$ = 40 V and maximum permissible collector current = 300 mA. The circuit
 to be designed for maximum output power at the smallest possible input signal.
ssuming a linear characteristic for the transistor, sketch the region of operation
hat is permitted and determine the effective a-c load resistance and the Q-point.
f R_s = 8 Ω, what is the required turns ratio of the coupling transformer?

The linear $i_C - v_{CE}$ characteristics are shown in Fig. 10-10. The maximum
oltage that can appear across the transistor is 40 V, occurring at $i_C = 0$. The
aximum permissible collector current is the other end of the a-c load line.

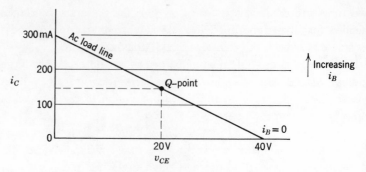

Fig. 10-10   Idealized common-emitter characteristics of a transistor.

Maximum power will be developed in the a-c load when the input signal permit operation from one end of the load line to the other. This requires that the Q-poir be located at the midpoint of the load line, at $i_C = 150$ mA and $v_{CE} = 20$ V The a-c load may be calculated from the slope of the load line

$$R_L{}' = \frac{40}{300} \times 10^3 = 133 \ \Omega$$

The required turns ratio is calculated,

$$\frac{n_p}{n_s} = \left(\frac{133}{8}\right)^{\frac{1}{2}} = 4.08$$

10.7   Parallel Operation of Power-Amplifier Tubes

Two or more tubes may be operated in parallel in order to obtain greate power output. The plates are connected in parallel and so are the grids a shown in Fig. 10-11. The optimum load resistance in the plate circuit, whe

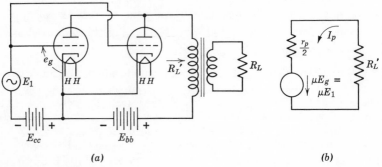

(a)   (b)

Fig. 10-11   (a) Parallel operation of triodes in an amplifier. (b) Equivalent circui for (a).

two tubes are used, is approximately half the optimum resistance recommended for single-tube operation. If the current is doubled and the load resistance is halved, the power will be doubled; $P = I^2R$.

High-power audio-output amplifiers such as those used in large public-address systems and in some radio receivers often use tubes in parallel in push-pull circuits. This practice is also common in the final amplifiers of radio broadcast transmitters.

It is logical to conclude that electronic-control applications may indicate the desirability of operating two or more tubes in parallel when the output of one tube is too small to operate a control device.

10.8   Theoretical Limit on Power Efficiency of Class A

The maximum possible efficiency of a single-ended vacuum-tube amplifier operating Class A may be deduced from Fig. 10-5 or 10-7 if we make a few assumptions. The analysis holds also for pentode and beam-power tubes, as well as for transistors.

Assume that the characteristic curves are straight lines and equally spaced, and that shunt feed of d-c plate current is used. Also assume the tube is driven between $E_c = 0$ and complete plate-current cutoff. The amplitude of the a-c component of plate current is then

$$I_{p\,\text{max}} = \sqrt{2}I_p = \frac{i_{b\,\text{max}} - i_{b\,\text{min}}}{2} = I_{bo}$$

This means that the negative half-cycle of i_p dips all the way down to the voltage axis, i.e., to $i_b = 0$. The a-c power output is then $I_p^2R_L$ and the input power is $E_{bb}I_{bo}$. The theoretical plate-current efficiency is then

$$\text{Theoretical } \eta_p = \frac{I_p^2R_L}{E_{bb}I_{bo}} = \left(\frac{I_{bo}}{\sqrt{2}}\right)^2 \frac{R_L}{E_{bb}I_{bo}}$$

$$= \frac{I_{bo}R_L}{2E_{bb}} \times 100\% \tag{10-17}$$

For *maximum* theoretical η_p, the maximum value of $I_{bo}R_L$ is required. R_L must be large enough for the load line to be practically horizontal. When that is true, the a-c output-voltage swing ($I_{bo}R_L$) approaches E_{bb} in value. It is evident from Fig. 10-4 that this can be more nearly accomplished with transistors and pentodes than with triodes. If $I_{bo}R_L$ is *theoretically* equal to E_{bb}, the *maximum theoretical* efficiency becomes

$$\text{Maximum theoretical } \eta_p = \tfrac{1}{2} \times 100\% = 50\%$$

10.9   The Push-Pull Principle

When two tubes of the same type are connected in an amplifier circuit as
shown in Fig. 10-12*b*, they are said to be connected in push-pull. Their
alternating components of plate current flow in the same direction in the
center-tapped primary winding of the transformer, *both either up or down*.

When the instantaneous plate current of one tube is at its maximum, the
other tube has minimum plate current. When the plate current in the first
tube is rising (increasing in value), the plate current in the second tube is
falling (decreasing in value). The a-c component of current of tube 1 is
flowing into its plate while the a-c component of current of tube 2 is flowing
out of its plate. Thus *both tubes contribute current in the same direction in the
primary winding*. Both tubes carry their own d-c components of plate current,
but these are equal in magnitude and flow in opposite directions through the
output-transformer primary. The effect of this is the same as if there were
no direct current in the transformer at all. This is a big advantage, as will be
seen shortly.

Figure 10-13 shows the relation between the number of magnetic lines
of flux in the core of a transformer and the current in its primary winding.
With single-ended operation, as shown in Fig. 10-13*a*, there is a comparatively
large d-c component of current (I_{bo}) which magnetizes the core in the amount
ϕ_o. With no signal voltage on the grid, the flux is constant at this value. When
signal voltage is applied, the plate current alternately rises above and falls

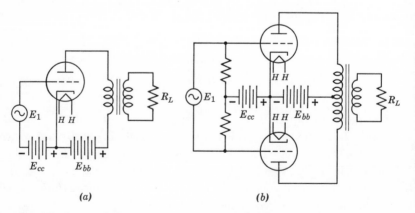

(a)   (b)

Fig. 10-12   Power-amplifier circuits. (*a*) Single-ended operation; (*b*) push-pull
operation.

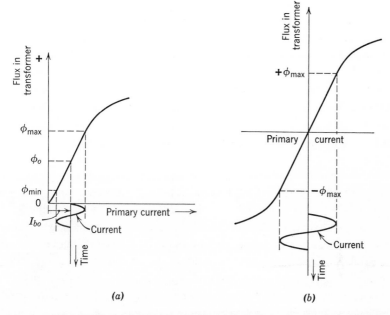

(a)   (b)

Fig. 10-13 Flux-current relations in output transformers of power amplifiers. (*a*) Single-ended operation. (*b*) Push-pull operation. There is no net d-c component of current in the transformer.

below the I_{bo} value. The corresponding variations in flux above and below the ϕ_o value induce voltage in the secondary winding of the transformer and cause secondary current to flow in the load.

In push-pull action the absence of a net effective direct current in the primary winding makes it possible to use an a-c component of current that is larger than twice the amount obtained with a single tube. This advantage, coupled with the automatic elimination of second harmonics (all even-numbered harmonics, in fact) of plate current, results in the realization of more than twice the power output obtainable from single-ended operation. Notice in Fig. 10-13*b* that the effective primary current, which is alternating current, is larger than twice the a-c component of primary current in Fig. 10-13*a*. A smaller output transformer may be used than would be needed for even a single-ended amplifier employing the same type of tube because *there is no d-c current in the windings to produce a large steady component of flux* in the transformer.

When the circuit is properly balanced, there is no signal-frequency component of current in the plate power supply. On the other hand, if there are any low-frequency components of current in the power supply, which are due to inadequate filtering, they will cancel out in the primary halves of the transformer and thus cause no objectionable effects in the output.

Cathode bias may be used; it is then unnecessary to use a bypass capacitor across the cathode resistor if the tubes are matched.

Transistors behave in much the same way as described for tubes in push-pull amplifiers. Their operation is discussed in Section 10.12.

10.10   Analysis of Push-Pull Amplifier Operation

The automatic elimination of all even-order harmonics of plate current (2nd, 4th, 6th, and so forth) permits the operation of push-pull power amplifiers in Classes AB_1, AB_2, B_1, B_2 and C. In public-address systems, AB_1 and AB_2 are extensively used. Class B is used where larger power output is desired than can be obtained with AB_2, but in both of these types of operation the driving unit (the amplifier or other device that feeds the grids) must be capable of supplying the required grid-circuit input power without distortion. Class C operation is used when only one frequency, or a narrow band of frequencies, must be amplified. Class B and Class C amplifiers are used in the high-frequency circuits of radio and television transmitters.

The manner in which the plate currents of the tubes in push-pull AB_1 operation are produced and combined to form a large effective primary current in the output transformer is shown in Fig. 10-14. Since the grid of one of the tubes is being driven more negative while the other is being driven more positive, it has been found convenient by many authors to place the transfer-characteristic charts back to back at the common bias voltage ($E_c = E_{c1} = E_{c2}$). This makes it convenient to draw the *composite dynamic transfer characteristic* (*b*), which is midway between the curves for the individual tubes, except when either tube's current is zero (in which case the composite characteristic coincides with the curve for the tube that is carrying current). Observe that the second-harmonic components, which flow simultaneously, are in phase opposition to each other and hence cancel in the transformer primary. That is also true of the other even harmonics.

There is reinforcement of the third-harmonic component effects, however; also of the fifth and other odd-numbered harmonics. Fortunately, the magnitude of even the third harmonic in each tube is appreciably less than that of the second, and the harmonic magnitudes higher than the third diminish rapidly. Nevertheless, the odd-harmonic distortion is a limiting factor in power output of push-pull stages. The symmetrical nature of the

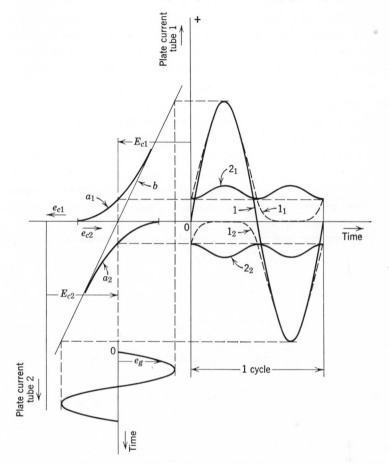

Fig. 10-14 Graphical construction of plate-current wave forms push-pull amplifier, Class AB_1, operation. 1_1 is current in tube 1; 2_1 is second harmonic in 1_1; 1_2 is current in tube 2; 2_2 is second harmonic in 1_2; 1 is a-c current in transformer primary.

combined dynamic transfer characteristic is the underlying cause of the absence of even-numbered harmonics and the presence of odd-numbered harmonics.

A composite load line, drawn on a pair of plate-characteristic-curve charts placed back to back, is used to analyze the performance of push-pull amplifiers. As an example we have chosen two type 6L6 tubes operating Class AB_1 at a plate-supply voltage of 350 V and with 22.5 V fixed bias. We did not choose 360 V, merely because the graphical representation is somewhat clearer with 350 V.

Figure 10-15 shows two plate-characteristic charts placed back to back (or more precisely, bottom to bottom) so that the 350-V points on the plate-voltage axes coincide. This causes the other major divisions on the voltage scales to line up, and this is advantageous for studying the process. This would not occur if the operation were at 360 V on the plates, for then the light vertical lines on the graph sheet would not line up. Nevertheless, operation at 360 V is recommended and the analysis for that condition would require lining up the charts at the 360-V points.

The load line for the two tubes in push-pull operation, called the composite load line, is drawn so that its slope (tangent θ) is $+1/R_L$ and is equal to one-fourth of the effective load resistance. The effective load resistance, called the plate-to-plate resistance in the tube manual, is specified as 6,600 Ω for Class AB_1 operation at highest possible output power using fixed bias. The load line is then drawn so that the acute angle θ with the voltage axis has a tangent value obtained by using

$$6,600 \div 4 = 1,650 \ \Omega = R_L'$$

$$\tan \theta = \frac{1}{R_L'} = \frac{1}{1,650} = \frac{\Delta i_b}{\Delta e_b} = \frac{\Delta i_b}{350}$$

$$\Delta i_b = \frac{350}{1,650} = 0.212 \ \text{A} = 212 \ \text{mA}$$

The composite load line is therefore drawn through points $e_b = 350$, $i_b = 0$ and $e_b = 0$, $i_b = 212$ mA, extending over both charts.

When the grid-driving voltage is zero, the voltage on each plate is 350 V and each tube carries 30 mA (this is the reading at 22.5 V bias, and, although not exact, it is quite accurate). When a driving voltage of 22.5 V peak value is applied, one grid is driven to zero volts with respect to its cathode, and at the same time the other is driven to -45 V with respect to its cathode. The curves in the tube manual show that the plate current is practically zero when the grid voltage is driven below -35 V. This means that the current of each tube will be cut off during that part of the grid-voltage cycle in which the grid voltage is dropping from about -35 V to its lowest value (-45 V), and while it is coming up again to -35 V to start current flow.

The plate current in each tube rises to a peak value of 165 mA once each cycle. The plate voltage drops to about 77 V at the same time. With sine-wave grid excitation the a-c components of plate current and plate voltage are nearly exact sine waves in this operation (total harmonic distortion is listed at only 2 per cent in the tube manual). We may therefore calculate

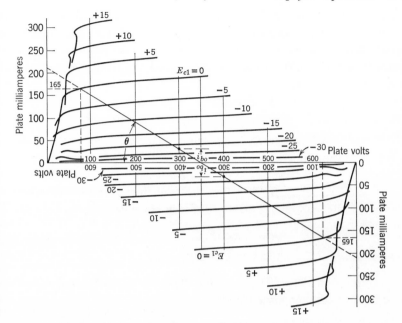

Fig. 10-15   Push-pull power-amplifier analysis; two 6L6 tubes. $E_b = 350$ V; $E_c = -22.5$ V fixed bias; $E_{gm} = 22.5$ V.

the approximate power output on the basis of pure sine waves of current and voltage, and use their effective values.

$$P_o = E_p I_p = \frac{(350 - 77)}{\sqrt{2}} \times \frac{0.165}{\sqrt{2}} = 22.5 \text{ W}$$

The push-pull circuit may be represented by the equivalent circuit of Fig. 10-16a. R_L is the load resistance after its reflection into the primary

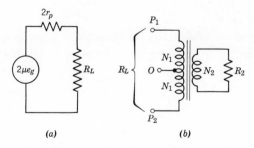

(a)   (b)

Fig. 10-16   (a) Equivalent circuit for push-pull amplifier. (b) Output transformer and load, push-pull operation.

circuit of the output transformer. It is the plate-to-plate load resistance and is called R_{pp}. The transformer puts R_L into the circuit. The transformer turns ratio must be selected so that a given actual load (R_2) on its secondary will be reflected into the primary as the desired (R_{pp}) value. In Fig. 10-16b,

$$R_L = \left(\frac{2N_1}{N_2}\right)^2 R_2$$

and this must be equal to R_{pp}, the plate-to-plate resistance of the push-pull tubes, for proper operation. Looking into only half the primary (between P_1 and O), a single tube "sees" $(N_1/N_2)^2 R_2$, which is only one-fourth of R_L (i.e., one-fourth of R_{pp}). Therefore the load line for push-pull operation (the composite load line) is drawn so that its slope is

$$-\frac{\Delta i_b}{\Delta e_b} = \frac{1}{\frac{1}{4}R_{pp}} = \frac{1}{R_L{}'}$$

In the discussion above, R_{pp} was 6,600 Ω, so $R_L{}' = 6{,}600/4 = 1{,}650$ Ω was used to draw the composite load line.

10.11 Class B Push-Pull Amplification

Operation of a push-pull amplifier in Class B is very similar to that of Class A and AB push-pull operation, except that the tubes are biased to cut-off at the Q point. The plate current in each tube has a large swing; as a result, the tube has more power output than in Class A. The efficiency is also improved. Figure 10-17 shows instantaneous grid-driving voltage and plate-current relations. Since the individual grids are driven alternately positive and grid current must flow, the driving amplifier must be capable of supplying the necessary power without distorting the signal voltage on the grids. The power tubes are usually designed with dynamic characteristics that are linear well within the positive-grid region.

Theoretical analysis of Class B performance is possible if the *composite dynamic characteristic is assumed to be linear*; this will make the current wave form sinusoidal for sinusoidal input voltage. Assuming that the plate current of each tube is a half sine wave, the equivalent d-c current *per tube* is

$$I_{\text{d-c}} = \frac{I_m}{\pi} \tag{10-18}$$

and the *total* d-c plate current is twice this value. The effective a-c load

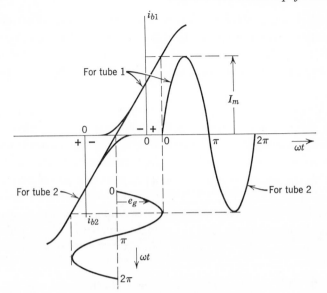

Fig. 10-17   Class B push-pull action.

current is then

$$I_{\text{rms}} = \frac{I_m}{\sqrt{2}} \ \text{A} \tag{10-19}$$

The output power is

$$P_o = \left(\frac{I_m}{\sqrt{2}}\right)^2 R_L' = \frac{I_m{}^2 R_L'}{2} \ \text{W} \tag{10-20}$$

in which R_L' is the *load resistance on each tube*, which is one-fourth the plate-to-plate value. Using Equation (10-18), the d-c input power to both tubes is

$$P_{\text{d-c}} = \frac{2}{\pi} I_m E_{bb} \tag{10-21}$$

The ratio of output power to input power gives the plate-circuit efficiency:

$$\eta_p = \frac{I_m{}^2 R_L'/2}{2 I_m E_{bb}/\pi} = \frac{\pi}{4} \frac{I_m R_L'}{E_{bb}} \tag{10-22}$$

$I_m R_L'$ is the amplitude of the a-c component of plate voltage and is equal to $E_{bb} - E_{b\,\text{min}}$ as always; therefore

$$\eta_p = \frac{\pi}{4}\left(\frac{E_{bb} - E_{b\,\text{min}}}{E_{bb}}\right) \tag{10-23}$$

Theoretically, $E_{b\,\text{min}}$ should be made zero; then the maximum theoretical plate-circuit efficiency of a Class B push-pull amplifier is

$$\text{Maximum theoretical } \eta_p = \frac{\pi}{4} = 0.785 = 78.5\% \tag{10-24}$$

This assumes that plate voltage may vary from E_{bb} to zero along the load line. For vacuum tubes, this variation is unattainable; it may be approached in transistor circuits. The low saturation voltage of many transistors permits an efficiency very close to this theoretical maximum.

The plate dissipation is equal to the input power minus the output power. Since the input power $(E_{bb}I_{\text{d-c}})$ is expressed in terms of the d-c current, Equation (10-18) may be used to express output power in terms of $I_{\text{d-c}}$. The *total d-c plate current* is $2I_m/\pi$, and I_m equals $\pi I_{\text{d-c}}/2$, in which $I_{\text{d-c}}$ is the d-c current *per tube*. The output power is

$$P_o = \left(\frac{\pi I_{\text{d-c}}}{2}\right)^2 \frac{R_L{'}}{2} = \frac{\pi^2}{8} I_{\text{d-c}}^2 R_L{'} \tag{10-25}$$

The plate dissipation is then

$$P_p = P_i - P_o = E_{bb}I_{\text{d-c}} - \frac{\pi^2}{8} R_L{'} I_{\text{d-c}}^2 \tag{10-26}$$

10.12   Transistor Class B Push-Pull Operation

Transistor operation in Class B push-pull power amplifiers is similar to tube operation. The transistors are connected in the circuit so the input signal drives the two transistors on alternate half-cycles, and the load passes the two output currents out of phase.

The fundamental circuit is shown in Fig. 10-18. The input transformer is usually the collector circuit of a Class A *driver stage*. The load on the power amplifier is connected to the transistors through the center-tapped output transformer. A voltage-divider network, R_1 and R_2, is connected across V_{CC}. It is used to eliminate *crossover distortion* that occurs because of the forward-biased emitter voltage at zero base current. The resistors R_E are temperature-compensating components.

For each input half-cycle to be equally amplified in the load, the input transformer and the circuit must be balanced. If the two transistors are equal in their operation, there will be no net average MMF in the core of the output transformer. Thus, a smaller transformer may be used than for a

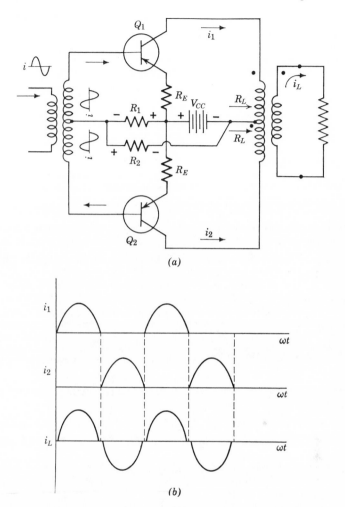

Fig. 10-18 Transistor push-pull amplifier. (*a*) Schematic diagram. (*b*) Crossover distortion in load current without R_1 and R_2.

comparable Class A amplifier; this is an advantage of the Class B push-pull operation of transistor power amplifiers.

The input signals cannot be applied to the power amplifier when their magnitudes are smaller than the potential difference across the emitter-base junction. During the small period of time when the input signals are going through zero, discontinuities in load current are shown in Fig. 10-18*b* as crossover distortion in the load current. The resistor R_1 is used to apply a potential in the base circuit that just equals the forward-biased junction potential. Thus, its effect is to eliminate the distortion.

DESIGN EXAMPLE. A push-pull Class B power amplifier is to be designed to deliver 10 W to a load of 8 Ω resistance. An output transformer is available that has 10 Ω primary resistance and 2 Ω secondary resistance. Its required turns ratio will be determined in the design.

The amplifier is to employ transistors that have the following ratings and specifications:

$$i_{C\,max} = 2\,\text{A}$$
$$P_{C\,max} = 20\,\text{W}$$
$$h_{fe} = 50\ (\text{assumed linear})$$
$$BV_{CEO} = 65\,\text{V}$$
$$\text{Emitter-base voltage (at } i_B = 0) = 0.6\,\text{V}$$

We will use the circuit of Fig. 10-18, and assume the input signal is a signal-frequency sine wave. Each half of the output primary winding delivers the same power to the load and dissipates the same power internally. Similarly, each transistor supplies half of the total power delivered to the secondary winding and the load. Some of the available a-c power will be dissipated in the resistance of the secondary winding.

For the assumed linear h_{fe}, one transistor will deliver power to the transformer primary given by

$$P_o = \frac{I_{c\,max}V_{ce\,max}}{2}, \qquad \text{during alternate half cycles} \tag{1}$$

The power supplied will divide between the primary winding resistance and the reflected load:

$$P_o = \frac{I_{c\,max}^2}{2} \cdot \frac{R_p}{2} + \frac{I_{c\,max}^2}{2} R_L' \tag{2}$$

The power delivered to the secondary is given by the second term in this expression, and the power delivered to the load will be less by an amount of dissipation in the secondary resistance:

$$P_L = \frac{I_{c\,max}^2}{2} R_L' - \frac{I_{s\,max}^2}{2} R_s$$

$$= \frac{I_{s\,max}^2}{2} (R_L) \tag{3}$$

To meet the specifications of the design and deliver 10 W to R_L, the secondary must be supplied with the power

$$P_{s\,min} = 10 + \frac{2}{8}(10) = 12.5\,\text{W}$$

The maximum value of $V_{ce\,max}$ must be less than $\frac{1}{2}BV_{CEO}$, or less than 32.5 V.

From Equations (1) and (2), the power supplied to the secondary is

$$P_{sec} = \frac{I_{c\,max}\,V_{ce\,max}}{2} - \frac{I_{c\,max}^2}{4}\,R_p \qquad (4)$$

Equation (4) is graphed in Fig. 10-19 for the known values of R_p and $V_{ce\,max}$. Note that two values of $I_{c\,max}$ will yield the power needed in the secondary. The two values may be calculated from Equation (4):

$$\frac{I_{c\,max}^2}{4} \cdot R_p - \frac{I_{c\,max}\,V_{ce\,max}}{2} + P_{sec} = 0$$

or

$$I_{c\,max} = \frac{\frac{1}{2}V_{ce\,max} \pm (\frac{1}{4}V_{ce\,max}^2 - R_p P_{sec})^{\frac{1}{2}}}{\frac{1}{2}R_p} \qquad (5)$$

Two values of $I_{c\,max}$ result, each of which will yield the required P_{sec}. However, only one of them is likely to meet the transistor specification of $i_{c\,max} = 2$ A.

Inspection of Equation (5) shows that some minimum collector voltage must be used, or the current expression will become a complex number, i.e.,

$$\frac{V_{ce\,max}^2}{4} \geq R_p P_{sec}$$

or the minimum voltage is

$$V_{ce\,max} \geq 2\sqrt{R_p P_{sec}} \qquad (6)$$

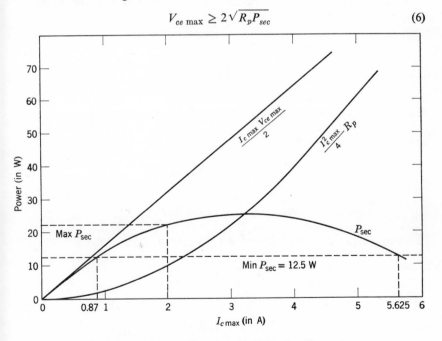

Fig. 10-19   Plot of Equation (4) in Design Example.

Substituting known values into Equation (6) yields,

$$\min V_{ce\,max} = 2\sqrt{10(12.5)} = 22.4 \text{ V},$$

which is less than the 32.5 V calculated as the maximum possible value. Any value of collector supply V_{CC} that is between these extremes may be used in this design.

The necessary turns ratio of the transformer can be calculated from the ratio of primary and secondary power requirements and from the ratio of load values in the primary and secondary windings:

$$\frac{P_{p\,max}}{P_{s\,max}} = \frac{I_c^2{}_{max}\,R_L'}{I_s^2{}_{max}\,(R_L + R_s)} = \left(\frac{n_s}{n_p}\right)^2 \cdot \frac{R_L'}{(R_L + R_s)} = 1$$

if the transformer is considered having no losses except those in the resistive components. Thus

$$\frac{n_p}{n_s} = \left(\frac{R_L'}{R_L + R_s}\right)^{\frac{1}{2}} \tag{7}$$

Since

$$V_{ce\,max} = I_{c\,max}\left(R_L' + \frac{R_p}{2}\right), \tag{8}$$

the value of reflected load resistance can be found to calculate the necessary turns ratio, when the voltage and current values have been chosen in the design.

To complete the design, we calculate the two values of $I_{c\,max}$ in the figure; using $V_{c\,max} = 32.5$ V

$$I_{c\,max} = \frac{16.25 \pm \sqrt{(16.25)^2 - 10(12.5)}}{5}$$

$$= \frac{16.25 \pm 11.8}{5} = 5.61, \text{ or } 0.89 \text{ A}$$

The value 0.87 A is well within the limiting value of 2 A set by the specifications.

Equation (8) and these limiting values of $I_{c\,max}$ may be used to calculate minimum and maximum values of R_L':

$$R_L'{}_{min} = \frac{32.5}{2} - 5 = 11.25 \ \Omega$$

$$R_L'{}_{max} = \frac{32.5}{0.87} - 5 = 32.4 \ \Omega$$

Choosing a value of collector current at 1.2 A then the value of R_L' is

$$R_L' = \frac{32.5}{1.2} - 5 = 22.1 \ \Omega$$

From Equation (4), the power supplied to the secondary will be

$$P_{sec} = \frac{(1.2)(32.5)}{2} - \frac{(1.2)^2(10)}{4} = 15.9 \text{ W}$$

Since only 12.5 W is required, this will provide more than the design requirements but may be used. The turns ratio is

$$\frac{n_p}{n_s} = \left(\frac{22.1}{8+2}\right)^{\frac{1}{2}} = 1.49$$

so that the power will be transformed to the load. A slightly smaller turns ratio could be used with the other calculated values, since the power available will be greater than the amount needed.

The transistor must be capable of dissipating 20 W of power. Since the load is taking 15.9 W and maximum collector current is 1.2 A, we may calculate the average power dissipated by the transistor. The average power over 1 cycle of input signal is

$$P_C = \frac{1}{2}\left[V_{CC}\frac{2I_{c\,max}}{\pi} - \frac{I^2_{c\,max}(R_L' + R_p/2)}{2}\right]$$

The first term must be less than 20 W:

$$\text{Power} = \left(\frac{1}{2}\right)(32.5)\frac{(2)(1.2)}{\pi}$$

$$= 12.4\ \text{W}$$

which is well within the specified maximum collector dissipation.

The R_1-R_2 network will be calculated to eliminate crossover distortion. The d-c current in R_1 must be larger than the largest signal current in the base circuit. Since h_{fe} is given as 50, and $I_{c\,max}$ is 1.2 A, the base current will have a maximum value of

$$i_{b\,max} = \frac{1.2}{50} = 24\ \text{mA}$$

Then R_1 must be

$$R_1 \le \frac{0.6\ \text{V}}{0.24\ \text{A}} = 2.5\ \Omega, \text{ to permit a d-c current equal to } 10 \times i_{b\,max}$$

Choosing its value at 2.5 Ω, then R_2 will be

$$R_2 = \frac{(32.5 - 0.6)\ \text{V}}{0.24\ \text{A}} = 132.9\ \Omega$$

Transformers designed for use as output devices for power amplifiers are usually rated for the load resistance and its corresponding *total* primary load resistance, i.e., the transformer used in the design example would be rated as 88.4 to 8 Ω. In terms of its turns ratio

$$88.4\ \Omega = \left(\frac{2n_p}{n_s}\right)^2 (8)\ \Omega$$

DRILL PROBLEM

D10-3 A certain output transformer is marked "400 Ω C.T. to 3.2 Ω." When used in a push-pull circuit, what will be the reflected load?

10.13 Complementary Symmetry

The transformers supplying, simultaneously, grid voltages of opposite polarities to tubes in push-pull and base currents of opposite direction of flow to transistors in push-pull perform *phase inversion* inherently. Note, in Fig. 10-18 that, although there is a single sine-wave current in the transformer primary, there are two secondary currents. We may call the current in the upper half of the secondary a positive-polarity current flowing into the base of Q_1 and the current in the lower half a negative polarity current flowing into the base of Q_2.

By complementary symmetry is meant a principle of assembling a Class B transistor amplifier without requiring the center-tapped phase-inverting input transformers at the input and output of the stage. Fortunately, *NPN* and *PNP* transistors require opposite polarized bias voltages; the *NPN* common-emitter transistor must have a *positive* base-current drive, while the grounded-emitter *PNP* requires a *negative* base-current drive. If two such transistors having similar characteristics are connected back-to-back and biased at cut-off in a push-pull circuit, the *NPN* will amplify the current in the positive half-cycles, and the *PNP* in the negative, thus obviously eliminating phase inversion. This arrangement is shown in Fig. 10-20.

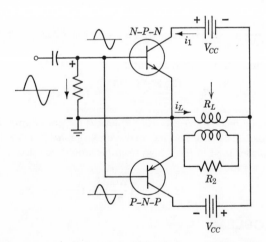

Fig. 10-20 Transistor push-pull amplifier using complementary symmetry. Current arrows show direction of flow during positive half-cycle of input current.

10.14   Class C Amplifier

It was mentioned earlier that Class C operation is employed only when a single frequency, or a narrow band of frequencies, is to be handled. The reason is that the amplifier has a parallel-resonant circuit for its load and exact resonance can be maintained at only one frequency. For all practical purposes, however, the circuit is resonant over a narrow band of frequencies. For this reason the name *tuned Class C amplifier* is often used.

Although the Class C amplifier is used almost entirely in radio-transmitter circuits, it is worthy of study here because its principles are applicable in electronic engineering. For example, an important type of generator of high-frequency currents and voltages—the Hartley oscillator—is in reality a Class C amplifier that is kept in operation by feedback of some a-c energy from the plate circuit to the grid circuit.

A comparatively high voltage is needed to drive a Class C amplifier because the grid is biased far below cutoff value for the tube. Bias voltages of two or more times the cutoff value are common. For example, a type 802 pentode power tube is recommended for operation at a bias of -100 V, and 500 V on its plate. In order to drive the grid positive, and that is always done in Class C operation, the a-c grid voltage must exceed 100 V peak. A peak value of 155 V is recommended for this service by the manufacturer.

The advantages of Class C operation are (1) low cost and small space requirements, (2) high plate-circuit efficiency, (3) comparative simplicity. Although its plate current flows only in short spurts, its resonant circuit, called the "tank circuit," makes possible the production of large output power and acceptable wave form.

The circuit is shown in Fig. 10-21. The "tank" consists of an air-core inductance coil and a capacitor. If the capacitor is variable, as shown, the tank may be tuned to resonance over a range of frequencies. For a given capacitor setting, the frequency of resonance is given very closely by the familiar relation

$$f = \frac{1}{2\pi\sqrt{LC}} \tag{10-27}$$

in which L is in henrys and C is in farads. As stated before, the tank will be practically in resonance over a narrow band of frequencies. This band lies on either side of, and includes, the center frequency given by Equation (10-27). Tuned loads are discussed in more detail in Chapter 11.

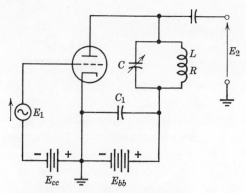

Fig. 10-21   Class C tuned amplifier circuit.

As an example of power amplification, it is common to obtain 100 W from a single-tube amplifier operating Class C and driven by a $7\frac{1}{2}$-W driver. The 100 W come from the plate-supply source and not from the $7\frac{1}{2}$-W input unit, of course. The output power is obtained from a d-c source by tube and circuit action, being delivered at high efficiency as a-c power. Briefly, then, large amounts of high-frequency power may be generated economically by starting with a small voltage at the desired frequency, amplifying it so that it can drive a low-power Class B or Class C amplifier, and using this in turn to drive a high-power Class C amplifier.

10.15   Voltages and Currents of a Class C Amplifier

The relationships of voltages and currents of a Class C amplifier are very interesting. The grid is driven far enough positive to cause substantial grid-current flow, for only thus may high efficiency be obtained.

Because the d-c resistance of the tank inductance coil is very low, there will be negligible voltage drop due to the d-c component of plate current. We may therefore consider the plate-supply voltage (E_{bb}) to be applied directly to the plate of the tube. The action of the resonant-tank circuit is to produce a sinusoidal voltage of large amplitude, because its characteristics are such that it acts like a high resistance (as explained in Section 10.16) to the flow of the fundamental component of plate current. This is an important principle that will be explained further. First, however, additional comment is desirable with reference to the plate current.

Figure 10-22a shows two cycles of plate current drawn to an enlarged scale. The fundamental component is shown in dash lines. Many other

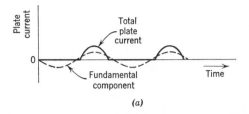

(a)

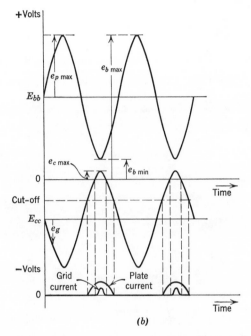

(b)

Fig. 10-22 (a) Plate current of Class C amplifier. (b) Voltages and currents in Class C power amplifier.

components exist, but they are smaller and are not shown. When the tank circuit is tuned to the frequency of the fundamental, it will offer high resistance at that frequency only. At the frequency of the second harmonic and at higher frequencies, the impedance of the tank is very low, so that practically no voltage at second-harmonic frequency, or at the higher frequencies, exists across the tank. Only at the fundamental frequency will there be an output voltage. The narrow band of frequencies heretofore mentioned would be made up of fundamentals only. An example is the band of frequencies put out by the tank circuit of a radio-transmitter Class C amplifier that feeds the antenna. A station operating on 1,000 kHz would have in its

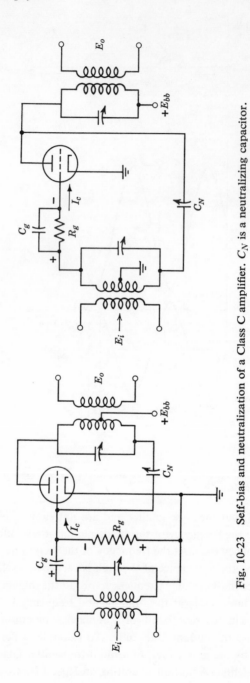

Fig. 10-23   Self-bias and neutralization of a Class C amplifier. C_N is a neutralizing capacitor.

tank circuit individual frequencies that lie in the band between about 995,000 Hz and 1,005,000 Hz. These are all fundamentals. The second harmonics would have frequencies twice these values, of course.

The high efficiency of the Class C amplifier is due to the fact that plate current flows only while the instantaneous plate-to-cathode voltage, e_b, is going through its lowest values (see Fig. 10-22b). The maximum value of the current occurs at the instant the plate-to-cathode voltage is a minimum. *During this brief period of each cycle the power stored in the tank circuit is large because its voltage is large.* It is important to note here that the varying component of plate-to-cathode voltage, which is shown to have maximum values $e_{p\,max}$, is also the a-c component of the tank voltage. There is negligible a-c drop across the power supply, i.e., across the capacitor C_1 (Fig. 10-21). Practically all the power stored in the tank circuit each cycle is delivered to the load. Because this is a large portion of the input power each cycle, the tube losses therefore being small, the efficiency is high. It exceeds 90 per cent under favorable conditions. The high voltage across the high-resistance tank circuit of Fig. 10-21 sends load current through the load which is coupled to the tank by the capacitor.

A Class C amplifier may be operated with self-bias produced by a grid-leak resistor shunted by a capacitor as shown in Fig. 10-23. When grid current flows, the capacitor is charged with the polarity indicated. The time constant $(R_g C)$ is long, compared with the time of 1 cycle of the grid-driving voltage, and therefore the capacitor discharges only slightly through R_g while grid current is not flowing. The required negative component of voltage thus remains nearly constant and is continuously applied to the grid while the tube receives grid-excitation voltage.

Many Class C amplifiers are operated with fixed bias. This is expensive because a special d-c voltage source is required, but it is safer than using only grid-leak bias, because the plate current rises to very high values when no grid-excitation voltage is being applied, and this current may damage the tube. A combination of fixed and grid-leak bias is commonly used.

10.16 Tank-Circuit Impedance at Unity Power Factor

It will now be shown how the tank circuit may be tuned so that, at the desired frequency, it will become an impedance that is a pure resistance. Let the inductance and the resistance of the tank coil be L and R, and let the value of the capacitance of the tank capacitor be C (Fig. 10-24). The equivalent series impedance (Z) at any frequency (f), is given by the product

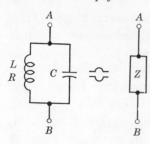

Fig. 10-24 Tank circuit presents an equivalent series impedance to its terminals.

of the two branch impedances divided by their sum:

$$Z = \frac{(R + j\omega L)(1/j\omega C)}{R + j\omega L + 1/j\omega C} \quad (10\text{-}28)$$

Multiplying numerator and denominator by $j\omega C$

$$Z = \frac{R + j\omega L}{j\omega CR - \omega^2 LC + 1}$$

$$= \frac{R + j\omega L}{(1 - \omega^2 LC) + j\omega CR} \quad (10\text{-}29)$$

Note that the denominator has a real term, $1 - \omega^2 LC$, and an imaginary term, $j\omega CR$. It should, therefore, be rationalized (i.e., converted to real terms only). Accordingly, we multiply both numerator and denominator by the conjugate of the denominator. The conjugate of a complex quantity is obtained by changing the sign of its imaginary term.

$$Z = \frac{R + j\omega L}{(1 - \omega^2 LC) + j\omega CR} \times \frac{(1 - \omega^2 LC) - j\omega CR}{(1 - \omega^2 LC) - j\omega CR}$$

$$= \frac{j\omega L - j\omega^3 L^2 C - j\omega CR^2 + R - \omega^2 LCR + \omega^2 LCR}{(1 - \omega^2 LC)^2 + \omega^2 C^2 R^2} \quad (10\text{-}30)$$

$$Z = \frac{R + j\omega(L - CR^2 - \omega^2 L^2 C)}{\omega^2 C^2 R^2 + (1 - \omega^2 LC)^2}$$

In order for this impedance to be *purely resistive*, the j term in the numerator must be *equal to zero*; this means that

$$L - CR^2 - \omega^2 L^2 C = 0 \quad (10\text{-}31)$$

From this is obtained

$$\omega = \sqrt{\frac{L - CR^2}{L^2 C}} = \sqrt{\frac{1}{LC} - \frac{R^2}{L^2}} = 2\pi f \quad (10\text{-}32)$$

The capacitance of C in Fig. 10-21 can be made the required value to satisfy this equation when R, L, and f are fixed. Then the criterion for the equivalent-tank impedance (Z) to be a pure resistance will be achieved, Equation (10-30) having no reactive term.

By means of substitution of numerical values into Equation (10-32), it may be shown that the second term under the radical sign is very small compared with the first term, and, therefore, Equation (10-27) may be used to determine the resonant frequency or to determine L or C when one of them and the frequency are given. A set of practical values is: $L = 200 \times 10^{-6}$ H; $f = 10^6$ Hz; $R = 50\ \Omega$. Using Equation (10-31) to determine C:

$$\frac{1}{C} = 4\pi^2 f^2 L + \frac{R^2}{L} = 39.5 \times 10^{12} \times 200 \times 10^{-6} + \frac{50^2}{200 \times 10^{-6}} = 7.9 \times 10^9$$

$$C = 126.5 \times 10^{-12}\ \text{F}$$

Comparing R^2/L^2 and $1/LC$:

$$\frac{R^2}{L^2} = \frac{50^2}{200^2 \times 10^{-12}} = 0.0625 \times 10^{12}$$

$$\frac{1}{LC} = \frac{1}{200 \times 10^{-6} \times 126.5 \times 10^{-12}} = 39.5 \times 10^{12}$$

It is thus seen that $1/LC$ is more than six hundred times R^2/L^2, for this circuit at $f = 10^6$ Hz. Equation (10-32) relating L and C to f may, therefore, be written in the simpler form when the resistance of the tank coil is small:

$$f = \frac{1}{2\pi\sqrt{LC}} \tag{10-33}$$

10.17   Quantitative Analysis of Class C Operation

The Class C amplifier is distinctly nonlinear in operation, giving an output voltage that is a complicated function of input voltage. In order to express the plate current and the grid current mathematically, they may first be plotted as functions of ωt from data obtained graphically. With grid voltage on the vertical scale and plate voltage on the horizontal, a set of curves each at a constant value of grid current is plotted, as well as a set of curves each at constant plate current. A straight line is then drawn at the Q-point located by data from two equations, one relating E_{cc} to e_c and the signal voltage, the other relating E_{bb} to e_b and the a-c component of plate voltage.

Using the plot and operating line, values may be taken from the graph and used in plotting i_b and i_c against ωt as stated above. Then, using graphical integration (getting the area under each curve), values for E_p, I_b, E_g, and I_g may be obtained. These may then be used to calculate input power, output power, and efficiency. The whole procedure is lengthy and tedious, and the results do not justify its inclusion here.

10.18   Neutralization of a Class C Amplifier

A triode tube with a plate load that is inductive conducts high-frequency energy from the plate end of the tank circuit to its grid through the grid-plate interelectrode capacitance. This can cause current surges back and forth through the tube, in which case the amplifier is said to *oscillate*. Oscillation ruins the amplifying action. To prevent oscillation a correction known as *neutralization* must be applied. This is done by connecting a small variable capacitor between the *bottom point* of the tank circuit and the grid terminal of the tube. It is then adjusted until oscillation stops. This is evident when high output voltage of good sine-wave form appears across the tank. It is also possible to connect the neutralizing capacitor between the plate of the tube, which is at the top point of the tank, and the bottom point of the parallel *LC* circuit that feeds the grid as shown in Fig. 10-23.

Push-pull circuits operating at high frequencies also require neutralization. They may be neutralized by a criss-cross connection of two capacitors, one from each tube plate, to the end of the input parallel circuit that is connected to the grid of the other tube.

10.19   Phase-Inversion Circuits

With the exception of transistor stages with complementary symmetry, push-pull amplifier stages require input signals that are equal but opposite in phase. To save in cost, weight, and space requirements and to remove the handicap of possible poor transformer frequency response, special circuits have been devised to provide the signals required.

Figure 10-25 is a simple circuit in which the load on the tube consists of two equal resistors, R_1 and R_2. The voltages E_1 and E_2 will be proved equal and opposite in phase. Before going into the mathematical analysis, we can easily see that, while the signal E_s on the grid is going in the positive direction and the plate current is increasing, the top point of R_2 is *rising* in potential with respect to ground and therefore E_2 is *increasing*. At the same time the top point of R_1 is *falling* in potential with respect to ground and E_1 is therefore decreasing. Furthermore, at the instant the a-c plate current is at its positive peak, E_2 is at its positive peak and E_1 is at its negative peak. E_1 and E_2 are thus 180 degrees out of phase.

By now we know that the constant-voltage equivalent circuit for Fig. 10-25 has a generator μE_g supplying I_p to R_1, R_2, and r_p in series. Symbols

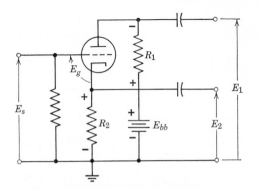

Fig. 10-25   Split-load phase inverter.

for rms values of voltage and current will be used in this analysis.

$$\mu E_g = I_p(R_1 + R_2 + r_p) \tag{10-34}$$

and, from the actual circuit

$$E_g = E_s - I_p R_2 \tag{10-35}$$

$$E_2 = I_p R_2 \tag{10-36}$$

$$E_1 = -I_p R_1 \tag{10-37}$$

Substituting for E_g in Equation (10-34) gives

$$\mu E_s = I_p[r_p + R_1 + (1 + \mu)R_2]$$

Using Equation (10-37) for I_p gives

$$E_1 = -\frac{\mu R_1 E_s}{r_p + R_1 + (1 + \mu)R_2} \tag{10-38}$$

Using Equation (10-36) for I_p gives

$$E_2 = +\frac{\mu R_2 E_s}{r_p + R_1 + (1 + \mu)R_2} \tag{10-39}$$

The opposite signs verify the phase reversal, and it can be seen that R_1 must equal R_2 if E_1 and E_2 are to have equal magnitudes.

A disadvantage of this circuit is low gain, in spite of the fact that large load resistances must be used to obtain output-voltage balance because the two output impedances are quite different.

Higher gain is obtainable with the two-tube circuit of Fig. 10-26. So far as operation is concerned, it can be seen that the grid of T_2 is actuated by

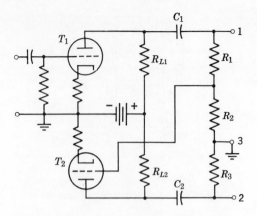

Fig. 10-26   Two-tube phase-inverter circuit.

a voltage 180 degrees out of phase with the input signal to T_1. R_1 and R_2 serve as a split a-c load on T_1. R_3 is made equal to $R_1 + R_2$, and, to give the output voltages $E_{1,3}$ and $E_{2,3}$ equal magnitudes, R_1 and R_2 must be so chosen that

$$\frac{R_2}{R_1 + R_2} = A_2$$

the voltage gain through T_2.

Transistor circuits may be constructed as split-load phase inverters, but biasing problems limit their practical use. Inexpensive transformers are available, and they are often used to obtain two signals that are out of phase and equal in amplitude by using a center-tapped secondary. Refer to Fig. 10-18 where the transistors are driven from such a transformer. In the interest of miniaturization of electronic equipment, class B push-pull operation may be achieved with complementary symmetry transistors, rather than with a phase inverter or transformer circuit.

10.20   The Magnetic Amplifier; Amplification by Change in Magnetization

Consider a load resistance to be supplied with alternating current that flows through two coils A and B wound on a three-legged iron core, as illustrated in Fig. 10-27.

A third coil C wound on the middle leg is supplied with direct current that can be varied in amount by adjustment of a control rheostat R_c. Note that coils A and B are so wound that the alternating flux produced by the load current will flow only in the outside legs. Any flux that coil A would tend to send through the middle leg of the core would be completely cancelled by an opposing flux due to coil B. The flux produced by coil C always has the direction shown and divides equally between the two outside legs of the core, flowing upward in both.

This core-and-coil structure is called a saturable reactor. The reactance which it provides may be varied over a wide range by changing the amount of direct current in coil C. This means that the load current may be varied over a wide range and thus the power (I^2R_L) delivered to the load may be varied, or controlled, over a corresponding range. This control may be accomplished with relatively small variations of the direct current, and therefore of the input power, in coil C. Thus power amplification results.

The reasons why changes in reactance occur when the direct current is changed may be understood after a study of Fig. 10-28. When a steady current I_1 flows through the control winding C, the magnetization of the core, which is called the magnetic bias, is such that the "operating point" is on the straight portion of the magnetization curve. The inductance of the AB coils is therefore large $[L = N(\Delta\phi/\Delta I)]$, and the inductive reactance in series with the load will be correspondingly large ($X_L = 2\pi fL$). This will allow only a small current to flow in the load.

When the direct current in C is increased to a value I_2, the core of the reactor becomes saturated, owing to the NI_2 bias, and the inductance is

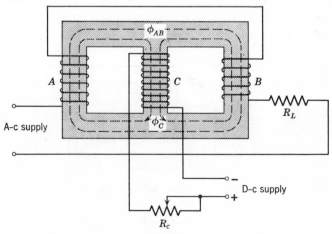

Fig. 10-27   Magnetic amplifier.

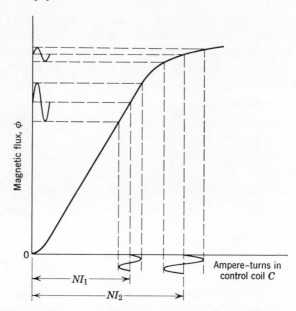

Fig. 10-28 Magnetization curve of a saturable reactor. The superimposed variations in NI_1 and NI_2 are due to the load current, alternating in nature, flowing through coils A and B.

greatly reduced. This is accompanied by a corresponding reduction in the reactance in series with the load, and therefore a much larger current will flow in the load. Because coil C has many turns of wire, the d-c value I_1 is small and the change in current from I_1 to I_2 is also small. Thus a small change in power input to the control coil C causes a large change in power output to the load, and power amplification results.

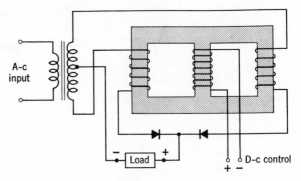

Fig. 10-29 Magnetic amplifier as a controlled full-wave rectifier.

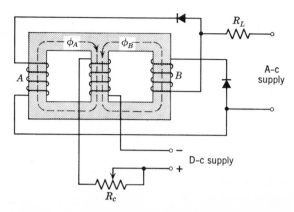

Fig. 10-30 Self-saturated magnetic-amplifier circuit.

The magnetic-amplifier circuit shown in Fig. 10-29 is a full-wave rectifier whose output voltage is controllable by the direct current in the control winding. Large values of the d-c load current may be obtained.

The control ampere-turns may be given assistance in saturating the core by an arrangement that provides a self-saturating effect. This is accomplished in Fig. 10-30 by connecting the load-current coils *A* and *B* in parallel and in such manner that their fluxes add in the center core instead of cancelling. A rectifier is connected in series with each of the two coils with such polarity that alternating current will flow in the load.

Feedback may be used in a magnetic amplifier. It is accomplished by introducing a small part of the a-c load current into a control winding with the aid of a rectifier. A discussion of the magnetic amplifier with feedback is presented after the general topic of feedback is discussed in Chapter 12.

There are many useful applications of the magnetic amplifier. Among them are voltage regulators for generators and supply lines, controls for lamps, furnaces, battery charges, and temperature regulators. Magnetic amplifiers are used with time-delay devices and control relays, and they serve as instrument amplifiers. They are rugged, trouble free, and comparatively simple in construction. Large amounts of power may be controlled by small amounts of input power, the ratio ranging from about 50 to 50,000 or higher.

10.21 Power Amplifiers in Servomechanisms

The block diagram of Fig. 10-31 represents what is called a *servomechanism.* It is also referred to as an automatic feedback control system. Its operation

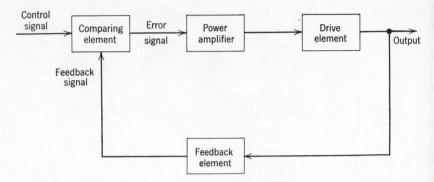

Fig. 10-31   Basic elements of a servomechanism, showing use of power amplifier.

will be discussed more fully in a subsequent chapter, where the principles of *feedback* are presented. It is shown here to point out that its operation may require a power amplifier for the drive element that is producing the output.

The system operates to generate an output only when there is some difference between the control signal and the feedback signal. The comparing element then yields an error signal to the power amplifier, whose output is applied to the drive element. Some parameter related to the system output, for example, voltage, power level, temperature, or position, is returned to the comparing element. When the error signal is reduced to zero, the system ceases operation until conditions permit the error signal to reappear. Any difference between the control signal and the feedback signal will generate an error signal.

As an example of a simple feedback control system, consider the operation of a home furnace. The room thermostat is the comparing element, which compares a preset temperature with the actual air temperature. Whenever the air temperature is below the preset value, an electric circuit is closed in the thermostat and an error signal is applied to a relay causing it to open a valve on the fuel line for the furnace. The furnace heats the circulating air and increases the temperature of the air at the thermostat. When the air temperature reaches the preset value, it opens the electric circuit and the de-energized relay allows the fuel valve to close. In this example, the power amplifier is a combination of an electric circuit and mechanical valve, and the system output is air temperature. Similarly, the comparing element, the thermostat, is a combination of a mechanical, temperature-sensitive device and a switch in an electric circuit.

It should be evident that many complex combinations of electrical, mechanical, pneumatic, and hydraulic components may be used in servo-mechanism systems. The study of these systems is too broad for detailed discussion here. The subject has been included to indicate that power amplifiers are necessary components in many applications, whether they are wholly electronic or combinations of other types of devices.

SUGGESTED REFERENCES

1. G. E. Happell and W. M. Hesselberth, *Engineering Electronics*, McGraw-Hill New York, 1953.

2. Herbert J. Reich, *Theory and Applications of Electron Tubes*, 2nd ed., McGraw-Hill, New York, 1944.

3. Frederick E. Terman, *Electronics and Radio Engineering*, 4th ed., McGraw-Hill, New York, 1955.

4. Samuel Seely, *Electron-Tube Circuits*, 2nd ed., McGraw-Hill, New York, 1958.

QUESTIONS

10-1 Sketch from memory a single-tube power-amplifier stage with a transformer-coupled load. Redraw the plate circuit only, but with shunt feed giving *L-C* coupling.

10-2 Compare the two kinds of coupling in the preceding question with respect to advantages.

10-3 For best performance, what size a-c load resistance should be used in a single-tube Class A power amplifier? How low should distortion be kept?

10-4 Explain the use of an output transformer in a power amplifier that feeds a loud speaker. Explain how to determine the required turns ratio.

10-5 State two ways to calculate the power output of a single-stage Class A amplifier. How is power input calculated? What assumption must be made to justify the use of these formulas? What correction should be made if the d-c resistance of the output transformer primary is not negligible?

10-6 Explain how to locate a load line, by trial construction, which will give maximum power output of an amplifier without excessive distortion.

10-7 When the dynamic transfer characteristic curve of an amplifier tube has the shape of a parabola, what are the three components present in the plate-current wave? Two of them have the same amplitude. Which are they?

10-8   Define second-harmonic distortion.

10-9   What harmonics are most prominent in pentode-tube amplifiers?

10-10   Discuss the importance of temperature in transistor-amplifier operation, telling where it becomes a critical matter and what should be done to prevent serious consequences.

10-11   Describe plate-current flow in the primary winding of the output transformer of a push-pull amplifier.

10-12   Explain how it happens that the iron core of an output transformer used in push-pull has no constant magnetic flux.

10-13   Explain why a bypass capacitor is not needed across a self-biasing resistor placed in the cathode circuit of a push-pull amplifier.

10-14   In Fig. 10-12*b* imagine a biasing resistor in place of the wire going from $-E_{bb}$ to the common-cathode point. E_{cc} is then no longer needed, but where should the point between the two grid-circuit resistors be connected?

10-15   How is the slope of the composite load line for a push-pull amplifier determined?

10-16   Considering that a certain amount of power output is obtainable from a single-ended power amplifier, why is it possible to get more than twice that amount out of two tubes in push-pull?

10-17   What are the advantages of Class B push-pull operation over Class A? How do their efficiencies compare?

10-18   What causes transistors to tend to overheat when used in amplifiers? How does the value of load resistance enter into the heat problem?

10-19   Describe, fully, complementary symmetry with reference to transistor amplifiers.

10-20   Compare the tube current in a single-ended Class C amplifier with that in a Class A amplifier.

10-21   How is it possible to get an output voltage of sine-wave form from a Class C amplifier when the current going into and out of the tank circuit flows in spurts?

10-22   Why are Class C amplifiers referred to as *tuned* amplifiers?

10-23   How do you account for the unusually high efficiency of a Class C amplifier?

10-24   Explain bias methods used with Class C amplifiers.

10-25   What determines the resonant frequency to which the tank circuit of a Class C amplifier should be tuned?

10-26   Of what *practical use* is the formula $f = 1/2\pi\sqrt{LC}$ in Class C operation?

10-27   Why must Class C amplifiers be neutralized, and what is the process of neutralization?

10-28 What is phase inversion in reference to voltages? To currents?

10-29 Draw from memory a split-load phase-inverter circuit using vacuum tubes. Explain in words how it produces phase inversion.

10-30 In what way is a two-tube phase-inversion circuit superior to a one-tube split-load circuit, and why?

10-31 Explain in detail how a small change in the d-c input to a three-legged iron-core reactor can cause a large change in alternating current flowing in coils of wire wrapped around two legs of the reactor.

PROBLEMS

10-1 A 6F6 tube, triode-connected, supplies power to an a-c load $R_L' = 4,000 \ \Omega$ coupled in by an output transformer. The transformer primary resistance is negligible, so that $E_{bo} = 250$ V is E_{bb} on the plate. A fixed bias of -20 V and a peak input-signal voltage of 20 V are used. Draw the load line and then determine the following: (*a*) output power, (*b*) second-harmonic distortion, (*c*) plate-circuit efficiency, (*d*) plate dissipation.

10-2 Repeat Problem 10-1 for the condition when the 4,000 Ω load is in series with the plate supply battery E_{bb}.

10-3 A triode for which $\mu = 6.8$, $g_m = 2,600 \ \mu$mho, is used in an amplifier with cathode bias (R_k in parallel with C_k) and shunt feed, the inductor of which has 250 Ω resistance. Capacitively coupled is $R_L = 3,750 \ \Omega$. $E_{bb} = 255$ V. The coupling capacitor has negligible reactance. (*a*) Calculate the power output when the input-signal voltage is 25 V peak. (*b*) Assume $R_k = 480 \ \Omega$ and calculate C_k for satisfactory operation down to 50 Hz.

10-4 A 2A3 triode vacuum tube operates at $E_{bo} = 250$ V, $E_{cc} = -20$ V, $R_b = 3,000 \ \Omega$, and the signal voltage is 20 V peak. (*a*) Determine graphically the power output, plate dissipation, and plate-circuit efficiency. (*b*) Find the value of plate-supply voltage needed.

10-5 The amplifier of Problem 10-4 is changed to shunt feed by using an inductance of negligible resistance and a coupling capacitor of negligible reactance. Find the quantities asked for in that problem.

10-6 A heater coil is shunted by a resistance $R = 10,000 \ \Omega$ as shown in Fig. P10-6. The relay, which has a resistance of 15,000 Ω, allows the spring on the contact arm to close the bell circuit at 4 mA and opens it at 7 mA. What value of I, the heater current, will cause the bell to ring? To what value must I be reduced to stop the ringing?

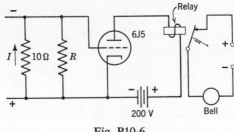

Fig. P10-6

10-7   A 6L6 tube is used with a transformer-coupled load in a single-ended power amplifier with a fixed bias of -10 V and $E_{bo} = 200$ V. Find by trial the a-c load resistance for maximum power output with not over 5% second-harmonic distortion. The plate current should never go below 10 mA. How much would the plate dissipation be?

10-8   For the load resistance found in Problem 10-7, determine the transformer turns ratio needed to match a speaker voice coil having 8 Ω resistance. Calculate the maximum power and rms current that would be delivered to the voice coil, assuming no loss in the transformer. Suppose the resistance of the transformer primary is 100 Ω, and the secondary 8 Ω. How much a-c power is lost in the windings?

10-9   Two 6F6 pentodes are operated in Class B push-pull. $E_{bb} = 350$ V, $E_{cc} = -38$ V, $e_g = 38$ V peak. The plate-to-plate resistance is 6,000 Ω. (*a*) Draw the composite load line. (*b*) Calculate the output power and plate-circuit efficiency.

10-10   A 6L6 tube in the second stage of an amplifier is to deliver 7.5 W to a 10-Ω load. An *R-C* coupled voltage amplifier using a 6J5 drives the 6L6 tube at frequencies between 160 and 5,100 hertz. Cathode bias is to be used on both tubes, and the plate-supply voltage is 250 V. Draw a complete circuit diagram. Design the complete circuit for good efficiency and low distortion.

10-11   Figure P10-11 shows common-base characteristics of an *NPN* power transistor. Choose a *Q*-point at $V_{CB} = 40$ V, $I_E = -25$ mA, and determine the optimum load resistance for maximum power output by constructing trial load lines not allowing the collector current to go below 2 mA. Calculate the power output, using your optimum load.

10-12   In Fig. 10-25, 2 A of direct current are supplied to coil *C* at 24 V. $R_L = 9$ Ω, and the total impedance of the series coils *A* and *B* is $1 + j20$ Ω. The a-c supply voltage is 240. When the direct current is increased to 2.2 A by cutting out more of R_c, the series impedance of *A* and *B* becomes $1 + j2$ Ω. (*a*) What could cause this drop in series impedance? (*b*) Compute the power gain: $\Delta P_o / \Delta P_i$.

10-13   A 6F6 triode connected tube is used in a power-amplifier circuit of the type in Fig. P10-13. The input voltage is sinusoidal with a peak value of 20.

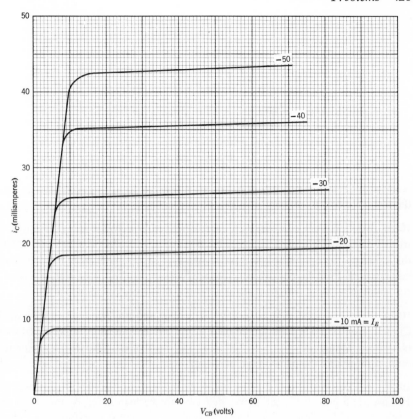

Fig. P10-11

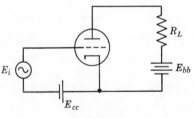

Fig. P10-13

$E_{bo} = 280$ V, $E_{cc} = -20$ V, $R_L = 2,500$ Ω. Compute the plate-circuit efficiency, η_p, and the plate dissipation P_p.

10-14 A triode-connected 6V6 is used in a power-amplifier circuit of the type in Fig. P10-14. The input voltage is sinusoidal with a peak value of 10 V. $E_{bo} = 225$ V, $E_{cc} = -10$ V, $R_L = 3,000$ Ω. Compute P_o, P_{po}, P_p, and η_p. Assume that the

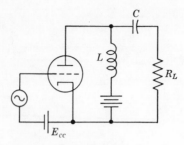

Fig. P10-14

reactance of the choke is very large and its resistance negligible. Also assume that the reactance of the coupling capacitor is very small.

10-15 (*a*) Construct composite characteristics for type 6V6 tubes connected as triodes operating at a plate voltage of 250 V and a grid bias of −20 V. (*b*) Draw the load line for optimum load, and determine the optimum plate-to-plate load resistance. (*c*) Find the power output for the two tubes. (*d*) Find the plate efficiency. (*e*) Find the required transformer turns ratio if the secondary load is 50 Ω.

10-16 Two 6F6 tubes are connected as triodes and are operated in push-pull Class A_1 from a 350-V plate source, with a grid bias of −30 V. A 30-V peak input is assumed. Assume $\mu = 4$, $r_p = 710$ Ω, and $I_{bo} = 40$ mA. Calculate the maximum output power and the plate-circuit efficiency.

10-17 A Class B amplifier has the following ratings: $E_{bb} = 290$ V, $E_{co} = -75$ V. For a certain grid drive, the power delivered to the load is 3.5 W and the plate efficiency is 65%. Find $i_{b\,max}$ and $e_{b\,min}$, assuming the load is transformer-coupled.

10-18 Determine L and C in the tank circuit of a Class C amplifier using a tube under the following conditions: $E_{bb} = 2,000$ V, $E_c = -160$ V, $E_p = 1,250$ V. Power delivered to the tank circuit $= 350$ W, $f_o = 2,000$ kHz, $Q_o = 15$.

10-19 Design a Class C amplifier using a 833 tube. The maximum ratings of this tube when used as a Class C amplifier (without modulation) are given by the manufacturer as $E_{bo} = 3,000$ V, $E_{co} = -500$ V, $I_b = 600$ mA, $I_c = 75$ mA, plate dissipation $= 300$ W. Assume $\eta_p = 75\%$ and the plate current conduction angle $\theta_p = 75°$.

10-20 A Class C amplifier uses an 852 tube and operates under the following conditions: d-c plate voltage $= 3,000$ V, d-c grid bias $= -600$ V, peak r-f signal $= 850$ V, d-c plate current $= 85$ mA, d-c grid current $= 15$ mA, fundamental component of plate current $= 120$ mA peak. Calculate the following, assuming $e_{b\,min} = e_{c\,max}$: (*a*) output power; (*b*) plate-circuit efficiency; (*c*) grid-driving power; (*d*) the amplifier is to operate at 1,500 kHz; specify the elements of the plate tank; choose $Q = 23$.

10-21   The current in either branch in the tank circuit of a certain properly designed Class C amplifier is 0.5 A. The reactance of the tank capacitor is 1,280 Ω, $E_{bb} = 1,000$ V, and $E_{cc} = -100$ V. If the amplifier is operating under conditions of maximum efficiency, what is the value of E_g, the rms grid-to-cathode voltage?

10-22   A 6V6 beam-power tube is used in a power-amplifier circuit. It should work into a load resistance of 5,000 Ω. What would be the turns ratio of the output transformer that would properly couple it to a 15-Ω resistance load?

10-23   A 6AQ5 beam-power tube is used in a single-ended amplifier employing an output transformer having negligible primary resistance. The a-c load resistance is 5,000 Ω, the plate-supply voltage is 250, and the grid bias is -12.5 V. The grid-signal voltage is a sine wave having a maximum value of 12.5 V. (*a*) Draw the load line. (*b*) Determine the maximum and minimum instantaneous values of plate current and plate voltage. (*c*) Determine the power output. (*d*) Calculate the percentage of second-harmonic distortion. (*e*) Calculate the power input and the plate dissipation.

10-24   A Class C amplifier operates at a frequency of 1 MHz. The bias resistor has a resistance of 10,000 Ω. What size capacitor should be connected across the resistor if the time constant should be twenty times the time of one cycle?

10-25   (*a*) In an undistorted Class A operation of a transistor, the collector dissipates a certain average power for no signal applied. When the input signal is applied, the collector dissipation decreases. For 35 per cent conversion efficiency, how large must the collector dissipation rating be for a load that takes 10 W? (*b*) The same transistor is used in Class B push-pull operation. The load requires the same 10 W of power. What is the collector dissipation when this power is supplied to the load? What is the collector dissipation when the signal is disconnected?

10-26   An input signal to a Class B push-pull, transformer coupled amplifier can be expressed as $e_s = 20 \sin \omega t + 10 \sin 2\omega t + 5 \sin 3\omega t$. Assuming linear, undistorted amplification, what will be the frequency content of the output?

10-27   A power transistor having the following specifications is to be used in the design of a Class A power amplifier driving a resistance load of 150 Ω in the collector circuit and operating as a common-emitter amplifier:

$$BV_{CEO} = 60 \text{ V}$$
$$P_{C\,max} = 6 \text{ W}$$
$$h_{fe} = 50, \quad \text{assumed constant}$$

(*a*) Sketch its collector characteristic curves and locate the maximum power hyperbola. (*b*) Determine a Q-point that will permit 50 per cent efficiency in the collector circuit, hence, develop maximum load power. (*c*) Under the conditions of (*b*), what is the load power? collector dissipation?

10-28   An audio power amplifier is to be designed to supply power to a 10-Ω load using a Class B push-pull, transformer coupling. The transistors being used require a collector load of 400 Ω for maximum undistorted operation. What should be specified for the output transformer?

10-29   The circuit shown in Fig. P10-29 is used as a Class A driver for a Class B power amplifier. The diodes and 5-Ω resistors represent the input to the Class B stage. One diode will conduct when the collector current increases, the other when

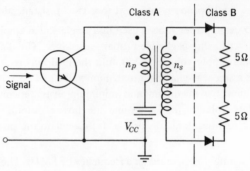

Fig. P10-29

it decreases. It is designed to yield the maximum possible power in the Class B stage, with the smallest possible base current. Determine the required values of V_{CC}, turns ratio, and Q-point, for these ratings of the transistor:

$$BV_{CEO} = 40 \text{ V}$$

$$i_{C \max} = 70 \text{ mA}$$

$$P_{C \max} = 0.5 \text{ W}$$

10-30   The coupling transformer shown in Fig. P10-30 is rated "400 Ω C.T. to 3.2 Ω." The transistors are rated at 120 mW maximum collector dissipation and

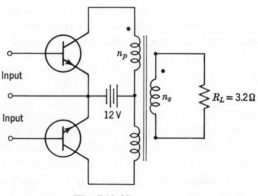

Fig. P10-30

100 mA maximum collector current. (*a*) What is the value of the reflected load in the collector circuit? (*b*) What maximum collector current can be used without exceeding any ratings? (*c*) Assuming linear operation of the transistors, calculate the turns ratio of the transformer for maximum efficiency, without exceeding any transistor ratings.

10-31 Show that the collector dissipation of a transistor operating in Class A common-emitter mode is minimum when power output is maximum.

10-32 A Class B push-pull circuit as shown in Fig. P10-32 uses 2N1485 transistors. Determine appropriate values for V_{CC}, n, R_1, and R_2, to provide maximum output

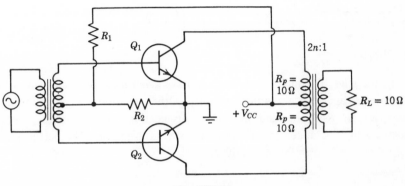

Fig. P10-32

power at minimum distortion. Assume operation at 25°C ambient temperature and the transistors mounted on efficient heat sinks.

10-33 Assuming the values found in Problem 10-32 permit sinusoidal operation, calculate the collector efficiency and compare it with the theoretical maximum value for Class B.

10-34 From the results of Problem 10-32, determine the input power required for maximum output power. What is the power gain of the amplifier? (Assume the emitter-base junction may be approximated by a 0.7-V source in series with 100 Ω resistance.)

10-35 Show that the maximum theoretical collector efficiency of a transistor amplifier operating in Class A is 50%; in Class B, 78.5%.

CHAPTER ELEVEN

Special-Purpose Amplifiers

In the study of amplifiers so far, we have given most of our attention to resistance-capacitance coupled circuits. Other methods of coupling are superior to *R-C* coupling in certain applications. Furthermore, special features may be added to conventional circuits to improve the performance of otherwise conventional amplifiers.

11.1 Direct-Coupled Amplifiers

When conventional amplifiers operate at *very low* frequencies the loss of amplification due to coupling elements such as capacitors, inductances, and transformers is a serious matter. They do not respond at all to d-c input voltages. By connecting the plate of one tube directly to the grid of the next or the collector of one transistor to the base of the next, *direct coupling* is achieved. When we consider this kind of connection the question of how to provide a negative bias on a grid when it is connected to a positive plate arises. One answer is: do it with a voltage-divider system.

The direct-coupled amplifier circuit of Fig. 11-1 is known as the Loftin-White circuit. A voltage-regulated power supply is generally used, although a battery is even more satisfactory if it has adequate capacity to maintain constant terminal voltage. The bleeder resistors should have values that are low enough so that changes of currents due to the input signal do not affect potentials at the resistor taps more than a minute amount. The resistance values of the bleeder sections are easily determined when the quiescent plate currents are known. Resistors R_2 and R_5 should be chosen first in the circuit design because of their high voltage drops.

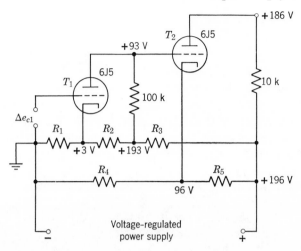

Fig. 11-1   Direct-coupled amplifier; Loftin-White circuit.

In analyzing the performance of this d-c amplifier we shall need to consider that the input voltage may be slowly alternating or that it may be merely a slow change in potential of one polarity only. Therefore a small incremental change in bias Δe_c will be the signal. Then the conventional e_g becomes Δe_c, $i_p = \Delta i_b$ and voltage gain $A_v = \Delta e_b / \Delta e_c$. These symbols are used in the equivalent circuit of Fig. 11-2.

We have seen before that

$$\Delta i_{b1} = \frac{\mu \Delta e_{c1}}{r_{p1} + R_{b1}}$$

so that

$$\Delta e_{b1} = -\frac{\mu \Delta e_{c1} R_{b1}}{r_{p1} + R_{b1}} \tag{11-1}$$

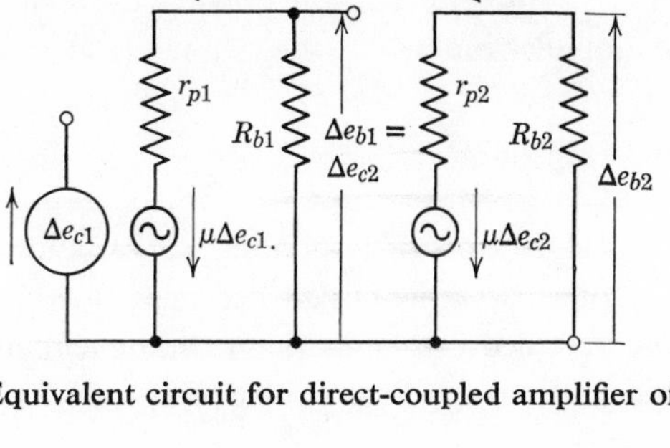

Fig. 11-2   Equivalent circuit for direct-coupled amplifier of Fig. 11-1.

The voltage gain of the first stage is

$$A_{v1} = \frac{\Delta e_{b1}}{\Delta e_{c1}} = - \frac{\mu R_{b1}}{r_p + R_{b1}} \tag{11-2}$$

It is significant that when vacuum triodes are operated at very low values of plate current and plate-to-cathode voltage their dynamic plate resistances are very much larger than when operation is more conventional. The tubes in this example normally have $r_p = 6,700 \ \Omega$ and $\mu = 20$, but measurements of these quantities with -3 V bias and $E_{bo} = 90$ V give about $15,500 \ \Omega$ for r_p and $\mu = 19$.

The first stage has a gain of

$$A_{v1} = -19 \times 10^5/(0.155 + 1.0)10^5 = -16.4$$

The second stage gain is

$$A_{v2} = -19 \times 10^4/(1.55 + 1)10^4 = -7.45$$

The overall gain is then $-16.4 \times (-7.45) = 122.$

The input signal required for a needed output (Δe_{b2}) may be determined. One-half milliampere increment (Δi_{b2}) in the final load resistor will give a 5-V output signal $(0.0005 \times 10,000)$. To produce this output effect the required input signal is

$$e_{c1} = \frac{5}{122} = 0.041 \text{ V} = 41 \text{ mV}$$

Note that a positive increment of voltage at the input will cause a drop in potential at the first grid. The plate current of the second tube will then decrease, and the potential at the plate of the second tube will rise, making Δe_{b2} positive.

Direct-current amplifiers are very sensitive to very small fluctuations in voltages, changes in rate of electron emission from a spot on a cathode, temperature changes, and grid current flowing through the impedance of the driver at the amplifier input.

11.2   Chopper Amplifiers

An amplifier that is capable of responding to variations in d-c quantities presents serious problems to the circuit designer. It might seem that a d-c response could be obtained by eliminating from the circuit all elements that restrict low-frequency response; the circuit of Fig. 11-1 is an example.

Such circuits are easy to construct, and they usually perform well for d-c inputs. They perform so well, in fact, that they may be almost useless for small inputs, because the circuit cannot distinguish between signals that originate at the input terminals and signals that are generated internally. Any drift in the potentials at various points in the circuit will be interpreted by the circuit as input signal variations. The amplifier will generate an output signal for internal drift signals as well as for the input signals.

An amplifier that responds to a-c inputs should be immune to the low-frequency drift signals. Consequently, a d-c amplifier may change the incoming d-c signal to an equivalent a-c signal for amplification and then rectify the output to form a d-c output. Such an arrangement may take the form of the circuit in Fig. 11-3, a *chopper amplifier*.

The chopper is shown as a mechanical switch that samples the input signal at the frequency of the alternating current driving the chopper. If a polarized a-c magnet is used to drive the chopper, the input will be connected to the amplifier in synchronism with the a-c supply. The input is connected during alternating half cycles, and, for a 60-Hz supply, the input is sampled 60 times per second. The a-c amplifier must be capable of amplifying signals at this frequency.

In order for the input samples to accurately follow any variations of the input signal, the switching frequency will have to be large compared with the input variation frequency. For some applications the 60-Hz sampling frequency may not be sufficient. For higher input frequencies the mechanical chopper is occasionally driven from a 400-Hz source.

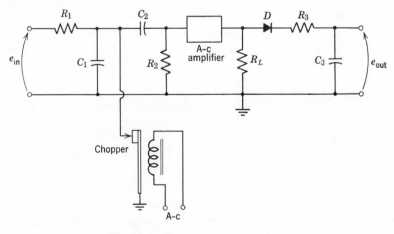

Fig. 11-3 Basic chopper amplifier.

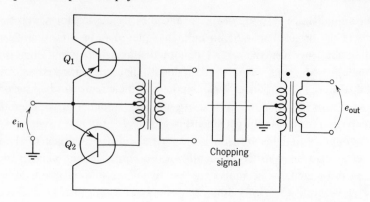

Fig. 11-4   Transistor chopper amplifier. (Bias supply omitted for convenience.)

As indicated in the circuit diagram, the input signal is passed through a low-pass filter consisting of R_1 and C_1. Its action will remove any high-frequency components of the input signal, and only an average of the input signal will be applied to the amplifier when the chopper switch is open. The diode at the output will rectify the amplified version of the average input to produce the required output voltage. The filter R_3C_3 removes the chopper frequency and its harmonics.

A transistor that operates between saturation and cutoff is a close approximation to the mechanical chopper. In the circuit of Fig. 11-4, the base of each transistor is alternately driven from saturation to cutoff by the square wave input. The d-c input is alternately switched from Q_1 to Q_2, producing an a-c signal at the transformer. Of course, the output voltage will be offset by the collector saturation voltage of the transistors, and the two transistors should be matched to secure a balanced output from the circuit.

11.3   Operational Amplifiers

An *operational amplifier* is a very-high-gain d-c amplifier designed for use with external feedback circuit elements. It is used in analog computers and data-handling equipment. It may be used with the proper feedback circuits to perform the mathematical operations of summation, subtraction, multiplication, division, integration, and differentiation. The amplifier circuit is usually designed to have a single input–single output configuration but may be constructed to have single and differential input and output. Since the single input–single output configuration is the most used, this is the one that will be discussed here. It is commercially available in either vacuum-tube or transistor versions and in either unstabilized or drift-stabilized modes.

The block diagram of a typical drift-stabilized operational amplifier is given in Fig. 11-5. The amplifier consists of two sections, the stabilizing amplifier and the direct-coupled amplifier. It is essentially the same as the combination of the circuits of Figs. 11-1 and 11-3.

One of the fundamental requirements of an operational amplifier is that the input terminal, or *summing junction*, be adjustable to within a few millivolts of zero (microvolts, for higher accuracy systems). Further, it must not be subject to drift on an instantaneous basis nor over long periods of time. This requires that the internal gain of the amplifier be very high,

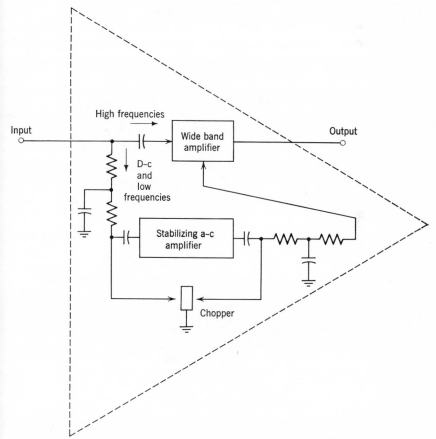

Fig. 11-5   Chopper-stabilized operational amplifier.

typically 10,000 or more. High input impedance is also desirable to minimize the power drain on the signal input circuits by the amplifier.

The stabilizing amplifier, or chopper amplifier, is used to virtually eliminate the current and voltage drift that might occur at the summing junction. The synchronous switch generates an a-c signal based on samples of the input signal. It is then amplified by the stabilizing amplifier and applied through a filter to the wideband, d-c amplifier. The sampling frequency is often 60 Hz for tube circuits and 400 Hz for transistor circuits. The use of the stabilizing amplifier also increases the total d-c gain of the operational amplifier by a factor of its own internal gain.

The high-frequency components of input signals and the d-c and low-frequency components from the chopper amplifier are summed at the input to the wideband amplifier. The combined signal is amplified by a large factor, 10,000 or more, to yield an output replica of the input signal. The gain and phase properties of the d-c amplifier will determine the behavior of the circuit over wide ranges of frequency, because the chopper is effectively isolated from any high-frequency components of the input signal.

11.4   Applications of Operational Amplifiers

The fundamental theorem that permits the operational amplifier to be used successfully to perform mathematical operations is Kirchhoff's law for electric currents: "the sum of currents entering a node equals the sum of currents leaving the node." If the amplifier of Fig. 11-6 is assumed to have an infinite input impedance, then the currents at the input node (the summing junction) can exist only in the input resistors and the feedback resistor. Thus, the currents entering from the inputs must just equal the current leaving

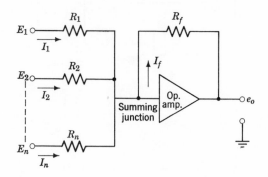

Fig. 11-6   General form of a summing amplifier.

through the feedback path. This means that when an input voltage increases, the output voltage must decrease. This equality of currents will be maintained within the bandwidth of the amplifier, and permits the summing of a-c as well as d-c input signals. Since the summing junction is ideally maintained at zero potential, inputs of zero will result in an output of zero. When several input paths are provided at the summing junction, the amplifier will cause the feedback current to equal the algebraic sum of all input currents, and the output voltage will be proportional to the sum of the input voltages. Written in terms of the voltages and resistances, the general equation for the circuit is

$$e_o = -R_f\left(\frac{E_1}{R_1} + \frac{E_2}{R_2} + \cdots \frac{E_n}{R_n}\right)$$

This basic equation can also be used to analyze the behavior of the amplifier when the resistors are general impedance elements. For example, if the feedback resistor is replaced with a capacitor, the circuit will produce an output that is the integral of the inputs. If a current flows into a capacitor, the voltage across it will change at a rate that is proportional to the current

$$e_C = \frac{1}{C}\int i\, dt$$

Since the summing junction is at zero potential, the voltage across the feedback capacitor will be e_o, and the current summation may be written in terms of the voltages and impedance elements as

$$e_o = -\frac{1}{C}\int_0^t \left(\frac{E_1}{R_1} + \frac{E_2}{R_2} + \cdots \frac{E_n}{R_n}\right) dt + E_o(t = 0)$$

Similarly, the circuit may be used to subtract input signals. This only requires that the signals be of opposite phase and the ratio of feedback and input resistors be the same for the two signals. Of course, it can be seen that by a proper choice of resistance ratios, the circuit can multiply an input signal by any desired factor, which will be set by the resistance ratio. The process of division by a constant can be handled in the same way, since division is inverse multiplication.

The many applications of operational amplifiers are too broad for our discussion here. The reader is referred to the multitude of technical papers and textbooks on the use of these circuits in analog computers and data systems. This introduction should be sufficient to indicate the importance of these circuits in electronic applications.

EXAMPLE 11-1. An operational amplifier is used to sum three voltages, using a circuit similar to that of Fig. 11-6. The amplifier has a gain of $-1,000$ and may be considered to have infinite input impedance. Then the sum of currents entering the summing junction from the input sources equals the current in the feedback impedance.

Considering the summing junction to be at ground potential, we may write the currents as

$$\frac{E_1}{R_1} + \frac{E_2}{R_2} + \frac{E_3}{R_3} = \frac{-e_o}{R_f} \tag{1}$$

Suppose each of the resistors is chosen to be 1 MΩ. Then the output voltage will be the sum of the three input voltages, provided the sum does not exceed the saturation voltage of the operational amplifier. The saturation voltage will be determined by the d-c supply used to power the operational amplifier. In some cases, it may be about 100 V, in others, 10 V or so. Higher saturation voltages are permissible in circuits that use vacuum tubes, whereas those using transistors typically have lower permissible values.

This circuit may also be used to scale certain input voltages, or apply "weighting factors," before the addition process. This is easily accomplished by adjusting the ratios of input and feedback resistors. Suppose it is desired that the input E_1 be reduced by a factor of $\frac{1}{10}$ compared with the other two inputs before addition. Equation (1) may be rewritten,

$$\frac{R_f}{R_1} E_1 + \frac{R_f}{R_2} E_2 + \frac{R_f}{R_3} E_3 = -e_0 \tag{2}$$

For R_f equal to 1 MΩ, R_1 would equal 10 MΩ, and R_2 and R_3 would be chosen equal to R_f, 1 MΩ. Then the sum of the voltages would include E_1 reduced by a factor of $\frac{1}{10}$.

In effect, by choosing the ratio $R_f/R_1 = \frac{1}{10}$, the input voltage E_1 was divided by 10. Similarly, multiplication of the input voltage may be obtained by choosing a ratio necessary for the required multiplication factor.

11.5   Transformer Impedance Matching in Amplifiers

The transformer as a coupling device in amplifier circuits has been introduced in earlier chapters. Its ability to transform an impedance is an important property, and we shall discuss it further here.

The *impedance ratio* of a transformer, also called the *impedance-transformation ratio*, is an important quantity. Figure 11-7 illustrates a transformer, the N's representing numbers of turns. N_1 is nearly always different from N_2.

It was explained in Chapter 10 that for a transformer the following relations are true

$$\frac{E_1}{E_2} = \frac{N_1}{N_2} = \frac{I_2}{I_1} \tag{11-3}$$

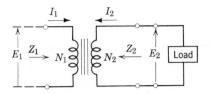

Fig. 11-7   Interstage transformer.

In Fig. 11-7

$$Z_1 = \frac{E_1}{I_1}$$

and   (11-4)

$$Z_2 = \frac{E_2}{I_2}$$

From Equation (11-3)

$$\left(\frac{E_1}{E_2}\right)\left(\frac{I_2}{I_1}\right) = \left(\frac{N_1}{N_2}\right)^2$$

Using Equation (11-4)

$$\frac{Z_1}{Z_2} = \left(\frac{N_1}{N_2}\right)^2 \qquad (11-5)$$

Thus *the impedance ratio of a transformer is equal to the square of the turns ratio*, and, conversely, the *turns ratio is equal to the square root of the impedance ratio*. As an example, to match 250,000 Ω to 10,000 Ω a turns ratio of $\sqrt{25}$ is required; this is a ratio of 5 to 1. If the output impedance of a first stage of an amplifier is 10,000 Ω and the input impedance of a second stage is 250,000 Ω, the matching transformer would contribute a voltage gain of 5 while doing the matching. Current gain may be obtained in a "step-down" arrangement.

In addition to providing added voltage (or current) gain, transformer coupling eliminates the necessity for a voltage-blocking (coupling) capacitor and provides a *low-resistance* path for direct current to the tube plate. This materially reduces d-c power loss and thus improves operating efficiency.

The basic circuit of Fig. 11-8 has an *interstage transformer*, Tr 1, and an output transformer, Tr 2. The output transformer is almost always used to couple a low-impedance voice coil of a loud speaker (4, 8, or 16 Ω, usually) to the relatively high-impedance plate circuit of the *power amplifier* stage feeding it. The first stage and the interstage transformer will be treated for what it is—a *voltage* amplifier.

A transformer has *an equivalent circuit of its own*. The detailed theory of

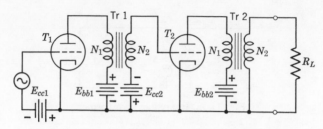

Fig. 11-8 Transformer-coupled amplifier.

its development will be left for reference reading by the student.* A few pertinent facts will be evident from brief explanations.

We have seen that *a secondary impedance may be expressed as a primary impedance* if the secondary impedance is multiplied by the square of the turns ratio. The secondary impedance is thereby *reflected into the primary circuit.* By this means the resistance and reactance of the secondary winding may be reflected into an *equivalent circuit* for the transformer, consisting of a series-parallel circuit all on the primary side of an *ideal transformer* having *no losses* and a turns ratio of n chosen to fit the situation.

A transformer with an iron core (to make the mutual inductance between windings, and therefore the coupling coefficient, a maximum) has not only I^2R losses in both windings, but also *core loss* due to flux changes in the iron. The symbol R_c will designate a fictitious resistance which will represent the core loss in watts by the equivalent I^2R_c when current I flows through it. R_e is the sum of the primary-winding resistance and the secondary-winding resistance after it has been reflected into the primary.

The transformer also has capacitance between windings and distributed capacitance between each turn and all other turns of the same winding. The distributed capacitance of the primary winding is represented by a shunt capacitor between the plate and cathode points of the equivalent circuit of the amplifier stage. The capacitance between windings, C_M, may be represented by a shunt capacitor across the secondary of the ideal coupling transformer in the equivalent circuit. It can be shown that this capacitance representing C_M is equal to $C_M(1 \pm N_1/N_2)$, the sign of the turns ratio depending on the relative polarity and connection of the primary and secondary windings. Notice that the bottom ends of the windings of a transformer are effectively tied together for alternating current by the bias supplies. The distributed capacitance, C_2, of the secondary winding is

---

* See Terman, reference 2 at end of chapter.

included in C_o, which is placed across the secondary of the ideal transformer in the equivalent circuit.

$$C_o = C_2 + C_M \left(1 \pm \frac{N_1}{N_2}\right) + C_{i2}$$

where C_{i2} is the input capacitance of the next stage.

Inductances must also be represented. There is a primary winding inductance due to *leakage flux* which produces the very important *leakage reactance* of the primary. There is also a secondary leakage reactance represented by a secondary winding inductance. A single inductance symbol, L_e, in the equivalent circuit represents the sum of the primary leakage inductance and the additional primary inductance needed to represent the secondary leakage reactance *after it has been reflected* into the primary. It is pertinent to state here that an *inductive reactance* on one side of a transformer becomes a *capacitive reactance* on the other side, through the reflection process, and thus an inductance on one side becomes a *negative* inductance (or a capacitance) on the other side. Also, a capacitive reactance reflects as an inductive reactance on the other side, so that a reflected capacitance becomes an inductance or a negative capacitance after reflection. The reflected inductance or capacitance is placed in series with the *transformer winding*. A *reflected resistance* is still a resistance when it gets on the other side of the transformer.

Let us diverge from our main path for a moment to illustrate this point. It would take a little more time than we wish to spend to show that the impedance after reflection, Z_{ref}, is given by $\omega^2 M^2$ divided by the impedance to be reflected;* thus

$$Z_{\text{ref}} = \frac{\omega^2 M^2}{R + jX} \tag{11-6}$$

M is the mutual inductance of the transformer.

After rationalizing the denominator, we see how the reflected impedance has a capacitive component rather than an inductive one:

$$Z_{\text{ref}} = \frac{\omega^2 M^2}{R + jX} \cdot \frac{R - jX}{R - jX} = \frac{\omega^2 M^2 R}{R^2 + X^2} - j\frac{\omega^2 M^2 X}{R^2 + X^2}$$

Of course, if we reflect $R - jX$, the j term of the reflected impedance is positive. We see, therefore, that

$$Z_{\text{ref}} = R_{\text{ref}} + jX_{\text{ref}} \tag{11-7}$$

* The derivation is given in the Appendix.

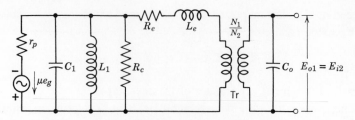

Fig. 11-9   Approximate equivalent circuit for the first stage of Fig. 11-8.

C_1 = primary winding cap. + output cap. of T_1
L_1 = primary winding magnetizing inductance
R_c = power-consuming resistance representing transformer core losses
R_e = primary res. + res. reflected from secondary
L_e = primary leakage ind. + reflected secondary leakage inductance
C_o = output capacitance of transformer + input capacitance of T_2
Tr = ideal transformer with N_1/N_2 turns ratio

where

$$R_{\text{ref}} = \omega^2 M^2 \frac{R}{|Z|^2} \quad \text{and} \quad X_{\text{ref}} = -\frac{\omega^2 M^2 X}{|Z|^2} \tag{11-8}$$

There is one more inductance in the primary, and it is the self-inductance due to the *magnetizing* component of the primary current. It is shown in parallel with the shunt capacitor representing the distributed capacitance of the primary winding. All these elements are shown in the equivalent circuit of a transformer-coupled amplifier stage in Fig. 11-9.

At low-audio frequencies the reactances of C_1 and C_o are so high that their effects may be neglected. Also, the reactance of L_1 is so small compared with R_c that R_c has a negligible effect. With the transformer secondary virtually open-circuited, the current through the primary is extremely small owing to the high impedance of the coil on an iron core, so the effects of R_e and L_e are negligible. These approximations simplify the all-frequency equivalent circuit to that of Fig. 11-10.

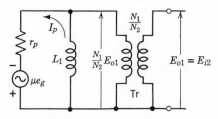

Fig. 11-10   Approximate low-audio-frequency equivalent circuit of a transformer-coupled voltage amplifier.

Recalling that the current in the transformer primary is practically zero, and therefore negligible, the voltage across L_1 is given by its current times its reactance, $j\omega L_1$:

$$\left(\frac{N_1}{N_2}\right) E_{o1} = \mu e_g \frac{j\omega L_1}{r_p + j\omega L_1} \tag{11-9}$$

The signal voltage e_g will be called E_{i1}. The voltage gain in the low-audio range is then

$$A_{vl} = \frac{E_{o1}}{E_{i1}} = \frac{N_2}{N_1} \times \frac{\mu}{1 - jr_p/\omega L_1} \tag{11-10}$$

This expression may be positive or negative, depending on the way the secondary of the transformer is connected relative to the primary.

As the frequency increases from a very low-audio value, ωL_1 increases until it gets large enough to make $r_p/\omega L_1$ so small that it has little effect. When $\omega L_1 \geq 10 r_p$, $A_{vl} = \mu N_2/N_1$. The gain remains practically constant as the frequency is further increased until the capacitances begin to have an effect. The region in the frequency spectrum *between the value at which $r_p/\omega L_1$ is regarded as zero in Equation* (11-10) and the value where ω is so high that the gain is reduced by capacitance effects is called the midband, where the magnitude of the gain is

$$|A_{vm}| = \frac{\mu N_2}{N_1} \tag{11-11}$$

The approximate mid-audio-frequency circuit is shown in Fig. 11-11.

It is important here to note that the gain $|A_{vm}|$ as expressed in this equation includes the voltage transformation ratio of the transformer. If the gain through the tube is 20, and the transformer has a step-up ratio of 3, the stage gain is 60.

From theory learned in the study of R-C coupled amplifiers, we conclude

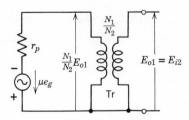

Fig. 11-11   Approximate mid-audio-frequency equivalent circuit of a transformer-coupled voltage amplifier.

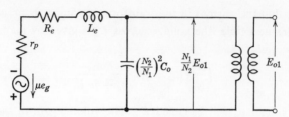

Fig. 11-12 Approximate high-frequency equivalent circuit of a transformer-coupled voltage amplifier.

that the lower half-power frequency must satisfy the condition that $r_p = \omega L_1$. Calling this frequency f_1,

$$f_1 = \frac{r_p}{2\pi L_1} \tag{11-12}$$

The phase angle is 45 degrees leading with respect to the phase at mid-frequency. Because values of r_p for pentodes are so high that they increase f_1 so greatly, they are not used in transformer-coupled audio-frequency voltage amplifiers. High-gain triodes are also not suitable for the same reason.

High-frequency (*audio-range*) *operation* of a transformer-coupled voltage amplifier may be studied with reference to the approximate equivalent circuit of Fig. 11-12. The distributed capacitance of the primary winding (C_1) is small enough that its effect as a shunting reactance may be neglected throughout the entire audio range. Remember that a very high-reactance parallel path takes negligible current.

The output capacitance of the transformer and its load is a very important quantity because it is largely responsible for the decrease in amplification as the frequency is increased, as we would expect from our experience with R-C coupled voltage amplifiers. C_o represents the total secondary circuit capacitance, which is the sum of the input capacitance of the next stage, the distributed capacitance of the secondary winding, and $C_M(1 + N_1/N_2)$, where C_M is the "mutual capacitance" between the windings. Reflection of the *reactance* of C_o into the primary circuit requires dividing the total *secondary* reactance by the square of the primary-to-secondary turns ratio.

$$X_1 = \left(\frac{N_1}{N_2}\right)^2 X_2$$

$$X_2 = \frac{-1}{\omega C_o}$$

$$X_1 = -\left(\frac{N_1}{N_2}\right)^2 \frac{1}{\omega C_o} = -\frac{1}{\omega C_o/(N_1/N_2)^2}$$

Therefore a value C_o in the secondary is represented in the primary by

$$\frac{C_o}{(N_1/N_2)^2} = \left(\frac{N_2}{N_1}\right)^2 C_o$$

The current through the transformer primary is negligible because its impedance with the secondary virtually open-circuited is so high compared with the parallel capacitive reactance. The result is that the current is given by

$$I_p = \frac{\mu e_g}{r_p + R_e + j[\omega L_e - (N_1/N_2)^2/\omega C_o]} \tag{11-13}$$

The voltage represented by $(N_1/N_2)E_{o1}$ is

$$\left(\frac{N_1}{N_2}\right)E_{o1} = I_p X_C = \frac{\mu e_g[-j(N_1/N_2)^2/\omega C_o]}{r_p + R_e + j[\omega L_e - (N_1/N_2)^2/\omega C_o]} \tag{11-14}$$

making the high-frequency gain

$$A_{vh} = \frac{E_{o1}}{E_{i1}} = \mu \frac{N_2}{N_1} \frac{-j(N_1/N_2)^2/\omega C_o}{r_p + R_e + j[\omega L_e - (N_1/N_2)^2/\omega C_o]} \tag{11-15}$$

As the high frequency decreases toward the midband range of values, $(N_1/N_2)^2/\omega C_o$ should be compared with the remainder of the denominator, $r_p + R_e + j\omega L_e$. When the former has become very much greater than the latter, the gain expression reduces to $\mu(N_2/N_1)$ as it should. The signs of all gain expressions depend on the relative polarities of the transformer winding connections, as previously stated. Using the 10:1 ratio as before, the following relation can be used to determine the approximate upper limit of the midfrequency range.

$$(N_1/N_2)^2/\omega C_o = |10(r_p + R_e + j\omega L_e)| \tag{11-16}$$

Observe that, if we consider the transformer primary as an open circuit, which it practically is, there is a possibility of series resonance of L_e with the shunt capacitor. This does happen, and as a result the gain becomes much greater than the midfrequency-range value. From series-resonance theory this resonant frequency is given by

$$f_r = \frac{1}{2\pi(N_2/N_1)\sqrt{L_e C_o}} \tag{11-17}$$

We might think that the maximum gain occurs at f_r, but it does not because of the presence of r_p in series with the resonant pair. It occurs at a frequency

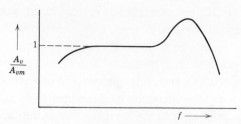

Fig. 11-13   Typical frequency-response curve of a transformer-coupled audio-frequency amplifier.

well below f_r, which may be shown to be expressed by

$$\text{Frequency for maximum gain} = f_{\max} = \frac{1}{2\pi}\sqrt{\frac{(N_1/N_2)^2}{L_e C_o} - \frac{(R_e + r_p)^2}{L_e^{\,2}}}$$

$$(11\text{-}18)$$

A typical response curve for the transformer-coupled audio-frequency voltage amplifier is shown in Fig. 11-13. Although there are ways of reducing the peak and flattening the whole curve if a wider range of uniform gain is desired, the superiority of the *R-C* coupled voltage amplifier when pentodes are used makes the improvement additions hardly worthwhile.

DRILL PROBLEMS

D11-1   A certain transformer has a mutual inductance of 1 mH. An impedance of $5 + j15\ \Omega$ in its secondary circuit is to be reflected in the primary circuit. Determine the impedance appearing in the primary when the frequency is 10 kHz.

D11-2   A transformer-coupled amplifier uses a triode having $r_p = 10\ \text{k}\Omega$ and and $\mu = 20$. The transformer turns ratio is $N_2/N_1 = 3$ and primary magnetizing inductance $L_1 = 50\ \text{H}$. (*a*) Assuming operation in the low-audio-frequency range, calculate the frequency at which A_{v1} is down 3 dB from the midband gain. (*b*) How much error in calculated gain occurs when the reactance term is considered negligible? (*c*) Plot the frequency response curve for the amplifier in the low-frequency region of operation.

11.6   The Grounded-Grid Amplifier

By grounding the grid of the tube in an amplifier, some shielding is produced between the plate and cathode which reduces the interelectrode capacitance and makes a grounded-grid triode useful in very-high-frequency (VHF) circuits. Because in such applications the load on the tube is a parallel-resonant circuit, the load will be considered purely resistive in our analysis.

The circuit is shown in Fig. 11-14a. As the input-signal voltage goes positive, the grid becomes more negative with respect to the cathode, thus decreasing the plate current. This causes the potential at the plate to rise, giving a positive output voltage. This means that there is no phase reversal through the amplifier, i.e., E_o and E_i are in phase. Furthermore, we must add the signal voltage to the voltage E_i of the generator in the equivalent circuit (Fig. 11-14b). Because the input signal is making the grid more negative with respect to the plate, i_p is in its *negative* half-cycle and is therefore shown flowing away from the plate point P. Since the grid is in the process of becoming more negative with respect to the plate, e_g is directed as shown in the actual circuit diagram (Fig. 11-14a). Neglecting d-c let us write the equation for the voltage loop in the actual circuit first. Summing the voltages clockwise, we get

$$E_i - i_p R_g - e_g = 0 \qquad (11\text{-}19)$$

$$e_g = E_i - i_p R_g \qquad (11\text{-}20)$$

and, in the equivalent circuit,

$$E_i + \mu e_g - (r_p + R_g + R_L)i_p = 0 \qquad (11\text{-}21)$$

Substituting for e_g,

$$E_i(1 + \mu) - \mu i_p R_g = (r_p + R_g + R_L)i_p \qquad (11\text{-}22)$$

$$i_p = \frac{(1 + \mu)E_i}{r_p + (1 + \mu)R_g + R_L} \qquad (11\text{-}23)$$

and

$$E_o = \frac{(1 + \mu)R_L E_i}{r_p + (1 + \mu)R_g + R_L}$$

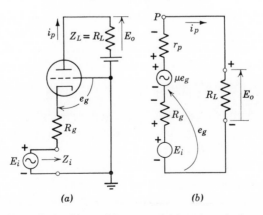

(a)   (b)

Fig. 11-14   Grounded-grid amplifier, actual (a) and equivalent (b) circuits.

The voltage gain is

$$A_v = \frac{E_o}{E_i} = \frac{(1 + \mu)R_L}{r_p + (1 + \mu)R_g + R_L} \tag{11-24}$$

The effective generator in the plate circuit is seen to have a voltage $(1 + \mu)E_i$, (Equation 11-23), which has the same effect as an increase in amplification factor of the tube.

The input impedance of the amplifier is obtained from Equations (11-23) and (11-24):

$$Z_i = \frac{E_i}{i_p} = \frac{r_p + (1 + \mu)R_g + R_L}{(1 + \mu)} = R_g + \frac{r_p + R_L}{1 + \mu} \tag{11-25}$$

After saying that the grounded-grid amplifier is especially useful in very high-frequency circuits when interelectrode capacitances cannot be neglected, it seems out of order to ignore them in the analysis of the amplifier's performance. To include the interelectrode capacitances, however, would add much complexity to the process, and this is hardly justifiable because it is found that their major effect is to add a capacitance C_{gk} in parallel with Z_i of Equation (11-25).

A useful characteristic of the grounded-grid amplifier is its impedance-matching property. An ordinary type 6J5 tube with $\mu = 20$, $R_g = 500$, $r_p = 7{,}700\ \Omega$ and a recommended plate load $R_L = 47{,}000\ \Omega$ provides an input impedance $Z_i = 3{,}100\ \Omega$. This gives a forward-impedance transformation ratio Z_i/R_L of 1 to 15 for the stage.

DRILL PROBLEM

D11-3 The grounded-grid amplifier of Fig. 11-14 has $R_g = 1\ \text{k}\Omega$, $R_L = 20\ \text{k}\Omega$, and the tube has $r_p = 6{,}000\ \Omega$ and $\mu = 40$. Compute the voltage gain and input impedance of the amplifier, for $E_i = 1\ \underline{/0°}$ at 1 kHz.

11.7 Cathode-Follower (Grounded-Plate) Amplifier

An amplifier circuit that delivers output voltage from across an unbypassed cathode resistor and has its a-c input signal applied between grid and plate is shown in Fig. 11-15. It is called a cathode follower because the output voltage, which is taken off the cathode resistor, is practically equal to the input signal voltage and they are in phase. The most important feature and value of the cathode follower is its *low output impedance*. Its voltage gain is less than unity (actually a loss), but its ability to match a high-impedance

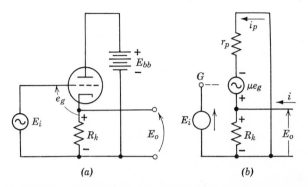

Fig. 11-15   Actual (*a*) and equivalent (*b*) circuits for a cathode-follower amplifier.

source to a low-impedance load has made it very useful and therefore popular.

In the grid circuit, neglecting d-c,

$$E_i - e_g - i_p R_k = 0$$

$$e_g = E_i - i_p R_k \tag{11-26}$$

Evidently,

$$E_o = i_p R_k \tag{11-27}$$

$$e_g = E_i - E_o \tag{11-28}$$

By now we know

$$i_p = \frac{\mu e_g}{r_p + R_k}$$

Substituting Equation (11-27) and Equation (11-28),

$$i_p = \frac{\mu(E_i - i_p R_k)}{r_p + R_k}$$

From this, it follows that

$$i_p = \frac{\mu E_i}{r_p + (1 + \mu)R_k} \tag{11-29}$$

Using Equation (11-27),

$$E_o = \frac{\mu R_k E_i}{r_p + (1 + \mu)R_k} \tag{11-30}$$

The voltage gain is then

$$A_v = \frac{E_o}{E_i} = \frac{\mu R_k}{r_p + (1 + \mu)R_k} \tag{11-31}$$

This expression shows that the gain is always less than unity. The only way for it to become very, very near unity is for μR_k to be very much larger than the remainder of the denominator, which is $(r_p + R_k)$.

The output impedance, Z_o, of the cathode follower, which is, by way of review, the impedance seen when looking back from the load terminals with the source voltage short-circuited, has been designated an important quantity.

Looking back at Equation (11-29), let us divide numerator and denominator by $1 + \mu$:

$$i_p = \frac{[\mu/(1 + \mu)]E_i}{[r_p/(1 + \mu)] + R_k}$$

This suggests that in the cathode-follower circuit the tube has an effective amplification factor of $\mu/(1 + \mu)$ and an effective dynamic plate resistance of $r_p/(1 + \mu)$.

Let us assume that a generator applies E_o to the output terminals of the equivalent circuit and sends a current i downward through R_k. E_o is thus applied between cathode and grid and is therefore equal to $(E_i - e_g)$. With E_i reduced to zero [driving generator in (a) short-circuited], two voltage equations can be written.

$$iR_k + i_p R_k = E_o = -e_g \tag{11-32}$$

$$iR_k + (r_p + R_k)i_p = \mu e_g \tag{11-33}$$

Solving for i_p,

$$i_p = -\frac{(\mu + 1)R_k i}{(\mu + 1)R_k + r_p}$$

Substituting this into Equation (11-32) and dividing by i gives the *output impedance of the cathode follower*:

$$Z_o = \frac{E_o}{i} = \frac{r_p R_k}{r_p + (1 + \mu)R_k} \tag{11-34}$$

The *input impedance* of the cathode-follower amplifier can be shown to be a capacitive reactance. Note that the input signal is *applied between the grid and the plate*. The input capacitance is almost entirely C_{gp} of the tube. To show this we can use the constant-current generator type of equivalent circuit shown in Fig. 11-16. Because the reactance due to the plate-to-cathode capacitance inside the tube is so large compared with R_k, with which it is in parallel, we are neglecting C_{pk}.

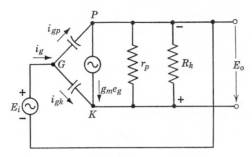

Fig. 11-16   Equivalent circuit for the cathode follower.

It is evident that

$$i_g = i_{gp} + i_{gk} \tag{11-35}$$

$$i_{gp} = \frac{E_i}{X_{C_{gp}}} = j\omega C_{gp} E_i$$

$$i_{gk} = j\omega C_{gk}(E_i - E_o)$$

Substituting the last two equations into Equation (11-35) and dividing by E_i gives the *admittance* of the grid circuit:

$$Y_g = \frac{i_g}{E_i} = j\omega \left[C_{gp} + \left(1 - \frac{E_o}{E_i} \right) C_{gk} \right] \tag{11-36}$$

The gain A_v is E_o/E_i, so the admittance may be expressed in terms of voltage gain, which has been found to approach unity.

$$Y_g = j\omega[C_{gp} + (1 - A_v)C_{gk}] \tag{11-37}$$

With resistance loading (R_k), the phase shift through the tube is negligible since the effect of C_{gp} is negligible in comparison with R_k. This makes A_v both real and positive. Equation (11-37) then shows that the input admittance is due to capacitance only, and that the input capacitance is given by

$$C_i = C_{gp} + (1 - A_v)C_{gk} \tag{11-38}$$

When A_v is near unity, the input capacitance is practically equal to the grid-plate capacitance of the tube. The cathode-follower circuit is superior to the grounded-cathode circuit for amplifying in a frequency range from zero (d-c) up to frequencies well beyond the upper limit for grounded-cathode amplifiers.

As an example of the impedance-matching possibilities of a cathode follower circuit, consider its output impedance when the following constant are known (type 6S4A tube).

$$R_k = 8,000 \ \Omega, \quad r_p = 3,600 \ \Omega, \quad \mu = 16$$

$$C_{gp} = 2.6 \text{ pF}, \quad C_{gk} = 4.2 \text{ pF}, \quad C_{pk} = 0.6 \text{ pF}$$

$$Z_o = \frac{r_p R_k}{r_p + (1 + \mu)R_k} \tag{11-39}$$

$$Z_o = \frac{3,600 \times 8,000}{3,600 + (1 + 16)8,000} = 206 \ \Omega$$

Should this be lower than needed for matching, another tube could be selected. For example, one with $r_p = 7,700 \ \Omega$, and $\mu = 20$ could be used with $R_k = 10,000 \ \Omega$. Z_o would then be 354 Ω.

The input impedance of the first tube, which has a calculated output impedance of 206 Ω, is dependent on the operating frequency since C_{gp} is the most important limiting factor. The amplifier is often used when the frequency band in which it operates is narrow enough so that changes in input impedance due to changes in frequency of the input-signal voltage do not disturb the matching situation.

Using $A_v = 0.917$ for this amplifier, as computed with Equation (11-31), the input impedance at 1 MHz is

$$Y_g = j2\pi \times 10^6[2.6 \times 10^{-12} + (1 - 0.917)4.2 \times 10^{-12}]$$

$$Y_g = j18.5 \times 10^{-6} \text{ mho}$$

$$|Z_i| = \frac{1}{18.5 \times 10^{-6}} = 54,100 \ \Omega$$

Obviously the input impedance at frequencies near 500 kHz is about twice this value.

11.8   Graphical Analysis of Cathode-Follower Performance

A numerical example of cathode follower action will now be given, using the load line of the amplifier for which Equation (11-34) was used to calculate 206 Ω output impedance. The 6S4A triode has relatively high current capacity, can stand a substantial grid-voltage swing without drawing grid current, and has a μ of 16, a medium value. With 400 V on the plate, a grid bias of about -32 V will reduce the plate current to zero. The average plate characteristics are shown in Fig. 11-17.

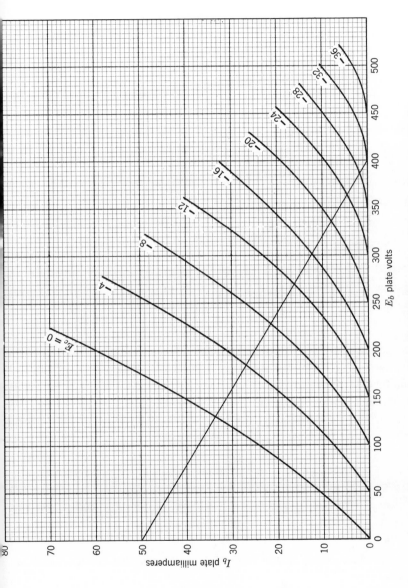

Fig. 11-17   Average plate characteristics for a 6S4A medium-mu triode.

Using $E_{bb} = 400$ V and a cathode-load resistance of 8,000 Ω, the load-line intercept on the current axis is $400/8,000 = 50$ mA. The load-line equation is written from examination of Fig. 11-15a:

$$E_{bb} = e_b + i_b R_k \tag{11-40}$$

and the input voltage is

$$E_i = e_g + i_b R_k \tag{11-41}$$

Note that in this circuit the instantaneous value of the e_g is the instantaneous bias voltage on the grid, and it can be determined at the instantaneous

TABLE 11-1

i_b	e_g	$i_b R_k$	e_b	E_i
34	0	272	128	272
27	-4	216	186	212
21	-8	168	232	160
16	-12	128	272	116
8	-20	64	336	44
0	-32	0	400	-32

operating point on the plate characteristics. At $i_b = 16$ mA, e_b is seen to be 272 V, $i_b R_k$ is 128 V, and $e_g = -12$ V. At this instant the input-signal voltage, E_i, is

$$E_i = -12 + 128 = 116 \text{ V}$$

A set of these voltages was determined at other values of instantaneous plate current, and they are shown in Table 11-1. Notice the wide range of instantaneous input voltage values that is possible. Without a fixed-bias arrangement an a-c swing of ± 32 volts is the limit without cutting off plate current and introducing severe distortion. If the input signal is super-imposed upon a fixed bias E_{cc}, it can have a much greater swing. The optimum value of E_{cc} is halfway between cut-off bias (-32 volts) and the maximum input voltage without grid-current flow ($+272$ volts). This value is (272 $-$ 32)/2 $= +120$ volts. With this value of fixed bias, E_i could swing 152 volts *negative* and give -32 volts bias (e_g) and cut off the plate current, and its 152-volt positive swing would make $E_i = 272$ volts and make the bias $e_g = 0$.

An input amplitude of 152 V is 107 V rms, which is quite a large signal voltage.

To compute the gain of the amplifier, note that an input voltage range from -32 to 272, which is 304 V, produces an output voltage range ($i_b R_k$)

from 0 to 272 V *and in phase*. The ratio of these two values is the gain.

$$A_v = \frac{272 - 0}{272 - (-32)} = 0.895$$

The plate of the cathode follower is at a-c ground potential. It is therefore common to both the cathode and the grid. As the common-collector transistor circuit is therefore analogous to the cathode follower, the graphical analysis of the common-collector transistor amplifier would be developed in a manner similar to the development here.

The common-collector and common-base circuits are discussed in detail in Chapter 9. The reader may want to review that material at this point, for a comparison of similar vacuum-tube and transistor circuits.

11.9 Amplifiers with Tuned Loads

When it is necessary to transmit signals over long distances, amplifiers with tuned loads are used. The circuits that are discussed in this chapter are adequate for audio amplifiers used in home entertainment and public address systems and for servomechanisms, electronic instruments, and a host of other confined applications. However, when the signals are transmitted from an earth satellite or manned space probe to a receiving station on the earth, or from city to city, the distances involved require the use of high-frequency carriers, radio frequencies, for effective use of the transmitting medium. The signals may include only a narrow band of frequencies. Tuned loads are used to amplify the narrow-band signals with the carrier and to separate them from other spurious frequencies.

Various systems of this kind operate throughout the frequency spectrum. The commercial broadcast industry, for example, operates within the range from 550 kHz to 1,600 kHz to transmit voice and music signals. Each commercial transmitter is limited to a narrow-band spectrum of 5 kHz on either side of its assigned carrier frequency or a total bandwidth of 10 kHz which it may use to transmit its signals. A transmitter of this signal is required to operate within this restricted range, using tuned loads that resonate at the center of the bandwidth. Its operation must confine the signals to the region of 5 kHz above and below the carrier frequency. Similarly, receivers for the signals require frequency-selective circuits in order to distinguish between the many signals that may be present. Frequency-selective amplification can be achieved by using tuned loads, because their operation provides a form of filtering to eliminate unwanted signals.

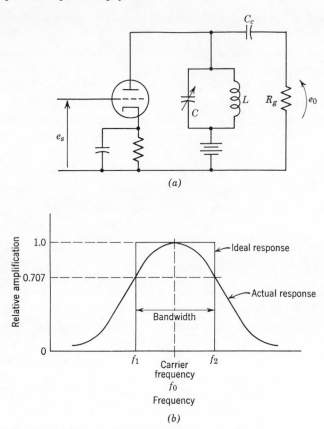

Fig. 11-18 Single-tuned amplifier and its response. (*a*) Amplifier with tuned load. (*b*) Actual amplification compared with ideal response.

The response of a single-tuned amplifier is compared in Fig. 11-18 with the ideal response needed to completely eliminate signals that are outside the desired passband. The circuit shown is one of the simplest narrow-band amplifiers, because it uses a single-tuned load. Double-tuning and stagger-tuning will be discussed after we have analyzed the behavior of the single-tuned amplifier.

Some frequency of the input signal will permit the load of the circuit to be in resonance. At this frequency the amplification will be greatest, and it will be less than maximum at frequencies above and below this value. This is shown by the curve of actual response in the figure. The resonant frequency is tunable within limits, by varying the value of the capacitance in the plate load. The parallel connection *L-C* is often called a *tank circuit*, because of its ability to store energy in the form of a circulating current. The principal disadvantage of this circuit is that the amplification decreases slowly with frequency about the carrier frequency, but the desired passband may be relatively narrow. Thus, it may not reject signals that are near the desired frequency range.

The voltage amplification of the circuit may be expressed as

$$A_v = -g_m Z_{\text{load}} = -g_m \frac{1}{j\omega C + 1/R + 1/j\omega L}$$

where R is the parallel combination of r_p and R_g. Let us examine only the expression for Z_{load}. This factor may be written,

$$\frac{1}{Z_{\text{load}}} = Y_{\text{load}} = \frac{1}{R} + j\left(\omega C - \frac{1}{\omega L}\right)$$

This may be put into a more useful form by using the relation that $1/\omega_0 L = \omega_0 C$, and factoring $1/R$:

$$Y = \frac{1}{R}\left[1 + j\left(\frac{R\omega C\omega_o}{\omega_o} - \frac{R\omega_o}{\omega L\omega_o}\right)\right]$$

$$= \frac{1}{R}\left[1 + j\omega_o RC\left(\frac{\omega}{\omega_o} - \frac{\omega_o}{\omega}\right)\right]$$

$$= \frac{1}{R}\left[1 - jQ_o'\left(\frac{f}{f_o} - \frac{f_o}{f}\right)\right] \tag{11-42}$$

The parameter Q_o' is convenient because it includes the resonant frequency and all circuit components:

$$Q_o' = \frac{R}{\omega_o L} = R\omega_o C$$

We may also express any frequency f in terms of the resonant frequency

$$f = f_o + \alpha f_o$$

where α is a fractional part of f_o, so it follows that

$$\alpha = \frac{f - f_o}{f_o} = \frac{f}{f_o} - 1$$

Then Equation (11-42) may be written

$$Y = \frac{1}{R}\left[1 + jQ_o'\left([1 + \alpha] - \frac{1}{1 + \alpha}\right)\right]$$

$$= \frac{1}{R}\left[1 + jQ_o'\alpha\left(\frac{2 + \alpha}{1 + \alpha}\right)\right] \tag{11-43}$$

The fraction $(2 + \alpha)/(1 + \alpha) = 2 - \alpha + \alpha^2 - \alpha^3 + \cdots$, or is ≈ 2 for $\alpha \ll 1$. Then the load admittance for $\alpha \ll 1$ can be written,

$$Y_{\text{load}} = \frac{1}{R}[1 + jQ_o'2\alpha] \tag{11-44}$$

and load impedance is

$$Z_{\text{load}} = \frac{1}{Y_{\text{load}}} = \frac{R}{1 + jQ_o'2\alpha} \tag{11-45}$$

The voltage amplification can be expressed

$$A_v = -g_m R \frac{1}{1 + jQ_o'2\alpha} \tag{11-46}$$

The gain will be "down 3 dB" when $Q_o'2\alpha = 1$, or at $\alpha = 1/2Q_o'$. Then the bandwidth of this amplifier is between the frequencies $f_o - (f_o/2Q_o')$ and $f_o + (f_o/2Q_o')$. These may be expressed as

$$f_o \pm f_o\left(\frac{1}{2Q_o'}\right) = f_o \pm \frac{1}{4\pi RC}$$

and the bandwidth, B, is $2\alpha f_o$, or

$$\frac{f_o}{2Q_o'} = \left(\frac{B/2}{f_o}\right)^{-1}$$

Then Equation (11-46) becomes

$$A_v = -g_m R\left(1 + j2Q_o'\frac{B/2}{f_o}\right)^{-1}$$

$$= A_{v\,\text{max}}\left(1 + j2Q_o'\frac{B/2}{f_o}\right)^{-1} \tag{11-47}$$

Thus, the half-power frequencies are a measure of the circuit Q_o' : $Q_o' = f_o/B$.

EXAMPLE 11-2.   A pentode having $g_m = 5{,}000$ μmho is to be used in a single-tuned amplifier. The signal to be amplified is centered at 1 MHz and has a half-power bandwidth of 10 kHz. Determine required values of R, L, and C, and the amplification at resonance.

For the specified center frequency, $\omega_o{}^2 = 1/LC$, and the bandwidth in terms of R and C is $B = 1/2\pi RC$,

$$\omega_o{}^2 = (2\pi)^2(10^6)^2 = \frac{1}{LC}$$

$$2\pi B = (2\pi)(10^4) = \frac{1}{RC}$$

Once the value of C is chosen, the other two components are fixed by the specified frequencies. The value of C will depend on parasitic wiring* and interelectrode capacitances as well as on the actual component used for C in the circuit. Suppose its value is taken as 200 pF for calculation, then the other components are

$$R = \frac{1}{2\pi(10^4)(200)(10^{-12})} = 80 \text{ k}\Omega$$

$$L = \frac{1}{(2\pi)^2(200)} = 0.126 \text{ mH}$$

The amplification at resonance is $A_v = -g_m R = -(5)(10^{-3})(80)(10^3) = -400$.

11.10 Double and Stagger Tuning

Figure 11-19 shows a double-tuned amplifier, which is probably the most widely used. The two tuned circuits are coupled by the mutual inductance between the coils. Each tank circuit is tuned to the same center frequency, and if each has the same amplification and filtering characteristics in the bandwidth of incoming signals, the overall performance is superior to that of the single-tuned circuit. The double-tuned circuit provides (1) selective filtering and amplification, and (2) impedance matching for maximum amplification of signals.

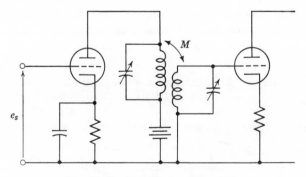

Fig. 11-19 Double-tuned amplifier.

* Stray capacitance is present between connecting wires in the amplifier circuit.

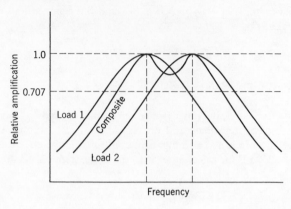

Fig. 11-20   Relative amplification response in a stagger-tuned amplifier.

When several tuned stages are cascaded for greater amplification, the bandwidth is reduced by the successive amplifications. It may be necessary or desirable to expand the bandwidth at some point in the cascaded amplification. This can be accomplished by *stagger tuning*, i.e., using a different center frequency for two successive tuned loads. Then the overall performance generates a composite amplification as shown in Fig. 11-20. Each stage of amplification may generate a sharply defined bandwidth, but the composite more closely resembles the ideal response of a frequency-selective amplifier.

11.11   Bandwidth Reduction in Cascaded Stages

When two or more stages of amplification are interconnected so that the output of one becomes the input of the next, they are said to be in *cascade* connection. If we denote the lower half-power frequency of one stage as f_1, the relative amplification of *n identical stages connected in cascade is,*

$$(A_{\text{rel}})^n = \left(\frac{1}{[1 + (f_1/f)^2]^{\frac{1}{2}}} \right)^n = \frac{1}{[1 + (f_1/f)^2]^{n/2}} \qquad (11\text{-}48)$$

The half-power frequency for the composite amplification is also determined where the relative gain is reduced to $1/\sqrt{2}$ of its maximum value. This requires that the denominator of Equation (11-48) equal $\sqrt{2}$

$$\left[1 + \left(\frac{f_1}{f} \right)^2 \right]^{n/2} = \sqrt{2}$$

This expression may be rearranged as

$$\frac{f_1}{f} = \sqrt{2^{1/n} - 1} \qquad (11\text{-}49)$$

A similar analysis of the upper half-power frequency, f_2, yields a similar expression for the bandwidth reduction

$$\frac{f}{f_2} = \sqrt{2^{1/n} - 1} \qquad (11\text{-}50)$$

TABLE 11-2   Bandwidth Reduction Factors

n =	1	2	3	4	5	6
$\sqrt{2^{1/n} - 1}$ =	1.000	0.643	0.510	0.435	0.387	0.350

The bandwidth reduction factor is given in Table 11-2 for several values of n. It is seen that if an overall bandwidth is necessary for an amplifier, each stage of a cascaded circuit must have a materially greater individual bandwidth. For example, three identical stages will require that each stage have a bandwidth that is approximately twice the desired bandwidth for the whole system.

SUGGESTED REFERENCES

1. Samuel Seely, *Electron-Tube Circuits*, 2nd ed., McGraw-Hill, New York, 1958.

2. Frederick E. Terman, *Electronic and Radio Engineering*, 4th ed., McGraw-Hill, New York, 1955.

3. Paul D. Ankrum, *Principles and Applications of Electron Devices*, International Textbook Co., Scranton, Pa., 1959.

4. R. B. Hurley, *Junction Transistor Electronics*, Wiley, New York, 1958.

QUESTIONS

11-1 How is negative bias on the grid of a second stage provided in a direct coupled amplifier even though a positive plate is directly connected to the grid?

11-2 What is the purpose of having the resistances as small as possible in a voltage divider used with a d-c amplifier?

11-3 Is there a phase shift in the voltage when amplified by a d-c amplifier?

11-4 To what conditions, inherent in amplifiers, are d-c amplifiers sensitive? In what kind of amplifying situation would these conditions make the d-c amplifier operate in an erratic manner?

11-5   What two useful purposes does an interstage transformer serve?

11-6   What is meant by impedance ratio, and how is it related to the turns ratio of a transformer?

11-7   Name two circuit advantages transformer coupling has when compared with *R-C* coupling.

11-8   When a resistance in the secondary circuit of a transformer is to be accounted for in an equivalent circuit representing the whole system as *primary quantities*, by what means is the resistance "reflected" into the primary?

11-9   When an impedance $R_s + jX_s$ in the secondary circuit of a transformer is to be reflected into the primary circuit, how is it done? Comment on what happens to reactances when they are reflected to the other side of a transformer.

11-10   What inductances are present in a transformer? What capacitances are present?

11-11   Which capacitances and inductance are negligible when a transformer-coupled amplifier is operated in the low audio range?

11-12   How is the gain of a transformer coupled amplifier affected by the turns ratio?

11-13   Why are pentode tubes not suitable for use in transformer-coupled audio-frequency amplifiers?

11-14   Which inductance and capacitances must be included in the equivalent circuit of a transformer-coupled amplifier operated above the midband audio range?

11-15   What phenomenon causes the appreciable hump at the high frequency end of the response curve in Fig. 11-13?

11-16   Draw the basic circuit of a grounded-grid amplifier.

11-17   What special effect has the grid in a grounded-grid amplifier, and what advantage does it provide?

11-18   Explain the phase relation between input and output voltages of a grounded-grid amplifier.

11-19   Discuss the impedance-matching ability of a grounded-grid amplifier.

11-20   Draw the basic circuit of a cathode-follower amplifier. Why is it called a cathode follower?

11-21   Name four distinctive features of the cathode-follower amplifier that relate to its use in a circuit.

11-22   What limits the magnitude of the input voltage of a cathode-follower amplifier when no fixed bias is provided?

PROBLEMS

11-1 Refer to the d-c amplifier circuit of Fig. 11-1. The quiescent plate currents are 1 mA in each tube. It is desired to load all resistors as near their full wattage capacity as possible. Calculate the resistance and wattage values you would specify.

11-2 An input voltage of 25 mV at 2 Hz is applied to the d-c amplifier of Fig. 11-1. The tubes are type 6J5. Determine (a) the signal voltage on the second grid, (b) the voltage at the output of the amplifier and its phase with respect to that of the input voltage, (c) the a-c current in the load resistor of T_2.

11-3 The 0.1 MΩ output impedance of an amplifier stage is matched by an inter-stage transformer to the 0.75 MΩ input impedance of the following stage. (a) What is the required turns ratio? (b) What is the voltage gain through the matching transformer?

11-4 A transformer has a mutual inductance of 1 mH. An impedance $10 + j30 \ \Omega$ in its secondary circuit is to be reflected into the primary circuit. Determine the impedance after reflection when the frequency is 10 kHz.

11-5 In the gain equation for low-audio-frequency operation of a transformer-coupled voltage amplifier (Equation 11-10) assume $\mu = 20$, $N_2/N_1 = 3$, $r_p = 7,700 \ \Omega$, $L_1 = 40$ H. (a) At what frequency is $r_p/\omega L_1 = \frac{1}{10}$? (b) Find the per cent error in calculating the gain when the j term is neglected. (c) At what frequency is A_{vl} 3 dB down?

11-6 In the amplifier for which constants are given in the statement of Problem 11-5, additional data are: $C_o = 80$ pF, $R_e = 500 \ \Omega$, $L_e = 0.3$ H. (a) Calculate the voltage gain at 10 kHz. (b) Plot a curve of A_{vh} vs. f in the high-frequency range. (c) Calculate the frequency at which the gain is maximum.

11-7 Plot the response curve of the amplifier of Problem 11-5 in the low-frequency range.

11-8 The grounded-grid amplifier of Fig. P11-8 has a tube for which $r_p = 6,300 \ \Omega$ $\mu = 43$. Compute the input impedance and the voltage gain.

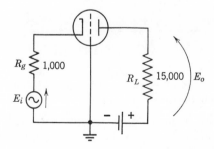

Fig. P11-8 Grounded-grid amplifier.

11-9   A cathode-follower amplifier uses a 6S4A tube having average plate characteristics as shown in Fig. 11-17. The load resistor $R_k = 10,000\ \Omega$ and $E_{bb} = 350$ V. Tube constants are given in the numerical example in Section 11.7. Find (a) the input and the output impedances at 500 kHz, (b) the voltage gain.

11-10   Compute value and construct a table similar to Table 11-1 for the amplifier of Problem 11-9. Determine optimum bias, maximum rms input voltage, and voltage gain by graphical analysis.

11-11   A transformer-coupled amplifier is to be constant within 3 dB over the frequency range from 100 to 8,400 Hz. Specify the required value of primary inductance. The tube is a 6J5 with $r_p = 7,700\ \Omega$. Neglect the winding resistance in the calculations.

11-12   In the cathode-follower amplifier circuit of Fig. P11-12, find $A = E_o/E_i$, assuming midband frequencies. $R_{g2} \gg R_L$ and the cathode resistors, R_k, are adequately bypassed.

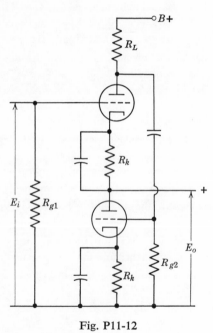

Fig. P11-12

11-13   For the cathode follower circuit in Fig. P11-13, find the input and output impedances, neglecting the tube capacitances.

11-14   An operational amplifier is being used as a simple voltage adder. The three input voltages are 10 V d-c, 5 V d-c, and $5 \sin \omega t$, and the feedback resistor $R_f = 1$ MΩ. What must be the values of input resistors?

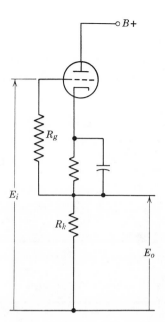

Fig. P11-13 $\mu = 20$; $r_p = 7,700\ \Omega$; $R_g = 1\ M\Omega$, $R_k = 10,000\ \Omega$.

11-15 Sketch the waveform of output voltage from the amplifier in Problem 11-14. Compare it with the output waveform if all input resistors are chosen to be equal to $0.5R_f$.

11-16 An operational amplifier is to be used to change the algebraic sign of an input voltage, but must not alter its amplitude. Show the circuit that is used, and specify the ratio of input and feedback impedances.

11-17 The three input voltages of Problem 11-14 are to be added together and appear at the output with the same phase as the input. Show a circuit that may be used to sum the voltages and maintain them in the same phase at the output.

11-18 A single tuned-circuit amplifier has its maximum amplification at 1 MHz and a half-power bandwidth of 10 kHz. A pentode having $g_m = 5,000\ \mu$mho and $r_p = 1\ M\Omega$ is used to drive the tank circuit. The output voltage is taken across a resistor $R_g = 100,000\ \Omega$, as in Fig. 11-18. Determine the required values of L and C and the amplification at the half-power frequencies.

11-19 If the value of C in Problem 11-18 is variable by ± 50 per cent, how will its variation affect circuit operation? When it is adjusted to its extreme values, what will be the circuit bandwidth? Through what range of frequencies will the variable C change the resonant frequency?

11-20 Three identical stages of amplification are connected in cascade. The desired bandwidth of the circuit is from 82 MHz to 88 MHz. What must be the bandwidth of each individual amplifier stage?

11-21   For the direct-coupled emitter-follower amplifier in Fig. P11-21, determine the quiescent operating points for both transistors.

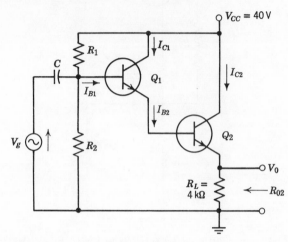

Fig. P11-21

11-22   From the results of Problem 11-21, estimate the input resistance of Q_2. Compare this value with a calculation using h-parameters and the value of R_L.

11-23   Calculate the output resistance of the amplifier in Fig. P11-21.

11-24   In the circuit of Problem 11-21, what value of C will permit the voltage gain to remain within 3 dB from 100 Hz to higher frequencies?

CHAPTER TWELVE

Feedback Amplifiers and Oscillators

In the use of cathode bias we found that part of the d-c voltage drop in the plate circuit was applied as negative voltage to the grid. If the operating frequency were reduced to the value where the bypass capacitor across the cathode resistor, R_k, no longer had negligible reactance, an a-c component of voltage would be fed to the grid. The phase of this "feedback" voltage is such that, while the grid potential is going in the positive direction (grid becoming less negative) and the a-c component of plate current is increasing, the bottom of R_k is becoming *more negative with respect to the cathode* and thus the feedback voltage opposes the grid-signal voltage. This is *negative, inverse,* or *degenerative feedback.*

It is reasonable to expect the gain of the amplifier to be reduced by negative feedback because it decreases the input voltage to the amplifier.

The reduction in gain caused by inverse feedback does not prevent its use, because it has beneficial effects, such as decrease of distortion and of unwanted signals (called noise), improvement of amplifier stability, and broadening of the constant-gain (midband) section of the amplifier's frequency-response curve.

When the phase of a feedback voltage is the same as the phase of the input signal to the amplifier, *positive feedback* exists. It will be shown that *negative feedback decreases the gain* of an amplifier while *positive feedback increases the gain.* If the amount of positive feedback is made (or allowed to become) large enough, it will *overdrive* an amplifier tube, with the result that the plate current will vary appreciably between large and then small values at a frequency determined by circuit and tube constants. This is called *oscillation.* This phenomenon is ruinous to amplifier performance, but it is the desired

463

situation in *feedback oscillators*, where the action is such that the tubes are not "overdriven."

Feedback is the necessary element in error-correction systems and in servomechanism control circuits.

12.1   Basic Considerations of Feedback

The process of applying to the input of an amplifier a signal that is proportional to either current or voltage in the output circuit of the amplifier is called *feedback*. Feedback affects the gain, input and output impedances, and frequency response of the amplifier, and under certain circumstances it will cause the amplifier to sustain an oscillation that is independent of any applied input signal.

Figure 12-1 shows the effects of feedback in its simplest form. A_v represents the voltage amplification of the amplifier without feedback, and the block designated as β is the feedback network.* The input signal, E_i, is to be amplified to form an output signal, E_o'. However, a feedback signal, E_{fb}, is introduced at the amplifier in addition to E_i. The feedback signal is a fraction of the output voltage E_o':

$$E_{fb} = \beta E_o' \tag{12-1}$$

and the signal at the output can be expressed as the sum of E_i and E_{fb}. The output signal becomes

$$E_o' = (E_i + E_{fb})A_v$$

$$= (E_i + \beta E_o')A_v \tag{12-2}$$

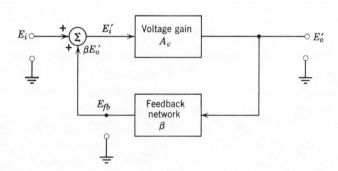

Fig. 12-1   Basic block diagram of a feedback amplifier.

* This symbol for feedback circuits does not refer to the beta\of transistors in the circuits.

Rearrangement of this equation shows that the gain with feedback is related to the gain without feedback in this way:

$$\text{Feedback gain, } A_v' = \frac{E_0'}{E_i} = \frac{A_v}{1 - \beta A_v} \qquad (12\text{-}3)$$

The term βA_v is called the *feedback factor*. We can see that its value compared to unity is important in determining the feedback gain.

The amount of feedback is often expressed in decibels and is usually given by

$$\text{decibel of feedback} = 20 \log_{10} \frac{1}{|1 - \beta A_v|}$$

$$= -20 \log_{10} |1 - \beta A_v|$$

Observe that the ratio of the gain with feedback to the gain without it is

$$\frac{A_v'}{A_v} = \frac{1}{1 - \beta A_v}$$

The *decibel of feedback* is given by $20 \log_{10}$ of this voltage ratio.

EXAMPLE 12-1. An amplifier has a gain without feedback expressed as 20 dB. What value of feedback factor is required to reduce the gain (with feedback) to 10 dB?

The amount of feedback is obviously 10 dB (20 dB − 10 dB = 10 dB). Then

$$10 \text{ dB} = -20 \log_{10} |1 - \beta A_v|$$

$$-\tfrac{1}{2} = \log_{10} |1 - \beta A_v| = \log_{10} 0.317$$

This requires that the feedback factor be equal to 1.000 − 0.317 = 0.683.

Since the gain without feedback is 20 dB, then its numerical value is $A_v = 10$. This means that the value of β is therefore negative, because the gain was reduced.

The feedback is termed *negative* or *degenerative* when it reduces the gain and is called *positive* or *regenerative* when it increases the gain. These statements can be symbolized

$$|1 - \beta A_v| > 1, \quad \text{for negative feedback}$$

$$|1 - \beta A_v| < 1, \quad \text{for positive feedback}$$

However, it is seen that a special case arises when $|1 - \beta A_v| = 0$, or $\beta A_v = 1$. The feedback gain under this condition becomes infinitely large, from Equation (12-3). This means, in a practical sense, that the amplifier provides its own input signal from its operation and breaks into self-sustaining oscillation. This condition is purposely introduced in a class of circuit called *oscillators*, the subject of the latter parts of this chapter. It is important to note here that an amplifier circuit employing feedback is subject to oscillation, and its design must take this possibility into account. A circuit design should be checked closely for the presence of feedback loops that might initiate oscillation where it is not wanted.

12.2   Negative (Inverse) Feedback in Amplifiers

The circuit designer may choose the phase of the feedback voltage that is required. If it is selected to be the same as the phase of the input signal, then β in Equation (12-3) is a *positive* number and *positive* or *regenerative feedback* results. When the feedback voltage is selected out of phase with the input signal, the value of β will be negative and *negative feedback* results.

Let us demonstrate that when feedback is added to an amplifier, its gain with frequency changes, i.e., its frequency response is affected. Consider an amplifier that has a voltage gain of 10 in the midband range without feedback, so that $E_i = 1\underline{/0°}$ when $E_o = 10\underline{/180°}$

$$E_i = 1 + j0, \qquad E_o = -10 + j0$$

If $\beta = -0.1$, from Equation (12-3)

$$A_v' = \frac{10}{1 - (-0.1)(10)}$$
$$= 5$$

This means a 50 per cent drop in midband gain caused by adding feedback $\beta = -0.1$.

We now investigate the effect of feedback at the half-power frequencies. The output voltage *without feedback* is down to 7.07 V and, since there is a $+45°$ phase shift E_o is represented at low frequencies as

$$E_o = 7.07\underline{/225°} = -5 - j5 \text{ V}$$

The feedback voltage is $\beta E_o = 0.5 + j0.5$ V and the *new input voltage* is $E_i + \beta E_o$.

$$E_i' = E_i + \beta E_o = 1 + j0 + 0.5 + j0.5 = 1.5 + j0.5 \text{ V}$$

The gain with feedback is E_o/E_i':

$$A_v' = \frac{-5 - j5}{1.5 + j0.5} = \frac{7.07\underline{/225°}}{1.58\underline{/18.4°}} = 4.47\underline{/206.6°}$$

This shows that the gain at the low-frequency *half-power point* has fallen only to 89.4 per cent of the midband gain of 5, whereas without feedback the gain fell 29.3 per cent, from 10 to 7.07. The phase shift is an angle of $206.6° − 180° = 26.6°$ instead of $45°$ without feedback. The same amount of shift occurs at the upper half-power frequency. This means that the bandwidth is extended at a sacrifice in gain, when negative feedback is used.

With high-mu triodes and pentodes the sacrifice in midband gain due to feedback is not a problem, because even with feedback the gain is very satisfactory. That the gain holds up so well at low and high frequencies means that the new midband range is much wider than the old one.

The decrease in phase-shift effect means that *phase distortion* in amplifiers is appreciably diminished by feedback. Whether there is amplitude or phase distortion in an amplifier without feedback, we can visualize how it is diminished by considering the following. Assume that an amplifier distorts a sine wave and delivers an output wave E_o as in Fig. 12-2a. The wave shape indicates that the gain of the amplifier is below normal from θ_1 to θ_2 on the ωt axis. The wave βE_o, which is a fraction of E_o, is picked off at the feedback tap and fed back in inverse phase to the grid where it combines with the pure sine wave from the source (Fig. 12-2b), modifying it to the shape in Fig. 12-2c of $E_i + (−\beta E_o)$. When the amplifier performs its "faulty-gain"

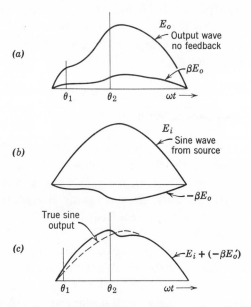

Fig. 12-2 Voltages of negative-feedback amplifier. $E_i + (−\beta E_o')$ is the input voltage with feedback.

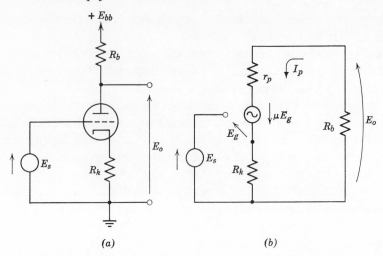

Fig. 12-3   Simple feedback circuit and its equivalent. (*a*) Feedback amplifier. (*b*) Equivalent circuit.

procedures on this wave, its below-normal effects from θ_1 to θ_2 have larger than normal values to work on, and so these are reduced to normal and the result is a true sine-wave output. It is true that Equation (12-1) has β applied to E_o', but let us imagine switch S in Fig. 12-3 open at first. The output is now E_o in Fig. 12-2*a*. Closing the switch results in the feedback circuit "picking off" a fractional part of E_o *in the first cycle*, and, by the time a few cycles have passed, a miniature E_o', i.e., $\beta E_o'$, is fed back. This justifies Equation (12-1).

12.3   Inverse-Feedback Circuits

When the voltage fed back from the output of an amplifier to the input is *proportional to voltage*, the circuit has *voltage feedback*; when the voltage fed back is *proportional to current*, the circuit has *current feedback*. Both types occur in practical circuits. There are many different types of feedback circuit. Figure 12-3 is a simple current feedback circuit.

Another designation for type of feedback is *series feedback*, used to show that the input signal and feedback signal are additive in series, and *shunt feedback* is used to show that the signals are additive in parallel. The circuit we are analyzing here could be called series feedback, because the feedback voltage is in series with the input signal, as far as their influence on the grid and cathode voltages are concerned.

From the small-signal equivalent circuit, we can write equations to describe the circuit operation:

$$E_g = E_s - I_p R_k$$
$$E_o = -I_p R_b \qquad (12\text{-}4)$$
$$\mu E_g = (r_p + R_k + R_b)I_p$$

Solving for the plate current I_p yields

$$I_p = \frac{\mu E_s}{r_p + R_b + (\mu + 1)R_k}$$

and the gain with feedback is

$$A_v' = -I_p R_b = -\frac{\mu R_b}{r_p + R_b + (\mu + 1)R_k} \qquad (12\text{-}5)$$

DRILL PROBLEM

D12-1 Determine the expression for β in Equation (12-5), and the feedback factor. Is the feedback negative or positive?

Voltage-stabilizer circuits using degenerative feedback are common. The one in Fig. 12-4 employs a gas diode which will accommodate small changes in its current without a change in its terminal voltage. The purpose is to maintain constant voltage across the load resistance R_L. If the input voltage rises a little, the current through R_1 and the diode immediately increases. All the rise in voltage shows up across R_1 because the diode voltage does not change. This increases the negative bias on the triode and shifts the static Q-point to a new position where the tube has a larger voltage drop of just the right value to absorb the increase in input voltage from the rectifier filter. This leaves the voltage across R_L unchanged.

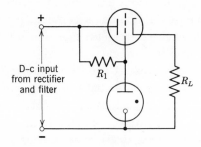

D–c input
from rectifier
and filter

R_1

R_L

Fig. 12-4 Degenerative voltage-stabilizing circuit using gas diode.

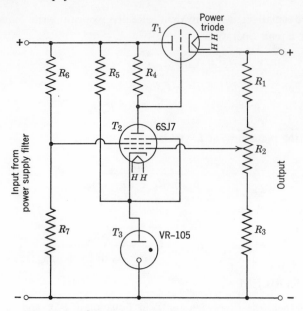

Fig. 12-5 Voltage regulator for a d-c power supply.

When the input voltage from the rectifier filter decreases, the reverse takes place. The current through R_1 decreases, reducing the bias on the triode with the result that the static Q-point is shifted the other way, causing the tube to have a smaller internal voltage drop and leaving the same voltage across R_L as before.

The circuit of Fig. 12-5 is an example of amplified degenerative feedback for stabilization of load voltage. It has one control, the potentiometer (R_2) by means of which the output voltage may be set at a fixed value over a fairly wide range. That is, a circuit of this type that will deliver a steady d-c voltage of 250 V with the slider of R_2 at its midposition may be set to deliver any voltage desired between approximately 200 and 300 V, and it will stabilize the output voltage very well. The amount of output current is limited by the current capacity of T_1. A power-amplifier tube should be used because of its high current capacity and transconductance.

Observe that the plate current of the control tube (T_2) flows through R_4 and thus determines the potential on the grid of T_1 with respect to its plate. The output voltage is always greater than the potential difference between the plate of T_2 and the negative line (bottom of R_3), and so the grid of T_1 is always negative with respect to its cathode. This grid bias controls the plate current of T_1, and hence the output current.

The cathode of T_2 is held by T_3 at a constant potential with respect to the negative line, and because of this any change in voltage between the slider of R_2 and the negative line will show up as a change in the grid bias of T_2.

Suppose some of the load is removed from the output. The load voltage will tend to rise because the decrease in load current will reduce the voltage drop across T_1. The grid bias of T_2 rises simultaneously (becomes less negative) and its plate current, flowing through R_4, increases. The resulting fall in grid potential of T_1 (its bias becomes more negative) increases the voltage drop from plate to cathode and prevents a rise in output voltage.

If more load current is drawn through T_1, its voltage drop will tend to increase and the result will be a momentary drop in output voltage. This increases the amount of negative bias on T_2 because the current through R_1, R_2, R_3 decreases, and thus the positive voltage contribution to the grid by R_3 and part of R_2 has become less. The resulting decrease in current through R_4 raises the potential of the grid of T_1, allowing it to carry the new, higher value of load current without an increase in voltage drop between its plate and cathode.

It takes appreciable time to explain the action of one of these regulating circuits, but it should be understood that the regulating action itself requires only a few microseconds of time. A very slight change in the grid potential of T_2 sets off the compensating action of T_1, so that even the most minute change in output voltage is met with opposition.

A mechanical analogy that may illustrate the action of the voltage regulator follows. Suppose that a water tank has an outlet valve at the bottom and a supply pipe that enters somewhat below the water surface at the top. Consider that a float on the water surface controls an inlet valve in the supply pipe in a highly efficient manner, being so sensitive that a change of one hundredth of an inch in water level will result in a compensating adjustment of the intake. Let it be true also that the maximum possible change in rate of outflow (from "off" position to "wide open" position of the outlet valve) can be compensated by moving the float only a few hundredths of an inch from its middle position. Changes in rate of flow of outlet water, whether sudden or slow, large or small, will then have a negligible effect on the water level in the tank because the first hundredth of an inch change in level sets up a compensating action on the input valve, and only a few hundredths are needed for full-range control.

The circuit of Fig. 12-5 compensates to some extent for changes in input voltage. If the input voltage drops, the screen voltage on T_2 drops, and this

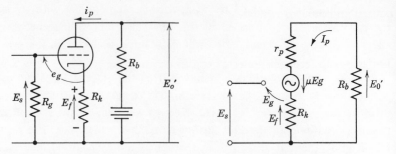

Fig. 12-6   Amplifier with simple current feedback and its equivalent.

decreases its plate current. The resulting rise in the grid voltage of T_1, caused by the reduction of current in R_4, lowers the potential drop through T_1, and thus output voltage is held up because it is getting a larger fraction of the lowered input voltage. It should be observed that with this type of voltage-regulator circuit the series tube (T_1) "accepts the changes in input voltage" and thus allows the constant load resistance to have constant terminal voltage. When the load resistance is changed, T_1 "accepts the required changes in load current with practically no change in its plate-to-cathode voltage," thus allowing the load voltage to remain constant.

Figure 12-6 shows an amplifier with simple current feedback. The feedback voltage E_f in the amplifier circuit is proportional to i_p, obviously, because i_p flows through R_k. For current feedback, the voltage that is fed back is proportional to the current in the a-c load impedance, Z_L. The value of β is, in general, related to Z_L by a constant that is independent of Z_L

$$\beta = \frac{k}{Z_L} \tag{12-6}$$

In the circuit of Fig. 12-6, the a-c load impedance is

$$Z_L = R_b$$

and the value of β is

$$\beta = \frac{E_f}{-E_0'} = \frac{I_p R_k}{I_p R_b} = \frac{R_k}{R_b} \tag{12-7}$$

The voltage gain from grid input to load voltage is easily derived from the equivalent circuit:

$$A_v' = \frac{E_o'}{E_g} = \frac{-\mu R_b}{r_p + R_b + R_k} \tag{12-8}$$

but the grid input is related to E_s

$$E_g = E_s - I_p R_k$$

The gain with feedback, referred to the source E_s, is

$$A_v' = \frac{E_o'}{E_s} = \frac{-\mu R_b}{r_p + R_b + (\mu + 1)R_k} \tag{12-9}$$

The plate current with feedback then is

$$I_p' = \frac{-E_o'}{R_b} = \frac{-A_v' E_s}{R_b} = \frac{\mu E_s}{r_p + R_b + (\mu + 1)R_k} \tag{12-10}$$

The plate current *without feedback* is given for comparison:

$$I_p = \frac{\mu E_s}{r_p + R_b} \tag{12-11}$$

Note that feedback increases the denominator by the term $(\mu + 1)R_k$. This means that a change in R_b will have less effect on the a-c plate current when *current feedback* is used (because the denominator is larger) than on the plate current without feedback.

When voltage feedback is carried across one or more stages of an amplifier, the matter of inherent phase reversal of each stage must be taken into account. If the feedback resistor is connected to the plate end of R_b, the voltage will already be inverse in phase with respect to the signal voltage on the grid; therefore the grid end of the feedback resistor is connected so it will be in series with the signal voltage to the grid. This connection must not be made to ground. A temptation to do this would appear when cathode bias is used. In this case another resistor is placed between the feedback resistor and ground, their junction point then serving as the lower terminal for the signal voltage. This type of inverse-voltage-feedback connection is correct for feeding back across any odd number of stages. In practice the number seldom exceeds three.

Inverse-voltage feedback across an even number of stages is illustrated in Fig. 12-7. With the switch S closed, the fraction of the output voltage returned in series with the signal entering the input end is $800/100,800 = \beta$. The phase reversal of voltage is 180 degrees in each stage and makes the final output voltage in phase with the voltage on the input grid. In order for the feedback to be *inverse* or *negative*, the connection must be made to the cathode instead of to the grid. Because this is not easy to understand at first, let us consider the fact that, when the grid at the input is receiving

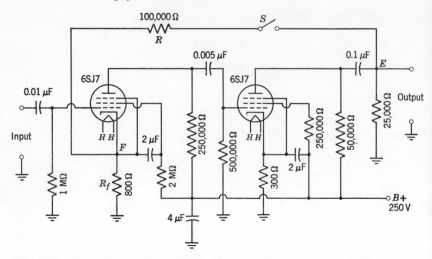

Fig. 12-7   Two-stage *R-C* coupled voltage amplifier. Closing switch *S* inserts inverse feedback. Gain without feedback 60 dB, with feedback 40 dB.

increasing potential in the positive half-cycle of the input-signal voltage, point *E* at the output is growing more positive with respect to ground and so is point *F*, the cathode of the first tube. This means that the voltage fed back is making the *cathode* more positive with respect to ground and therefore *more negative with respect to the grid*. Thus the connection to *F* makes the feedback inverse. Point *F* is growing more positive with respect to ground only β times as fast as is point *E*. The value of β given above can be expressed numerically as $\frac{1}{126}$.

Figure 12-8 illustrates current feedback. It could be an output-amplifier stage feeding a loud speaker, where R_L represents the voice coil.

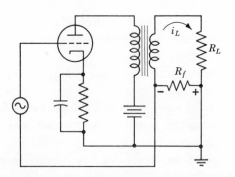

Fig. 12-8   Negative-current feedback circuit. Transformer secondary terminals must be properly connected.

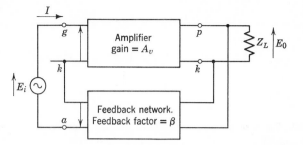

Fig. 12-9 General block diagram for voltage-feedback amplifier.

12.4 Effect of Voltage Feedback on Input and Output Impedances

In Fig. 12-9, the *input impedance* of the amplifier *with feedback* is

$$Z_{if} = \frac{E_i}{I}$$

But $E_i = E_{gk} + E_{ka}$, so that

$$Z_{if} = \frac{E_{gk} + E_{ka}}{I}$$

Note that $E_{ka} = -E_{gk} \times A_v' \times \beta$, so that *the input impedance of an amplifier with negative feedback* is

$$Z_{if} = \frac{E_{gk}(1 - \beta A_v')}{I} \qquad (12\text{-}12)$$

Now the input impedance of the amplifier *without feedback* is

$$Z_i = \frac{E_{gk}}{I} \qquad (12\text{-}13)$$

so that the relation between the two impedances is

$$Z_{if} = Z_i(1 - \beta A_v') \qquad (12\text{-}14)$$

Since β is always negative for inverse, or negative, feedback, it is evident that *adding inverse-voltage feedback increases the input impedance.*

As an example, suppose an amplifier has an input impedance without feedback of 10,000 Ω and a voltage gain of 40. Adding inverse feedback for which $\beta = -0.15$ increases the input impedance to 70,000 Ω. Negative-voltage feedback is responsible for the high input impedance of the cathode follower.

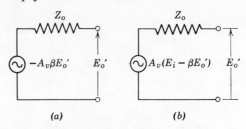

Fig. 12-10   Equivalent-output circuits of voltage amplifier with inverse feedback.
(a) No input signal, but E_o' is applied at output terminals. (b) Input signal E_i at
front end; this produces E_o' at output.

Negative-current feedback also increases the input impedance of an
amplifier. Since voltage, proportional to current, is fed back, it does not
matter where or what the source is.

To obtain an expression for Z_{of}, *output impedance with feedback*, it is
necessary to recall that Thévenin's theorem enables us to represent an
amplifier stage as shown in Fig. 12-10. In Fig. 12-10a, instead of looking
back from the output terminals and seeing a series impedance, formerly
called Z' but now Z_o, which is the output impedance of the amplifier itself
without feedback, we now *apply a voltage E_o'* to the output terminals. The
feedback branch inside the amplifier picks off the β fraction of the terminal
voltage E_o' and sends it back to the input in inverse phase as $-\beta E_o'$. The
gain of the amplifier (used here like μ of the tube) then makes the generated
voltage in the equivalent circuit become $-\beta A_v E_o'$. This is all *without an
input signal* at the front end of the stage.

When an input signal E_i is applied at the front end, it too is amplified,
and the equivalent circuit becomes that in Fig. 12-10b. E_o' shows up of
its own accord; i.e., we need not apply an external voltage at the output,
of course. But let us use this value of E_o' and apply it to the output terminals
after first removing the input signal and its source.

The current in the series circuit of Fig. 12-10a is then

$$I_L' = \frac{(E_o' - A_v \beta E_o')}{Z_o}$$

from which we can get $E_o'/I_L' = Z_o'$, the output impedance with feedback.
The *output impedance of an amplifier with negative voltage feedback is*

$$Z_0' = \frac{E_0'}{I_L'} = \frac{Z_0}{1 - \beta A_v} \tag{12-15}$$

This means that the impedance looking back into an amplifier from the
output terminals is reduced (remember that β is negative) by adding voltage

feedback. Changes in external-load impedance, and therefore in current, will not cause as much change in load terminal voltage when inverse (negative) feedback is used, and the larger the product βA_v, the greater the reduction in Z_o' and the more stable the output voltage.

As an example, suppose an amplifier has an output impedance of 35,000 Ω and a voltage gain of 40 without feedback. Adding inverse feedback for which $\beta = -0.15$ decreases the output impedance to 5,000 Ω. We will note later that current feedback increases the output impedance.

DRILL PROBLEMS

D12-2 A voltage amplifier has a gain of -100 without feedback. What value of β will increase the input impedance from 100 kΩ to 500 kΩ?

D12-3 What value of β in the amplifier of D12-2 will decrease the output impedance from 50 kΩ to 10 kΩ?

D12-4 (*a*) How much change occurs in the output impedance in D12-2? (*b*) How much does the input impedance change in D12-3?

EXAMPLE 12-2. We are interested in the voltage gain and output impedance of the amplifier shown in Fig. 12-11 and in particular, the parameters that may be varied to match the output and load impedances. The bias and supply sources have been omitted for convenience.

We will calculate the voltage gain first, by writing the circuit equations:

$$E_g = E_s + E_{fb} \tag{1}$$

$$\mu E_g = (r_p + Z)I_p - E_p \tag{2}$$

$$E_o = -Z_L I_o \tag{3}$$

$$E_o = \frac{N_3}{N_1 + N_2} E_p \tag{4}$$

$$E_{fb} = \frac{N_2}{N_3} E_o - ZI_p \tag{5}$$

$$I_o = \frac{N_1 + N_2}{N_3} I_p \tag{6}$$

Eliminating E_g; E_{fb}, and I_o leaves a system of these three equations, where the turns ratio of the transformer has been defined as $a = n_s/n_p = N_3/(N_1 + N_2)$:

$$I_p[r_p + (\mu + 1)Z] - E_p - \mu \frac{N_2}{N_3} E_o = \mu E_s \tag{7}$$

$$I_p\left(\frac{1}{a}\right) + \frac{1}{Z_L} E_o = 0 \tag{8}$$

$$aE_p - E_o = 0 \tag{9}$$

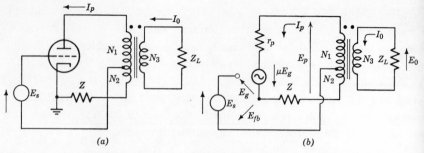

Fig. 12-11   (*a*) Feedback amplifier. (*b*) Equivalent circuit for midband frequencies.

A determinant may be set up for these equations:

$$
\begin{array}{ccc}
I_p & E_p & E_o \\
\end{array}
$$

$$
\begin{vmatrix}
r_p + (\mu + 1)Z & -1 & -\mu \dfrac{N_2}{N_3} \\[2ex]
\dfrac{1}{a} & 0 & \dfrac{1}{Z_L} \\[2ex]
0 & a & -1
\end{vmatrix}
\begin{matrix}
\mu E_s \\[2ex]
0 \\[2ex]
0
\end{matrix}
$$

Solving for E_o:

$$
E_o = \frac{\begin{vmatrix}
r_p + (\mu + 1)Z & -1 & \mu E_s \\[2ex]
\dfrac{1}{a} & 0 & 0 \\[2ex]
0 & a & 0
\end{vmatrix}}{\begin{vmatrix}
r_p + (\mu + 1)Z & -1 & -\mu \dfrac{N_2}{N_3} \\[2ex]
\dfrac{1}{a} & 0 & \dfrac{1}{Z_L} \\[2ex]
0 & a & -1
\end{vmatrix}}
$$

The gain with feedback is

$$
A_v' = \frac{E_o}{E_s}
$$

and can be written

$$
A_v' = -\frac{\mu}{\mu \dfrac{N_2}{N_3} + \dfrac{1}{a} + a \left[\dfrac{r_p + (\mu + 1)Z}{Z_L} \right]}
$$

The gain without feedback may be found by setting $Z = 0$ and $N_2 = 0$:

$$A_v = -\frac{\mu}{\dfrac{1}{a} + a\dfrac{r_p}{Z_L}} = -\frac{\mu a Z_L}{Z_L + a^2 r_p}$$

Manipulation to put the feedback gain expression in the form of Equation (12-3) yields

$$A_v' = \frac{A_v}{1 - \beta A_v} = -\frac{\mu a Z_L}{Z_L + a^2 r_p + \mu a \dfrac{N_2}{N_3} Z_L + a^2(\mu + 1)Z}$$

$$= -\frac{\mu a Z_L}{Z_L + a^2 r_p} \cdot \left[1 + \frac{\mu a (N_2/N_3)Z_L + a^2(\mu + 1)Z}{Z_L + a^2 r_p}\right]^{-1}$$

$$= \frac{A_v}{1 - A_v\left[\dfrac{N_2}{N_3} + \dfrac{a(\mu + 1)Z}{\mu Z_L}\right]}$$

Thus, β is the bracketed factor

$$\beta = \frac{N_2}{N_3} + \frac{a(\mu + 1)Z}{\mu Z_L}$$

Feedback will be positive if a negative real part of Z/Z_L becomes large enough to counteract N_2/N_3. We see that feedback will be negative if $Z = 0$.

To determine the output impedance, E_o/I_o, the input signal E_s is set to zero and Z_L is removed. The circuit equations are

$$E_g = E_{fb} \tag{10}$$

$$\mu E_g = (r_p + Z)I_p - E_p \tag{11}$$

$$E_o = \frac{N_3}{N_1 + N_2} E_p = a E_p \tag{12}$$

$$E_{fb} = \frac{N_2}{N_3} E_o - ZI_p \tag{13}$$

$$I_o = \frac{N_1 + N_2}{N_3} I_p = \frac{1}{a} I_p \tag{14}$$

Eliminating E_{fb} and E_g yields the determinant

	I_p	E_p	E_o	
	$\dfrac{1}{a}$	0	0	I_o
	0	a	-1	0
	$-[r_p + (\mu + 1)Z]$	1	$\mu\dfrac{N_2}{N_3}$	0

Then solving for E_o in terms of I_o, the output impedance may be written

$$Z_{\text{out}} = \frac{(N_3/N_2)[r_p + (\mu + 1)Z]}{\dfrac{N_3}{N_2}\left(\dfrac{1}{a^2}\right) + \dfrac{\mu}{a}}$$

$$= \frac{a^2[r_p + (\mu + 1)Z]}{1 + \mu a(N_2/N_3)}$$

This means that matching the output impedance to a load impedance can be done by varying any combination of the three parameters Z, a, and N_2/N_3.

Suppose the tube has $r_p = 1\,\text{k}\Omega$ and $\mu = 40$; the transformer has $a = \frac{1}{4}$ and $N_2/N_3 = \frac{1}{5}$; $Z = 100\text{-}\Omega$ resistor; and $Z_L = 100\text{-}\Omega$ resistor. The feedback network will have

$$\beta = \frac{N_2}{N_3} + \frac{a(\mu + 1)Z}{\mu Z_L}$$

$$= \frac{1}{5} + \frac{\frac{1}{4}(40 + 1)(100)}{40(100)}$$

$$= \frac{1}{5} + \frac{41}{160}$$

$$= 0.200 + 0.256$$

$$= 0.456$$

The gain without feedback will be

$$A_v = \frac{-\mu a Z_L}{Z_L + a^2 r_p}$$

$$= \frac{-(40)(\frac{1}{4})(100)}{100 + (1/4)^2(1000)}$$

$$= \frac{-1000}{100 + 1000/16}$$

$$= -6.15$$

and the gain with feedback

$$A_v' = \frac{A_v}{1 - A_v \beta}$$

$$= \frac{-6.15}{1 - (-6.15)(0.456)}$$

$$= -\frac{6.15}{3.80}$$

$$= -1.62$$

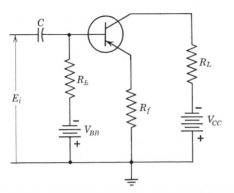

Fig. 12-12   Transistor amplifier stage using negative current feedback.

12.5   Feedback in Transistor Amplifiers

The effects of inverse feedback on the impedances of an amplifier are much more important in transistor circuits than in tube circuits. The inherently low input impedance of transistor amplifiers, particularly the common-base and common-emitter circuits, is a handicap in amplifier design. Inverse-voltage feedback therefore helps solve that problem. Similarly, the reduction in output impedance of common-emitter and common-base transistor amplifiers through the use of inverse feedback is conducive to better matching possibilities in *R-C* coupled circuits.

Transistor circuits using negative feedback are shown in Figs. 12-12 and 12-13. R_f is the feedback resistor in both circuits. Current feedback, generally termed *series* feedback, is achieved in Fig. 12-12 in the following manner. First we assume static conditions; this means there is no input signal ($E_i = 0$).

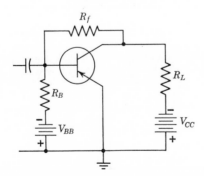

Fig. 12-13   Transistor amplifier using negative voltage feedback.

The d-c current direction in R_f is then *upward*, making the emitter terminal negative with respect to the positive terminals of the power sources. The collector current flows *out* of the collector terminal.

A positive input signal, E_i, will be in opposition to this polarity and will reduce the emitter and collector current. This causes the a-c component of the collector current (i.e., the output current in R_L) to flow *into* the collector terminal and *downward* through R_f. Therefore the net input signal from emitter to base has been reduced and the feedback is negative. Because it is proportional to the a-c current in R_f, it is termed *current* or *series* feedback.

To explain negative-voltage (*shunt*) feedback in a transistor amplifier stage, we refer to Fig. 12-13. When a positive input signal is applied, an *amplified current* (a-c) appears in the output resistor R_L which is due to the current amplification. This current flows into the collector terminal, as explained for the circuit of Fig. 12-12, with the result that the potential of the upper end of R_L, and of the collector terminal, is given a negative increment. This drop in potential is fed to the base of the transistor and thus opposes the positive-input signal.

EXAMPLE 12-3.   We will analyze the circuit of Fig. 12-12 to determine the effects of feedback on gain and input impedance. The equivalent circuit for Fig. 12-12 is shown in Fig. 12-14.

From Chapter 9 the voltage gain of the common-emitter amplifier without the feedback across R_f is

$$A_{ve} = \frac{-h_{fe}R_L}{h_{ie}(1 + h_{oe}R_L) - h_{re}h_{fe}R_L}$$

and its input impedance is

$$Z_{\text{in}} = h_{ie} - \frac{h_{re}h_{fe}R_L}{1 + h_{oe}R_L}$$

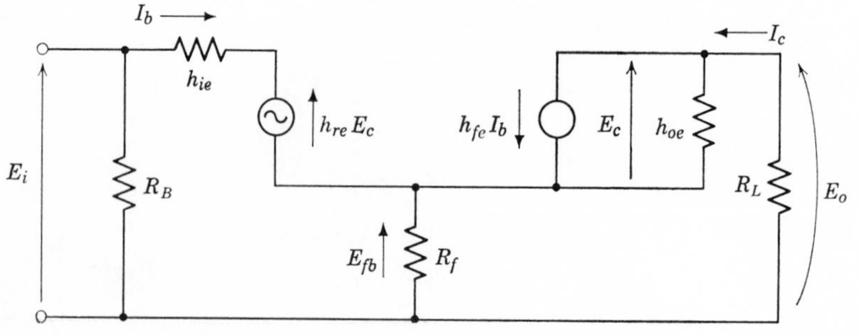

Fig. 12-14

We will calculate the voltage gain of the circuit by first writing the equations that define its operation, based on the equivalent circuit:

$$E_i = h_{ie}I_b + h_{re}E_c + E_{fb} \tag{1}$$

$$E_{fb} = (I_b + I_c)R_f \tag{2}$$

$$E_c = \frac{1}{h_{oe}}(I_c - h_{fe}I_b) \tag{3}$$

$$E_o = E_{fb} + E_c \tag{4}$$

$$E_o = -I_c R_L \tag{5}$$

By eliminating E_{fb}, E_c, and E_o, we form a determinant from these equations:

$$E_i = I_b h_{ie} + h_{re}\left[\frac{I_c - h_{fe}I_b}{h_{oe}}\right] + (I_b + I_c)R_f$$

$$0 = (I_b + I_c)R_f + \left[\frac{I_c - h_{fe}I_b}{h_{oe}}\right] + I_c R_L$$

$$\begin{array}{cc} I_b & I_c \end{array}$$

$$\begin{vmatrix} h_{ie} - \dfrac{h_{re}h_{fe}}{h_{oe}} + R_f & \dfrac{h_{re}}{h_{oe}} + R_f & E_i \\[3mm] -\dfrac{h_{fe}}{h_{oe}} + R_f & \dfrac{1}{h_{oe}} + R_f + R_L & 0 \end{vmatrix}$$

The output voltage is

$$E_o = -I_c R_L = -\frac{\begin{vmatrix} h_{ie} - \dfrac{h_{re}h_{fe}}{h_{oe}} + R_f & E_i \\[3mm] -\dfrac{h_{fe}}{h_{oe}} + R_f & 0 \end{vmatrix}}{\begin{vmatrix} h_{ie} - \dfrac{h_{re}h_{fe}}{h_{oe}} + R_f & \dfrac{h_{re}}{h_{oe}} + R_f \\[3mm] -\dfrac{h_{fe}}{h_{oe}} + R_f & \dfrac{1}{h_{oe}} + R_f + R_L \end{vmatrix}} \times R_L$$

and the feedback gain is, after considerable manipulation,

$$A_{ve}' = \frac{E_o}{E_i} = -\frac{(h_{fe}R_L)\left(\dfrac{1}{h_{oe}}\right)\left(1 - \dfrac{R_f h_{oe}}{h_{fe}}\right)}{\dfrac{1}{h_{oe}}[h_{ie}(1 + h_{oe}R_L) - h_{re}h_{fe}R_L + R_f(1 + h_{oe}R_L + h_{oe}h_{ie} - h_{re}h_{fe} - h_{re} + h_{fe})]} \tag{6}$$

For $R_f = 0$, this reduces to the expression for A_{ve}. We should put this in the form of Equation (12-3):

$$A_{ve}' = \frac{A_{ve}}{1 - \beta A_{ve}} = \frac{N/D}{1 - \beta(N/D)}$$

where N is the numerator and D the denominator of the expression for A_{ve}, without feedback. If we note the similarities of expressions for A_{ve} and A_{ve}', we may write A_{ve}' in a "shorthand" notation as

$$A_{ve}' = \frac{NB}{D + C}$$

or

$$= \frac{N/D}{1 + \left(\dfrac{C - BD + D}{NB}\right)\left(\dfrac{N}{D}\right)} \tag{7}$$

Then

$$\beta = -\left(\frac{C - BD + D}{NB}\right)$$

Substituting the circuit parameters from Equation (6) yields for A_{ve}', after considerable manipulation,

$$A_{ve}' = -\frac{\dfrac{h_{fe}R_L}{h_{ie}(1 + h_{oe}R_L) - h_{re}h_{fe}R_L}}{1 - \left[+ \dfrac{R_f}{R_L}\left(1 + \dfrac{1 + h_{oe}R_L}{h_{fe}}\right)\left(\dfrac{(h_{fe} + \Delta_h)}{1 - h_{oe}R_f}\right)\right]\left[\dfrac{h_{fe}R_L}{h_{ie}(1 + h_{oe}R_L)} - h_{re}h_{fe}R_L\right]}$$

where $\Delta_h = (h_{oe}h_{ie} - h_{re}h_{fe})$. Thus

$$\beta = + \frac{R_f}{R_L}\left(\frac{h_{fe} + \Delta_h}{1 - h_{oe}R_f}\right)\left(1 + \frac{1 + h_{oe}R_L}{h_{fe}}\right)$$

In a practical circuit, $h_{oe}R_L \ll 1$ and,

$$\beta \approx + \frac{R_f}{R_L}(h_{fe} + \Delta_h) \tag{8}$$

The input impedance is $Z_{in}' = E_i/I_b$, which may be calculated from the determinant above Equation (6):

$$Z_{in}' = \frac{E_i}{I_b} = \frac{\Delta_h}{\dfrac{1}{h_{oe}}(1 + h_{oe}R_L + h_{oe}R_f)}$$

$$= \frac{\left(h_{ie} - \dfrac{h_{re}h_{fe}}{h_{oe}} + R_f\right)(1 + h_{oe}R_f + h_{oe}R_L) + \left(\dfrac{h_{fe}}{h_{oe}} - R_f\right)(h_{re} + h_{oe}R_f)}{1 + h_{oe}R_L + h_{oe}R_f}$$

$$= \frac{[h_{ie}(1 + h_{oe}R_L) - h_{re}h_{fe}R_L] + R_f(1 + h_{oe}R_L - h_{re}h_{fe}) + R_f(h_{oe}h_{ie} - h_{re} + h_{fe})}{[1 + h_{oe}R_L] + h_{oe}R_f}$$

The bracketed terms are the input impedance *without feedback*. We note that Z_{in}' can be larger than Z_{in}, if the numerator increases more than the denominator. For the values of *h*-parameters expected for a practical transistor and circuit element values expected, e.g., $R_L \gg R_f$, $1/h_{oe} \gg R_f$, the input impedance with feedback may be approximated:

$$Z_{in}' \approx h_{ie} - \frac{h_{re}h_{fe}R_L}{1 + h_{oe}R_L} + \frac{R_f(1 + h_{fe})}{1 + h_{oe}R_L} \qquad (9)$$

A 2N334 transistor may have these typical values of common-emitter parameters: $h_{ie} = 2 \times 10^3 \, \Omega$, $h_{oe} = 16 \, \mu$mho, $h_{fe} = 30$, $h_{re} = 10^{-3}$. We will calculate the effect of R_f on voltage gain when the feedback resistor is increased from 0 to $100 \, \Omega$, and its effect on input impedance. For these calculations we will assume the load is $R_L = 5 \, k\Omega$.

The voltage gain without feedback is

$$A_{ve} = \frac{-h_{fe}R_L}{h_{ie} - R_L \Delta h}$$

where Δ_h is defined as $(h_{ie}h_{oe} - h_{fe}h_{re})$

$$A_{ve} = \frac{-30 \times 5 \times 10^3}{2 \times 10^3 - 5 \times 10^3(2 \times 10^3 \times 16 \times 10^{-6} - 30 \times 10^{-3})}$$

$$= \frac{-150 \times 10^3}{2 \times 10^3 - 10} = -\frac{150,000}{1990}$$

$$= -75.4$$

From Equation (8), when $R_f = 100 \, \Omega$

$$\beta \approx \frac{R_f}{R_L} (h_{fe} + \Delta_h)$$

$$\approx \frac{100}{5 \times 10^3} (30 + 2 \times 10^{-3})$$

$$\approx 0.6$$

The voltage gain will be reduced because of the feedback

$$A_{ve}' = \frac{A_{ve}}{1 - \beta A_{ve}}$$

$$= \frac{-75.4}{1 - 0.6(-75.4)}$$

$$= -\frac{75.4}{46.24}$$

$$= -1.63$$

The emitter resistor has a significant effect on the voltage gain of the common-emitter amplifier. Even if R_f were increased to only 2 Ω, the feedback would affect the gain:

$$\beta \approx \frac{2 \times 30}{5 \times 10^3} = 0.012$$

and the gain would reduce to

$$A_{ve}' = \frac{-75.4}{1 + 0.012(75.4)}$$

$$= \frac{-75.4}{1.905}$$

$$= -39.5$$

A resistor is often used in the emitter circuit to stabilize circuit operation, occurring because of the negative feedback. However, the value of the emitter resistor will be exceedingly small compared with other circuit components, because of its great effect on amplifier gain.

The input impedance without feedback ($R_f = 0$) may be calculated

$$Z_{\text{in}} = h_{ie} - \frac{h_{re}h_{fe}R_L}{1 + h_{oe}R_L}$$

$$= 2 \times 10^3 - \frac{30(10^{-3})(5 \times 10^3)}{1 + 16 \times 10^{-6}(5 \times 10^3)}$$

$$= 2 \times 10^3 - \frac{150}{1 + 80 \times 10^{-3}}$$

$$Z_{\text{in}} \approx 2000 - 150 = 1850 \ \Omega$$

Because of the feedback in the emitter circuit when $R_f = 100 \ \Omega$, the input impedance will increase, according to Equation (9)

$$Z_{\text{in}}' \approx Z_{\text{in}} + \frac{R_f(1 + h_{fe})}{1 + h_{oe}R_L}$$

$$= 1850 + \frac{100(1 + 30)}{1 + 16 \times 10^{-6}(5 \times 10^3)}$$

$$Z_{\text{in}}' \approx 1850 + 3100 = 4950 \ \Omega$$

DRILL PROBLEMS

D12-5 A transistor having these parameter values is used in a common-emitter amplifier circuit that has a voltage gain of 100 without feedback: $h_{ie} = 500 \ \Omega$, $h_{re} = 3 \times 10^{-4}$, $h_{fe} = 200$, $h_{oe} = 15 \ \mu$mho. The load is $R_L = 5$ kΩ and emitter feedback resistor $R_f = 100 \ \Omega$. Calculate the voltage gain with feedback and the value of β.

D12-6 For the amplifier of D12-5, calculate the change in input impedance caused by R_f.

12.6   Feedback in Magnetic Amplifiers

Before discussing feedback in magnetic amplifiers, it is desirable to examine further the theory of magnetic amplifiers without feedback. The circuit of Fig. 12-15, in which variations in the amount of a small direct current can change the value of a reactance in series with a load, is the basic magnetic amplifier circuit. Rectification of the alternating current after it passes through the reactor makes possible a d-c output which can be used as the d-c control for a second stage of the magnetic amplifier. The output of the second stage may be used as *controlled alternating current*, or it may be rectified and become the *controlling direct current* in a second stage.

Feedback in magnetic amplifiers may be either *internal* or *external*. Modern practice requires that maximum gain be obtained in a single stage, where possible. Feeding part of the output power back into the input circuit in proper phase so that the control input is assisted is *positive* or *regenerative* feedback. It increases the gain of the amplifier.

Most positive feedbacks are inductively coupled in by a winding which is added to the basic amplifier circuit. Figure 12-15 is the circuit of a magnetic amplifier with positive feedback and supplying controlled a-c power to a load.

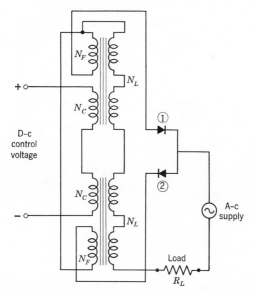

Fig. 12-15   Magnetic amplifier with a-c load and external positive feedback.

For external feedback a third winding N_F is added to each leg of the reactor and so connected that its flux must aid the control winding flux. The N_F are the feedback windings, the N_C are the control windings and the N_L are the load-circuit windings, sometimes called the *gate windings*.

Assume the a-c supply is momentarily positive at the top. Current flows through rectifier 2, the feedback winding at the bottom, both sections of the load winding, and *to the right* through R_L. On the next half-cycle, current flows *to the left* through R_L, both sections of the load winding, the feedback winding at the top, and then through the top rectifier 1. The current paths of the d-c output amplifier of Fig. 12-16 are easily traced, and it will be found that the current flows only to the right in the load.

There are external-feedback circuits in which a transformer with a center-tapped secondary supplies the amplifier and load, and others in which a bridge-type full-wave rectifier can be arranged to deliver direct current to one load while another load is receiving alternating current, both subject to the same d-c control, however.

Control characteristics of the magnetic amplifier with positive external feedback are shown in Fig. 12-17. Since current gain is the ratio of I_L to I_C, it is evidently increased by adding positive feedback. Furthermore, we should expect the gain to increase when the *feedback ratio*, N_F/N_L, increases.

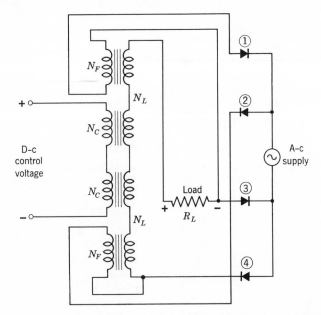

Fig. 12-16   Magnetic amplifier with d-c load and external positive feedback.

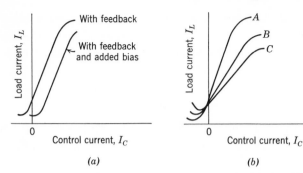

Fig. 12-17 Control characteristics of magnetic amplifiers with positive feedback. In (*b*) *A* represents largest feedback ratio and *C* smallest feedback ratio.

N_F and N_L are the numbers of turns on the feedback and load circuit windings, respectively.

Current gain, A_C, in a magnetic amplifier without feedback, is defined as

$$A_C = \frac{I_L}{I_C} = \frac{N_C}{N_L} \tag{12-16}$$

where N_C is the number of turns on the control winding and I_L is the average load current over a half-cycle. From Equation (12-16),

$$N_C I_C = N_L I_L \tag{12-17}$$

When feedback is added,

$$N_C I_C + N_F I_F = N_L I_L$$
$$N_C I_C = N_L I_L - N_F I_F \tag{12-18}$$

$$N_C I_C = N_L I_L (1 - N_F I_F / N_L I_L) \tag{12-19}$$

$$A_C = \frac{I_L}{I_C} = \frac{N_C}{N_L} \left(\frac{1}{1 - N_F I_F / N_L I_L} \right) \tag{12-20}$$

In Figs. 12-15 and 12-16 the load windings are series-connected and in series with N_F, so that $I_F = I_L$, which reduces Equation (12-20) to

$$A_C = \frac{N_C}{N_L} \left(\frac{1}{1 - N_F / N_L} \right) \tag{12-21}$$

N_F / N_L is the feedback ratio. If it is made equal to unity, the gain is *theoretically* infinite, as in the case of positive-feedback tube and transistor amplifiers. This is not, of course, a practical possibility. If $N_F = 0$, the gain is that without feedback, as expected. When the feedback is *negative*, or *degenerative*, in which case the feedback ampere-turns oppose the control ampere-turns, the gain is reduced.

Voltage gain, A_v, of a series-connected magnetic amplifier is derived as follows.

$$A_v = \frac{E_L}{E_C} = \frac{R_L I_L}{R_C I_C} = \frac{N_C R_L}{N_L R_C} \tag{12-22}$$

where voltages and currents are average values. R_C is the sum of the resistances of control windings and the series resistance regulating the control current.

Direct-current power gain, $A_{Pd\text{-}c}$ is

$$A_{Pd\text{-}c} = \frac{E_L I_L}{E_C I_C} = \frac{N_C^2 R_L}{N_L^2 R_C} \tag{12-23}$$

where E_L and I_L are quantities at the load after rectification and filtering, if any.

Alternating-current power gain, $A_{Pa\text{-}c}$ is the a-c power at the load divided by the d-c power in the control circuit. It may be expressed in terms of turn and resistance ratios, if the so-called *form factor*, F, of the a-c wave in the load is determined.

$$A_{Pa\text{-}c} = \frac{N_C^2}{N_L^2} \frac{R_L}{R_C} F^2 \tag{12-24}$$

Form factor is the ratio of the *effective* to the average value of a half-cycle of an a-c wave. It is equal to 1.11 for a pure sine-wave form.

Equations (12-22) to (12-24) are gains *without* feedback. The gains *with* feedback have the added factor $1/(1 - N_F/N_L)$. Voltage gain of a *feedback magnetic amplifier* is

$$A_{vf} = \frac{N_C R_L}{N_L R_C}\left(\frac{1}{1 - N_F/N_L}\right) \tag{12-25}$$

Direct-current power gain of a *feedback magnetic amplifier* is

$$A_{Pd\text{-}cf} = \frac{N_C^2 R_L}{N_L^2 R_C}\left(\frac{1}{1 - N_F/N_L}\right)^2 \tag{12-26}$$

Alternating-current power gain of a *feedback* magnetic amplifier is

$$A_{Pa\text{-}cf} = \frac{N_C^2}{N_L^2} \frac{R_L}{R_C} F^2 \left(\frac{1}{1 - N_F/N_C}\right)^2 \tag{12-27}$$

A magnetic amplifier with externally supplied regenerative feedback may be controlled by alternating current. See Fig. 12-16. The control voltage (a-c instead of d-c) is applied to the control windings in series, and all the other windings, with R_L in series, are served from the a-c power source through a bridge-type rectifier. The connections are such that the load current flows through the feedback windings in the same direction for both polarities of the power source that supplies the load. It is not necessary that the control voltage and power-source voltage be in phase, but when they are the load voltage is a maximum. This is fortunate because only one pair of input terminals is needed, and, as a consequence, no phase-shift network is required to establish maximum output conditions.

12.7   Bias in an External-Feedback Magnetic Amplifier

Figure 12-17a shows that when the control current is zero the load current has an appreciable value. This quiescent current should be minimized. It can be reduced to only one or two per cent of the value it has when no bias is used. The principle is simply to introduce a core flux in opposition to the flux produced by the quiescent current in the feedback windings. This has the effect of demagnetizing the core almost completely and amounts to moving the characteristic to the right until its lowest point is on the load-current axis just slightly above the origin. The slope of the straight part of the curve is unchanged.

It would be a simple matter to add bias to the circuit of Fig. 12-16. A bias winding would be put on each core, and they would be supplied with current in the proper direction from the terminals of R_L through a controlling series resistance. For the a-c load circuit, additional rectifiers would be needed.

12.8   Negative Feedback in Magnetic Amplifiers

Negative-feedback magnetic-amplifier circuits use internal feedback and they do not require separate feedback windings. Figures 12-18 and 12-19 are full-wave internal-feedback circuits, the first supplied by a transformer with a center-tapped secondary and the second through a bridge rectifier. Their operating features are identical. The rectifiers cause each load winding to conduct current on alternate half-cycles. The d-c component of the half-wave of current added to the d-c control current produces near saturation flux in the core, reducing the reactor impedance to a low value. This happens in each load winding in the alternate half-cycles. The d-c component of the half-wave constitutes the feedback.

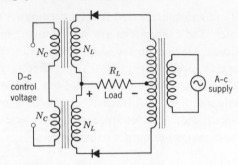

Fig. 12-18   Full-wave magnetic amplifier with internal feedback.

The bridge circuit is more reliable and more stable when the load is inductive, because counter EMF's built up in an inductive load would find a path through the load windings in the circuit containing a transformer. This would have undesirable effects on gain and stability.

Magnetic amplifiers with either external or internal feedback may be operated in push-pull. The push-pull circuits have advantages comparable to those of push-pull tube and transistor amplifiers when comparisons are made with single-ended circuits.

12.9   The Tuned-Plate Oscillator

Electronic-tube circuits have largely displaced rotating machines as generators of alternating voltages, except at very low frequencies. They have even been used in place of motor-generator sets in the "inversion" of d-c power to a-c power at 60 Hz.

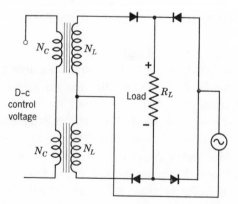

Fig. 12-19   Same as Fig. 12-18, but bridge rectifier replaces transformer.

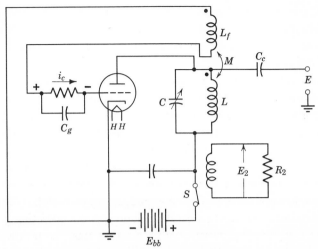

Fig. 12-20   Tuned-plate oscillator.

The advantages of vacuum-tube circuits used to generate alternating currents and voltages are: (1) the frequency may be changed readily over wide ranges; (2) no driving unit is needed; (3) first cost is comparatively low, weight and size small; (4) high voltages are easily handled. Some disadvantages are: (1) a d-c power source is needed (readily available for inverter operation, of course); (2) power capacity is limited; (3) tubes are subject to burnout; (4) efficiency is low.

Electronic oscillators have extensive application: as test and service instruments; in radio transmitters and receivers; in industrial service for high-frequency heating; and where high-frequency power is needed.

If the grid excitation voltage of a Class C amplifier is supplied in proper phase by transformer action from the tank circuit, the amplifier will oscillate. This means that surges of plate current continue to flow, the surges being maintained by energy transferred from the plate circuit to the grid circuit.

Large currents surge through the tank elements (L and C) of Fig. 12-20. They will readily produce an alternating voltage (E_2) in the coil below L, and power will be delivered to the load (R_2). The top of the tank circuit will rise well above—and fall equally far below—ground potential during each cycle of oscillation. If a capacitor is connected to the top of the tank as shown, an alternating output voltage (E) is obtained. The frequency of E and of E_2 is, of course, the frequency of oscillation. This frequency is given very closely by

$$f = \frac{1}{2\pi\sqrt{LC}} \qquad (12\text{-}28)$$

in which L is in henrys, and C is in farads. The equivalent-series resistance (R_s) of the tank circuit must be such that $2\pi fL$ is large compared to R_s. R_s is made up of the resistance of the coil L and some resistance coupled in from the load.

The amount of coupling between the feedback coil (L_f) and L may be made too small to produce oscillations. If it is made larger than necessary, excessive grid current may flow. The amount of coupling (mutual inductance between L_f and L) depends on the plate resistance and transconductance of the tube and on the tank-circuit constants (L, C and R). In fact, when an oscillator begins to operate, its grid bias automatically increases until the decreasing transconductance of the tube is just right to allow oscillations to continue at low grid current. The transconductance is the slope of the dynamic transfer characteristic at the operating point. As the operating point slides down the transfer characteristic with increasing bias, the transconductance becomes smaller.

Self-bias is used in feedback oscillator circuits. It allows oscillations to start and to build up. It is self-adjusting and thus helps to protect the tube from excessive plate current and allows variations in average plate current to accommodate changes in load.

12.10   Physical Description of the Start of Oscillations in a Feedback Oscillator

The starting up of a feedback oscillator is generally of interest. Assume that the switch S in the $B+$ lead in Fig. 12-20 is open but that the heater current has brought the cathode of the tube up to operating temperature. All other currents are zero. From theory studied earlier we know that, upon closing the switch, plate current goes right on through the capacitor C and that the initial current through L is zero. See Fig. 12-21. The tube is immediately driven to temperature saturation and the top of the current wave becomes flat. *Immediately* after the switch S in Fig. 12-20 is closed, the current in L *begins to build up* and the current through C *begins to decrease*. The magnetic flux linking L_f induces a voltage in that coil which makes the grid positive and causes grid current to flow. At first the grid current flows mostly through the grid-leak capacitor C_g, but as it becomes charged it establishes negative bias on the tube and the grid current begins to fall. This causes the plate current to decrease, and the decreasing magnetic field due to the decreasing current in L induces a *reverse* voltage in L_f which sends

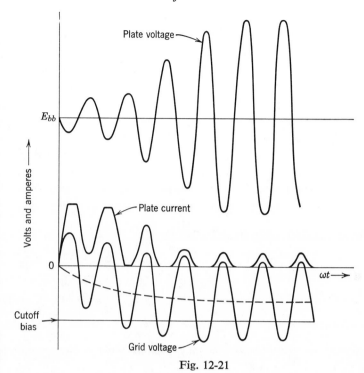

Fig. 12-21

negative potential to the grid on top of the negative bias already created by C_g.

The plate current decreases rapidly with an accompanying rise of plate potential. The negative-grid potential approaches cutoff value, but let us assume that it is not reached until the next cycle of grid voltage. Actually there may be several cycles in which the current goes from saturation value to zero, giving a waveform that is flat at both top and bottom. Soon, however, the situation settles down to that represented at the time of the fifth current pulse (in Fig. 12-21) and thereafter.

The oscillatory nature of the plate-voltage wave indicates that the capacitor of the tank is reversing polarity with periodic regularity in accordance with the natural behavior of a parallel-resonant circuit. Alternating current in L induces the alternating voltage in L_f, maintaining the large amplitude of grid-voltage swing, thus keeping the oscillations going. During the short time interval when the grid potential is in the tube conduction region and also while the grid is so briefly positive, the plate voltage is seen to be very low. This accounts for lower plate-current values than we might expect to find. Furthermore, the decrease in operating transconductance of the tube with the large increase in negative grid bias is conducive to stable operation with limitation of plate current.

The *tuned-grid* oscillator has the tank circuit connected to the grid instead of to the plate. Energy from the plate circuit is coupled into the tank coil by means of a small coil in series with the plate. The load receives its energy from the tank by means of inductive or capacitive coupling, as in the case of the tuned-plate oscillator.

A combination of the two tank-circuit arrangements just described makes up the *tuned-plate tuned-grid* oscillator; thus there are two parallel-resonant circuits, one connected to the grid and one to the plate of the tube.

12.11   Criterion for Sustained Oscillations

Figure 12-22 is a block diagram of a feedback oscillator with gain $A_v = E_2/E_1$ and a feedback voltage $E_f = \beta E_2$. Since E_1 is the required input voltage to produce the output voltage E_2, the following is a requirement for maintaining operation:

$$\beta E_2 = E_f = E_1 \tag{12-29}$$

This means that the feedback network must produce this required magnitude and its phase must be the same as the phase of E_1. This can be achieved in a practical circuit, so we can write

$$A_v = \frac{E_2}{\beta E_2} = \frac{1}{\beta}$$
$$\beta A_v = 1 \tag{12-30}$$

a requirement for sustained oscillation.

Since β can never be greater than unity, A_v must not be less than unity in a feedback oscillator.

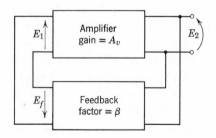

Fig. 12-22   Block diagram of a feedback oscillator.

Using the familiar gain expression for a single-stage amplifier,

$$A_v = \frac{-\mu Z_L}{r_p + Z_L}$$

in Equation (12-30), we obtain

$$\frac{-\mu \beta Z_L}{r_p + Z_L} = 1$$

from which

$$\beta = -\left(\frac{1}{\mu} + \frac{1}{g_m Z_L}\right) \tag{12-31}$$

This is called the *Barkhausen criterion for sustained oscillation*.

Because reactance elements are required in oscillator circuits, we use Z_L instead of R_L in the gain equation. There will be more than 180 degrees phase reversal unless Z is resistive.

Recalling that the condition $\beta A_v = 1$ in the expression for gain with feedback (Equation 12-3) makes it theoretically infinite, we conclude that this has been made impossible by temperature saturation and cutoff conditions in the tube.

Detailed mathematical analysis of feedback-oscillator performance is quite lengthy and involved for presentation here, but a study of the results is pertinent. In the tuned-plate oscillator circuit of Fig. 12-20 let the *mutual inductance* between L and L_f be M, and let R be the resistance of the coil L plus any coupled-in resistance. Analysis shows that

$$-\frac{M}{L} = -\left(\frac{1}{\mu} + \frac{1}{g_m R_L}\right) \tag{12-32}$$

in which R_L represents the impedance of the tank at resonant frequency with the load on the oscillator already coupled in $R_L = L/RC$. The circuit will satisfy this *Barkhausen criterion if M is negative*. M/L is recognized as equal to β.

The expression for the natural oscillation frequency also comes out of the analysis:

$$= \frac{1}{2\pi}\left[\frac{1}{LC}\left(1 + \frac{R}{r_p}\right)\right]^{\frac{1}{2}} \tag{12-33}$$

In practice R is small with respect to r_p, so the operating frequency is very close to the resonant frequency of the tuned circuit alone. Because R includes the coupled-in load resistance, the operating frequency is a function of the loading on the oscillator. Consequently, it is necessary to avoid variable loading effects if good frequency stability is required.

EXAMPLE 12-4. Calculate the necessary conditions and frequency of oscillation in the tuned-collector oscillator, shown in Fig. 12-23.

Set up circuit equations

$$E_i = I_b h_{ie} + h_{re} E_c \tag{1}$$

$$I_c = I_b h_{fe} + h_{oe} E_c \tag{2}$$

For the transformer

$$E_c = (j\omega L_1 + R_1) I_1 - j\omega M I_2 \tag{3}$$

$$E_2 = -j\omega M I_1 + (j\omega L_2 + R_2) I_2 \tag{4}$$

The currents I_1 and I_2 are

$$I_1 = -I_c - j\omega C E_c \tag{5}$$

$$I_2 = -\frac{E_2}{Z_i} = -\frac{E_2}{h_{ie}} \tag{6}$$

Substituting (2), (5), and (6) into (3) and (4) yields

$$E_c = -(R_1 + j\omega L_1)(I_b h_{fe} + h_{oe} E_c + j\omega C E_c) + j\omega M \left(\frac{E_2}{h_{ie}}\right) \tag{7}$$

$$E_2 = j\omega M (I_b h_{fe} + h_{oe} E_c + j\omega C E_c) - (R_2 + j\omega L_2)\left(\frac{E_2}{h_{ie}}\right) \tag{8}$$

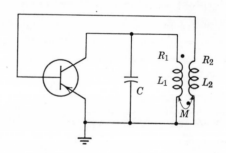

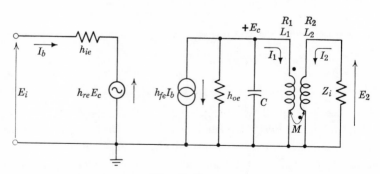

Fig. 12-23

and with (1)

$$E_i = I_b h_{ie} + h_{re} E_c \tag{1}$$

a determinant may be written for the coefficients of E_2, E_c, and I_b.

	E_2	E_c	I_b	
Equation (1):	0	h_{re}	h_{ie}	E_i
Equation (7):	$\dfrac{j\omega M}{h_{ie}}$	$\begin{array}{c}-1 - (R_1 + j\omega L_1) \\ \times (h_{oe} + j\omega C)\end{array}$	$-h_{fe}(R_1 + j\omega L_1)$	0
Equation (8):	$-1 - \dfrac{R_2 + j\omega L_2}{h_{ie}}$	$j\omega M(h_{oe} + j\omega C)$	$j\omega M h_{fe}$	0

We then calculate the value of the determinant D; after a lengthy manipulation, it is

$$
\begin{aligned}
D = &-h_{ie} - R_2 + R_1 \Delta \left(1 - \frac{R_2}{h_{ie}}\right) \\
&+ \omega^2 \left[\frac{\Delta}{h_{ie}} (L_1 L_2 - M^2) + C(h_{ie}L_1 + R_2 L_1 + R_1 L_2)\right] \\
&+ j\omega \left[\omega^2 (L_1 L_2 - M^2)C + L_2 \left(R_1 \frac{\Delta}{h_{ie}} - 1\right)\right]
\end{aligned}
$$

where $\Delta = h_{ie}h_{oe} - h_{fe}h_{re}$.

Similarly, we calculate E_2:

$$E_2 = -\frac{j\omega M h_{fe} E_i}{D}$$

The Barkhausen criterion (Section 12.11) requires that $E_2/E_1 = 1$, so we may set the real parts equal and the imaginary parts equal. The real parts determine the frequency of oscillation, and the imaginary parts set the conditions necessary for sustained oscillation. Thus, from the real part

$$
\begin{aligned}
0 = &-h_{ie} - R_2 + R_1 \Delta \left(1 - \frac{R_2}{h_{ie}}\right) \\
&+ \omega^2 \left[\frac{\Delta}{h_{ie}} (L_1 L_2 - M^2) + C(h_{ie}L_1 + R_1 L_2 + R_2 L_1)\right]
\end{aligned}
$$

The frequency of oscillation is

$$f = \frac{1}{2\pi} \left[\frac{h_{ie} + R_2 + R_1 \Delta \left(\dfrac{R_2}{h_{ie}} - 1\right)}{\dfrac{\Delta}{h_{ie}} (L_1 L_2 - M_2) + C(h_{ie}L_1 + R_1 L_2 + R_2 L_1)}\right]^{\frac{1}{2}}$$

In a practical circuit, the resistances R_1 and R_2 will be small compared to the reactances of L_1 and L_2, i.e., the Q will be high. Further, if the two windings are closely coupled ($k \approx 1.0$), since $M = k\sqrt{L_1 L_2}$, the frequency may be approximated by

$$f \approx \frac{1}{2\pi} \left(\frac{1}{L_1 C} \right)^{\frac{1}{2}} \tag{9}$$

Thus, the oscillation frequency will be determined by the tuned circuit in the collector and primary of the transformer.

The condition for oscillation derives from the imaginary parts:

$$-j\omega M h_{fe} = j\omega \left[\omega^2 (L_1 L_2 - M^2) C + L_2 \left(R_1 \frac{\Delta}{h_{ie}} - 1 \right) \right]$$

Substituting from Equation (9) yields,

$$-M h_{fe} = \frac{1}{L_1 C} (L_1 L_2 - M^2) C + L_2 \left(R_1 \frac{\Delta}{h_{ie}} - 1 \right)$$

Assuming R_1 small compared to the reactance of L_1, as before, this reduces to:

$$h_{fe} = \frac{M}{L_1} \tag{10}$$

If the circuit is to oscillate at a frequency of 1 MHz, using an *R-F* transformer having $L_1 = 1$ mH, the required value of C may be calculated by Equation (9), assuming high Q and close coupling in the transformer:

$$f = \frac{1}{2\pi} \left(\frac{1}{L_1 C} \right)^{\frac{1}{2}}$$

$$\sqrt{C} = \frac{1}{2\pi f} \left(\frac{1}{L_1} \right)^{\frac{1}{2}}$$

$$= \frac{1}{2\pi \times 10^6} \left(\frac{1}{10^{-3}} \right)^{\frac{1}{2}}$$

then

$$C = \frac{1}{4\pi^2 \times 10^{12}} \left(\frac{1}{10^{-3}} \right)$$

$$\approx 25 \text{ pF}$$

DRILL PROBLEMS

D12-7 A tuned-collector oscillator is designed to have a variable frequency from 200 kHz to 1 MHz, using an *R-F* transformer having $L_1 = 1$ mH. Assuming high Q and close coupling, determine the necessary variable range of C.

D12-8 In the oscillator of D12-7, the coefficient of coupling between primary and secondary windings of the transformer is $k = 0.95$ and $L_2 = 10$ mH. (*a*) If the transistor has $h_{fe} = 10$ or more, will the circuit generate oscillations? (*b*) As h_{fe} is increased, say by using a different transistor in the circuit, what must be done at the transformer to continue oscillations?

12.12 Other Feedback Oscillator Circuits

There are other arrangements of the L, C, and R components which produce sustained oscillations and which have been named after their originators. The Hartley oscillator, which is very popular in applications where substantial radio-frequency power is needed, as in radio transmitters, induction and dielectric heating, and remote-control systems, is shown in two forms in Fig. 12-24. RFC is a *radio-frequency choke* which prevents the power supply from short-circuiting the output R-F energy to ground.

The Colpitts oscillator is also well known. It is so similar to the Hartley shunt-feed circuit, Fig. 12-24b, that it will not be illustrated here. Instead, we refer to Fig. 12-24b and state that to convert it to the Colpitts oscillator we (1) remove the variable capacitor and replace it with an inductance coil, and (2) remove the center-tapped inductance coil and replace each half with a variable capacitor so that the wire from the grounded cathode goes to the junction point between the two capacitors.

Improved frequency stability is an important characteristic of the *electron-coupled* oscillator in Fig. 12-25. It will be recalled that changes in load conditions contribute to instability of frequency, since the load resistance is in the exact equation that defines the frequency value. If the load section of the oscillator can be isolated from the main oscillating resonant circuit, the effects of load change will be greatly reduced.

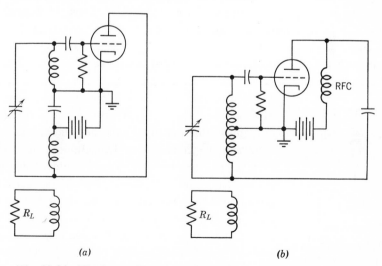

(a)   (b)

Fig. 12-24 Hartley oscillator circuits. (a) Series feed; (b) shunt feed.

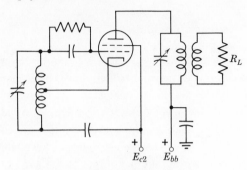

Fig. 12-25   Electron-coupled oscillator.

In Fig. 12-25 the cathode, control grid, and R-F grounded screen grid serve as the three elements of the tube in the Hartley circuit of Fig. 12-24*b*. During the oscillations, electrons pass through the screen to the plate of the tube and set up a voltage across the tuned circuit connected to the plate. Hence the name *electron-coupled*. The screen potential determines the plate current in a well-screened tetrode (or a pentode with its suppressor and screen grids tied together), and changes of load have a negligible effect on the grid circuit containing the frequency-determining resonant circuit. The tube actually amplifies the a-c voltage between the control grid and cathode.

Crystal control of the oscillating frequency is a still more effective means of acquiring stability of frequency. A small, flat plate of quartz, properly cut, has the property of developing positive and negative charges of electricity on opposite faces if mechanical stress is applied to another pair of faces. And, conversely, if an alternating voltage is applied to flat electrodes in contact with the opposite faces of the crystal plate, the plate will vibrate mechanically, There is a resonant frequency of mechanical vibration at which the vibrations have appreciable amplitude. This *natural frequency* is almost constant. Effects of temperature change can be eliminated by mounting the crystal in a temperature-controlled oven when extremes of stability are required, as in radio and television transmitters. Certain orientations of the axes of the natural crystal make possible a temperature coefficient that is zero over a wide range. When all possible measures are taken in an effort to keep the oscillating frequency constant, a 100-kHz crystal oscillator can be made

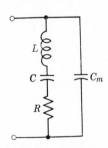

Fig. 12-26   Equivalent circuit of piezoelectric crystal.

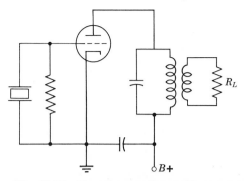

Fig. 12-27   Crystal-controlled oscillator.

so stable in frequency that it will not vary as much as 1 cycle in a million over long periods of time.

A piezoelectric crystal with plated electrodes has an equivalent circuit, as shown in Fig. 12-26. C_M is the capacitance of the metallic mounting, which may be metal surfaces plated on the crystal. The Q of the crystal equivalent circuit is extremely high—20,000 in the lower range—and it may go as high as half a million when special care is taken in the mounting. This is the main reason for such excellent frequency stability.

The circuit of Fig. 12-27 is actually that of a tuned-plate-tuned-grid oscillator with the crystal and holder forming the resonant circuit that determines the frequency. Feedback in this type of oscillator circuit is through the grid-to-plate internal capacitance of the tube.

12.13   Resistance-Capacitance Oscillators

These oscillators are used as *signal generators* from the lowest audio range up to around 10 MHz. Before studying the circuit, we should consider the basic *phase-shift oscillator* in which sustained oscillations are produced without the aid of inductive-capacitive resonance.

We have found that in a series path containing capacitance and resistance a shift in phase takes place, and the voltage between the terminal common to the two units and either one of the ends has a phase that is different from the phase of the voltage across the ends. In the circuit of Fig. 12-28, it has been found that with three *R-C* networks a total phase shift of 180 degrees can be achieved without losing the signal. This means that the output voltage, which is actually a small fraction of the a-c voltage across R_L, is applied to the grid of the tube in proper phase to maintain oscillation.

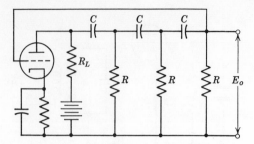

Fig. 12-28 Basic circuit of a phase-shift oscillator.

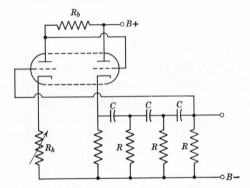

Fig. 12-29 Practical circuit for a phase-shift oscillator.

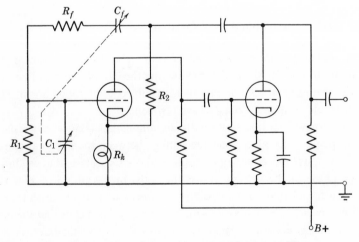

Fig. 12-30 Circuit of resistance-capacitance oscillator.

A practical circuit for a phase-shift oscillator is that of Fig. 12-29. The tube is a twin triode with a large amplification factor, such as a 6SC7 or a 6SL7. The variable R_k adjusts the gain of the first stage so that oscillation may be started or controlled. The second stage is a cathode follower which isolates the phase-shift network from the first stage.

The *resistance-capacitance* oscillator is a two-stage R-C coupled amplifier with special features for feeding voltage back to the first grid through a phase-sensitive network. In Fig. 12-30, R_f and C_f are the phase-sensitive feedback components, and they form a *coupling network* with R_1 and C_1. The net phase shift through the amplifier is zero; therefore it is evident that for oscillation to be maintained the *phase shift through the coupling network must be zero*. It can be shown that this occurs at the frequency given by

$$f = \frac{1}{2\pi\sqrt{R_f C_f R_1 C_1}} \qquad (12\text{-}34)$$

It is practical to make $R_f = R_1$ and $C_f = C_1$. Calling them simply R and C, we have

$$f = \frac{1}{2\pi RC} \qquad (12\text{-}35)$$

For these conditions, $\beta = \frac{1}{3}$ for the feedback and this means the amplifier must have a gain of at least 3.*

In addition to this positive feedback, a *negative* feedback voltage is developed across a cathode resistor R_k which has a special feature that is very valuable here. The required ohms value of R_k *increases rapidly* as its current increases, and therefore the negative bias increases *much more rapidly than the current*. An ordinary 3-W 115-V light bulb with a tungsten filament may be used as R_k. Its resistance depends on temperature, whether developed by $I_{\text{d-c}}^2 R_k$ or $I_{\text{eff}}^2 R_k$.

A change in amplitude, and therefore in effective value, of signal current through R_k will change the negative feedback and the gain of the amplifier in a direction to oppose the change in signal level. The signal current in R_k has nearly the same effective value as the d-c plate current. The result is extremely stable output voltage, good frequency stability, and excellent output-voltage waveform.

The resistance-capacitance oscillator is capable of *continuous variation of frequency* by means of a single tuning arrangement consisting of C_f and

---

* The proof is given in the Appendix.

C_1 as variable air capacitors mounted on a single shaft carrying a frequency dial. Multiplication of the frequency values of scale divisions of the dial by 10, 100, 1,000, etc., is made possible by switching in different pairs of R_f and R_1 which are fixed resistors.

DRILL PROBLEM

D12-9   The circuit of Fig. 12-30 is used to generate audio frequencies from 20 Hz to 40 kHz in 3 ranges: 20 to 400 Hz, 200 to 4,000 Hz, and 2 to 40 kHz. The frequency is varied by adjusting C_1 and C_f, and the range is chosen by switching in different values for R_1 and R_f. If C_1 and C_f can vary from 100 to 2,000 pF, what are the required values of R_1 and R_f in each of the three ranges, assuming $R_1 = R_f$?

12.14   Feedback Transistor Oscillators

Although the circuits of feedback transistor oscillators are similar in appearance to vacuum-tube oscillator circuits, their operating characteristics differ because the transistor is inherently a low-impedance device when compared with the vacuum tube. When the parallel-resonant circuit (tank) is connected to a transistor terminal, the Q of the tank is reduced too much, so it is frequently desirable to connect the transistor-electrode terminal to a tap somewhat near the R-F ground point of the tank inductor L, and thus leave the bulk of L to maintain a fairly high Q.

Figure 12-31 shows the Hartley circuit with a transistor The Colpitts, and other, oscillators have transistor circuits that are analogous to their vacuum-tube circuits. This is also true of transistor phase-shift, crystal-controlled, and other oscillator circuits.

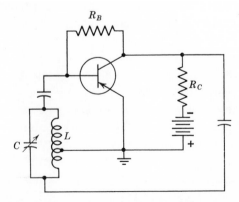

Fig. 12-31   Transistor oscillator, Hartley circuit.

12.15   Negative-Resistance Oscillators

An oscillator circuit that depends on a *negative resistance* appearing across a tuned circuit, rather than on amplification, to supply its input through feedback is an interesting device. A negative resistance is, in effect, a generator of power; a positive resistance is an *absorber* of power, of course.

The plate-characteristic curve of a pentode connected as a tetrode (its screen and suppressor grids tied together) is shown in Fig. 12-32a. The section of the curve between the crest and valley points 1 and 3 exhibits negative resistance, because *when the plate voltage increases* in that region *the current decreases*. The slope of the curve ($\Delta i_b / \Delta e_b$) is *negative* because, as has just been pointed out, when Δe_b is positive, Δi_b is negative. Since $-\Delta i_b / \Delta e_b$ is the reciprocal of resistance, the resistance here must be negative.

To understand how the *dynatron oscillator* works, consider that the Q-point is at 2, the middle point of the negative-resistance portion of the curve. E_{T_o} is the instantaneous voltage of the tank at this instant. A small *increase* in tube current can occur for a number of reasons, such as a slight change in cathode emission, in bias voltage, in tube temperature. This would cause the plate voltage to fall slightly and the load voltage to rise. The fall

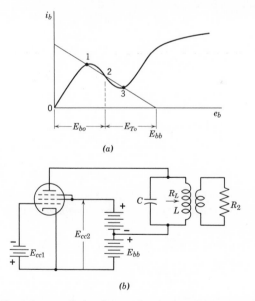

Fig. 12-32   (*a*) Plate characteristic of pentode connected as a tetrode, with d-c load line for tuned circuit of oscillator in (*b*). (*b*) Negative-resistance (dynatron)-oscillator circuit.

in plate voltage causes an additional increase in plate current, the process continuing until the plate current reaches the maximum value at 1, making the load voltage a maximum also. The tank capacitor voltage is at its maximum and its polarity is + at the bottom, − at the top. The capacitor immediately begins to discharge through L, and this raises the plate voltage because the voltage drop across the capacitor is decreasing. This process continues until point 3 is reached, where the load voltage is at its minimum value. Operation of a practical amplifier will carry on to the right of point 3 and to the left of point 1 to new points located by the intersection of the a-c load line and the characteristic curve. The tuned circuit assures good waveform, even though operation extends outside the linear region of the characteristic curve.

12.16   Magnetic-Deflection Current Generator

A saw-tooth voltage wave, which has an important application in providing horizontal deflection in a cathode-ray oscilloscope tube, may be generated and synchronized in an arc-tube relaxation oscillator circuit. In very-high frequency applications, for example in television receiver circuits, a saw-tooth-voltage wave may be obtained conveniently from a blocking oscillator if a pentode tube is used. The control grid and screen grid are made to serve in a Hartley oscillator circuit, and voltage pulses electron-coupled to the plate produce a rectangular wave (very unsymmetrical) at the plate. This is applied to a differentiating circuit to produce a saw-tooth output waveform.

A *saw-tooth current wave* is needed for producing the sweep in *magnetic* deflection. The reason is that the displacement of the cathode spot on the screen is proportional to B, the magnetic flux density, and, to have the spot displaced *more* and *more*, the density must get *larger* and *larger*. In fact, it must increase in direct proportion to time; this means that the *current* producing it must have saw-tooth waveform. This waveform of current is obtainable most easily from a saw-tooth voltage waveform.

The rectangular output voltage of the *multivibrator* is frequently used as the starting waveform for the production of saw-tooth current waves in very-high frequency work. By making the time constants of the R-C circuits, which drive the grids negative, very much unequal, a waveform as shown in Fig. 12-33a may be obtained. When this is fed into the integrating circuit, the saw-tooth voltage (Fig. 12-33c) is formed. The time required for the voltage to decrease to minimum in each cycle may be made very short in comparison with the rise time.

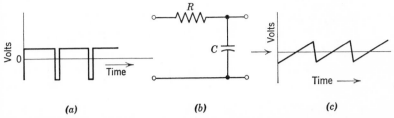

Fig. 12-33   Saw-tooth voltage from multivibrator. (*a*) Unsymmetrical rectangular wave from multivibrator; (*b*) integrating circuit; (*c*) saw-tooth output.

To obtain a current in the magnetic-deflection coil of a cathode-ray tube, whether desired for television or other use, it is necessary to apply a *voltage* which is different from saw-tooth shape because the coil has *inductance* as well as resistance. We already know that the current through a resistance has the same waveform as the applied voltage. An inductance acts differently. Remembering that the current cannot change in zero time but requires time to make a change, we can see without much additional consideration that the saw-tooth *current waveform* in Fig. 12-34*c* requires the unsymmetrical rectangular voltage input shown.

It is reasonable to conclude that, in order to get a saw-tooth current to flow in an *RL circuit*, the magnetic-deflection coil for example, the applied voltage should have a saw-tooth component, and an unsymmetrical rectangular component like the one in Fig. 12-34*b*. Such a waveform is shown in Fig. 12-35*b*. In order to show that this waveshape is obtainable by adding a saw-tooth voltage to the rectangular wave of Fig. 12-34*b*, we reproduce the saw-tooth at reduced amplitude and combine the two in Fig. 12-36, showing how the required *trapezoidal waveform* is synthesized graphically.

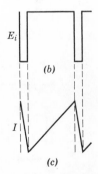

To actually produce this trapezoidal form of voltage wave, only an unsymmetrical rectangular wave, as obtained from a multivibrator, is required if it is applied to the circuit of Fig. 12-37*c* with properly adjusted R and C values. We already understand how E_o is produced in Figs. 12-37*a* and 12-37*b*. All we are doing, then, is adding them together in Fig. 12-37*c* by taking E_o across *both* R_2 and C. If a coil has considerable resistance, more saw-tooth component than rectangular is needed. If the resistance is small, less saw-tooth component is needed.

Fig. 12-34   Voltage and current waveforms of a pure inductance.

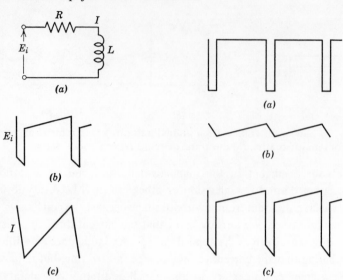

Fig. 12-35 Voltage and current waveforms for saw-tooth current in *RL* circuit.

Fig. 12-36 Combination of saw-tooth wave and rectangular wave produces trapezoidal wave in Fig. 12-37*c*.

12.17 Saw-Tooth Wave Generators

A voltage wave of saw-tooth form is required in the cathode-ray oscilloscope which uses an "electrostatic" sweep to move the electron beam horizontally across the tube. The intensity of the electric field between the deflection plates must increase linearly with time so that the force on the

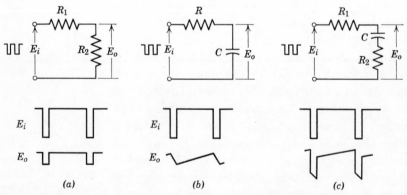

Fig. 12-37 Synthesis of trapezoidal output using combination *R* and *RC* circuit arrangement.

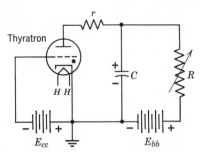

Fig. 12-38 Simplified saw-tooth voltage-generating circuit using an arc-discharge tube (arc-tube relaxation oscillator).

electrons of the beam will cause the spot on the screen to have a lateral component of velocity that is constant. This requires a voltage wave of saw-tooth shape.

The ability of a gas diode and of an *arc-discharge tube* (thyratron) to have *zero current* until the anode voltage reaches a certain positive value and then *suddenly conduct maximum circuit current* is an important feature and a useful one in a saw-tooth voltage generator.

A simplified circuit is shown in Fig 12-38. It is also called an *arc-tube relaxation* oscillator circuit. The action of the circuit is easily understood. When the plate-supply voltage is first connected, the capacitor (C) charges, its voltage rising exponentially as shown in Fig. 12-39. The tube is biased at a fixed value such that a plate voltage (E_b) is required for breakdown. When the capacitor voltage reaches this value, the thyratron suddenly conducts, allowing the capacitor to discharge rapidly through the tube. The capacitor voltage falls almost instantaneously to the extinction voltage (E_e) of the tube.

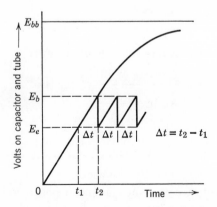

Fig. 12-39   Saw-tooth voltage developed across capacitor in Fig. 12-38.

At the instant the tube ceases to conduct, the capacitor begins to charge, and the cycle just described starts over again. The time of one cycle is $\Delta t = t_2 - t_1$, and this is almost entirely devoted to recharging the capacitor. The time of discharge in each cycle is considered sufficiently short to be neglected.

The equation for the instantaneous voltage at any time (t) of a capacitor charging through a resistor (R) with constant applied circuit voltage (E_{bb}) is

$$e_C = E_{bb}(1 - \epsilon^{-t/RC}) \tag{12-36}$$

Solving for t, we have

$$\epsilon^{-t/RC} = 1 - \frac{e_C}{E_{bb}}$$

$$t = RC \ln \frac{E_{bb}}{E_{bb} - e_C} \tag{12-37}$$

The symbol ln stands for natural logarithm. When $t = t_2$, the instantaneous voltage is the breakdown voltage, so that

$$t_2 = RC \ln \left(\frac{E_{bb}}{E_{bb} - E_b} \right)$$

When $t = t_1$, the instantaneous voltage is the extinction voltage (E_e), and

$$t_1 = RC \ln \left(\frac{E_{bb}}{E_{bb} - E_e} \right)$$

The time of one cycle is then

$$t_2 - t_1 = RC \left[\ln \frac{E_{bb}}{E_{bb} - E_b} - \ln \frac{E_{bb}}{E_{bb} - E_e} \right]$$

$$= RC \ln \frac{E_{bb} - E_e}{E_{bb} - E_b} \quad \left(\text{since } \ln x - \ln y = \ln \frac{x}{y} \right)$$

The frequency is the reciprocal of the time of one cycle,

$$f = \frac{1}{RC \ln \dfrac{E_{bb} - E_e}{E_{bb} - E_b}} \text{ c/s, or Hz} \tag{12-38}$$

In Equation (12-38) the E's are in volts, R is in ohms, and C is in farads. The frequency of oscillation of the saw-tooth-wave generator is given only approximately by this equation. Certain conditions that affect the frequency have not been taken into account; for example, the breakdown and extinction voltages change somewhat with tube current and temperature as well as with frequency itself. The equation does, however, show that the frequency varies inversely with R and with C. An increase in the supply voltage (E_{bb}) will increase the frequency because the capacitor will charge at a faster rate. The logarithm factor in the equation will decrease if E_{bb} is increased, because its denominator will increase faster than its numerator. E_b is several times as large as E_e in a grid-controlled gas tube. The result is an increase in the value of f in the equation.

The rising part of the saw-tooth-voltage wave will be a straight line only if the capacitor-charging current is constant. The lower region of the charging voltage curve in Fig. 12-39 is practically straight if E_{bb} is fairly high. Sometimes a pentode tube is used in place of R to provide constant charging current. It will be recalled that the plate current of a pentode is constant, for all practical purposes, over a wide range of plate voltage, particularly at low current values.

A number of capacitors of graduated sizes and a selector switch are used, instead of one capacitor as shown. Each capacitor provides a separate range of frequencies, and R serves as a fine-frequency adjustment for each range. The plate-circuit resistor r is necessary to prevent excessive tube current when the capacitor discharges.

A cathode-ray oscilloscope has provisions for introducing synchronizing signals into the sweep-frequency oscillator from either the vertical input circuit, which amplifies the signal to be seen on the cathode-ray-tube screen, or an external circuit. The synchronizing signal is fed to the grid of the gas triode as a positive pulse which will cause the tube to conduct at the proper instant in each cycle so that the sweep of the spot across the tube will be started at the proper instant in each cycle of signal voltage. Of course, if the spot requires the time of two cycles to cross the screen, two cycles will be shown on the screen and the synchronizing signal will be injected every second cycle.

The rectangular wave from the plate of a multivibrator may be used as a synchronizing signal, or it may be sent through a "pulse-sharpening circuit," before it is used as a synchronizing signal.

A sinusoidal voltage serves very well as a synchronizing signal. Inserted between E_{cc} and the grid in Fig. 12-38, *it has the effect of raising and lowering the plate potential at which the tube will suddenly conduct.* Look at Fig. 12-39 and observe that, with a given fixed bias on the grid, the tube cannot "fire" until the plate voltage reaches E_b each cycle. The nature of the tube is such that, when the grid is made less negative by reducing the bias, the tube will "fire" at a lower plate potential. A small a-c signal on the grid will

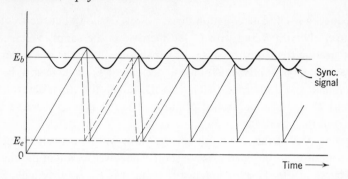

Fig. 12-40   Sine-wave synchronization of saw-tooth generator of Fig. 12-38.

thus *have the effect of a plate voltage pulsating* above and below the value E_b *for no synchronizing signal*. This is illustrated in Fig. 12-40.

This synchronizing voltage happens to have such phase that the capacitor must charge to a voltage above E_b in order to fire the tube at first. In the second charging cycle the capacitor voltage must go only slightly above E_b, and from the third charging cycle on the firing voltage is constant at about half the synchronizing-signal amplitude below E_b.

In using the generator it is found that a certain frequency of the saw-tooth voltage is desired. The proper capacitor and setting of the variable resistance R are chosen to provide a wave pattern at that frequency on the oscilloscope screen. As a synchronizing signal is usually required to hold the image stationary on the screen, a synchronizing signal having the same frequency as the wave being observed is applied in the grid circuit. This synchronizing signal is either taken from the observed signal itself or from another source associated with that signal.

SUGGESTED REFERENCES

1. Samuel Seely, *Electron-Tube Circuits*, 2nd ed., McGraw-Hill, New York, 1958.

2. Marlin P. Ristenbatt and Robert L. Riddle, *Transistor Physics and Circuits*, 2nd ed., Prentice-Hall, Englewood Cliffs, N.J., 1966.

3. Paul D. Ankrum, *Principles and Applications of Electron Devices*, International Textbook Co., Scranton, Pa., 1959.

4. H. F. Storm, *Magnetic Amplifiers*, Wiley, New York, 1955.

QUESTIONS

12-1 What is meant by *feedback* in an amplifier?

12-2 Distinguish between negative and positive feedback.

12-3 Why is negative feedback called degenerative?

12-4 Why is positive feedback called regenerative?

12-5 What does β represent? When is it a negative number and when is it a positive number?

12-6 What is the significance of the product of β and the gain without feedback?

12-7 What is the effect of inverse feedback on the gain of an amplifier? On the frequency-response curve?

12-8 How does inverse feedback affect phase shift through an amplifier stage?

12-9 Explain the negative-feedback action in a cathode-follower amplifier. What is the numerical value and sign of β?

12-10 Draw from memory a simple voltage-stabilizer circuit using a gas diode and a triode tube that carries output current.

12-11 Explain how the voltage-regulator circuit of Fig. 12-5 counteracts a rise in input voltage.

12-12 The voltage-regulator action of Fig. 12-5 responds to a tendency of the load voltage to change. Assume that the load *resistance* is suddenly *increased* somewhat. Explain what happens in the stabilizing circuit.

12-13 Show by comparison of the equations for plate current of an amplifier with and without current feedback that feedback makes the amplifier less sensitive to *changes* in loading.

12-14 How does the connection of the voltage-feedback circuit of an amplifier with an odd number of stages differ from that for an even number of stages?

12-15 How does inverse feedback affect the input impedance and the output impedance of an amplifier? Are these effects advantageous in transistor amplifiers? Explain.

12-16 How is *external* feedback added to a magnetic-amplifier circuit? Is it usually the positive or negative kind? Explain its action.

12-17 What is meant by feedback ratio? What is its sign ($+$ or $-$) in external positive feedback?

12-18 What is bias in a magnetic amplifier? How is it introduced in an external-feedback amplifier that is delivering d-c power?

12-19 Explain the action of a magnetic amplifier that has internal negative feedback.

12-20 Draw from memory the circuit of a tuned-plate oscillator. How does it differ from the circuit of a Class C amplifier?

12-21 What kind of bias is used in the tuned-plate oscillator? Why do you suppose fixed bias cannot be used?

12-22 What must be the value of βA_v in a feedback oscillator for sustained oscillation?

12-23 Draw from memory the circuit of an electron-coupled oscillator and describe how it works.

12-24 In what way is a crystal-controlled oscillator superior to other types not using a crystal?

12-25 Draw the equivalent circuit of a piezoelectric crystal. What are the electrical properties of a quartz plate crystal?

12-26 How is feedback provided in a resistance capacitance oscillator? What about phase shift?

12-27 What makes possible a *continuous frequency band* of operation of a resistance-capacitance oscillator? How is the frequency band made so extensive, i.e., from low-audio to the megacycle range?

12-28 What gives the negative resistance oscillator its negative resistance feature?

12-29 Explain the operation of the dynatron oscillator, using the plate characteristic of the tube.

12-30 What is responsible for good sine-wave form of the dynatron, even though the plate characteristic is non-linear?

PROBLEMS

12-1 The circuit of Fig. 12-3 has negative feedback for which $R_k = 1,000 \ \Omega$, $R_b = 30,000 \ \Omega$, $\mu = 40$, $r_p = 10,000 \ \Omega$. (*a*) Calculate the voltage gain without feedback ($R_k = 0$). (*b*) Calculate the voltage gain with feedback for $\beta = -0.005$, -0.01, -0.015, -0.02, -0.05, -0.1, -0.2. (*c*) Plot a curve of gain with feedback on the vertical vs. β expressed as per cent on the horizontal.

12-2 To observe the effect of higher load resistance on gain, repeat Problem 12-1 with $R_b = 100,000 \ \Omega$, with and without feedback.

12-3 The amplifier of Fig. 12-3 has negative feedback with $\beta = -0.3$ and gain with feedback $A_v' = 4.5$. What is the gain when $R_k = 0$?

12-4 An amplifier stage has negative current feedback. $R_L = 20 \ \mathrm{k\Omega}$, $R_k = 1 \ \mathrm{k\Omega}$, $r_p = 8 \ \mathrm{k\Omega}$, $\mu = 20$, $e_g = 1 \ \mathrm{V}$ rms. (*a*) Calculate the a-c plate current. (*b*) Calculate the a-c plate current without feedback, considering that R_k is left in the plate circuit. (*c*) Assume R_L is decreased 10% by adding another load in parallel.

Calculate the *per cent change* in a-c plate current in the two cases caused by the change in load.

12-5 The gain of an amplifier with *negative*-voltage feedback is 100. Assume the input impedance without feedback to be 50 kΩ. Calculate data for plotting a curve of input impedance with feedback vs. β from $\beta = -0.005$ to $\beta = -0.15$.

12-6 On the same curve sheet used for Problem 12-5 plot the output impedance of the amplifier vs. β over the same range, using the output impedance *without* feedback as $Z_o = 500$ kΩ.

12-7 Calculate the midband gain of the two-stage amplifier of Fig. 12-7: (*a*) without feedback, (*b*) with feedback.

12-8 The following data are given for a magnetic amplifier: $N_L = 100$, $N_F = 90$, $N_C = 1,000$, $R_L = 20$, $R_C = 200$. Calculate the current, voltage, and power gains with feedback.

12-9 What should be the capacitance of the capacitor to be used in a tuned-plate oscillator in which the inductance of the tank is $L = 100\ \mu H$ and the frequency is to be 1.5 MHz? The total a-c resistance of the L branch is 225 Ω, and the dynamic plate resistance of the triode tube is 2,300 Ω.

12-10 A resistance-capacitance oscillator uses two variable capacitors mounted on the same shaft for tuning. At one setting of the decade multiplier the frequency range is from 200 to 2,000 Hz. (*a*) What values of the product RC in Equation (12-35) are required for the two extremes of the frequency range? (*b*) Assume that 0.0001 μF is the minimum value possible for C. Calculate the value of R. (*c*) Since R is fixed, C must be changed to obtain a frequency at the other end of the range. Calculate the value of C.

12-11 A 100-μV noise voltage of 120 Hz frequency is inserted at point P in the circuit of Fig. P12-11. Determine the output voltage due to this noise when: (*a*) $R_1 = 0$, (*b*) $R_1 = 1,000$ Ω.

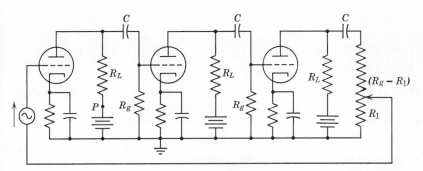

Fig. P12-11 $R_L = 30,000$ Ω; $R_g = 1$ MΩ; $r_p = 10,000$ Ω; $g_m = 1,000$ μmho.

12-12   A crystal used in a crystal oscillator has the following constants: $L = 0.29$ H, $C = 0.069$ pF, $R = 2,900 \ \Omega$. What is the Q of the crystal? If the shunting capacitance of the holder together with the input capacitance of the tube is 7 pF, at what frequency will the crystal and associated circuit be resonant?

12-13   A single-stage amplifier has a gain of $A_v = -8$ and an output of 100 V with 5% harmonic distortion. If negative-voltage feedback is to be used to reduce the distortion to 1%, what value of β (feedback ratio) should be used and what input voltage will be required if the output is to be maintained at 100 V?

12-14   An amplifier without feedback has a voltage gain of $3{,}000 \ \underline{/180^\circ}$. When feedback is effective, the gain is $96.8 \ \underline{/180^\circ}$. What is βA_v, the loop gain?

12-15   A one-stage amplifier with voltage feedback uses a tube with $\mu = 20$, $r_p = 12{,}000 \ \Omega$. The load is $R_L = 100{,}000 \ \Omega$, $\beta = 0.1 \ \underline{/0^\circ}$. What μ and r_p values would a tube need to have in a nonfeedback amplifier having the same R_L and the same gain?

12-16   Determine the conditions for oscillation and the frequency of the tuned-plate oscillator circuit in Fig. P12-16. Bias and d-c sources are omitted for convenience. Assume operation in Class A_1, and the tube coefficients are constant

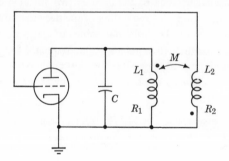

Fig. P12-16

and have the values $\mu = 20$, $r_p = 10 \ \text{k}\Omega$. Also assume the transformer windings have a high Q at the operating frequency, $L_1 = L_2 = 10 \ \text{mH}$, $C = 100 \ \text{pF}$, and $M = 9.6 \ \text{mH}$.

12-17   The circuit of Example 12-2 has $Z = 0$, $a = 1$, and $N_1 = N_2$. The tube constants are $\mu = 20$, $r_p = 10{,}000 \ \Omega$. (*a*) Calculate the feedback gain and output impedance for $Z_L = 100 \ \Omega$. (*b*) What changes should be made in the circuit to make $Z_o = Z_L$ for maximum power transfer?

12-18   Calculate an expression for the oscillation frequency of the circuit in Fig. 12-28, in terms of R and C, assuming $R_L = R$ and no grid current.

12-19   In the circuit of Fig. 12-11, $V_{CC} = 20$ V, $R_L = 10$ kΩ, and the circuit has a voltage gain of -100 when $R_f = 0$. Parameters h_{ie} 500 Ω, h_{re} 3 × 10^{-4}, h_{fe} 200, h_{oe} 15 μmho. Calculate the change in voltage gain when $R_f = 10$ Ω. What is the value of feedback β?

12-20   In the circuit of Problem 12-19, what is the change in input resistance caused by R_f?

12-21   Examine qualitatively the effects of R_f on voltage gain and input resistance in the circuit of Problem 12-19.

CHAPTER THIRTEEN

Modulation, Detection, Frequency Translation

An alternating current, or voltage, that has an unchanging waveshape cannot convey any information other than that which is related to its frequency and magnitude. Some physical property of the wave, e.g., its amplitude, frequency, or phase, must be varied if it is to convey information (intelligence). The variation must be caused by, and therefore correspond to, the intelligence that is to be transmitted. When such variations are produced in the waveform the wave is said to be *modulated.*

Let us visualize a current or a voltage that has simple sine-wave form (Fig. 13-1*a*). It may be called a *carrier wave* whose shape can be changed to conform to the tones of the voice or to some other kind of intelligence. When this is done the waveform may change to something like that in Fig. 13-1*b*. It is then called a *modulated carrier wave.*

13.1 Amplitude Modulation (A-M)

As the name implies, amplitude modulation is the process of changing the instantaneous values of the voltage or current from those of the steady sine-wave contour to other values with the result shown in Fig. 13-1*b*. The amplitude of each cycle of the wave might thus be changed and *amplitude modulation* would exist.

The frequency of the carrier wave is much higher than the frequency of the voice wave, or other modulating wave. The *contour* of the modulated carrier (shown as a dash-line along the envelope) varies at *modulation*

520

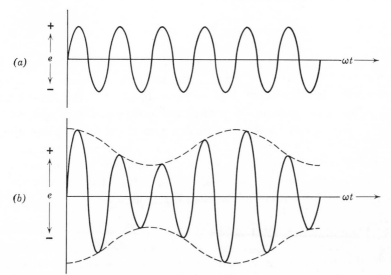

Fig. 13-1   Voltage waves radiated by the antenna of an amplitude-modulated radio transmitter. (*a*) Carrier wave, (*b*) Modulated carrier wave.

frequency. For example, if modulation is done by a 256 Hz sine wave (the frequency of middle *C* on the musical scale) the *variations in the envelope* occur at 256 Hz.

13.2   Frequency Modulation (F-M)

The electromagnetic wave radiated by commercial television antennas and by some radio antennas is *frequency modulated.* The instantaneous frequency of the antenna current and, simultaneously, of the antenna voltage is varied in accordance with the intelligence transmitted. Only the frequency of the carrier is changed; the amplitude remains practically constant, as shown in Fig. 13-2.

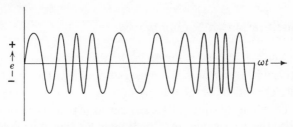

Fig. 13-2   Voltage wave radiated by the antenna of a frequency-modulated radio transmitter.

13.3   Pulse-Code Modulation

The modulating signal may be in the form of pulses (sudden bursts of voltage) that are coded to represent the elements of information to be transmitted. The information may be the temperature inside an orbiting space vehicle, for example. A prearranged *code* would be employed to identify variations in the pulses that represent temperature values which can be *telemetered* by radio to an earth station. The *telemetry* transmission system converts temperature readings into electrical signals and arranges them into a *pulse code*. In this form they alter the high-frequency carrier wave, by varying either the amplitude or the frequency.

13.4   Demodulation (Detection)

After the modulated carrier wave is received and converted into an amplified signal, the intelligence must be recovered from it. The process of information recovery is called *demodulation* or *detection*.

The transmission of information by modulation and subsequent detection is usually achieved by *frequency translation:* i.e., the modulated carrier being a *complex wave* has components with frequencies different from the frequency of the original, unmodulated carrier. Some components have frequencies higher than carrier frequency and some have lower frequencies. These components are related to the lower frequency components in the original modulating signal that carries the intelligence. For example, an A-M carrier at 100-kHz frequency modulated by a 1-kHz signal will have frequency components at 99 and 101 kHz in addition to the 100-kHz carrier. The 1-kHz signal has been *translated in frequency* so that it appears in the form of two much higher frequencies, one above carrier frequency and one below. In the detection process the original 1-kHz signal is recovered in its original form from the modulated components.

13.5   Classification of Radio Waves

Electrical energy travels in space in the form of electromagnetic waves. *Radio waves* are those whose frequencies are within the 30 kHz to 30,000 MHz range. They travel at the velocity of light: 3×10^8 m/s (186,000 miles). The distance, along the line of travel, between two successive crests of a wave is called the wavelength, usually denoted by λ, the Greek letter *lambda*. Physicists usually use the letter c to represent the velocity of light. We shall use v to represent the velocity of a radio wave in free space, or in air. The relation between wavelength, velocity and frequency (f) is

$$v = f\lambda \tag{13-1}$$

Frequencies (cycles per second) are now expressed in hertz, but the frequencies of radio waves are in the kilohertz (kHz) or megahertz (MHz) ranges.

TABLE 13-1   Frequency Ranges and Classification

Classification	Frequency Range	Wavelength Range	Typical Uses
Low frequency (LF)	30–300 kHz	10,000–1,000 m	Long-distance point-to-point, marine, and navigation aid systems
Medium frequency (MF)	300–3,000 kHz	1,000–100 m	Commercial broadcast, marine communication, navigation, harbor telephone
High frequency (HF)	3–30 MHz	100–10 m	Moderate and long distance communications of various sorts
Very high frequency (VHF)	30–300 MHz	10–1 m	Television, F-M radio broadcast, radar, aircraft navigation
Ultra-high frequency* (UHF)	300–3,000 MHz	100–10 cm	Television, radar, telemetry, telephone relay systems
Super-high frequency (SHF)	3,000–30,000 MHz	10–1 cm	Radio relay, radar, space communications

* Waves with frequencies higher than 2,000 MHz are usually called *microwaves*.

Wavelength is expressed in meters, or in centimeters, when the frequency is extremely high (see Table 13-1).

Radio waves are classified into groups (or bands) according to their frequencies. Classification and typical applications are listed in Table 13-1.

Radio waves in the various frequency ranges are important in the study of modulation and detection, because they are commonly used as carriers. From the table it is seen that typical uses include commercial A-M and F-M broadcasting and television.

DRILL PROBLEM

D13-1   Calculate (*a*) the wavelength of a 60-Hz power line signal, (*b*) the frequency of a 40-m wave.

13.6   Separation of Radio Carriers

Radio waves are sent out into space from radiating systems called *antennas*. An antenna is designed with specific dimensions that are related to the wavelength of the carrier wave that is radiated, although an alternating current radiates a certain amount of energy in the form of electromagnetic waves from an open-wire circuit of any configuration. The amount of radiated energy is much greater, however, if all dimensions of the circuit approach the wavelength dimension. In the same way, a radio wave is captured by an antenna having dimensions that approach the wavelength of passing waves.

The electromagnetic energy of a radio wave, in passing over a conducting antenna, induces a voltage in the antenna that varies in the same manner as the passing wave, and produces a current at the frequency of the wave. The induced voltage and associated current carry energy that is absorbed from the wave. Since every radio wave passing over an antenna produces its own voltage in the antenna, the receiving equipment must be capable of separating the signal of a desired frequency from all the others that are present. The separation of one signal from the others is made on the basis that each transmitting source uses a different carrier frequency. In commercial A-M broadcast systems, each station is allotted a *bandwidth* of 10 kHz about its carrier frequency, and carrier frequencies are separated by 10 kHz. For example, one broadcast station may be assigned a carrier frequency of 700 kHz. Other stations may be assigned 710 kHz, 720 kHz, etc. Television stations in the United States are assigned carrier frequencies that are 6 MHz apart in frequency because of the larger bandwidth needed in the much higher frequency range to transmit information.

Resonant circuits can be made to discriminate between radio waves that have different frequencies. Resonant circuits respond very strongly in favor of a particular frequency. The ability of a system to separate waves of different frequencies is referred to as *selectivity*. A high selectivity means that frequency separation is achieved for signals that are close together in frequency. The process of adjusting a resonant circuit to select first one frequency and then another is called *tuning*. For example, a radio broadcast receiver can be tuned to pick up any station in the A-M range, from 550 to 1,600 kHz, by varying a capacitance in a resonant circuit of the receiver.

13.7 Amplification of Modulated Carriers

The voltage induced in a receiving antenna may be very small, either because the transmitted signal is weak or because the signal at the receiving antenna is small. Much more satisfactory operation of a receiving detector is possible from relatively large signals. Modulated carriers may be amplified by circuits in the receiver before detection, or the information signal may be amplified after detection. The first case is referred to as *radio-frequency* amplification, performed by amplifiers specifically designed to handle the high frequencies. Amplification of the detected information is handled by *audio amplifiers*.

Sometimes it is necessary to amplify signals within the receiver that are at some *intermediate frequency* between the carrier and the information frequencies. These intermediate frequencies (I-F) are generated in the frequency translation process, combined with the information frequencies conveyed by the carrier. The reasons for amplification of modulated carriers will be understood better as we progress through the study of modulation and detection processes.

13.8 Fundamental Modulation Process

A sinusoidal* voltage may be written as

$$e = E_c \cos \phi = E_c \cos (\omega t + \phi_o) \tag{13-2}$$

and represented on the complex plane as a phasor rotating in time at a constant angular velocity. Its instantaneous amplitude is a projection on the real axis of the complex plane, as shown in Fig. 13-3.

The concept that a sinusoid can be considered as a rotating phasor is sometimes lost in repeated study of steady-state, constant-frequency networks. We will examine the notation of Equation (13-2) to show that the argument of the cosine function, $(\omega t + \phi_o)$, can be used to modulate the voltage waveform. We will also see that the amplitude, E_c, may be altered to achieve modulation.

---

* Because a cosine wave has the same shape as a sine wave, and because it is more convenient to work with cosine terms in this application, we are using cosines. The only difference between sine and cosine functions of time is the choice of the instant at $t = 0$.

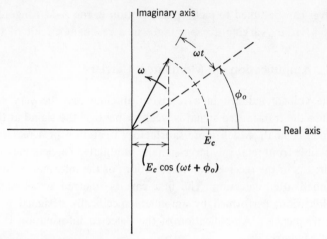

Fig. 13-3   Rotating phasor on complex plane representing a time-varying sinusoid.

The instantaneous angle of the phasor in Fig. 13-3, referred to the real axis, is changing in time because of the constant angular velocity, ω. Then we may write,

$$\omega = \frac{d\phi}{dt}$$

$$\phi = \int \omega \, dt + \text{a constant}$$

The constant may be zero, or an angle ϕ_o at $t = 0$. In general, then

$$\phi = \int \omega \, dt + \phi_o \qquad (13\text{-}3)$$

This may be used in the form of Equation (13-2) to express the voltage as

$$e = E_c \cos \left(\int \omega \, dt + \phi_0 \right) \qquad (13\text{-}4)$$

Equation (13-4) indicates that two factors may be used to modulate the sinusoidal voltage waveform: (1) the amplitude, E_c, and (2) the instantaneous angle, $\int \omega \, dt + \phi_o$. Variations in these two factors result in the two basic types of modulation. Variation of amplitude, E_c, with time is amplitude modulation (A-M). Variation of the angle with time results in two types of angle modulation: (1) time variation of $\int \omega \, dt$ yields frequency modulation (F-M), and (2) time variation of ϕ_o yields *phase modulation*. In our discussions of this chapter, we shall be interested only in A-M and F-M. Phase-modulation systems are used in special applications, but A-M and F-M are by far the most prevalent.

13.9   Analysis of Amplitude Modulation (A-M)

We shall first consider the process of *altering the amplitude* of a sinusoidal waveform to generate A-M signals for transmission. We shall rewrite Equation (13-2) in a form to show a sinusoidal variation of its amplitude:

$$e = (E_c + E_s \cos \omega_a t) \cos (\omega_o t + \phi_o) \qquad (13\text{-}5)$$

in which the angular velocity symbol ω_o represents the constant frequency of the carrier wave, and ω_a is 2π times the frequency of the modulating signal. The amplitude expression is usually written as

$$A = E_c + E_s \cos \omega_a t$$
$$= E_c\left[1 + \left(\frac{E_s}{E_c}\right) \cos \omega_a t\right]$$
$$= E_c(1 + m_a \cos \omega_a t)$$

The factor m_a is the ratio of the modulating signal and carrier amplitudes. It is called the *degree of modulation* or *modulation factor*. Why it has this name may be better explained by reference to Fig. 13-4, where Equation (13-5) is graphed as a function of time. It can be seen that the amplitude of the modulating signal is important in determining the resultant waveshape of the modulated carrier. The alterations in carrier amplitude are directly related to the changes in modulating signal. The modulating signal produces an *envelope* on both positive and negative portions of the carrier. As stated earlier, it is this envelope that represents the information being conveyed by the modulated carrier.

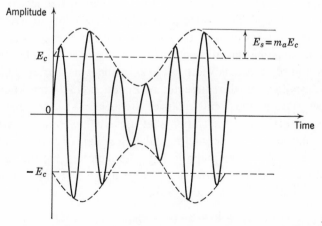

Fig. 13-4   Sketch of A-M wave, showing envelope developed by modulating signal.

The factor m_a is often used to designate the degree of modulation as "modulation percentage." This number results when the value of m_a is multiplied by 100%. It is usually desirable to maintain $m_a < 1.0$, i.e., $<100\%$. If the modulation is allowed to become greater than this, distortion of the signal may result so that detector circuits cannot retrieve the original information.

The A-M voltage wave may be written as

$$e = E_c(1 + m_a \cos \omega_a t) \cos \omega_o t \qquad (13\text{-}6)$$

dropping the constant angle, ϕ_o, as having no significance in the variations. This equation may be shown to contain *three* frequency components, even though it was derived from the combination of only *two*, a modulating signal and a carrier frequency. A familiar trigonometric identity permits the A-M signal to be written as*

$$e = E_c \cos \omega_o t + \frac{m_a}{2} E_c \cos (\omega_o + \omega_a)t + \frac{m_a}{2} E_c \cos (\omega_o - \omega_a)t \quad (13\text{-}7)$$

This shows that the A-M signal is a composite of three separate component phasors, each at a different frequency. The first term of Equation (13-7) is the carrier-frequency component. Added to it are two other frequency components, one at a frequency equal to the *sum* of the carrier and modulating frequencies and the other at a frequency equal to their *difference*. Thus, *three* frequency components make up the total A-M wave.

As long as the carrier frequency is greater than the modulating frequency, Equation (13-7) represents the A-M signal. Should the modulating signal frequency become equal to or greater than the carrier frequency, then this expression is not valid. The final term would generate a "negative frequency." In typical applications of A-M the carrier frequency is many times greater than the modulating signal frequency. For example, a radio broadcast station is limited to a bandwidth of 10 kHz about its carrier. This means that its transmitted signal, represented by Equation (13-7), must not contain frequencies outside the limits (carrier frequency $+5$ kHz) and (carrier frequency -5 kHz). For the station that uses 1,000 kHz as a carrier, the ratio of frequencies of carrier and modulating signal is 1,000 to 5 (200 to 1) as a limit. The form of the graph in Fig. 13-4 indicates that the A-M signal will contain 200 cycles of carrier *for each cycle* of 5 kHz modulating signal. The graph actually shows a 4 to 1 ratio.

---

* $\cos A \cos B = \frac{1}{2}[\cos (A + B) + \cos (A - B)]$.

DRILL PROBLEMS

D13-2 Make graphical sketches for these two expressions, plotting at least one cycle of the lowest frequency, on a common time scale. Which one describes A-M? Why doesn't the other describe A-M? (*a*) $e = \cos t + \cos 10t$, (*b*) $e = \cos 10t + \cos t \cos 10t$.

D13-3 An A-M wave of frequency 1 kHz is modulated at 50 per cent by a signal frequency of 100 Hz. Sketch the waveform over one cycle of the modulating signal.

D13-4 An A-M broadcast station is assigned a carrier of 1 MHz and is limited to a bandwidth of 10 kHz. What range of information frequencies may it transmit?

13.10　Generation of Sidebands in A-M

A constant-frequency carrier of a broadcast transmitter is modulated by signals that originate as vocal or musical sounds. Seldom is the modulating signal a single-frequency sinusoid. It is usually a combination of several components having different frequencies and acting simultaneously.

The principle of superposition may be used to show that the A-M signal is composed of the *carrier*, a *sum component* and a *difference component* for *each individual frequency* that is a part of the modulating signal. The sum component is called the *upper sideband* component and the difference component is called the *lower sideband* component of the A-M wave. They are called sideband components because they exist in *bands of frequencies* above and below the carrier on the frequency scale. All of the information being transmitted is contained in the sideband components, none in the carrier. The carrier (or central) component of the complex wave acts only as the vehicle that "carries" the information-bearing sideband components. Because signals representing voice and music contain many frequencies and their harmonics, a band of frequencies on each side of the carrier frequency exists in the A-M wave.

An advantage of amplitude modulation is that low-frequency signals may be translated to higher frequencies for ease of transmission, and then retranslated at a receiver to the lower (original) frequency range for reclaiming the information. Because the carrier may be produced at any desired frequency, the information may be carried at a convenient frequency at any location in the frequency spectrum. Figure 13-5 shows the placement of the sidebands about a carrier in A-M transmission. The shaded sidebands about a second carrier frequency indicate that several transmitting stations may

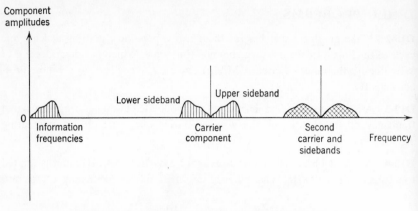

Fig. 13-5   Frequency translation in A-M system.

broadcast simultaneously without interference, provided the sidebands do not overlap. A detector must have a bandwidth large enough to include the sideband frequencies it is tuned to receive. If another carrier has sideband components that overlap those of one desired, the detector will accept both signals and produce a distorted output.

In a *superheterodyne* radio receiver, the carrier and its sidebands are translated from the carrier frequency to some intermediate frequency (I-F), amplified at the I-F, and then translated again to the information frequency band by a detector. This employment of an intermediate frequency makes the superheterodyne receiver more selective than any other kind. That is, its circuit is more efficient than that of any other receiver in rejecting unwanted signals that have sidebands very near to the one desired.

13.11   Bandwidth Requirements in A-M

Each frequency component of a modulating signal produces a pair of sideband frequencies that are spaced symmetrically about the carrier frequency. The new frequencies describe a bandwidth about the carrier that is twice the frequency of the modulating signal. It is apparent that the bandwidth of the A-M wave will be twice the highest frequency contained in the modulating signal. The bandwidth that is required in a given application to convey information by A-M depends on the type of intelligence involved. The bandwidth will be greater for rapidly varying signals (containing high-frequency components) than for signals made up of low-frequency components. Bandwidth requirements for voice transmission are less stringent than for television signals, because of the different ranges of frequencies involved.

From a listener's point of view the quality of a voice transmission can be measured in terms of two characteristics, intensity and intelligibility. These two parameters, although they are virtually independent of each other over a broad frequency range, together determine the quality of reproduced sound. Most of the *energy* in a voice signal, and hence the intensity, is contained in the lower frequencies, while the higher frequency harmonics and overtones contribute intelligibility.

The curves of Fig. 13-6 illustrate the roles of frequency and intensity in the intelligibility of voice signals. If all frequencies of a voice signal below 1,000 Hz are eliminated, about 85 per cent of the intelligence would be understood, even though the energy content would be only about 17 per cent of the original voice energy. Similarly, if the frequencies above 1,000 Hz were eliminated, 83 per cent of the energy would remain but only about 45 per cent of the signal would be understood. This means that voice transmission must include both low- and high-frequency components for acceptable operation.

Because bandwidths are limited in many A-M systems, some compromise is usually necessary. In a telephone relay system, in which voice signals are the primary information, a smaller bandwidth may be used than for A-M radio broadcast in which signals are both voice and music. The compromise in A-M radio is shown by the 5 kHz bandwidth allowed for each station

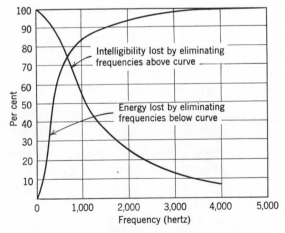

Fig. 13-6   Curves show that energy in voice signals is concentrated at low frequencies, and intelligibility is concentrated in high frequencies.

TABLE 13-2   Typical Signals and Amplitude-Modulation Frequencies

Type of Signal	Transmitted Frequency Range (hertz)
Telegraph	
Morse Code at 100 words per minute	0–170
Voice	
Typical broadcast program	100–5,000
Long-distance telephone	200–3,200
Intelligible but poor quality	500–2,000
Television	
Standard 525-line picture, interlaces, 30 frames per second rate	60–4,500,000
Pulses	
1 μs duration	0–1,000,000

(10 kHz about the carrier, because the total bandwidth is twice the highest frequency of information). This bandwidth is insufficient for high-fidelity reproduction of music signals, which may have frequency components above 15 kHz, but it is acceptable for most listeners. In a telephone transmission system designed to carry voice signals only, a smaller bandwidth may be used because the normal speaking voice generates significant frequencies up to about 3,500 Hz. Typically, the range of frequencies that is transmitted is from 200 to 3,200 Hz. Table 13-2 lists several signals and the bandwidth that is used for A-M transmission.

13.12   Power Contained in A-M Sidebands

When a modulating signal is a single-frequency sinusoid, the power in the A-M wave will consist of three parts, one for the carrier voltage and two for the sideband voltages. Equation (13-7), rearranged here, is useful in a study of the distribution of power in the A-M wave:

$$e = E_c\left[\cos\omega_o t + \frac{m_a}{2}\cos(\omega_o + \omega_a)t + \frac{m_a}{2}\cos(\omega_o - \omega_a)t\right] \quad (13\text{-}7)$$

Since power associated with a sinusoidal voltage is proportional to the square of the rms voltage, total power in the A-M wave is,

$$\text{Power} = kE_c^2\left(1 + \frac{m_a^2}{4} + \frac{m_a^2}{4}\right) = K\left(1 + \frac{m_a^2}{2}\right) \quad (13\text{-}8)$$

The power contained in the sidebands is $m_a^2/2$ times the carrier power. For a modulation percentage of 100 per cent ($m_a = 1$) sideband power is only $\frac{1}{2}$ the carrier power. For lower modulation percentages, the portion of total power contained in the sidebands is even less than this fraction. The ratio of sideband power to total power can be expressed as follows:

$$\frac{\text{Sideband power}}{\text{Total power}} = \frac{m_a{}^2/2}{1 + m_a{}^2/2} \tag{13-9}$$

When $m_a = 1.0$, the ratio of sideband power to total power is $\frac{1}{3}$, i.e., the sidebands contain only $\frac{1}{3}$ of the total power. At 10 per cent modulation, the ratio is $\frac{1}{201}$; the sideband power is less than half of 1 per cent of total power. The fraction of the total power that represents information carried by the sidebands decreases rapidly as the modulation percentage is reduced. For this reason a high percentage of modulation is generally desired.

13.13 Methods of Generating A-M

We have discussed the possibility of slowly varying the envelope of a sinusoidal carrier voltage in accordance with a modulating signal. We may now ask: How is A-M physically produced? To answer this question, we note that the mathematical expression for A-M (Equation 13-5) is a *product* function, in which two quantities are multiplied together. The simplest solution would be to devise a method for multiplying the two quantities and obtain an output that is their product. A circuit that performs this function is called a *product modulator*.

Before looking at some circuits that may be used to generate A-M, we need to discuss some characteristics that are commonly used as a basis for their operation. Modulating devices and circuits may be classified in one of two categories: (1) their terminal (voltage-current) characteristics are *nonlinear*, or (2) the device contains a *switch* independent of the signals that changes operation from one linear characteristic to another, *piecewise linear* characteristics.

We know that linear networks are characterized by two basic properties: the response to a sum of inputs is the sum of responses to each input applied separately, and multiplication of an input by a constant also multiplies the response by the same constant. The first property is a statement of the superposition principle. The second property reinforces the concept of linearity, implying that the terminal conditions are independent of any unique property of the network. As an example, consider the steady-state behavior of a simple RLC network in response to a sinusoidal input voltage.

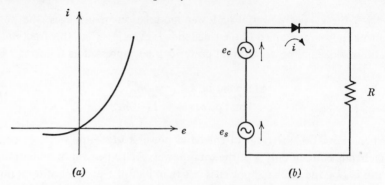

Fig. 13-7   Diode characteristic and circuit used to generate A-M from nonlinear operation. (*a*) Diode characteristic. (*b*) Circuit.

The network merely modifies the amplitude and phase of the voltage. *No new frequencies are generated in the linear network.* Thus, a linear network cannot be used to modulate a carrier signal.

A nonlinear network, on the other hand, distorts input signals and generates frequency components that are not present in the input. Similarly, a *piecewise linear* network or device, such as the diode rectifier studied in Chapter 5, can generate harmonic frequencies based on the input signal frequency. When it is operated so that its region of operation extends over two separate linear portions of its characteristic, harmonic frequencies are generated. Thus, its overall operation would be nonlinear.

As an example of A-M generation by a nonlinear device, we will use a semiconductor diode whose characteristic is given in Fig. 13-7. Assume that this characteristic curve can be expressed by

$$i = a_1 e + a_2 e^2 + a_3 e^3 + \cdots \tag{13-10}$$

The first term is a linear relationship between current and voltage, and the other terms show the presence of curvature in the graph of *i* vs. *e*. The first two terms represent the only significant portions of the current, because the higher powered terms represent components that have frequencies too far from those of the side-band components. They are filtered out of the system and thus do not affect the shape of the modulated output wave.

The voltage contains a carrier component and a signal-frequency component. Its equation is

$$e = E_c \cos \omega_o t + E_s \cos \omega_a t$$

where $E_s < E_c$. Substituting this into Equation (13-10) we get the expression for the current

$$i = a_1 E_c \cos \omega_o t + a_1 E_s \cos \omega_a t + a_2 E_c^2 \cos^2 \omega_o t$$

$$+ 2a_2 E_c E_s \cos \omega_a t \cos \omega_o t + a_2 E_s^2 \cos^2 \omega_a t + \cdots \quad (13\text{-}11)$$

The components represented by the second, third, and fifth terms are eliminated by filters in the modulating unit. The remaining components, represented by the first and fourth terms, give the following equation for the current:

$$i = a_1 E_c \left(1 + \frac{2a_2}{a_1} E_s \cos \omega_a t \right) \cos \omega_o t \quad (13\text{-}12)$$

This has the form of Equation (13-6) and describes an A-M current wave. The current in an antenna will radiate an electromagnetic wave that has components in the sidebands and a carrier component.

13.14   Plate-Modulated Class C Amplifier

Commercial broadcast transmitters commonly use plate-modulated Class C amplifiers to generate A-M signals for transmission. The basic method is shown in Fig. 13-8a, and the circuit in Fig. 13-8b. The modulation process involves the extreme nonlinearity of Class C operation of the amplifier, because of the high negative grid bias. The tuned circuit at the plate provides the necessary filtering to allow only the carrier and sidebands in the output.

Conduction in the tube of a Class C amplifier occurs only at the peaks of the carrier input signal. Pulses of current at the carrier frequency are fed to the tank circuit. Thus, the tube acts as a switch to turn on and off the carrier signal. Since the amplitudes of the current pulses are nearly linearly related to the plate supply voltage, varying the effective value of E_{bb} by means of the modulating signal generates current pulses that are proportional to the modulating signal amplitude.

If the plate supply voltage is varied slowly compared to a cycle of carrier frequency (the resonant frequency of the tuned load), then the oscillating current in the tank circuit will have an envelope corresponding to the waveshape of the modulating signal. Thus, modulation of the carrier will occur at the rate and amplitude of the modulating signal, generating the required waveshape for A-M.

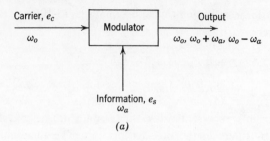

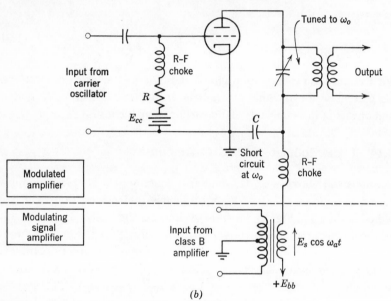

(b)

Fig. 13-8   Basic concept and circuit to generate A-M. (a) Basic method for A-M. (b) Plate modulated Class C amplifier.

The effective plate supply voltage will be a combination of E_{bb} and modulating signal

$$E_b = E_{bb} + E_s \cos \omega_a t$$

or

$$= E_{bb}(1 + m_a \cos \omega_a t)$$

which is the usual form representing the amplitude of the A-M wave.

13.15   Balanced Modulators and Single-Sideband (SSB) Operation

We have seen that the information being carried in A-M operation is contained entirely in either one of the two sidebands. The carrier itself conveys no information, and the information in one of the sidebands is rejected. Only one sideband needs to be transmitted to convey all of the information contained in the modulating signal. Further, transmission of the carrier and a second sideband is a waste of power, particularly carrier power. It is also a waste of frequency space because transmission of both sidebands requires double the bandwidth needed for a single sideband. This has resulted in the use of single-sideband (SSB) transmission for radio-telephony in fixed and mobile radio services at frequencies below 25 MHz, by regulations issued by the Federal Communications Commission (FCC). In this type of transmission the carrier and one sideband are suppressed, and the remaining sideband is transmitted.

A modulator that generates the two sidebands but suppresses the carrier is the *balanced modulator*. This method uses two nonlinear components of the type previously discussed. Simplified circuit diagrams are given in Fig. 13-9.

In the bridge modulator, the diodes act as switches to turn on and off the output signal, operating at the carrier rate. With identical diodes (balanced operation) the bridge is balanced when the modulating signal is zero, and there will be no output. When the carrier signal has the polarity shown the diodes are essentially short-circuits, and the output is again zero. But when the carrier signal changes polarity, the diodes are essentially open circuits, and the output will equal the modulating signal. Thus, the output is alternately zero and equal to the modulating signal, and is switched at the carrier frequency rate. The mathematical analysis of this type of modulator is complex and will not be attempted here. The important result is that the carrier frequency will not be a component of the output; only the two sidebands remain. The output signal is sometimes referred to as a *double-sideband, suppressed carrier* signal (DSB/SC).

The balanced modulator using vacuum triodes is essentially a Class B, push-pull amplifier. The carrier signal in the grid circuit has the modulating signal superimposed on it. Plate current in each of the tubes is determined by the combination of the two signals. Assuming the triodes identical and balanced, the plate currents may be written,

$$i_1 = k(1 + m_a \cos \omega_a t) \cos \omega_o t$$
$$i_2 = k(1 - m_a \cos \omega_a t) \cos \omega_o t \tag{13-13}$$

The load voltage is proportional to the difference $i_1 - i_2$, or

$$e_o = K \cos \omega_a t \cos \omega_o t \tag{13-14}$$

which leads to the sum and difference frequencies in the two sidebands. In

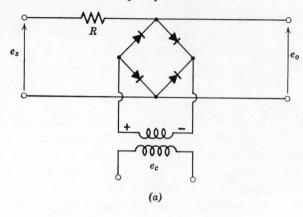

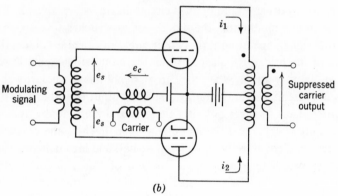

Fig. 13-9   Balanced modulators for suppressed carrier operation. (*a*) Diode bridge modulator. (*b*) Triode modulator.

the circuit operation the carrier has been suppressed, leaving only the two sidebands.

The unwanted sideband may be removed by a bandpass filter that passes only the desired sideband. The balanced modulator and sideband filter are the heart of SSB transmitters commonly used. The modulator is operated at low power levels in SSB operation, then followed by linear power amplifiers to produce the level necessary for transmission. This is in contrast to conventional A-M, where modulation is frequently performed at high power levels in Class C amplifiers.

EXAMPLE 13-1.   We will calculate the peak power required in single-sideband operation compared to that in conventional A-M, where each carries equal intelligence at 100 per cent modulation.

In ordinary A-M operation, the peak A-M voltage at 100 per cent modulation will be

$$\text{Peak voltage} = 2E_c$$

and the peak power will be proportional to the voltage squared, or

$$\text{Peak power} = k(4E_c{}^2)$$

The peak voltage in one sideband is $E_c/2$, and the peak power at 100 per cent modulation is $k(E_c{}^2/4)$. Then the peak power required in conventional A-M is 16 times as great as the peak power required in SSB operation, or

$$\frac{k4E_c{}^2}{kE_c{}^2/4} = 16$$

13.16   Frequency Conversion and Mixing

The block diagram of a superheterodyne A-M receiver is illustrated in Fig. 13-10. The block labeled *mixer* represents a circuit that is used to translate the carrier and sidebands to a new frequency, the I-F frequency of the receiver. The local oscillator and R-F amplifier are simultaneously tuned so that the output of the mixer occurs at the I-F frequency, regardless of the incoming carrier frequency. Such receivers usually have the I-F frequency at about 455 kHz.

The basic operation of a frequency converter may be understood in terms of the circuit in Fig. 13-11. The vacuum tube is usually referred to as a *pentagrid converter*, and the circuit operates as a *frequency mixer*. The

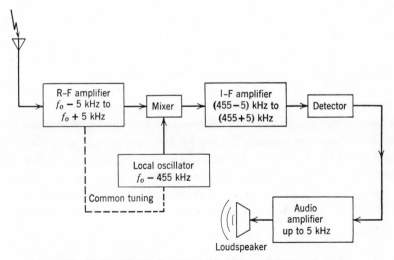

Fig. 13-10   Block diagram of A-M broadcast receiver.

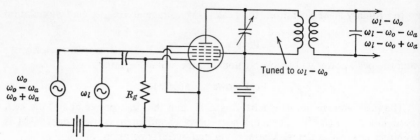

Fig. 13-11   Pentagrid mixer as a frequency converter.

cathode, first grid, and second grid act as a triode to alter the plate current at a rate corresponding to the local oscillator frequency, ω_l. The third grid acts as the normal control grid of a conventional triode, altering the plate current at a rate corresponding to the R-F carrier frequency and sidebands. The tuned load is designed to resonate at the I-F frequency and is broad enough to include the sideband frequencies. The mixer acts as a frequency converter, shifting the incoming signal down to the intermediate frequency of the receiver.

Several stages of amplification at the I-F level may be used to increase the magnitudes of information signals. They are then detected to recover the original information. (Some inexpensive portable receivers use only one I-F stage.)

13.17   Single-Sideband Receiver

Because of the economy of power and frequency space offered by SSB, it may be wondered why it has not been used for standard radio broadcasting. Part of the reason is the complexity required in a receiver. Mass production of inexpensive home receivers has not been possible.

A block diagram of a typical SSB receiver is given in Fig. 13-12. The complexity arises because the carrier must be reinserted at the receiver in order that the detector circuits may operate properly to retrieve the transmitted information signals. The receiver contains a local oscillator that is used to reinsert the carrier. Deviations of its output cause distortion in the sideband signals and in the information being conveyed. The precision frequency control that is required in the receiver, as well as in the transmitter, is the basic deterrent to the widespread use of SSB in radio broadcasting.

A basic difference in the SSB receiver compared with the conventional A-M receiver is the method of detection. The SSB system uses frequency mixing and translation of the information frequencies by shifting them down to the audio range. A conventional A-M system uses *envelope detection*, a process that will be discussed in subsequent sections of the chapter. Generally, the A-M wave has an envelope that varies in time with the information signal, but an SSB wave bears no relationship in time with the information signal.

13.18   The Envelope Detector for A-M Signals

The process of retrieving information from a modulated carrier is called *demodulation*, or *detection*. A detector operates basically as the inverse of a modulator; frequency translation brings the information signals down in frequency to their original locations in the frequency space. It requires nonlinear circuit elements for proper operation. Two general types of nonlinear detectors are used, the so-called *square-law detector* and *piecewise linear detector*. First, we shall discuss the piecewise linear type.

In conventional A-M detection the information is recovered by applying the A-M wave to a half-wave rectifier, as in Fig. 13-13. The output is then

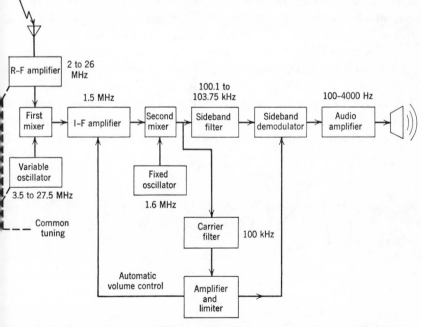

Fig. 13-12   Block diagram of SSB receiver. (Adapted from "Information Transmission, Modulation, and Noise," by M. Schwartz, McGraw-Hill, 1959.)

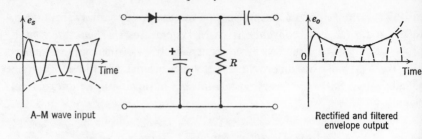

Fig. 13-13   Linear diode envelope detector for A-M.

filtered to recover the envelope. Such a detector-filter combination is called an *envelope detector*, for obvious reasons.

The output of an envelope detector can be a close replica of the original information signal. Depending on the time constants of the R-C network that acts as a filter to remove the carrier component, the amount of distortion can be kept small. Figure 13-14 illustrates the effect of the R-C time constant on the distortion that may be introduced by the filter. If the RC product is too small, the time constant will be too short and the output waveform will have a ragged edge that essentially follows the envelope. Some frequency components will be present that are not portions of the original information signal. Conversely, if the time constant is too long, the filter will not be able to follow rapidly varying envelopes. Some compromise will be designed in the filter to permit good reproduction of the envelope, without introducing extraneous frequencies or "clipping" some wanted frequencies. A short time constant introduces higher frequency components than are contained in the envelope, and a long time constant eliminates some of the high frequency components of the original signal. Either, of course, is a form of distortion and should be avoided where possible.

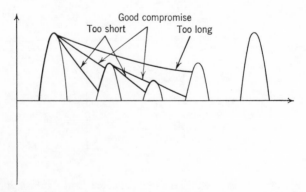

Fig. 13-14   Effect of time constant on the output waveform of envelope detector.

In the design of the filter, the value of C is chosen to have a small reactance ompared to the value of R at the carrier frequency. This will generate a lowly varying voltage across R at the modulating frequency. Compared o the variation of an individual cycle of carrier signal, the output voltage emains constant except for variations at the modulating signal frequency. This is especially true if the carrier frequency is many times greater than the modulating frequency, which is usually true in conventional A-M radio roadcasting. The value of R should be large compared to the forward esistance of the diode, so that the charging time constant permits the utput voltage to follow the envelope variations. However, its value is also mportant in the discharging time constant.

EXAMPLE 13-2. It is desirable to know how the values of R and C in the nvelope detector are related to the highest modulation frequency and modulation ercentage. Both of these factors affect the discharge rate of C so that the output oltage may follow the envelope variations. The problem is most serious at the ighest frequency and modulation percentage, because the discharge rate of C nust be less than the slope of the envelope at this instant.

For the A-M wave the rate of change of the envelope is

$$\frac{de}{dt} = \frac{d}{dt}[E_c(1 + m_a \cos \omega_a t)]$$

$$= -E_c m_a \omega_a \sin \omega_a t \tag{1}$$

The capacitor C will discharge from a voltage E according to the relation

$$e_C = E \epsilon^{-\frac{t}{RC}}$$

nd the rate of change of capacitor voltage will be

$$\frac{de_C}{dt} = -\frac{E}{RC} \epsilon^{-\frac{t}{RC}}$$

$$= -\frac{e_C}{RC} \tag{2}$$

When the capacitor voltage is at the peak of one of the rectified carrier pulses, ne voltage rate of change is

$$\frac{de_C}{dt} = -\frac{E_c(1 + m_a \cos \omega_a t)}{RC} \tag{3}$$

In order that the voltage across C may be able to follow the envelope variations, ne envelope should decrease slower than the capacitor voltage. A critical condition occurs when the two voltages have the same rate of decrease

$$\frac{de}{dt} = \frac{de_C}{dt}$$

$$E_c m_a \omega_a \sin \omega_a t = \frac{E_c(1 + m_a \cos \omega_a t)}{RC}$$

or

$$RC = \frac{1}{\omega_a}\left[\frac{1 + m_a \cos \omega_a t}{m_a \sin \omega_a t}\right] \qquad (4$$

Because the time constant should be equal to or less than this value, this equation describes a maximum value of RC. In order to eliminate the variable t, methods of calculus may be used to determine the maximum value of the expression on the right with respect to t. This process yields,

$$\cos \omega_a t = -m_a$$

and   (5)

$$\sin \omega_a t = \sqrt{1 - m_a^2}$$

Substituting these in Equation (4) yields the relationship of RC, m_a, and ω to be satisfied in the design of the filter circuit

$$RC \le \frac{1}{\omega_a}\frac{\sqrt{1 - m_a^2}}{m_a} = \frac{\sqrt{1 - m_a^2}}{2\pi f_a m_a} \qquad (6$$

Even though the RC time constant should approach zero for 100 per cent modula tion, a value not greater than $1/2\pi f_a m_a$ has been found satisfactory in practice

As an example of the required limiting time constant in A-M detection, consider the broadcast signal of 5 kHz and 50 per cent modulation. For this case the RC product should be

$$RC \le \frac{\sqrt{1 - 0.5^2}}{2\pi(5)(0.5)10^3} = \frac{0.866}{5\pi} \times 10^{-3}\,\text{s}$$

$$\le 0.055\,\text{ms}$$

Using the approximation resulting from practice, the value is

$$RC \le \frac{1}{2\pi(5)(0.5)10^3} = 0.064\,\text{ms}$$

For a 1,000-Ω resistor R, the value of C should be less than 0.064 μF. Larger values of resistance require smaller values of capacitance, of course.

13.19   Square-Law Detection of A-M Signals

The *square-law detector* operates on the principle that new component frequencies are generated because diodes and triodes have current-voltage characteristic curves that are nonlinear in the operating region. The plate irrent of a square-law triode detector is represented by an equation ontaining the following: two d-c terms (direct-current components), an iformation-bearing term containing modulation frequency, and an in-ormation-bearing term representing sideband components. After detection nd filtering only two component frequencies remain, one at the modulation equency containing the original information and a second-harmonic omponent that causes distortion.

Square-law detectors are used in small-signal operation when the per cent iodulation may be small and thus the distortion may be kept to a minimum. inear diode detectors are preferred in most broadcast-receiver designs.

3.20 Analysis of Frequency-Modulation (F-M)

In Section 13.8 we saw that the mathematical expression for a sinusoidal oltage can be written as

$$e = E_c \cos \left(\int \omega \, dt + \phi_o \right) \qquad (13\text{-}4)$$

:peated here for convenience. It was stated that time variation of the ngular frequency results in frequency modulation. Since an analysis of this juation is too complex for the purposes of this chapter, we will look at ome of the similarities and differences of A-M and F-M signals without gorous mathematical analyses.

To compare the waveshapes of A-M and F-M signals that are modulated y a single-frequency sine wave, refer to Fig. 13-15. The A-M wave varies s amplitude in time according to the changing amplitude of the modulating gnal, and *the F-M wave varies its frequency* according to the changing mplitude of the modulating signal. The *amplitude* of the F-M "carrier" :mains unchanged, however; carrier frequency occurs in the sketch at the istants when the modulating signal amplitude is zero.

Frequency modulation is a nonlinear process, and we should expect that ew frequencies will be generated in the process. Further, since the frequency scillates about a nominal "carrier" frequency, the new frequencies will efine a bandwidth for the F-M signal.

3.21 Bandwidth of F-M Signals

Suppose the modulating signal, a single-frequency cosine wave, is made to iry the angular frequency according to

$$\omega = \omega_o [1 + k_f \cos \omega_f t]$$

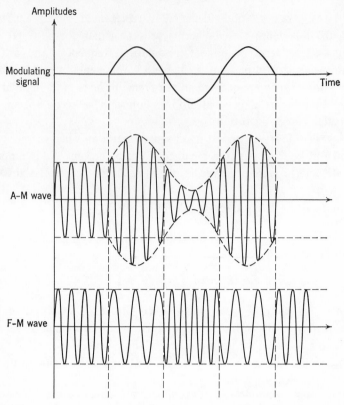

Fig. 13-15   Comparison of A-M and F-M waveshapes when modulating signal a sine wave.

where k_f is the amount of modulation of ω_o, the *center frequency*, and ω_f the modulating frequency. Substituting into Equation (13-4) and simplifyin yields this result:

$$e = E_c \cos \left(\omega_o t + \frac{\omega_o k_f}{\omega_f} \sin \omega_f t \right) \tag{13-1}$$

The coefficient of the sine term represents the deviation of the F-M sign from its center frequency, and in terms of frequency it can be written a

$$\frac{\omega_o k_f}{2\pi} = \Delta f, \quad \text{the maximum *frequency deviation*}$$

A *modulation factor*, or *deviation ratio*, is also defined for the F-M signal

$$\frac{\omega_o k_f}{\omega_f} = \frac{\Delta f}{f_f} = m_f, \quad \text{the deviation ratio} \tag{13-16}$$

The deviation ratio will have a *different value for every modulating frequency.* Because of this nonlinear behavior of F-M signals, the principle of super-position cannot be used when the modulating signal contains more than a single frequency. Of course, voice and music signals contain many frequency components, and an F-M signal that transmits them is very complex. We shall see that the F-M signal is much more complicated than that of the A-M signal, even for a signal of only one frequency.

The general expression for the F-M signal can be written in terms of the deviation ratio:

$$e = E_c \cos (\omega_o t + m_f \sin \omega_f t) \qquad (13\text{-}17)$$

The familiar identity* for the cosine of the sum of two angles may be used to express the F-M signal as

$$e = E_c[\cos \omega_o t \cos (m_f \sin \omega_f t) - \sin \omega_o t \sin (m_f \sin \omega_f t)] \quad (13\text{-}18)$$

This "simplification" results in unusual forms of trigonometric functions. In fact, it leads to what mathematicians call "Bessel functions," the use of which is beyond the scope of this book. For our purposes, we will write the F-M signal in its final form and work an example to show how to determine the bandwidth of a typical signal.

The F-M signal is expressed in terms of Bessel functions (the J factors) in this form

$$
\begin{aligned}
e = E_c[&J_0(m_f) \cos \omega_o t - J_1(m_f) \cos (\omega_o - \omega_f)t - \cos (\omega_o + \omega_f)t \\
&+ J_2(m_f) \cos (\omega_o - 2\omega_f)t + \cos (\omega_o + 2\omega_f)t \\
&- J_3(m_f) \cos (\omega_o - 3\omega_f)t - \cos (\omega_o + 3\omega_f)t \\
&+ \cdots]
\end{aligned}
\qquad (13\text{-}19)
$$

This expression continues as an *infinite series*, but, in general, higher order terms contribute small components. The important result is the generation of *sideband frequencies at all of the modulating signal harmonics.* This means that the F-M bandwidth must be considerably greater than that for an A-M signal carrying the same information. This fact is a fundamental difference between the operations of A-M and F-M systems.

In Equation (13-19), the coefficients $J_n(m_f)$ may have widely different values, depending on the deviation ratio, m_f. The coefficients are listed in the Appendix. These numbers are important in determining the relative amplitudes of the sideband frequencies and the required bandwidth. It can

* Cos $(A + B) = \cos A \cos B - \sin A \sin B$.

be seen that for small values of m_f, the amplitudes decrease rapidly. Therefore, if the center frequency does not change very much, the bandwidth will be small in order to include significant frequencies.

EXAMPLE 13-3. It will be instructive to determine the bandwidth required for a 5 kHz modulating signal, for two different deviation ratios. Suppose the modulating signal has such amplitudes that it causes a frequency deviation of first 50 and then 10 kHz, about a center frequency of 100 mHz in an F-M transmission system. This corresponds to a 5 kHz tone that decreases in amplitude.

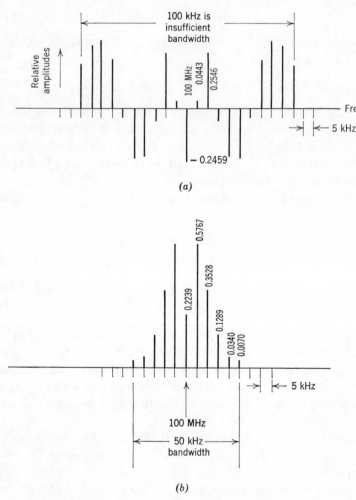

(a)

(b)

Fig. 13-16 Comparison of bandwidth requirements in F-M for deviation ratios of (a) $m_f = 10$. (b) $m_f = 2$.

Equation (13-20) is used for each of the two conditions separately, because the bandwidth is different for the two cases, even though the modulating frequency is the same in both. In order to compare the two cases directly, the F-M coefficients $(J_n(m_f))$ will be graphed on a frequency scale in Fig. 13-16.

In the case of 50 kHz frequency deviation, the value of m_f is

$$m_f = \frac{50\text{ kHz}}{5\text{ kHz}} = 10$$

The coefficients may be read from the Appendix for $m_f = 10$: -0.2459, 0.0443, etc. The minus sign for a coefficient simply means that the frequency component is shifted by 180° from those with plus sign. These coefficients are plotted in the figure.

Similarly, the second case has $m_f = 10/5 = 2$. The coefficients can be read from the Appendix and plotted as shown in the figure. It will be noted that, even though the sideband components are spaced 5 kHz apart, the bandwidth required for the larger m_f (the larger frequency deviation) is much greater than for the smaller m_f. In fact, more than the 100-kHz bandwidth shown should be used, because the coefficients have not decreased significantly to permit other higher sideband components to be eliminated. To remove these higher harmonics would generate distortion in the transmission. Conversely, the bandwidth required for $m_f = 2$ could be limited to about 50 kHz without significant distortion. The transmission system should then be able to pass frequencies from $(100 - 0.025)$ MHz to $(100 + 0.025)$ MHz, in the second case.

DRILL PROBLEM

D13-5   An F-M broadcast station is assigned a channel from 92.1 to 92.3 MHz. (*a*) What is its center frequency? (*b*) What is its permissible bandwidth for transmission? (*c*) What is the maximum permissible deviation ratio for a modulating frequency of 10 kHz? (*d*) For this f_f, how many sideband components could exist on each side of the center frequency and still remain in the bandwidth? (*e*) Plot the frequency spectrum, similar to Fig. 13-16, for this application, using the maximum deviation ratio calculated above.

13.22   Basic F-M Radio Receiver

A block diagram of a typical F-M radio receiver for the normal broadcast band from 88 to 108 MHz is given in Fig. 13-17. Except for the blocks labeled *limiter* and *discriminator*, the receiver is similar to the conventional A-M receiver. The I-F frequency is commonly 10.7 MHz in such a receiver. The audio amplifier has a much wider bandwidth than that of the A-M system, because F-M transmission can carry high-fidelity music signals within the assigned bandwidth, whereas the A-M signal is limited to signals up to 5 kHz.

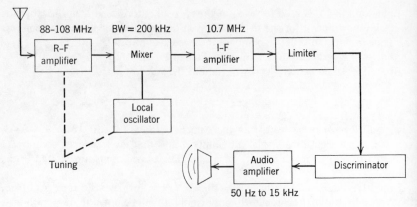

Fig. 13-17 Typical F-M receiver.

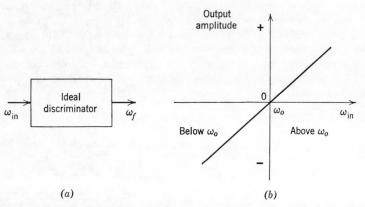

Fig. 13-18 Desired response of a frequency discriminator. (*a*) Ideal discriminator. (*b*) Output characteristics.

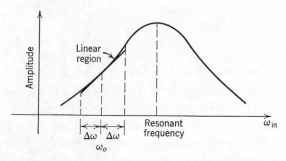

Fig. 13-19 Detuned resonant circuit characteristic for use as a frequency demodulator.

550

The limiter is essentially a saturated amplifier that clips signals whose amplitudes exceed the level permitted by the operation of the limiter. Since amplitude variations carry no information in the F-M system, but may exist because of noise pulses and atmospheric disturbances, the limiter may remove them without removing any information. Another purpose is to eliminate them before the I-F signal is applied to the discriminator, where they could cause distortion of the information signals.

13.23 The F-M Discriminator

Demodulation, or detection, of F-M signals is the process of converting them to their original location in the frequency spectrum. It may be performed by a circuit whose output voltage (or current) amplitude varies linearly with the frequency deviation of the F-M signal. The name *frequency discriminator* is commonly used to describe the circuit that *converts frequency to amplitude*. The desired characteristic of an ideal discriminator is shown in Fig. 13-18. The output amplitude varies linearly, plus and minus values, for frequency variations above and below the center frequency.

A tuned circuit may be used to approximate the operation of the ideal discriminator. If the tank is detuned so that the center frequency occurs below the resonant frequency, as shown in Fig. 13-19, the sloping part of the amplitude-frequency curve is essentially linear over a small region that may include the center frequency and the maximum frequency deviation (± 75 kHz for commercial F-M broadcasting). As the modulated F-M carrier is applied to the tuned circuit, the output amplitude will vary according to the frequency variations. This circuit is not very practical, though, because amplitude variations are superimposed on the large amplitude representing the center frequency, which carries no information. Of course, two circuits might be used in such a way that this amplitude is balanced out of the final output, similar to the operation of the balanced modulator.

A more practical arrangement is the circuit of Fig. 13-20. This circuit is known as the *Foster-Seeley balanced discriminator*. It uses a double-tuned input with transformer coupling to the diode and capacitor combinations that act as peak detectors. The output resistors operate with the associated capacitors to generate an output voltage that follows the envelope of voltage variations occurring on the two capacitors. The upper capacitor charges to follow the peak values of $E_1 + E_2$, and the lower capacitor follows $E_1 - E_2$. The output signal is then

$$e_o = |E_1 + E_2| - |E_1 - E_2| \qquad (13\text{-}20)$$

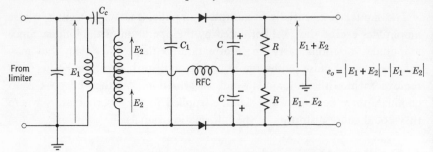

Fig. 13-20   Foster-Seeley phase discriminator.

so that it is a low-frequency signal having an amplitude proportional to the input frequency variation. When the voltage amplitudes in Equation (13-20) are equal (for the center frequency), the output is zero.

Both the primary and secondary of the input transformer are tuned to the unmodulated, center frequency. The coupling capacitor C_c is chosen to have negligible reactance at the carrier frequency, and the voltage E_2 will lead the voltage E_1 by about 90°. As the F-M signal varies, however, the phase difference varies. This occurs because the phase characteristic of a tuned circuit varies in the vicinity of its resonant frequency as the frequency varies. For small deviations about the resonant frequency, the phase-frequency curve is almost linear. The phase change that occurs for modulated signals is shown in Fig. 13-21, and its effect on the output voltage is evident. The output voltage will be nearly proportional to the frequency deviation.

The discriminator circuit is sensitive to amplitude changes of the input F-M signal, because of the way voltage E_1 enters its operation to generate

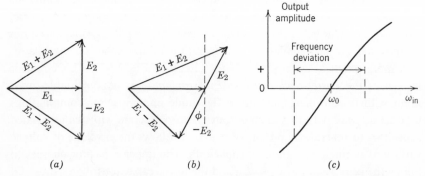

Fig. 13-21   Effect of frequency deviation on output of frequency discriminator. (*a*) Phase diagram, no modulation. (*b*) Phase diagram, below ω_0. (*c*) Discriminator characteristic.

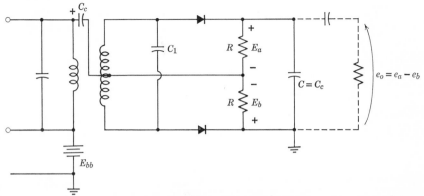

Fig. 13-22   Modified phase discriminator.

an output. For this reason, the amplifier immediately ahead of a discriminator is operated as a limiter. It clips positive and negative amplitudes of the signal and produces square-topped pulses of uniform amplitudes but of varying frequency. Since the frequency variation carries the modulating information, the limiter has no effect on the information content but eliminates amplitude variations caused by atmospheric noise.

The circuit of Fig. 13-20 is suitable for discussing the operation of the discriminator, but more practical arrangements are possible. The fact that the output voltage is "floating" about the circuit ground connection may lead to problems in the receiver. The circuit is rearranged in Fig. 13-22 to eliminate the floating ground. It will also be noted that fewer circuit components are needed. One of the output capacitors and the RFC have been removed, because they are unnecessary in the modified circuit.

It is also customary to obtain the output voltage without its d-c component, so that it will represent the original modulating signal. The R-C network shown by dashed lines may be used in practice. The capacitor C blocks the d-c component and passes the slowly-varying envelope corresponding to the original signal information.

13.24   Pulse-Code Modulation (PCM)

Many information-transmitting systems cannot operate from continuous time-varying signals either because they are not available or it is not feasible. For example, a radar system obtains target information only when the antenna scans it once each revolution. In the measurement of slowly varying quantities, such as the temperature of a chemical process, it is only necessary

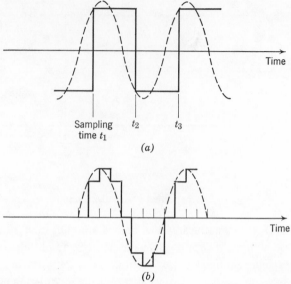

Fig. 13-23   Sample rate effect on reconstructed waveform. (*a*) Two samples per cycle approximation to a sine wave. (*b*) 8-sample approximation to a sine wave.

to sample the variable periodically rather than continuously. Continuous sampling would not yield much more information than readings taken at widely spaced points in time, particularly if the variable changes very slowly.

The famous *sampling theorem** of modern communication theory states that any two independent pieces of information about a single period of a periodically recurring variable will completely characterize the variable. For a single-frequency sinusoid, this means that two independent samples of its amplitude will completely describe its information content. This is shown in Fig. 13-23*a*, where the sine wave is approximated by two samples, one for the positive half-cycle and another for the negative half-cycle. It is seen that only two samples during one period yield a very rough approximation to the original sine wave, but according to the theorem, they are sufficient to characterize the wave. In Fig. 13-23*b*, several samples show that a much closer relationship exists between the samples and the sine wave.

In a pulse-code modulation system waveforms are sampled at intervals and the amplitude samples are arranged in a *binary code* for transmission. As an example of a simple code based on measured amplitude samples, consider the sketch of Fig. 13-24. The signal amplitude range is divided into 8 levels, including zero level.

* H. S. Black, *Modulation Theory*, D. Van Nostrand, New Jersey, 1953.

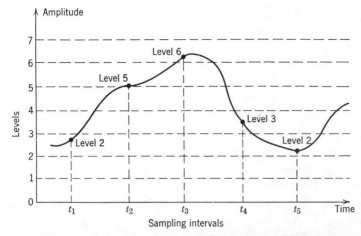

Fig. 13-24   Amplitude levels preselected to form the 8-level pulse code.

In order to have a unique pulse code for each amplitude level, a *binary* system consisting of a series of pulses of uniform height can be devised. Each separate signal amplitude level is then characterized by a coded sequence of pulses. In the binary system, the amplitudes are coded to represent the powers of 2 according to the arrangement of Table 13-3. As an example, the level 3 equals $2^1 + 2^0 = 2 + 1 = 3$; level 4 is $2^2 = 4$, and has code 100.

It should be obvious that a code based on many more levels would more closely resemble the original waveform, if it is desired to reconstruct the waveform. A continuous waveform has an infinite number of discrete amplitudes. In order to represent it by a pulse code, it is necessary to limit the number of levels used. Sampling the amplitudes according to preselected levels is called "quantizing" the waveform. The difference between the

TABLE 13-3   8-Level Binary Code and Pulse Sequence

Level	Powers of 2 $2^2 + 2^1 + 2^0$			Pulse Sequence for Pulse Code	Levels
0	0	0	0		0
1	0	0	1		
2	0	1	0		1
3	0	1	1		
4	1	0	0		2
5	1	0	1		
6	1	1	0		3
7	1	1	1		etc.

quantized level and the actual amplitude of the signal waveform leads to distortion at a receiver, called "quantizing noise." It may be minimized by using as many levels as possible, since this reduces the difference between the actual waveform and its reconstruction based on quantized samples. This was pointed out in connection with the sketch of Fig. 13-23. This increases the number of code pulses, however, because the number of levels equals 2^n, where n is the number of code pulses. In the example above, the 8-level code required 3 pulses: $2^3 = 8$.

As the number of code pulses increases, so does the bandwidth needed to transmit them. Since a waveform needs to be sampled at least twice during each cycle, the bandwidth will be twice that for the original signal. A compensating advantage of PCM, however, is its ability to sense the presence or absence of a pulse in time, even though the pulse may be severely distorted. Regenerative repeaters that are capable of reshaping distorted pulses on long-distance channels can restore the code to its original state for effective communication over great distances. The same distortion applied to the original continuous signal would mask its character so completely that communication would be impossible. Even though bandwidth requirements are greater for PCM than for conventional A-M, the ability to recognize and regenerate a pulse code more than compensates for this disadvantage in many applications.

For a more complete discussion of PCM and its many and varied applications, the reader is referred to the vast literature that is available. It is included in this chapter to show only an introduction to the use of sampling and coding as another method for transmitting information from one point to another.

SUGGESTED REFERENCES

1. Special Issue on Single Sideband (many contributors), *Proc. I.R.E.*, no. 12, **44** (1956).

2. H. S. Black, *Modulation Theory*, D. Van Nostrand, N.J., 1953 (an advanced level presentation).

3. M. Schwartz, *Information Transmission, Modulation, and Noise*, McGraw-Hill, N.Y., 1959 (advanced level).

4. F. E. Terman, *Electronic and Radio Engineering*, 4th edition, McGraw-Hill, N.Y., 1955.

5. J. D. Ryder, *Electronic Fundamentals and Applications*, 3rd edition, Prentice-Hall, N.J., 1964.

QUESTIONS

13-1   Compare the two fundamental methods of modulating a carrier, insofar as they alter the waveform of the carrier over a period of time.

13-2   What is "frequency translation" as it occurs in the process of modulation?

13-3   How is "frequency translation" manifested in the detection process?

13-4   What fundamental characteristic of a radio wave is independent of its frequency or wavelength?

13-5   How are the frequency and wavelength of a radio wave related?

13-6   Why do you think television transmission uses the VHF and UHF frequency bands rather than the MF band?

13-7   What is the significance of "selectivity" in a radio receiver?

13-8   Explain the process of "tuning" a radio receiver to separate a desired signal from all that are present at its antenna.

13-9   What is the need for I-F in a radio receiver?

13-10   Sketch from memory your concept of the waveshape of a general A-M wave that is modulated by a sine wave.

13-11   Why is it important that the carrier be much higher in frequency than the highest-frequency component of its modulated *envelope*?

13-12   What happens to the waveshape of the A-M wave if modulation exceeds 100 per cent?

13-13   Why is the bandwidth of A-M transmission required to be *twice* the highest frequency of modulating signals?

13-14   What are upper and lower *sidebands* of the A-M carrier? How are their frequencies related to the modulating signal frequency?

13-15   Why must voice transmission include both low- and high-frequency components?

13-16   Explain why the carrier power in the A-M wave is so much greater than the power of sidebands.

13-17   In a plate-modulated Class C amplifier, how does the tuned circuit at the plate provide *filtering* so that the output is only the carrier and sidebands?

13-18   Although SSB is an efficient method of conveying information, what are some of the problems that arise in the operation of an SSB receiver?

13-19   In a superheterodyne radio receiver, what is the purpose of the mixer?

13-20   What should be the bandwidth of mixer and I-F stages in a superheterodyne receiver that is designed for commercial A-M reception?

13-21   Explain the basic operation of an envelope detector.

13-22   Compare the time-varying alterations in A-M and F-M waves that are modulated by the same signal.

13-23   Why must the bandwidth of an F-M system be greater than that of an A-M system that is modulated by the same information signal?

13-24   Some of the coefficients of F-M frequency components (Appendix) are negative. What does this signify?

13-25   What is the purpose of the "limiter" in an F-M receiver that uses a frequency discriminator as a detector?

13-26   Explain the basic operation of a frequency discriminator as a detector for F-M.

13-27   What is "quantizing noise" in a pulse code?

13-28   Show that 4 amplitude levels of a signal, including zero, require 2 pulse positions to describe the signal levels, whereas 16 levels require only 4 pulse positions.

13-29   What is the primary advantage of PCM over continuous signal transmission, particularly for long-distance communication?

PROBLEMS

13-1   The amplifier in Fig. P13-1 generates a load voltage that is related to the grid signal by $e_L = a_1 e_g + a_2 e_g{}^2$. (*a*) Determine all frequencies present in the load voltage when $e_g = \cos \omega_a t$. (*b*) What changes or additions to the circuit would permit it to be used as a second-harmonic generator for the input frequency?

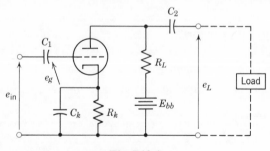

Fig. P13-1

(*c*) Assume $a_1 = 10$ and $a_2 = 1$. Calculate the ratio of second-harmonic amplitude to fundamental amplitude in the output. (This is "second harmonic distortion.")

13-2   In the circuit of Problem 13-1, the input signal is: (*a*) $2 \cos 100t$. Calculate the frequency components and their amplitudes in the load. (*b*) $2 \cos 100t + \cos 500t$. Calculate the frequency components and amplitudes. What new frequencies are present that were not in (*a*)? How did they arise?

13-3 A nonlinear amplifier has an output expressed by

$$e_o = 20(1 + 0.5 \cos 2\pi \times 1000t) \cos 2\pi \times 10^5 t$$

(*a*) Determine all frequency components of the wave. (*b*) What is the degree of modulation? (*c*) Calculate the relative power levels for all frequency components.

13-4 An A-M wave of $f_o = 1$ kHz is modulated at 50 per cent by $f_a = 500$ Hz. Plot a time graph of three frequency components over one cycle of the lowest frequency and add them point-by-point, to obtain the A-M wave. Consider ways of measuring m_a, e.g., with an oscilloscope.

13-5 Calculate the wavelength of a 100 Hz wave. Of a 1 MHz wave.

13-6 Calculate the frequency range of waves having wavelengths from 1 to 5 m.

13-7 What are the wavelengths of frequencies known as "microwaves"?

13-8 An A-M signal consists of a carrier voltage 10 sin $\omega_o t$ and sidebands generated by the modulating signal 5 sin $\omega_a t$. (*a*) What is the modulation precentage? (*b*) Sketch the envelope over a modulation cycle. Label with numerical values of significant points. (*c*) What average power will be delivered to a 100-Ω load? What peak power?

13-9 Assume that the carrier in Problem 13-8 is suppressed after modulation. How much may the average power of the sidebands be increased and not exceed the peak power level of Problem 13-8?

13-10 An A-M broadcast station is assigned a carrier of 1 MHz and is limited to a 10 kHz bandwidth. (*a*) What range of information signal frequencies may it transmit? (*b*) If its total radiated power is 1 kW, how much of this represents signal information?

13-11 The frequency spectrum of an A-M wave may be plotted by vertical bars representing amplitudes placed at the frequency component on a frequency scale. A certain A-M wave has an amplitude of 3 at 10 kHz, 1 at both 8 and 12 kHz, and 0.5 at 6 and 14 kHz. Sketch the frequency spectrum and write the time equation for this waveform.

13-12 A 5 kHz signal is used to modulate a 100 kHz carrier, and the upper sideband is transmitted in a SSB system. At a receiver the incoming signal is mixed with a local oscillator operating at 95 kHz before detection. Assuming the mixing and detecting operations regenerate the lower sideband, discuss the output signal from an audio amplifier following the detector. (What happens if the local oscillator varies from 95 to 100 kHz in a random manner?)

13-13 Show that the time constant RC defined for an envelope detector in Example 13-2 will be satisfactory for all frequencies lower than the maximum signal frequency and for all modulation percentages, if the filter is designed for maximum frequency and 100 per cent modulation as the "worst-case" condition.

13-14   An A-M transmitter is tested as shown in Fig. P13-14. The receiver includes a linear detector having a bandwidth of 1 kHz. The R-F amplifier in the receiver is swept successively through the range from 100 kHz to 1 MHz. With no audio input to the transmitter, the wattmeter reads 1 kW average power. The peak-reading VTVM indicates 10 V at 800 kHz. (*a*) For an audio signal of 10 V peak at

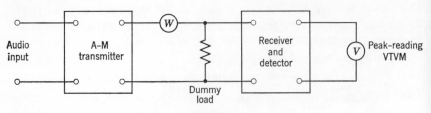

Fig. P13-14

2 kHz the wattmeter indicates 1.5 kW average power. What are the voltage peaks indicated by the VTVM? (*b*) At what frequencies will the voltmeter indicate an output?

13-15   An F-M system uses 1 MHz as a center frequency. The modulating signal at 1 kHz has an amplitude such as to generate a maximum frequency deviation of 2 kHz. (*a*) Find the required bandwidth to pass the F-M signal. (*b*) Repeat (*a*) for a 2 kHz signal of the same amplitude. (*c*) Find the bandwidth necessary to pass the signal for a modulating signal having twice the amplitude at 2 kHz.

13-16   A modulating signal of amplitude 1 V at 1 kHz is used to modulate a 1 MHz carrier in both A-M and F-M systems. The 1-V amplitude produces a 2 kHz frequency deviation in the F-M system. (*a*) Compare the bandwidths required in the R-F amplifiers and the audio amplifier in both A-M and F-M receivers. (*b*) Repeat (*a*) for signal having 3-V amplitude.

13-17   Two F-M signals are tuned into a receiver alternately. One has an audio modulating signal at 2 kHz and requires a bandwidth of 20 kHz. The other requires a bandwidth of 100 kHz for a 10 kHz modulating signal. The carriers of both have the same intensity. Compare the amplitudes of the audio signals.

13-18   In a general F-M transmission system, how many sideband frequency components are generated? In the frequency spectrum for the F-M wave, what determines the spacing between adjacent spectral amplitudes? What determines their amplitudes? Under what conditions could the carrier amplitude be zero?

13-19   An F-M signal has a maximum frequency deviation of 50 kHz and is modulated by a 10 kHz signal. If a receiver has a bandwidth of 50 kHz in its R-F amplifiers, is this adequate for good reproduction of the 10 kHz audio signal?

13-20   A "slope detector" utilizes a detuned tank circuit to obtain a modulating signal from an incoming F-M signal, as given in the block diagram of Fig. P13-20.

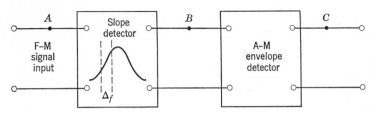

Fig. P13-20

Assuming a sine-wave modulating signal, sketch the waveforms at points *A*, *B*, and *C*.

13-21 Show that a general waveform can be approximated by taking samples of its amplitude at equal time increments. Suppose that 10 quantizing levels are used to generate a pulse code for the waveform samples. How many pulses are needed in the code, assuming it to be the simple binary code discussed in Section 13.24?

CHAPTER FOURTEEN

Glow and Arc-Discharge Tubes

Electron tubes containing mercury vapor or gases like argon and helium at very low pressures are able to conduct much larger currents than are vacuum tubes. When the cathode of such a tube is heated, it is capable of emitting a very large supply of electrons which, in turn, separate other electrons from atoms of gas within the tube, with the result that large currents may flow in the tube and through the circuit in which the tube is connected. The tube has the ability to act as a sensitive high-speed switch across which is a low, constant voltage drop and through which current flows according to the amount of resistance and voltage *in the circuit external to the tube.*

A grid-controlled type of arc-discharge tube, the *thyratron*, and an igniter-rod-controlled type, the *ignitron*, are the subjects of discussion in this chapter. The thyratron is used in a vast number of electronic-control circuits. The ignitron, more suitable for heavy duty, is used in power rectification and in control circuits where relatively large amounts of power are handled.

Glow discharges and arc discharges differ in many respects. A glow often precedes an arc discharge. Glow tubes are not used so extensively as arc tubes. Their principal use is in illumination, as voltage regulators, and as signal lamps in various forms. Presentation of some basic material on glow discharges preceding a study of arcs is appropriate.

14.1 Electrical Conduction in Gases

A glow tube is a sealed container, usually made of glass, from which air has been evacuated as completely as possible and into which another gas is fed under carefully controlled conditions so that the final pressure is very

562

low. Ions are formed when a few gas atoms receive energy from cosmic rays, which are extremely high-energy waves from outer space, or from other ionizing agents. When a potential difference is applied between two electrodes in the tube, these ions are driven by the established field and constitute current flow. This current is so small, perhaps less than a micro-ampere, that its presence is not revealed in any noticeable way. It may therefore be called the *dark current*. *Glow discharges*, as the name implies, are accompanied by the emission of light which is due to the changes in energy levels of electrons in the gas atoms. The currents are much larger (milliamperes or amperes), and therefore much more energy is put into the discharge, than in the dark-current case.

14.2   Glow and Arc Formation; Current and Voltage Relations

Figure 14-1 shows the relations between the terminal voltage of a cold-cathode diode (glow tube such as the types used in reference-voltage applications) and the tube current, while the anode voltage is being increased from zero to glow-discharge value. As the voltage is increased from zero, the dark current (called Townsend current after J. S. Townsend, an early re-searcher in gaseous conduction) increases slowly until the voltage approaches that at which the tube operates with a glow discharge. Protective resistance is used in the anode circuit. When the anode voltage reaches a high value,

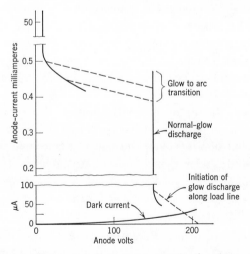

Fig. 14-1   Voltage-current curves for gaseous conduction between cold cathode and anode. Scales different for glow and arc currents.

ions are forced by the electric field to strike the cathode with enough energy to eject electrons which immediately travel under acceleration toward the anode. A process called *gas amplification* exists when high-velocity electrons travel through a gas. An electron strikes a gas atom and knocks an electron out of it. The two of them go on toward the anode while the ionized gas atom, now a *positive ion*, travels toward the cathode. If the two electrons can gain sufficient energy to enable each to knock another electron out of an atom before they reach the anode, two more ions will be formed and there will be four electrons reaching the anode instead of only the first one. Thus current increases rapidly at the higher voltages.

Soon there will be so much current that a glow discharge develops, and the current, which is now in the milliampere range, is determined by the applied voltage and the external resistance in accordance with conventional load-line theory. The glow is due to radiation in the visible band of the spectrum sent out by electrons of *excited atoms* of the gas as they fall back to lower-energy states. The *ionization potential* of a gas is the energy, in volts, which an electron must have in order to eject an electron from a neutral gas atom upon impact. If an electron with a smaller amount of energy strikes the atom, *excitation* will result without ionization. An *excited atom* has one or more electrons in energy states that are higher than their normal energies. The higher-energy electrons soon fall back into their original energy states or to other states of energy lower than their excitation energies.

We have seen in the study of gas diodes how the voltage regulator tubes (0A3, 0C3, etc.) can accommodate changes in their current with practically no change in their terminal voltage. The vertical portion of the curve in Fig. 14-1 from the 0.2 to the 0.4 mA region represents this type of operation. The tubes just mentioned have larger cathodes and anodes than the one these curves represent.

Somewhere near the top of the vertical portion of the curve, the cathode gets hot enough from positive-ion bombardment to begin emitting electrons by thermionic emission. Gas amplification continues and in a few microseconds the glow discharge changes into an arc discharge. In both types the tube has no limiting action on the current; *only the applied voltage and circuit resistance determine the amount of current through the tube.*

14.3   Physical Nature and Potentials of a Glow Discharge

When a glow discharge takes place in a tube containing gas at very low pressure (less than 1 mm of mercury) close observation detects distinct

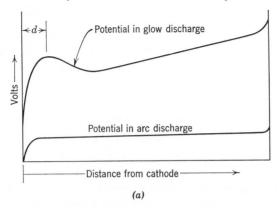

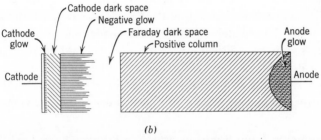

Fig. 14-2 Characteristics of a glow discharge between plane parallel electrodes. (a) Potential distributions in glow and arc tubes; (b) sections in glow discharge. [After Slepian and Mason, *Elec. Eng.*, **53**, 511 (1934).]

regions inside the tube. These are illustrated in Fig. 14-2b. The cathode glow and the negative glow are about all that can be seen in daylight in commercial electronic tubes of frequently used types, but in tubes built for research a very faint bluish light is seen in the cathode dark space, and a pink hue is seen in the positive column, both when the tube contains air. Elements have their characteristic colors of emitted light, and this accounts for different colors in the clear glass tubing used in glow-discharge electric signs, popularly called *neon signs*. Neon, helium, and argon are some of the elements used. Colored glass is used in tubing to produce other color effects.

The potential distribution curve for the glow discharge shows a very rapid rise of voltage with increase of distance from the cathode, almost reaching the anode value. Positive ions going down this high potential hill (they actually travel only the short distance *d*) acquire sufficient energy to knock electrons out of the unheated cathode by positive-ion bombardment.

Ions starting over near the anode move naturally down the more gentle slope in the positive column, and most of them acquire enough energy to coast up the reverse slope in the Faraday dark space and negative-glow regions. The dip in the potential line is considered to be due, not to electron space charge, but rather to the fact that there is a larger positive-ion density in the region where the hump is located than in the region where the dip is located. Ions are very heavy compared to electrons and they move much more slowly. This is natural because exactly the same field force that acts on an electron acts on the singly ionized gas atom as they have the same amount of net electric charge. The ion is thousands of times heavier than the electron. Electrons do not stay in the region of the dip but immediately move in the direction of the anode. Adjacent to the anode the potential curve shows a rise, but there may be a fall. The anode is not generally a positive-ion source. The electric field there tends to drive positive ions away from the region immediately in front of the anode. The result may require a potential rise to get sufficient electrons per second into the plate to satisfy the current demands of the circuit. However, a large anode area may exist and then a slight dip in potential may be needed to regulate the rate of electron arrival to that required by the current demand of the circuit.

14.4   Paschen's Law; Ignition and Breakdown

Breakdown occurs in a gas tube when the current has increased sufficiently to cause a substantial drop in an increasing voltage applied to two electrodes in the gas. In the breakdown state the current is limited in value only by the circuit conditions outside the conducting-gas region.

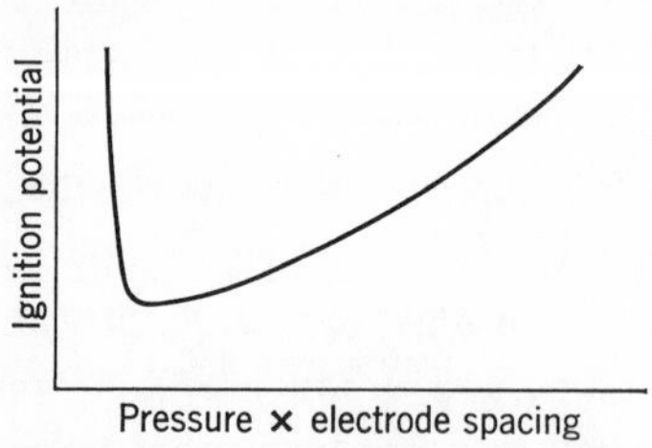

Fig. 14-3   Paschen's law curve. Relations for ignition in gas diodes.

Ignition has been defined as *the advent of a self-sustaining discharge.* Breakdown does not necessarily follow ignition, as will be shown in the discussion of *Paschen's law.*

Ignition potential has been found experimentally to be a function not only of the kind of gas and the cathode material but also of the product of the gas pressure and the distance between the electrodes. This dependence is shown in Fig. 14-3, as suggested by Paschen, an experimenter in physics in the late nineteenth century. The relation between ignition potential and the product of pressure and electrode spacing, as shown by this curve, is known as *Paschen's law*. To the left of the minimum point the increase in voltage required to produce ignition is due to these facts: (1) there are comparatively few gas molecules in the shortened space, and (2) the short distance electrons have for travel materially reduces their chances of having ionizing collisions before they arrive at the anode. The increase to the right of the minimum point is also caused by two facts: (1) increase of pressure reduces the spacing between molecules, resulting in more *collisions* with electrons *per centimeter of electron advance* up the potential hill, and this means fewer electrons can reach ionizing energy before the next collision; and (2) increase of spacing decreases the intensity of the electric field so that the force accelerating an electron up to ionizing values of kinetic energy is reduced. Because of this, fewer electrons acquire ionizing energy between collisions with molecules and atoms.

14.5   Arc Discharges

The curve near the bottom of Fig. 14-2a shows the potential along the path of an *arc discharge* in a tube to be much lower than that of a glow discharge through same type of gas. The main reason for the much lower voltage is that in an arc discharge electrons are *thermionically emitted* from the cathode, while in a glow discharge a high potential hill must build up in front of the cathode to get electrons out *by the mechanism of positive-ion bombardment*. Every arc discharge has a cathode fall of potential approximately equal to the ionizing potential of the gas or vapor conducting the current.

The main current-carrying region of both arc and glow discharges is called the *plasma*. The plasma is similar to the interior of a metal conductor in that it has positive ions and electrons in equal concentrations and hence no net space charge. A substantial *drift current* can be driven through a plasma by a moderate rise in potential across it.

Some ions leave the plasma of an arc-discharge tube by moving sidewise and entering the walls of the tube. There are weak radial electric fields that cause this. To supply this energy loss, compensating energy must be put into the plasma. The voltage drop lengthwise through the plasma multiplied by the plasma current provides power to produce this energy.

The temperatures *of electrons* in the plasma of an arc discharge are very high—of the order of 20,000°C—because of their extremely high random velocities. Some electrons must ionize gas particles to make up for the loss of ions to the lateral boundaries. High electron temperatures accompany these high electron energies.

14.6 Comparison of Glow and Arc Discharges

A summary of the differences and similarities of glow and arc discharges is given in Table 14-1.

TABLE 14-1

Features	Glow	Arc
Mechanism of electron emission at cathode	Positive-ion bombardment	Thermionic
Amount of current	Tens or hundreds of milliamperes	Few hundred milli-amperes to tens of thousands of amperes
Voltage drop	Approx. 70 to 1,000 V	Usually less than 100 V
Cathode fall space voltage drop	Large	Small
Plasma present	Yes	Yes
Gas pressures	Approx. 0.03 to 3 mm Hg	Approx. 0.03 mm Hg to a great many times atmospheric pressure

14.7 The Thyratron

The thyratron has a heated cathode and a control grid, which may be a wire mesh or a flat metal disk, with one or more holes in it, supported inside a metal cylinder. The anode of a small thyratron is similar in construction to the anode of a high-vacuum tube, but in medium and in some large sizes it has the form of a horizontal metal disk with a short metal cylinder extending upward from its circumference. A sketch of the internal construction of a medium-sized thyratron is shown in Fig. 14-4.

The cathode of a thyratron may be either directly or indirectly heated. In order to reduce cathode-heating power and at the same time attain sufficient operating temperature, cathodes are closely wound to reduce heat radiation. Even with this kind of construction a medium-sized thyratron takes 10 A of heating current at 2.5 V. The purpose of the low voltage is to keep low the peak value of potential difference between anode and end-turns of the cathode during conduction. This prevents destruction of parts of the cathode by positive-ion bombardment. Gas ions are accelerated toward

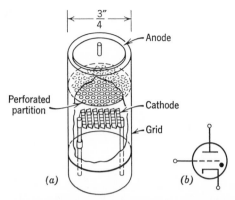

Fig. 14-4 (*a*) Internal structure of a medium-sized thyratron; (*b*) schematic symbol. The grid is a metal cylinder with a horizontal plate (perforated partition) located between the cathode and the anode.

the cathode by the electric field. The acceleration caused by the normal voltage drop of the arc discharge does not produce ion velocities that do harm. The normal voltage drop from anode to cathode of a mercury-vapor arc tube is about 10 to 15 V. If, for any reason this should rise to 20 to 25 V, it is practically certain that the cathode has been damaged by positive-ion bombardment.

The cathodes of thermionic gas tubes must be brought up to operating temperature before plate voltage is applied. This may require from 5 to 10 s or from 5 to 10 min, depending on the size and newness of the tube. When plate voltage is applied before the cathode is hot enough to emit electrons fast enough for the operating plate current, the plate voltage rises to abnormal values because the insufficient current does not cause the voltage drop across the plate-circuit resistor to be as high as it should. The higher-than-normal plate voltage accelerates the positive ions to such high velocities that they strike the cathode with destructive force and thus permanently damage the emitting surfaces. This process is cumulative in that the more cathode-surface destruction there is, the less the current will be and the higher the voltage drop across the tube. The positive ions are thus driven into the cathode with still greater force.

14.8   Grid-Control Features

Current does not flow (except for a few microamperes, which are insignificant) in a thyratron until there is *breakdown* in the anode-cathode space. Breakdown is the formation of an arc discharge. It is *permitted to take*

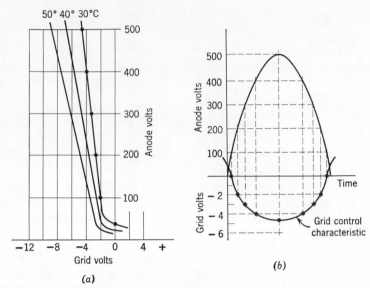

Fig. 14-5   (*a*) Grid-control curves of a mercury-vapor thyratron. (*b*) Grid voltage needed to fire at 30°C and sine-wave voltage on the anode.

place by the action of the grid. The grid may be made sufficiently negative to prevent conduction even at the highest positive anode voltages that the tube is built to handle. When the amount of negative bias on the grid is reduced to a critical value, however, the tube will conduct or *fire*. A set of curves relating fixed values of anode voltage to corresponding grid voltages at which conduction, or firing, takes place is shown in Fig. 14-5*a*. The following explanation may make the meaning of this statement clear. Suppose the temperature is 30°C, the anode voltage is 400 V, and the grid voltage is −10 V. The tube is not conducting. As the grid voltage is raised (made less negative), the tube remains nonconducting until −4 V is reached, at which instant it fires. These are important curves because they tell not only how much bias is needed to keep the tube dark (nonconducting) but also how large a positive voltage must be superimposed upon the bias to fire the tube *at the existing anode potential.*

When an a-c voltage is applied to the anode, there are minimum values of instantaneous negative grid voltage that will just keep the tube from firing. These are indicated by the grid-control characteristic curve (drawn for 30°C operation) in Fig. 14-5*b*. If an a-c voltage of 500 V maximum amplitude is applied between anode and cathode, a steady negative grid voltage larger than about 4.5 V will keep the tube from firing.

Some thyratrons require a small positive voltage on the grid before they will fire. Tubes of this type are necessary in some control circuits. They are built with a type of grid which so thoroughly shields the anode from the cathode that the electric field potential in front of the cathode requires positive charge on the grid in order to increase sufficiently to allow adequate electron flow for firing of the tube.

After a thyratron has fired, the grid completely loses control. The tube has then become merely an arc-discharge *diode*, having the characteristics described for arc-type diodes in Chapter 4. Like the gas diode, *the conducting thyratron can place no limitation on the current passing through it.* The current is determined *entirely* by the resistance in the circuit external to the tube and the voltage applied to the circuit. The grid regains control when the anode current falls to zero. Several ways of causing this to happen will be discussed later. A highly negative grid will succeed in shutting off a conducting thyratron if the anode current is very small. This is a trivial situation, however.

A negative grid in the path of an arc discharge quickly becomes surrounded by a collection of positive ions. A *positive-ion sheath* is thus formed which causes the grid to lose its effectiveness in controlling anode current. The sheath is not the result of a higher-than-usual concentration of positive ions but rather an absence of electrons. Ions are much larger and heavier than electrons and consequently move much slower. Their effect around a grid wire, or just outside the surfaces of other types of grid structure, is to neutralize the field effects of negative charges on the grid itself. The electric field in the *plasma*, which is the main body of the arc discharge, is consequently unaffected by the negative potential on the grid, and so the grid does not have control of the arc current.

DRILL PROBLEMS

D14-1 Sketch the grid-control characteristic of the thyratron of Fig 14-5, for anode voltage = 400 V d-c and operating temperature 30°C. Why is it different for a-c voltage on the anode?

D14-2 A sine wave of voltage having crest value of 400 V is connected between anode and cathode of the thyratron of Fig. 14-5. If grid bias is −6 V, will the thyratron fire at 30°C temperature? At 40°C? At 50°C?

14.9   Grid Current in a Conducting Thyratron

The variation of grid current with grid voltage in a conducting thyratron tube is interesting and important. It is shown in Fig. 14-6. Over a large

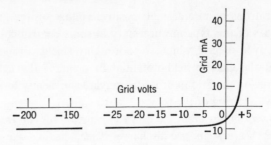

Fig. 14-6   Grid current plotted against grid voltage in a conducting thyratron.

range of negative grid voltages the grid current is practically constant. There is only a slight increase in current when the grid is made more negative by as much as 300 V. The positive-ion sheath increases in thickness, and neutralizes so effectively the effort of the grid to lower the potential in the region around it, that there is very little increase in ion penetration into the grid surface.

In the regions of Fig. 14-6 where the grid current is practically constant (grid more than 5 V negative), the grid current is all *positive-ion current*. Points on the bend of the curve, but below the horizontal axis, indicate current values that are the resultant of positive-ion current and *electron current*. Both kinds of particles—ions and electrons—find their way to the grid in this region of grid voltage. When the grid is slightly negative, the kinetic energies of some of the electrons, all of which are flying around at high random velocities, are great enough to force them through the sheath, which is very thin at such low grid voltage. When the grid is slightly positive, there is no positive-ion sheath and electrons are readily received, but the high random energies of some positive ions carry them into the slightly positive grid in spite of the opposing electric field due to the positive grid surfaces.

At the point where the curve crosses the voltage axis, the ion and electron currents are equal and cancel. Above this point the ion current rapidly becomes zero, and only electron current flows. This current rises very rapidly with grid voltage. In fact, an arc discharge between cathode and grid will take place at relatively small positive grid voltage because the grid is already in a plasma crowded with electrons and ions. An arc discharge to the grid would destroy it quickly.

We can readily understand the necessity of having sufficient resistance in series with the grid to limit the grid current to safe values. Such limiting action comes about because at safe values of grid current the *IR* drop through the grid-circuit resistance prevents the rise in potential of the grid with respect to the plasma, to a dangerous value. It should be noted that, although positive-ion current flow through a grid resistor adds a positive value to the grid voltage, there is little danger that a grid will be damaged while it is negative. Grids are usually able to carry safely the currents that flow when they are negative. However, grids swing positive when operated on alternating voltages, and they must be protected from resulting excessive electron current. It is fortunate that electron current flow provides some negative-bias voltage, as in the case of a high-vacuum tube operating with grid-leak bias.

14.10   The Thyratron in a D-C Circuit

A thyratron may be used in a d-c circuit to control the flow of current through a circuit branch, either by conducting the current itself or by conducting current through a relay coil which closes another independent circuit. After the thyratron has fired, it must be rendered nonconducting, i.e., extinguished.

A common method of extinguishing a thyratron is employed in the d-c control circuit of Fig. 14-7. This circuit is shown in order to illustrate how

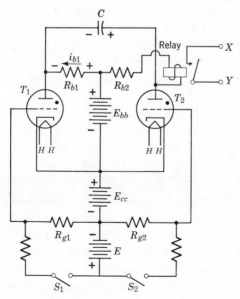

Fig. 14-7   Direct-current thyratron control circuit, T_1 conducting.

a relay, serving as a switch in a line (X-Y), could be operated by thyratron control when only direct current is available. It will be seen later, however, that a much simpler circuit, employing only one thyratron, can do the same thing when operated with alternating current.

Suppose tube T_1 is conducting and T_2 is not. This may be accomplished by momentarily closing the spring switch S_1. Current i_{b1} flows through R_{b1}, causing it to have positive polarity on the right and negative on the left. The capacitor C is simultaneously charged through R_{b2} and the relay coil, positive on the right and negative on the left. The capacitor-charging current is not large enough to operate the relay. The resistance of the relay coil is very small compared to the resistances of R_{b2} and R_{b1}, which are equal. The capacitor quickly becomes fully charged and R_{b2} then carries zero current. Both spring switches (S_1 and S_2) are open. T_1 is now conducting, T_2 is not, and the capacitor is charged with the polarity indicated in the drawing. The plate of T_2 is at full battery potential above its cathode because there is no voltage drop in R_{b2} or the relay coil.

Now imagine an object passing from left to right and closing spring switches S_1 and S_2 momentarily in that order. Tube T_1 is already conducting, so closing S_1 again has no effect. At the instant the moving object closes S_2, current flows from the battery (E) through R_{g2} and raises the grid potential of T_2, allowing it to fire. Its plate current flows through the relay coil and closes the relay contacts connecting X and Y.

T_2 acts as a switch to connect the right-hand side of capacitor C to the cathode of T_1, making it positive with respect to its plate. The plate is thus *negative with respect to its cathode*, and so T_1 is extinguished because with its plate negative the current is reduced to zero. The grid of T_1 immediately regains control. T_2 has acted as a switch with approximately 15 V drop across it. But C had been charged up to, say, 100 V $= E_{bb}$. Thus the firing of T_2 caused about $100 - 15 = 85$ V negative to be applied to the plate of T_1.

At the instant T_2 fires, the capacitor C starts to discharge through T_2, E_{bb}, and R_{b1}. In a very short time C is completely discharged but immediately begins to charge again, this time with its polarity reversed. It charges through R_{b1}; the plates connected to T_2 are negative. Soon the capacitor attains full battery voltage, and its charging current through R_{b1} is reduced to zero. The plate of T_1 is now at positive battery potential, but its negative grid prevents firing. When the moving object again closes S_1 and then S_2, the whole action described is repeated, but this time closing S_1 fires T_1, and C discharges through T_1, extinguishing T_2 because of negative polarity on its plate.

The relay contact circuit X-Y may be used to operate a counter, energize another control circuit, or do many other things. The circuit will operate with moving objects closing S_1 and S_2 at high speed. The resistances R_{g1} and R_{g2} are large, and E may be only a few volts. As a result, the current through S_1 and S_2 is small, and there will be no arcing or burning of their contacts.

14.11   The Thyratron in an A-C Circuit

When a thyratron is operated in an a-c circuit, the phase angle separating the plate voltage and grid voltage is of great importance. It determines the instant at which the tube fires. Firing can occur *only during the positive half-cycle of plate voltage*, of course. As the plate voltage goes through its positive-half-cycle instantaneous values, there are corresponding instantaneous values of negative grid bias that will *just prevent* the tube from firing. For example, when the plate voltage has the value A in Fig. 14-8, a negative bias value (e_c) will just prevent firing. The value of e_c may be obtained, when e_b is known, by locating e_b on the curve of Fig. 14-5a, corresponding to the

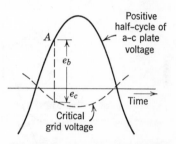

Fig. 14-8   Critical grid voltage curve for a thyratron.

operating temperature. It is obvious that a set of values of e_c may be obtained for the complete half-cycle of plate voltage and a *critical grid voltage curve* plotted as shown in Fig. 14-8. This is an important curve in the analysis of a-c operation of a thyratron.

A thyratron will operate satisfactorily in an a-c circuit without a d-c bias voltage. The grid must be protected from damage, since it goes substantially positive in most cases. When the grid voltage lags the plate voltage, it is possible to vary the average plate current flowing through the tube over the full range from maximum to zero by *simply varying the phase angle by which the grid voltage lags the plate voltage*. That is a very important matter because many useful control circuits owe their success to this phase-control feature, made possible by the operation of a thyratron on alternating current.

Figure 14-9 shows two conditions of firing of a thyratron, in which the grid voltage lags the plate voltage (*a*) by a small angle, and (*b*) by a large angle. In each case the tube fires at the instant the grid voltage comes up to the critical value, as indicated at the intersection of the grid voltage wave and the critical grid voltage curve. After the tube fires, it conducts throughout the remainder of the half-cycle, except for a very short time at the end, when the plate voltage becomes so low that it cannot maintain the arc discharge. This effect is of no practical significance, however. At the instant the tube fires, the plate voltage drops as shown in the sketch to a very low, constant value instead of continuing at sine-wave values.

It is readily seen that when the grid voltage is in phase with the plate voltage, or when the angle of lag is very small, the tube will conduct during practically the entire half of the cycle, *and the average plate current must then have its maximum value.* By gradually increasing the angle of lag, the

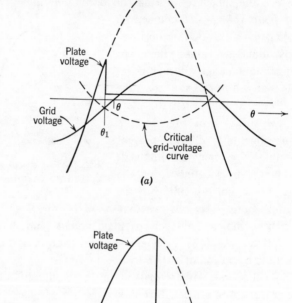

(a)

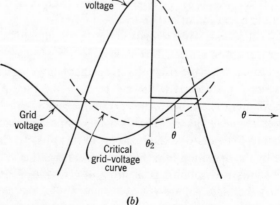

(b)

Fig. 14-9   Phase control of firing of a thyratron. (*a*) Grid voltage lags plate voltage by small angle θ. Tube fires and plate voltage drops at θ_1. (*b*) Grid voltage lags plate voltage by large angle θ. Tube fires and plate voltage drops at θ_2.

firing of the tube may be delayed more and more, until at about a 170-degree lag the tube does not fire at all. Thus the average plate current may be varied over a wide range, as stated before.

EXAMPLE 14-1.   The thyratron whose characteristics are given in Fig. 14-5 has a sine-wave voltage impressed between anode and cathode, and it is operating at 30°C. It has a variable d-c bias in the grid circuit, that can be varied from 0 to −5 V. The anode voltage can be expressed as 400 sin ωt. We want to determine the waveform of anode voltage as the grid bias is varied through its range.

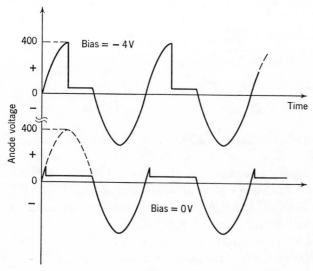

Fig. 14-10

From the characteristics shown in Fig. 14-5, it is seen that the thyratron will not conduct until the grid bias is less negative than about 4 V. Bias values between 0 and −4 V will permit conduction. The two extremes are shown in Fig. 14-10.

14.12   Phase-Control Circuits

We shall now use some a-c circuit fundamentals. In a circuit consisting of a resistor and a capacitor in series, the current *leads* the applied voltage by an angle that is determined by the resistance of the resistor and the reactance of the capacitor. If the frequency is kept constant, the angle of lead of the current will be reduced if either the resistance or the capacitance is increased or if both are increased. It will be recalled that increasing the capacitance *reduces* the capacitive reactance. The circuit and phasor diagram for two values of R are shown in Fig. 14-11.

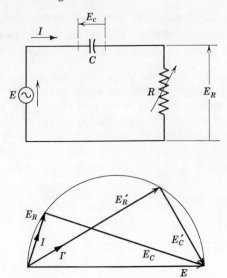

Fig. 14-11   Phase-control circuit and phasor diagram.

There are a number of practical phase-control circuits employing either capacitors or inductances with variable resistors. A sample one is shown in Fig. 14-12. The capacitance need not be variable if its reactance is not greater than about $R/10$ Ω, and if R is large. Such a combination will give a wide range of phase-angle variation, and as a result a wide range of average current variation. The transformer should have low d-c resistance in its windings and be capable of carrying the load current. The current

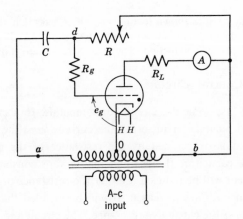

Fig. 14-12   Thyratron phase-control circuit.

through R and C will be small. If $C = 1\ \mu\mathrm{F}$, its reactance is a little over 2,650 Ω on 60 Hz. With this value of C, R should be about 26,500 Ω.

Before the thyratron fires, current flows only through C and R, and it is produced by the secondary terminal voltage (E_{ab}) of the transformer. It is seen in the diagram that the amount of voltage that the grid is above the cathode is the amount from point o to point a plus the amount from point a through the capacitor to point d. That is,

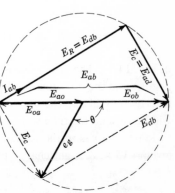

$$\vec{e}_g = \vec{E}_{oa} + \vec{E}_{ad} \qquad (14\text{-}1)$$

The arrows above the symbols mean that this is a phasor equation, and so the two voltages on the right must be added as phasors. We are interested in the phase of e_g with respect to the *plate voltage before firing*, which is E_{ob}. This may be ascertained by means of the phasor diagram of the circuit as it operates *before the tube fires*.

Fig. 14-13   Phasor diagram for Fig. 14-12 before the tube fires. I_{ab} flows from a through C and R to b.

The phasor diagram (Fig. 14-13) is drawn in the following steps:

1. $E_{ab} = E_{ao} + E_{ob}$ on the horizontal.

2. I_{ab} leads E_{ab} by any convenient angle. (The tangent of the angle is, however, X_c/R from Fig. 14-12.)

3. Since $\vec{E}_{ab} = \vec{E}_{ad} + \vec{E}_{db}$, as seen in the actual circuit diagram, the upper right triangle is formed with $E_R = E_{db}$ in phase with I_{ab} and $E_C = E_{ad}$ lagging I_{ab} by 90 degrees. The current leads the capacitor voltage.

4. Since $E_{oa} = -E_{ao}$, we have from Equation (14-1)

$$\vec{e}_g = -\vec{E}_{ao} + \vec{E}_{ad}$$

where e_g is the grid voltage before conduction. In order to show this phasor subtraction to better advantage, the phasor E_C will be shifted; this is permissible if its length and angle are not changed. Show E_{oa} (dashed) opposite and equal to $+E_{ao}$ in Fig. 14-13. Shift $E_C = E_{ad}$ to the dashed position and draw the phasor e_g. The grid voltage e_g is seen to lag the plate voltage E_{ob} by the angle θ. Varying R will vary E_R, and hence the angle.

On-Off Control

It is possible to arrange the phase-control circuit so that the tube carries *either maximum current or none at all*. The principle of the method is shown

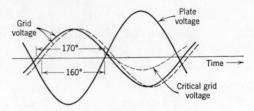

Fig. 14-14   Phase relations of voltages for on-off control.

in Fig. 14-14. The grid voltage leads the plate voltage by approximately 170 degrees and the tube does not fire. If the grid voltage is shifted slightly so that it leads only about 160 degrees (dash curve), the critical grid voltage curve will be cut and the tube will fire as soon as its plate voltage gets high enough to ionize the gas in the tube. For mercury-vapor thyratrons that voltage is around 15 V. A slight change in one of the constants of the circuit will shift the phase of the grid voltage enough *to change from conduction to nonconduction.* This is called *on-off control.* The tube conducts maximum current for all angles of lead from near 165-degree lead down to zero-degree lead.

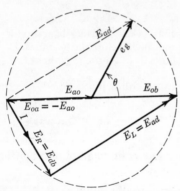

Fig. 14-15   Phasor diagram of *RL* phase-shift circuit that provides on-off control.

On-off control may be achieved by using an inductance of proper value in place of the capacitance of Fig. 14-12. A simpler arrangement, however, is to interchange the capacitance and resistance in that circuit.

The phasor diagram for a phase-shift circuit exactly like that in Fig. 14-12 except for a substitution of an inductance for the capacitor is shown in Fig. 14-15. The inductance is assumed to have negligible resistance. It is seen that the grid voltage e_g leads the plate voltage E_{ob} by the angle θ, which may be varied by changing the value of R.

It must be understood that the reason for using phase control on the grid of an a-c operated thyratron is to achieve variations in *average plate current.* The plate current may be the load current of the circuit, or it may operate another device or circuit which in turn controls load current. It is possible to control large amounts of current by the turn of a small knob on a control potentiometer that carries a few milliamperes and serves as R in a phase-control circuit similar to that of Fig. 14-12.

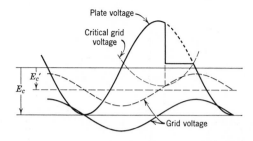

Fig. 14-16   Thyratron control by varying the d-c bias on the grid.

14.13   Variable D-C Grid Bias on an A-C-Operated Thyratron

A thyratron may be fired and its average plate current changed appreciably by providing a variable d-c bias in series with the a-c voltage on the grid. Many applications of this principle are in use in electronic control circuits. The grid voltage is usually given a constant angle of lag with respect to the plate voltage; this angle is often approximately 90 degrees.

The principle is illustrated in Fig. 14-16. The grid voltage lags the plate voltage by 90 degrees. With the bias set at E_c, the instantaneous grid voltage cannot reach the required value to fire the tube at any time. But, when the bias is reduced to E_c', the component of grid voltage, shown dashed, reaches the required firing voltage shortly after the plate voltage has reached its crest value. The firing condition is represented, as before, by the sudden drop of the positive plate-voltage wave beginning at the instant the dashed sine wave representing the a-c component intersects the critical grid-voltage curve. It should be pointed out again that, at the instant the tube conducts, the plate-to-cathode voltage of a mercury-vapor tube drops to about 15 V and remains at that value during conduction. With a grid-voltage wave of small amplitude, a fairly wide range of variation in average plate current may be achieved by this method.

EXAMPLE 14-2.   A thyratron tube, used in a phase-control circuit, is to be operated with on-off control. The 60-Hz voltages are to be 120 on the plate and 10 on the grid, both rms. Fig. 14-12 may be used *after the capacitance and resistance have been interchanged.* Adequate series resistance should be connected in series with the grid to protect it when the grid-input voltage gets large. The phasor diagram is shown in Fig. 14-17b. The grid-cathode voltage, e_g, is given by

$$e_g = E_{oa} + E_R$$

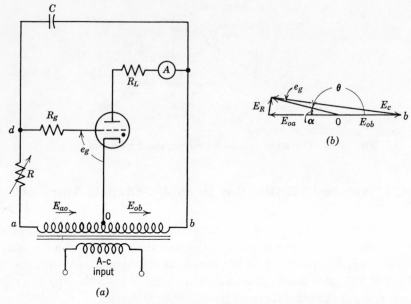

Fig. 14-17   (*a*) Circuit diagram for on-off control. (*b*) Phasor diagram.

and this leads the plate voltage angle E_{ob} by the large angle $\theta = 180° - \alpha$. Assume α to be 10 degrees and $E_{oa} = 60$ V rms. The angle at point b is $\alpha/2 = 5°$.

$$\sin 5° = 0.087 = \frac{E_R}{120}$$

$$E_R = 10.44 \text{ V}$$

$$E_c = 120 \cos 5° = 119.04 \text{ V}$$

Assume the current through R and C to be 10 mA.

$$R = \frac{10.44}{0.01} = 1044 \ \Omega \text{ for } \theta = 180° - 10° = 170°$$

$$X_C = \frac{119.04}{0.01} = 11904 = \frac{1}{2\pi fC}$$

$$C = \frac{1}{(377)(11904)} = 0.222 \ \mu\text{F}$$

Decreasing R increases θ, the angle of lead of e_g with respect to the phase of the anode voltage E_{ob}, which has the same phase as E_{ao}. Note that if R were zero, E_R would be zero, and θ would be 180°. On-off control is accomplished by varying the ohms value of R between a maximum value required by a leading angle of about 160° and a value of a few hundred ohms which will make θ large enough to prevent conduction.

DRILL PROBLEMS

D14-3 Find the values of E_R and R in Fig. 14-17 that that will make e_g lead the anode voltage by 160° when the preconduction current through R and C is 10 mA.

D14-4 Sketch a circuit showing a variable d-c bias in combination with an a-c bias that is in phase with the anode voltage, which is a sine wave of 400 V crest value. Show how the variable d-c bias can vary the starting time of conduction, by sketching the voltage waveforms that exist in the circuit. (You will need to assume a critical grid voltage characteristic or use the curves of Fig. 14-5.)

14.14 The Ignitron

Cathode emission is an important factor that limits the current-carrying capacity of thyratron tubes. The use of a mercury-pool cathode overcomes this limitation. However, the surface of liquid mercury will not give off electrons unless it is brought up to a high temperature. This is made possible in the ignitron tube by the use of a third electrode, the *igniter rod*, which conducts a short surge of current to the surface of the mercury and establishes a small localized arc (Fig. 14-18). An abundant supply of electrons is thus emitted from the hot surface of the mercury adjacent to the igniter rod, and breakdown occurs from the hot spot on the surface of the mercury pool (cathode) to the highly positive anode.

The igniter rod is made of carborundum, which has a rough surface that is favorable to the establishment of strong electrostatic fields where it depresses the mercury surface. After breakdown the igniter-rod current stops, but the cathode spot, which is created near the center of the mercury pool, continues to emit electrons as long as the anode potential is high enough to support conduction.

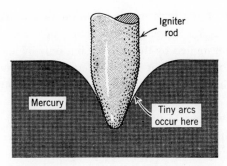

Fig. 14-18 Igniter rod establishes strong electric fields, where it depresses mercury surface, causing short arcs that initiate breakdown in the tube.

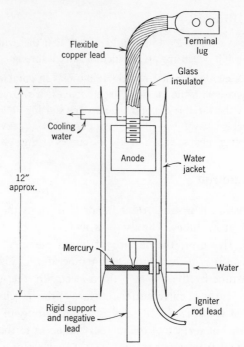

Fig. 14-19   Inside view of water-cooled ignitron. (Courtesy Westinghouse Electric Corp.)

There is a powerful electric field adjacent to the cathode spot caused by a positive-ion sheath. This is regarded as the most important cause of the tremendous rate of emission of electrons at the spot. The temperature is kept up by positive-ion bombardment. No separate supply of cathode-heating power is needed. Figure 14-19 is a sketch of the inside of a metal-clad ignitron.

The advantages of the ignitron, in addition to not requiring cathode-heating current, are: no danger of destruction of the cathode; maximum current capacity not limited by the cathode but only by anode temperature; instant readiness for use, since it does not require a warmup period. Water-cooled ignitrons that carry thousands of amperes of instantaneous current are in operation.

The great disadvantage of the ignitron when compared to the thyratron is the necessity of supplying a substantial pulse of current through the igniter rod. This may be accomplished by thyratron control of igniter-rod current or by a reactor-capacitor circuit. A simple circuit with thyratron phase control is shown in Fig. 14-20.

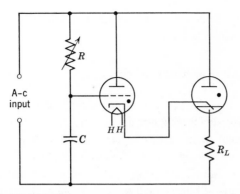

Fig. 14-20   Ignitron tube with thyratron control.

Attention should be called to the contrast between the principles of control inherent in thyratron and ignitron operation. The thyratron is prevented from firing by the negative potential on its grid. It must wait until it is allowed to fire. The ignitron cannot fire of its own free will; it *must be caused to fire* by a shot of current through its igniter rod.

14.15   Deionization

At the instant an arc-discharge tube is extinguished, there are billions of positive gas ions and electrons moving about with high energies in the tube. These charged particles must clear out from the space between the electrodes before the anode is again made positive. If they are not, the tube will fire before it is supposed to, and this means the tube is no longer controlled by its grid or igniter rod. This loss of control is more likely to occur in thyratrons than in ignitrons. Ions travel to the walls of the tube and to negatively charged electrodes and become gas atoms after they have taken on sufficient negative charge to become electrically neutral. Electrons also enter the walls and electrodes.

The time required for sufficient deionization to take place, so that control of tube conduction is not lost when the anode again becomes positive, is called *deionization time*. It varies with tube temperature, anode current, and grid voltage. In ordinary operation the deionization time of a thyratron may be of the order of fifty to several hundred microseconds. It is readily seen that deionization time is a limiting factor on the frequency at which thyratrons and ignitrons may be operated on alternating current. This is because the tube must be under control every time its anode voltage goes from negative polarity (nonconduction) to positive polarity. That occurs at the start of every positive half-cycle.

14.16 Saturable-Reactor Control

Figure 14-21 shows a circuit for supplying igniter-rod current through a *saturable* reactor using a magnetic amplifier for phase control. The capacitor C is charged through the linear reactor, which offers practically constant reactance to the charging current. When its voltage gets high enough, it will aid the small igniter-circuit current in *saturating* the saturable reactor and a sudden burst of current will be sent through the igniter rod to initiate conduction in the ignitron tube. The capacitor discharge current returns to the negative plates by being shunted around the resistance R through the diode rectifier 2, R is used to form a path with rectifiers 1 and 3 around the tube for control current of reverse polarity.

Phase shift for proper timing of the igniter-rod current pulse is provided by controlling saturation of the three-legged reactor core with direct current as explained in magnetic-amplifier theory.

The C_1, L_1 group serves as a compensating network to maintain nearly constant voltage to the firing circuit. At normal line voltage the leading current in C_1 is just balanced by the lagging current in L_1. If the line voltage gets too high, L_1 begins to saturate, its current increases and produces a 90-degree lagging voltage drop in the series coils of the three-legged phase-shift reactor. This reduces the firing-circuit voltage. A drop in line voltage

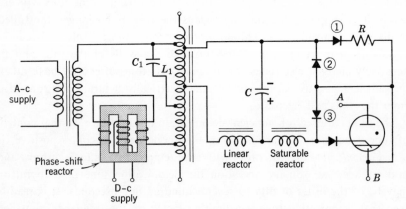

Fig. 14-21   Firing circuit for ignitron with line-voltage compensation and phase control.

below normal causes L_1 to fall below saturation with a resulting increase in its inductive reactance. The leading current in C_1 then predominates, and the effect in the phase-shift reactor is to introduce a 90-degree leading voltage which will increase the voltage applied to the firing circuit.

Large currents in the load circuit flow through the ignitron from A to B. The ignitron simply turns these currents on and off.

14.17 Applications of Thyratrons and Ignitrons

Some uses of the thyratron tube have already been mentioned. Its ability to serve as a high-speed switch that is sensitive to small and rapid changes in voltage is its most important feature. Thyratrons are available with sufficient current capacity for many control applications. They are able to carry igniter-rod current, field current for motors and generators, saturating current for transformers and reactors, and operating-coil current for relays and circuit breakers.

Ignitrons are useful in rectifiers, welding equipment, and other apparatus where medium and large currents are utilized. Figure 14-22 shows a simple welding control circuit in which ignitrons are used to control the flow of primary current in the welding transformer. The firing of the tubes is controlled by a timer that closes the circuit supplying igniter-rod current through copper oxide rectifiers.

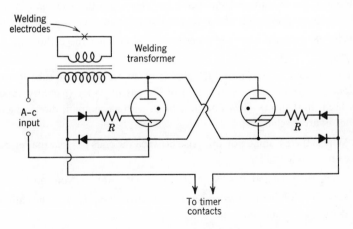

Fig. 14-22   Ignitron control circuit for a welding transformer.

SUGGESTED REFERENCES

1. W. G. Dow, *Fundamentals of Engineering Electronics*, 2nd ed., John Wiley & Sons, New York, 1952.

2. Herbert J. Reich, *Theory and Applications of Electron Tubes*, 2nd ed., McGraw-Hill Book Co., New York, 1944.

3. Paul D. Ankrum, *Principles and Applications of Electron Devices*, International Textbook Co., Scranton, Pa., 1959.

QUESTIONS

14-1   What is Townsend current in gaseous conduction?

14-2   Explain "gas amplification" in gaseous conduction.

14-3   What is the difference between excitation and ionization of atoms?

14-4   What determines the amount of current in a glow tube operating with glow discharge?

14-5   Explain in detail what happens when the circuit resistance in series with a glow-discharge tube is continuously reduced after the current is at maximum safe value.

14-6   What is the nature (magnitude and change with distance) of the potential at points near the cathode in a glow-discharge tube? What happens to ions and electrons in this region?

14-7   Explain the meaning of Paschen's law. How are the increases in ignition potential accounted for both above and below the minimum point of the curve?

14-8   What is the mechanism of electron emission from the cathode in an arc discharge?

14-9   The fall of potential near the cathode of an arc-discharge tube is approximately equal to the ionization potential of the gas or vapor (mercury vapor frequently used) in the tube. How does this tie in with gas amplification, which is so important in arc discharges?

14-10   Why is there a small potential rise between the ends of a plasma region, and why need it be only small?

14-11   Compare glow and arc discharges in as many ways as you can.

14-12   Describe the internal structure of a thyratron tube. How is the cathode, or heater, designed to conserve heat?

14-13   Discuss the reasons for allowing the cathode to reach full operating temperature before anode voltage is applied.

14-14 Draw a curve showing plate voltage required to fire a thyratron plotted against grid bias. Explain the meaning of the curve.

14-15 Why does the grid of a thyratron lose control of the tube after it starts conducting?

14-16 Draw a curve showing grid current in a conducting thyratron plotted against grid voltage. Explain the kinds of current flow involved.

14-17 In Fig. 14-7, T_1 is conducting, T_2 is not, and both switches are open. Explain fully what happens when S_2 is then closed momentarily.

14-18 If making the grid more negative will not extinguish a thyratron which is conducting appreciable current, what must be done to reduce its current to zero? Explain how this is accomplished in a-c operation of a thyratron.

14-19 Show by a sketch how variation of the phase angle of the grid voltage of a thyratron changes the average plate current. Explain the action of an RC phase-control circuit, showing the effect of varying R.

14-20 Repeat the preceding question for an RL phase-control circuit.

14-21 Illustrate on-off control by means of a sketch of voltage waves on the thyratron.

14-22 Illustrate with voltage wave sketches the principle of thyratron control by variable d-c grid bias.

14-23 Describe the physical structure of the ignitron tube.

14-24 How does the igniter rod manage to produce arcs at the mercury surface instead of allowing the pulse of current to pass to the mercury with much lower temperature rise than that of an arc?

14-25 Discuss the establishment of a cathode spot at the mercury surface in an ignitron, and the importance of the *positive-ion sheath* above the spot.

14-26 Does the ignitron have any disadvantages? If so, what are they?

14-27 What is meant by deionization in an arc-discharge tube? What is deionization time?

14-28 Why are arc-discharge tubes classified as high-speed switches? In what ways are they superior to high-vacuum tubes as switches? In what ways are they inferior?

PROBLEMS

14-1 In Fig. 14-1, consider that the load line along which the conduction changes from the dark-current phase to normal-glow discharge extends from the 220-V point to the 150-V point on the abscissa and strikes the characteristic curve at 70 μA. What is the load resistance?

14-2   Assume that the potential rise along the full length of the plasma (0.001 m) in an arc-discharge tube is 1 V. The current is 100 mA. (*a*) What is the average electric field intensity in the plasma parallel to its length? (*b*) How much power is lost in lateral current flow of ions which are lost from the plasma? (*c*) The total tube voltage drop is 15 V. How is the remainder of the input power utilized if there is no rise or fall of voltage adjacent to the anode?

14-3   In Fig. 14-7, the capacitor C is charged to 100 V, T_2 is dark, T_1 is conducting with a drop of 30 V from its plate to its cathode. Immediately after switch S_2 closes, what are the voltage and polarity from plate to cathode of T_1?

14-4   Draw a sine wave representing an a-c voltage, crest value of 500 V, and consider it to be impressed between the plate and cathode of a thyratron that is not conducting. Let Fig. 14-5 apply to the tube and consider the condensed-mercury temperature to be 50°C. Locate points for, and draw, the critical grid voltage curve for the tube. What is the maximum crest value that the a-c component of grid voltage may have, if it is in phase with the plate voltage, and still not fire the tube when the d-c bias voltage is -15 V?

14-5   For the conditions in Problem 14-4, assume that the d-c bias voltage is variable gradually from -15 V to zero. What range of firing angle of the thyratron would then be possible?

14-6   For the conditions in Problem 14-4, assume that d-c bias is held constant at -15 V and the amplitude of the sine wave a-c component of grid voltage is variable up to 15 V maximum. Determine the approximate minimum angle of delay of conduction of the thyratron.

14-7   In the phase-control circuit of Fig. 14-12, $C = 0.5$ μF, R is set at 1,200 Ω and the frequency is 60 Hz. (*a*) What is the phase angle between the plate and grid voltages? (*b*) What is the effect of increasing R on the starting time of tube conduction and on anode current?

14-8   Repeat Problem 14-7 using $C = 4$ μF, $R = 1,200$ Ω.

14-9   Solve Problem 14-7 using $C = 0.5$ μF, $R = 9,600$ Ω. Observe the effects on the angle θ, as shown in the conditions of Problems 7 and 8 compared with the conditions of this problem, when C is changed as R is held constant and when R is changed as C is held constant.

14-10   Redraw Fig. 14-12, interchanging the positions of C and R. Draw the phasor diagram for this circuit. What is the phase relationship of grid voltage and plate voltage?

14-11   Compute the phase angle between the grid and plate voltage of the circuit described in Problem 14-10, just before firing and using $C = 4$ μF, $R = 1,200$ Ω.

14-12   Redraw Fig. 14-12, showing an inductance (L) in place of R and putting R in place of C. Draw the phasor diagram for the phase-control circuit before conduction starts. Describe the type of control.

14-13 In Problem 14-12, let $L = 200$ mH and assume the inductance coil has no resistance. Compute the value of R that will make $\theta = 22.5°$ when operating from a source of 60 Hz.

14-14 Examine Fig. 14-9. (*a*) Describe fully how you would proceed in obtaining data from which to plot a curve of average plate current vs. angle of lag of grid voltage in a specific circuit. You would need the critical grid voltage curve for the tube. (*b*) In what way would the addition of a sharp positive pulse (a spike) to the sine-wave form of grid voltage be an advantage? Would an added d-c bias also be an advantage? Show how you would produce and inject the positive pulse.

14-15 When used with $C = 0.1$ μF and 60 Hz, what value of resistance in the circuit of Problem 14-10 will cause the grid voltage to lead the plate voltage by 170°, thus producing on-off control? Draw the phasor diagram.

14-16 The pulses of igniter-rod current in Fig. 14-22 turn on the primary current of a welding transformer at 30° following the point where the anode voltage passes through zero. The primary current will lag the input voltage somewhat but ignore this. Draw curves of a-c input voltage, voltage across the ignitrons, and voltage between igniter rods and cathodes. The waveform of the igniter-rod voltage is a tall spike with a small "back porch" like the hump on the letter *h*. This voltage should be plotted on a separate time axis just below the one for anode voltage.

14-17 A gas diode rated at 12 A peak plate current and 4 A average plate current is used in a half-wave circuit to charge a 6-V storage battery of 0.01 Ω resistance from a 24 V rms source. (*a*) Determine the maximum charging rate possible without exceeding any tube rating. (*b*) Assuming that sufficient resistance is added to prevent exceeding the rated peak current, how long will it take to add a charge of 120 A-hr?

CHAPTER FIFTEEN

Electronic Power Conversion

We have studied many situations in which the conversion of power from alternating to direct current is required. Electronic tubes and transistors need single-polarity current for both power input and bias supplies. Their needs are commonly met by half-wave and full-wave rectifiers of various circuit designs. These power converters can handle up to a few hundred milliamperes utilizing dry-plate rectifier elements, such as the copper oxide and selenium rectifiers that have been described, and several amperes using semiconductor diodes.

In circuits where the current demand exceeds a few amperes, it may not be practical to use the rectifier elements just referred to, except possibly stacks of semiconductor rectifiers. However, *semiconductor controlled rectifiers* (SCR), thyratrons, ignitrons, and the huge mercury-pool rectifier (enclosed in a steel tank) extend the power-handling capability into the several thousand-kilowatt range.

In this chapter we will study the SCR as a power converter, its operating characteristics, and some typical applications in practical circuits. Extensions of its characteristics in the construction of other similar devices will be discussed, to show the growing dependence on semiconductor rectifiers in electronic power conversion. The chapter concludes with a discussion of rectifiers in three-phase and multiphase applications in which load power may be several thousand kilowatts.

15.1 The Semiconductor Controlled Rectifier (SCR)

The SCR is one member of the more general class of semiconductor devices referred to as *P-N-P-N* or four-layer family. We have studied two-layer

592

diodes and three-layer triodes operating as rectifiers and transistors for controlling and converting power up to a few watts. The SCR can convert power up to thousands of watts. However, this capability is not its most useful property.

The family of *P-N-P-N* devices, including the SCR, *function as switches.* Their power-handling capability extends rectification of a-c power to much higher ranges than is possible using *P-N* devices, but their most important property is their ability to latch in either an ON or OFF state, similar to the action of a simple switch. They also require very little power to operate, often achieving 99 per cent conversion efficiency.

The SCR evolved as a practical device from the pioneering work of Bell Telephone Laboratories in 1954 and 1955, under the direction of J. L. Moll.* This group built the first working *silicon P-N-P-N* device that has served as the basis for all subsequent developments. Because it behaved as the thyratron but was an extension of transistor characteristics, the SCR that evolved from this work is often called a *thyristor*.

Although the SCR is conceptually a "solid-state thyratron," its forward voltage drop is about one-tenth that of a thyratron. This property makes it a more efficient device in power conversion applications. The SCR has other properties of solid-state devices, including long life, mechanical ruggedness, large power-handling capability, and fast switching action. The "turn-on" time of the SCR is a few microseconds, and its "turn-off" time (roughly similar to thyratron "deionization time") is several orders of magnitude faster than that of a typical thyratron.

Controlled rectifiers are used in a broad span of applications ranging from circuit protection to power switching and control. Examples are speed control of motors in hand tools, light dimming in theaters, light flashers, ignition systems for automobiles, and temperature controllers. Before looking in detail at some typical applications, we shall develop the underlying physical principles of operation.

15.2   Construction of SCR

Basically, the SCR consists of four alternating *P* and *N* layers of semiconductor, three junctions, and three external terminals for making connections in an operating circuit. The layer construction is shown in Fig. 15-1.

* J. L. Moll et al., "*p-n-p-n* Transistor Switches," *Proc. I.R.E.*, **44**, 1174–82 (1956).

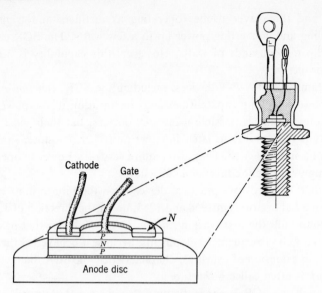

Fig. 15-1   Internal construction of SCR.

The terminals labeled *anode* and *cathode* perform the same function as their counterparts in a diode rectifier. The load in which power is to be controlled is connected in series with the anode-cathode terminations and the power source. Current will pass in the SCR only from anode to cathode as in the diode, but it may exist only when a bias current is applied to the *gate* terminal. Once conduction begins, the gate loses control of the current. This is similar to the loss of control by a thyratron grid. Anode current will stop only when its source is disconnected.

The SCR initially blocks both negative and positive currents at its anode. However, by applying a suitable gate current as a bias, the device can be "broken down" into conduction at any point on a positive voltage waveform at the anode.

15.3   Theory of Operation; Two Transistor Analogy

The four layers of semiconductor material making up the SCR may be regarded as two transistors as shown in Fig. 15-2. Three junctions are formed by the four layers, but their effects on one another are better understood from the two-transistor analogy.

The SCR has forward and reverse characteristics that are widely different, determined by the internal operation at the various junctions and their effects on one another. *Forward bias* for the composite device is defined as shown in Fig. 15-2. Junctions $J1$ and $J3$ are forward biased and $J2$ is reverse biased, i.e., both the *NPN* and *PNP* are biased in the conventional manner for common-emitter operation. When operated in this way, each transistor has associated with it a current gain α (alpha). Let us define the current gain of the *NPN* transistor as α_N, and that of the *PNP* unit as α_P. The current gain is the fraction of electron current (for the *NPN*), or hole current (for the *PNP*) injected at the emitter that reaches the collector.

At the junction that is common to both transistors ($J2$), current will be made up of three components—the electron current from the N-end region, the hole current from the P-end region, and the leakage current. The total current in an external circuit must be equal to the current at $J2$, thus,

$$I_{J2} = I = I\alpha_N + I\alpha_P + I_{CO} \tag{15-1}$$

Written in another form

$$I - I\alpha_N - I\alpha_P = I_{CO}$$

or

$$I(1 - \alpha_N - \alpha_P) = I_{CO}$$

so that the total current may be written in terms of I_{CO}

$$I = \frac{I_{CO}}{1 - (\alpha_N + \alpha_P)} \tag{15-2}$$

From this expression, it is seen that the circuit current can become very large as the sum of the current gains approaches unity. Circuit current will be limited only by any external resistance, when $\alpha_N + \alpha_P = 1$. Thus, increasing the current gains will "turn on" the SCR. When the current gains are very small, the current in $J2$ will be approximately the leakage current

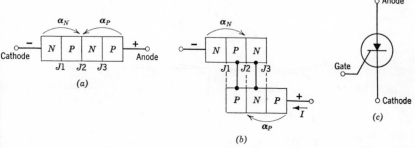

Fig. 15-2 Two-transistor analogy for four-layer SCR and its schematic circuit symbol. (*a*) Four-layer SCR. (*b*) Two-transistor analogy. (*c*) Circuit symbol for SCR.

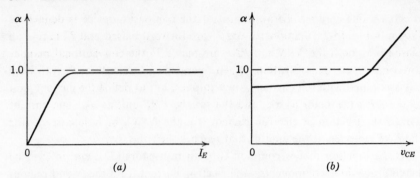

Fig. 15-3   Variations of α of a silicon-controlled rectifier with current and voltage (*a*) Current gain vs. emitter current. (*b*) Current gain vs. collector-emitter voltage.

I_{CO}. This condition corresponds to the "off" state, or blocking state, of the SCR.

When the emitter current of a transistor is small its current gain α is small. One way to increase the current gain is to increase the emitter current. Figure 15-3*a* shows the variation of current gain with emitter current. Another method of increasing the current gain is to increase the collector-emitter voltage. This is shown in Fig. 15-3*b*.

If the voltage between the collector and emitter is increased sufficiently, the junction *J*2 will break down and permit a large reverse current to pass through it. This effect is called *avalanche breakdown*. When *J*2 avalanches, the current in *J*2 increases and in turn increases both α_N and α_P. The SCR thus goes into its "on" state.

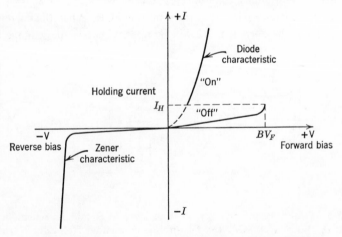

Fig. 15-4   Forward and reverse bias characteristics of SCR.

The anode-cathode voltage that causes these events in the SCR is called the *breakover voltage*, BV_F. After the device is turned on by the breakover voltage, it will remain in its *on* state as long as the current through $J2$ (external circuit current) is large enough to make $\alpha_N + \alpha_P$ equal to unity. The minimum current that will keep the SCR *on* is called the *holding current*, I_H. This is the current in $J2$, i.e., the *external* current.

When the polarity of the external bias is reversed, in order to cause $J1$ and $J3$ to be reverse-biased and $J2$ to be forward-biased, the SCR behaves much as a diode in reverse bias. The voltage-current curves for the SCR are shown in Fig. 15-4 for forward and reverse bias.

15.4   Effect of the Gate

The discussion so far has concerned the behavior of the SCR as a two-terminal device, as forward and reverse bias are applied to it. In a practical application of the SCR, however, the gate terminal is used to control the anode current over some range of anode voltage amplitudes. We shall now see how the gate current can be used to control the action of the SCR.

The two-transistor analogy of the SCR is used again to describe the action of the gate in controlling anode current. In Fig. 15-5, the gate terminal is shown connected to the *P*-region of the *NPN* transistor.

If a positive voltage is applied to the gate-cathode junction, $J1$, it is forward-biased and current will pass through it. Then many electrons appear at $J2$, and because of its reverse bias they will pass into the *N*-region between

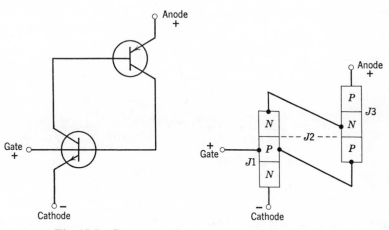

Fig. 15-5   Gate connection to two-transistor analogy.

*J*2 and *J*3. When these electrons reach *J*3, which is forward biased, they recombine with holes in the anode *P*-region. This will trigger the generation of an electron-hole pair at the anode terminal to restore equilibrium, and an electron leaves the anode to contribute to circuit current. Some of the holes that diffuse across the *J*3 junction during conduction will not recombine with electrons there. They drift on into the *N*-region between *J*2 and *J*3. As a hole reaches *J*2 it is swept into the *P*-region between *J*2 and *J*1. It then drifts until it reaches *J*1, where it combines with an electron from the emitter (cathode).

Thus, a current from the gate terminal into the junction *J*1 increases the current in the entire device. This occurs because the current gains, α_N and α_P, increase. As the gate current continues to increase, the breakover voltage, BV_F, is decreased significantly. This is shown in Fig. 15-6. The gate current can become large enough to cause the SCR to act as a *P-N* junction diode.

We see that the SCR remains in its "off" state when it is forward biased by a voltage less than the breakover voltage. Upon application of sufficient gate current, the breakover voltage will decrease and permit conduction in its "on" state. Once the device is turned on, the gate no longer has any control over the anode current. To return it to its "off" state, anode current must be reduced below the value of holding current. This may be accomplished either by reducing the anode current or by removing the forward anode voltage. If the anode voltage is a sine wave of voltage, it will reverse its polarity each

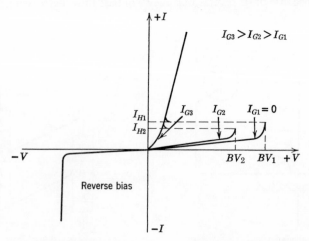

Fig. 15-6   Effect of gate current on breakover voltage and SCR behavior in forward bias.

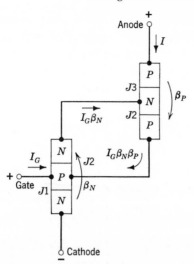

Fig. 15-7   Current regeneration in forward-biased SCR.

cycle and automatically return the SCR to its "off" state every cycle. It is this characteristic that is important in power control applications. It will be illustrated in connection with application of the SCR in following sections.

Since the gate current increases α_N, which in turn increases α_P, the gate current gain of the SCR is equivalent to that of a grounded-emitter *NPN* transistor connected to a *PNP* transistor, as illustrated in Fig. 15-7. As gate current flows into the base of the transistor, it is amplified and appears at the collector as $I_G\beta_N$, where $\beta = \alpha/(1 - \alpha)$. This amount of collector current enters the base of the *PNP* transistor and appears as a collector current of $I_G\beta_N\beta_P$. If α_P, the current gain of the *PNP* transistor, is large, a regenerative action occurs and the SCR anode current increases. A small gate current can then trigger the SCR into its "on" state, thus controlling a current that may be thousands of times larger than the gate current. To turn on the SCR, it is necessary that the gate current be large enough to trigger the regenerative action.

15.5   Ratings and Characteristics of SCR

The following ratings and characteristics of the SCR refer to their designations in Fig. 15-8a. Manufacturers' data sheets include numerical values for these parameters.

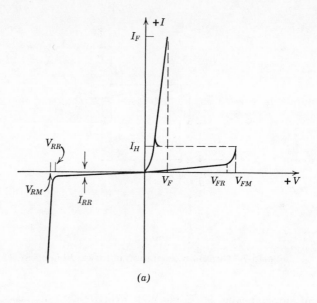

(a)

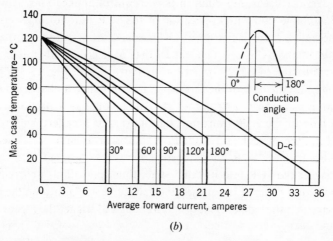

(b)

Fig. 15-8 Characteristics of the silicon controlled rectifier. (a) Identification of symbols. (b) Maximum case temperature for sinusoidal current waveforms, at various conduction angles, SCR type 2N685.

600

Ratings

V_{FM} Peak forward blocking voltage

The maximum instantaneous voltage that may be applied to the anode and will not switch on the SCR.

V_{FR} Continuous forward blocking voltage

The maximum continuous voltage that may be applied to the anode and not switch on the SCR.

V_{RM} Peak reverse blocking voltage

The maximum instantaneous reverse voltage that will not breakover the SCR in a reverse mode.

V_{RR} Continuous reverse blocking voltage

The maximum reverse voltage that may be applied continuously and not breakover in a reverse mode.

PFV Peak forward voltage

The maximum instantaneous forward voltage that may be used to switch on the SCR. If other specified maximum ratings are not exceeded and applied voltage is lower than *PFV*, the SCR will not be damaged.

I_F Maximum continuous anode current

The maximum current that may pass in the forward direction without exceeding rated junction temperatures.

I_O Maximum average forward current

Maximum current averaged over a full cycle that may be applied continuously under single-phase, 60 Hz half-sine-wave operation with a resistive load. Average current and peak reverse voltage ratings apply simultaneously.

Characteristics

V_F Static forward voltage	The forward voltage drop when the SCR is conducting in the forward direction.
V_{GT} Gate trigger voltage	The gate voltage required to produce the gate trigger current.
I_{GT} Gate trigger current	The minimum gate current required to switch the SCR from its "off" state to "on" state.
I_H Holding current	Minimum principal current required to maintain "on" state following conduction of steady-state forward current.
I_{RR} Static anode reverse current	Principal current for negative anode-to-cathode voltage.

As an example of ratings and characteristics furnished by a manufacturer, these numerical values are published for a 2N685 type of SCR:

Min. Fwd. Breakover Voltage, BV_F	200 V
Peak Reverse Voltage, V_{RM}	300 V
Continuous Reverse Voltage, V_{RR}	200 V

Maximum Allowable Ratings

Continuous Anode Current, I_F	35 A rms, all conduction angles
Maximum Average Anode Current, I_O	Depends on conduction angle (see Fig. 15-8*b*)
Peak One-Cycle Surge Current	150 A
Peak Forward Gate Voltage	10 V
Peak Reverse Gate Voltage	5 V

Characteristics

Gate Trigger Current, I_{GT}	15 mA typ., 40 mA max.
Gate Trigger Voltage, V_{GT}	1.5 V, typ.; 3.0 V d-c, max.
Holding Current, I_H	10 mA d-c, typ.; 100 mA d-c, max.

EXAMPLE 15-1. The type 2N685 silicon controlled rectifier is used in a single phase resistive-load circuit. If the maximum case temperature is maintained *at 80°C or less*, what average forward current can it safely carry?

From the curves of Fig. 15-8*b*, we see that different values of average current are permissible for different conduction angles. If the SCR is triggered as soon as its anode goes positive, current will flow during the entire 180° of a sinusoidal voltage. At this conduction angle the average current can be almost 12 A. For a conduction angle of 120°, the gate does not energize the anode until after the first 60° of the sinusoidal voltage, and average current can be about 10 A. At 30° conduction angle, the permissible average current is only 5 A.

The curves in Fig. 15-8*b* have definite endpoints (maximum *average current*) for each conduction angle. They occur because they represent the same rms values for each of the waveforms. For example, the 2N685 is rated at 35 A *rms*, at all conduction angles. For a conduction angle of 180° the average of the *half sine wave* is 0.318 times the maximum amplitude, and the rms value is 0.500 times the maximum amplitude. The ratio of rms to average values is 1.57, so that the average current is limited to $35/1.57 = 22.3$ A for 180° conduction angle. If the *average current* does not exceed 22.3 A, the *rms value* will not exceed the rated 35 A.

15.6   Gate Trigger Characteristics

The gate trigger characteristics of the SCR are usually presented in the graphical form in Fig. 15-9. These characteristics apply to the type 2N685. The SCR will trigger at a definite point on its gate characteristic. The shaded

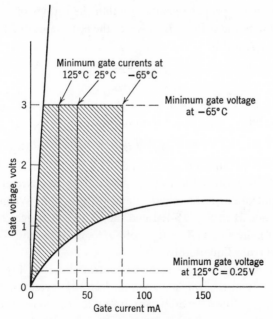

Fig. 15-9   Critical gate firing voltage and current for 2N685. Points for a firing condition (V_{GT}, I_{GT}) must fall *outside* the shaded area.

areas of the curve contains all of the possible trigger points, based on rated values of I_{GT} and V_{GT}. In order to achieve reliable triggering, the trigger circuit must simultaneously supply I_{GT} and V_{GT} values that lie *outside* the shaded area.

For the simple on-off switching of the SCR in a d-c circuit, a continuous gate signal of amplitude required to furnish V_{GT} and I_{GT} can be used. However, when d-c is used on the anode of the SCR, once the gate has triggered the device into conduction, the gate will have no more effect on the operation. Conduction will continue until the anode current falls below the holding current I_H. In order to stop anode current, the d-c anode supply may be disconnected. This can be done in an operating circuit by some mechanical switch (push-button, reed switch, etc.), actuated either manually or by transducers in response to heat, light sources, pressure, etc. If an a-c source is used on the anode, a d-c gate signal may be used and conduction will cease during negative half-cycles of the anode voltage. The gate can control conduction each time the anode voltage becomes positive.

A simple arrangement of a gate trigger circuit is shown in Fig. 15-10a, where $e_s = E_{\max} \sin \theta$. Gate current is supplied directly from the anode-voltage supply through a suitable current-limiting resistor, R. The resistor should be chosen for proper operation within the ratings of the SCR. For a worst-case design, the switch SW closes at the peak value of supply voltage, and the value of R will be

$$R \geq \frac{E_{\max}}{I_{G\max}} \tag{15-3}$$

The SCR will trigger into conduction when suitable values of V_{GT} and I_{GT} occur simultaneously or when the instantaneous anode voltage is

$$e = V_{GT} + I_{GT}R + V_F \tag{15-4}$$

where V_F is the forward drop of the diode D.

Figure 15-10b is another form of the anode triggering circuit. Resistor R is used as a control device for phase control of the SCR. When $R = 0$, the circuit is the same as in Fig. 15-10a and the SCR will trigger according to Equation (15-4), in which $R = R_{\min}$ of the control circuit. The value of $R_{\min}$ is selected from Equation (15-3).

As the value of R is increased, in the control circuit of Fig. 15-10b, the SCR will trigger at larger firing angles (θ in electrical degrees) until the anode voltage equals $E_{\max}$. Since the SCR cannot wait for I_{GT} to occur at some time *after* $E_{\max}$ is reached but will trigger the first time I_{GT} occurs, conduction can be controlled only for $0 \leq \theta \leq 90°$. Therefore, the circuit of Fig. 15-10b

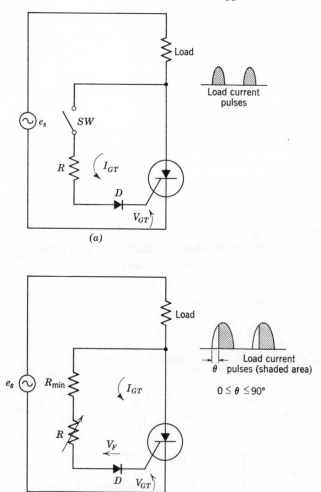

Fig. 15-10 Anode resistor trigger circuits. (*a*) Simple anode trigger. (*b*) Phase-control trigger.

provides continuous control of conduction from the full 180° of half-wave operation to 90° of half-wave.

DRILL PROBLEMS

D15-1 In the circuit of Fig. 15-10*a*, assume the diode forward voltage drop is 1 V, and the gate triggers at $V_{GT} = 1$ V, $I_{GT} = 10$ mA. For $R = 10$ kΩ, at what anode voltage will the SCR trigger? What must be the maximum gate current rating, I_{Gmax}?

D15-2 In the circuit of Fig. 15-10b, $I_{Gmax} = 10$ mA, $V_F = 1$ V, $V_{GT} = 1$ V, $I_{GT} = 1$ mA. When $R + R_{min} = 0$, what anode voltage will trigger the SCR? What will be the firing angle of the SCR?

The range of control provided by this simple circuit can be extended by the circuit in Fig. 15-11a. The resistor-capacitor combination in the gate circuit permits control of conduction through the entire 180° of half-wave operation. The action of the RC circuit delays the necessary V_{GT} for triggering to occur. During positive half-cycles of supply voltage, the capacitor will charge to the trigger voltage in a length of time determined by the RC time constant and the frequency of the supply voltage. On the negative half-cycle the capacitor charges to the negative maximum of supply voltage through diode D_2. This resets it for the next charging cycle.

Because of the spread of gate characteristics as in Fig. 15-9 an exact solution for values of firing angle θ is not practical. The value of resistor R is selected to allow trigger current I_{GT} when the capacitor voltage is large enough to trigger the SCR. When the capacitor voltage is equal to $V_{GT} + V_F$, resistor R must allow trigger current, I_{GT}, to flow and the anode voltage is

$$e \geq I_{GT}R + V_{GT} + V_F$$

requiring a maximum value for resistor R given by

$$R \leq \frac{e - V_F - V_{GT}}{I_{GT}} \tag{15-5}$$

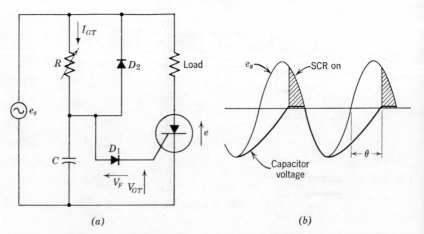

(a)	(b)

Fig. 15-11   Full half-cycle phase control of SCR. (a) RC phase control. (b) Voltage waveforms. Firing is delayed θ degrees after anode goes positive.

Figure 15-11*b* shows a typical waveform of capacitor voltage and the resulting firing angle of the SCR. The gate circuit will trigger the SCR into conduction whenever the gate voltage reaches its rated value.

By adjusting the value of the resistor R, the firing angle θ can be shifted to control the conduction angle anywhere in the range from 0° to 180°, giving full half-cycle control over the SCR and its load.

Phase control circuits that were discussed for the thyratron tube in Chapter 14 are applicable to the SCR as well.

15.7 Full-Wave Phase Control of SCR

The single-phase, half-wave circuits of Section 15.6 are often inadequate for supplying necessary power to the load of an SCR. Two devices may be used for full-wave control of load power in the circuit of Fig. 15-12*a*. In this circuit, two separate gate control circuits are required. If they provide equal conduction angles for the two SCR's, the load voltage will be symmetrical as shown in Fig. 15-12*b*.

An alternate form of full-wave control is shown in Fig. 15-12*c*. Only one gate control circuit is required to operate the circuit. This is sometimes an advantage, because the gates and cathodes have a common connection, either of which may be grounded in a practical application. The diodes prevent negative voltage across the SCR during alternate half-cycles of the input voltage.

15.8 Rms and Average Load Voltage Calculations

The SCR is rated in terms of average current, but most a-c loads require certain rms, or effective, currents and voltages to meet power requirements. Because the SCR is a switch that can proportion power according to its conduction angle, the load voltage will have average, rms, and peak values that depend on the conduction angle. The power in the load will be determined by the rms values of current and voltage. For a resistance load, the load power will be $I_{rms} \times V_{rms}$.

Figure 15-13 illustrates how the average, rms, and peak values of load voltage vary with conduction angle for both half-wave and full-wave phase control.

EXAMPLE 15-2. It is desired to supply 1 kW to a resistive load rated at 110 V from a supply of 220 V sinusoidal waveform. Since the power supplied is proportional to the square of the applied voltage, connecting it to the 220-V supply

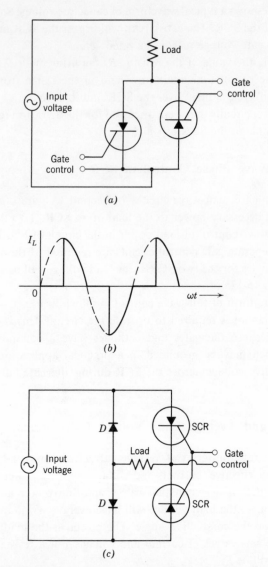

Fig. 15-12 Full-wave control of SCR load power. (*a*) Full-wave control of SCR's. (*b*) Symmetrical load voltage waveform, at 90° conduction angle. (*c*) Alternate full-wave control circuit.

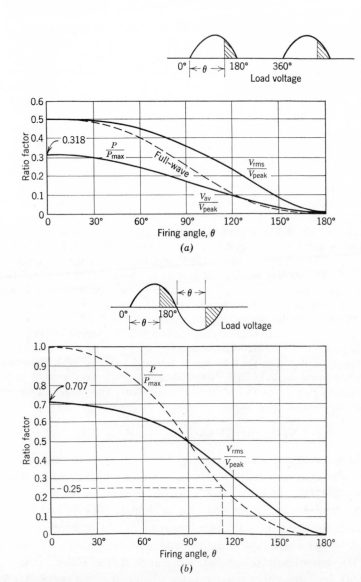

Fig. 15-13 Variations of rms and average voltages with changing firing angle. (a) Half-wave phase control. (b) Full-wave symmetrical phase control.

would result in 4 kW of power. In order to furnish only the required 1 kW, either half-wave or full-wave phase control may be used with an SCR.

Starting with the half-wave operation, we see from Fig. 15-13a that $\frac{1}{4}$ power (0.25 on the ordinate) occurs for a firing angle of 90°. The peak voltage is 1.414 × 220 = 312 V. Average voltage is 0.159 × 312 = 50 V and the rms voltage is 0.353 × 312 = 110 V. The ratios of rms to peak values and average to peak values are read from the graph at a conduction angle of 90°.

The load resistance has a value $110^2/1000 = 12.1\ \Omega$. Therefore the average load current is 50 V/12.1 Ω = 4.13 A. This value would be indicated by a d-c ammeter in series with the load. The SCR must therefore be rated to handle at least 4.13 A at 90° conduction angle. The load must also be capable of operating at the high values of peak voltage and peak current.

Now we will see how these values may be changed in a full-wave phase control connection of the SCR. In a symmetrical full-wave circuit, $\frac{1}{4}$ power occurs at a firing angle of about 115 degrees. This means that each half-cycle conducts for only (180° − 115°) = 65° in order to supply 1 kW to the load. The peak voltage is 312 × sin (115°) = 312 × 0.82 = 256 V, and the rms voltage is again 0.353 × 312 = 50 V. The average voltage will be zero for the symmetrical waveform. To determine the rating of each SCR, consider each as operating as a half-wave source at a firing angle of 115°. From Fig. 15-13a the average voltage is 0.095 × 312 = 29.6 V. Then the average current in each SCR is 29.6/12.1 = 2.44 A. Each SCR must be rated to carry this average current at a firing angle of 115°.

It is important to note that the first and last 30 degrees or so of each half-cycle contribute a small part of the total power (about 5 per cent only). Consequently a triggering range of firing angles from 30° to 150° will yield a range of power control from about 2.5 to 97.5 per cent of total power. This range of firing angles provides the most effective control of power in the load.

15.9   Bilateral SCR, the TRIAC

An important shortcoming of the SCR is its inability to conduct current through a load in both directions. In its "on" state it is limited to current in one direction. Of course, two SCR's can be connected in a full-wave bridge or center-tapped transformer mode, but it requires not only two SCR's but also makes the design of gate trigger circuits a further problem.

The fabrication techniques used to build four-layer *P-N-P-N* devices have led to the development of a novel, bilateral SCR, the TRIAC (triode, a-c operated). It is sometimes referred to as a bilateral triode switch. Its fabrication and internal behavior is beyond the scope of this book, and we will discuss briefly its features similar to the SCR in controlling load power.

The internal structure of this device permits it to trigger into conduction from a single gate terminal, from either positive or negative gate signals. Also, it is capable of conducting current in either direction. Trigger and phase-control circuits permit the power in a load to be varied from pure sinusoidal

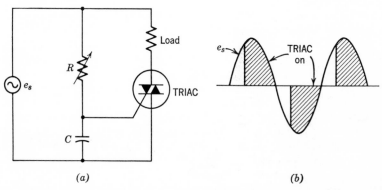

Fig. 15-14   Bilateral circuit using TRIAC. (*a*) Circuit. (*b*) Voltage waveforms.

input voltage to nonconduction in both half-cycles. Figure 15-14 is a typical circuit arrangement and load voltage waveform for the bilateral triode switch controlling power in a resistive load.

The curves developed in Section 15.8 for full-wave SCR control apply to the operation of the TRIAC. The difference is that the TRIAC will have average current and voltage values that are twice their counterparts in the SCR circuit, because they occur twice in each cycle in the same device, whereas two SCR's shared this behavior.

15.10   Three-Phase Power

Power conversion from a-c to d-c is usually done with *three-phase* power when the demand is greater than a few hundred watts. A few remarks about three-phase power are in order because some readers may be unfamiliar with it.

A three-phase generator has a stationary armature winding with taps to which three cables are connected to bring out current. The windings which produce the magnetic flux needed for the generation of voltage rotate on a shaft inside the armature. The three voltages differ in phase by 120 degrees, so their phasors form an equilateral triangle. Three transformers are used to change the voltage to a desired value. Their secondaries may be connected in a number of ways; the two most useful are the *delta* and the *wye* connections. These are shown, with phasors, in Fig. 15-15.

Assume that each transformer secondary coil has 120 V between its terminals. This amount of voltage will then be impressed across loads connected between any pair of the points 1, 2, and 3 of the delta connection.

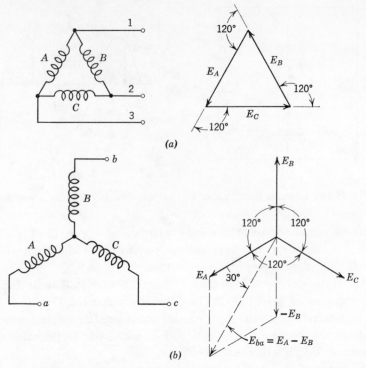

(a)

(b)

Fig. 15-15   Secondary windings of three single-phase transformers on a three-phase line. (*a*) Delta connection; (*b*) wye connection.

A load connected between points *a* and *b*, *b* and *c*, *c* and *a* of the wye connection would have $\sqrt{3} \times 120 = 208$ V impressed upon it. For example, the voltage E_{ba} (from *b* to *a*) is $-\vec{E}_B + \vec{E}_A = \sqrt{3}E_A$ at an angle of 30 degrees leading E_A, as shown. This is true with corresponding subscripts for the other two terminal pairs.

The *line currents* supplying a three-phase load are the same as the winding currents in a wye connection, whereas the *line currents* supplying a delta-connected load are equal to $\sqrt{3}$ *times the winding currents.*

The power in a three-phase load is equal to the sum of the powers taken by the three separate *single-phase* loads, each of which is given by $EI \cos \theta$, where E is the terminal voltage at the load and I is the current in the load. The power factor of that one load, is $\cos \theta$. When a three-phase load *is balanced*, i.e., all three parts have the same terminal voltage, current, and power factor values, the total taken by the load can be shown equal to $P_T = \sqrt{3}E_L I_L \cos \theta$, where E_L is the voltage between any two *line* wires,

I_L is the current in one *line* wire, and θ is the *angle between the current in* one of the three load sections *and the voltage across that load section.*

Three-phase power has several advantages over single-phase power. Generation, transmission, and utilization are all more efficient. Three-phase motors will start "directly across the line" without special features of design (required in single-phase motors) that contribute little to operation after the motor is at normal running speed. It will be found that in conversion of a-c to d-c power, three-phase operation is easier on the rectifiers and produces a smoother output wave form than does single-phase operation.

15.11   Three-Phase Half-Wave Rectification

Perhaps the simplest half-wave rectifier system for three-phase operation is the one in Fig. 15-16*a*. The voltages e_1, e_2, e_3 supplied by each secondary winding are 120 degrees apart in phase; this means that the SCR anodes are brought up to positive polarity and reach conduction potentials in succession. This is shown in Fig. 15-16*b*.

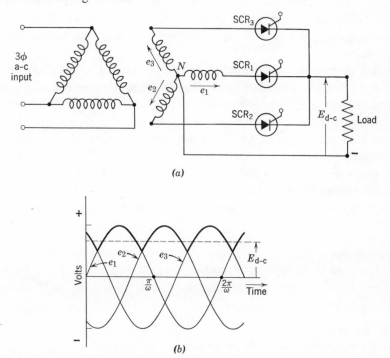

Fig. 15-16   (*a*) Three-phase half-wave rectifying system. (*b*) Three single-phase voltages supplied to SCR's and load of (*a*).

Although there is a small voltage drop through a conducting SCR, it will simplify the discussion if we consider the drop zero and thus, when the SCR fires, the full potential of the transformer secondary connected to that SCR is applied to the load. The voltage across the load is then represented by the heavy sections of the voltage waves in Fig. 15-16b. It is made up of parts of the three secondary terminal voltages, *each taking over the supply of load current at the instant it can supply more voltage than the one that is working.* We might surmise that the secondary winding which has just been relieved of the burden would continue to supply current as long as its voltage is positive, even though small. Such is not the case. To understand this, consider point P whose potential is always on the heavy line in Fig. 15-16b. At the instant e_2 can raise the potential of P *above the best e_1 can do*, SCR$_1$ is cut off because its cathode is immediately going *positive* with respect to its anode. Thus conduction shifts to the next SCR, and on to the next every 120 degrees of phase angle ($\frac{1}{3}$ cycle) which means every $\frac{1}{3}(2\pi/\omega)$ s.

For a resistive load, the "scalloped" curve is the exact waveform of the load current. These are three of these humps in the current wave, and in the voltage wave, per cycle; therefore *the ripple frequency is three times line frequency* as compared with twice line frequency in a full-wave single-phase rectifier.

The *average value* of the pulsating load voltage is $E_{\text{d-c}}$ and it is shown in Fig. 15-16b. If a graphical summation of the area under one of the loops and between straight lines perpendicular to the time axis is made, this area divided by the distance along the time axis will give the *average height* of one loop and will equal $E_{\text{d-c}}$ if the heights of individual thin vertical strips are called *volts* and the widths are called *seconds*. The total will then be volt-seconds and, when divided by the duration of the loop in *seconds*, the result will be in *volts*. The average turns out to be $1.17E_{\text{eff}}$, where E_{eff} is the rms value of the voltage supplied by each transformer secondary.

$$E_{\text{d-c}} = 1.17E_{\text{eff}} = 0.828E_m \qquad (15\text{-}6)$$

This compares with $E_{\text{d-c}} = 0.9E_{\text{eff}}$ for single-phase full-wave rectification.

In regard to the amount of ripple in the load voltage (no filter in between) it can be shown that the fundamental component E_{r1} of *ripple* voltage is 17.7 per cent of $E_{\text{d-c}}$.

$$E_{r1} = 0.177E_{\text{d-c}} \qquad (15\text{-}7)$$

In the single-phase case, $E_{r1} = 0.472E_{\text{d-c}}$.

The maximum peak-inverse voltage that the SCR can stand is one of its most important limiting ratings. The peak-inverse voltage with this three-phase half-wave rectifying system can be shown to occur, on SCR_3 for example, at the instant e_2 is zero and about to go positive. It can be deduced that e_3 is $-0.866E_m$ and e_1 is $+0.866E_m$ at that instant, making the *peak-inverse voltage on T_3 equal to $1.732E_m$*. Expressing this as a ratio to $E_{d\text{-}c}$, we have

$$\frac{1.732E_m}{E_{d\text{-}c}} = \frac{1.732\sqrt{2}E_{eff}}{1.17E_{eff}} = 2.09$$

Peak-inverse voltage on SCR $= 2.09E_{d\text{-}c}$

(15-8)

Therefore the maximum peak-inverse voltage on the SCR in the half-wave three-phase rectifier is 2.09 times the d-c load voltage. This compares with 3.14 for the single-phase full-wave rectifier.

The average anode current, i.e., the average current per SCR, is one-third the output current.

$$\text{Average tube current} = I_{d\text{-}c}/3 \qquad (15\text{-}9)$$

This is also very important in the rating of SCR's as it determines anode heating since the voltage drop is practically constant.

The *rms* value of SCR *current* can be shown to be

$$I_T = \frac{I_{d\text{-}c}}{\sqrt{3}}(\text{rms}) \qquad (15\text{-}10)$$

The relationships just presented are given in Table 15-1 (Section 15.13) which contains similar relationships for other systems of three-phase rectification.

15.12 Six-Phase Half-Wave Rectification

Three single-phase transformers, each having two identical but separate secondary windings, are used to provide a source of *six-phase power* for supplying six power converters in a rectification system. The arrangement of the secondaries is shown in Fig. 15-17. Observe that, if the second winding of the third transformer were connected *in reverse*, its terminal voltage would have the phase and direction denoted by the arrow labeled $-e_3'$. This voltage would lag e_1 by 60 degrees. In like manner, reversed connections of e_1' would give $-e_1'$, which lags e_2 by 60 degrees, and $-e_2'$ lags e_3 by 60 degrees. Figure 15-18 shows a six-phase rectifier system using these secondaries with reverse connections, as just explained, but with the voltages labeled in consecutive order. That is, what was formerly labeled $-e_3'$ is now e_2, what was formerly e_2 is now e_3, and so on, for simplicity.

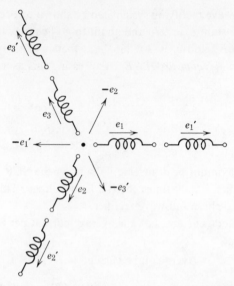

Fig. 15-17 Double secondary windings on a power three-phase transformer.

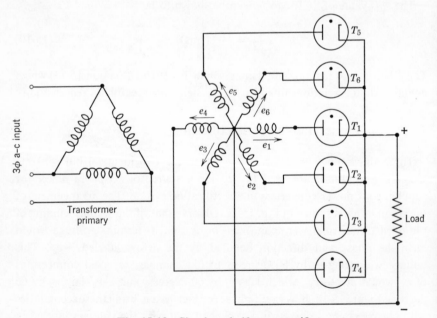

Fig. 15-18 Six-phase half-wave rectifier.

The six-phase half-wave rectifier operates in the same manner as the three-phase half-wave system. Although tubes are shown, the power converters may be SCR's. Each half of the transformer secondary supplies its own tube, and, when its falling output voltage becomes equal to the voltage which is rising on the next winding, its tube current is zero and the next tube takes over. As in the three-phase half-wave rectifier, *the incoming tube cannot start conducting*, although its supply voltage is getting more and more positive, *until* that supply voltage equals the voltage at the anode of the *tube presently conducting*, which is the potential at the *incoming tube's* cathode. It will be recalled that we consider the voltage drop through a conducting tube to be negligible. As this is not absolutely true, the second tube may start conducting a short time before the first tube ceases.

Six pulses of voltage per half-cycle are sent to the load; this means that a tube conducts for only *one-sixth* of each cycle. The d-c load voltage is the average of the top 60 degrees of each sine wave, which can be shown to be

$$E_{d\text{-}c} = 1.35E_{eff} \tag{15-11}$$

where E_{eff} is the rms value of the a-c voltage supplied to each tube.

The fundamental ripple frequency is six times the supply frequency, since the waveform repeats six times faster.

The average tube current is one-sixth of the d-c load current.

$$\text{Average tube current} = I_{d\text{-}c}/6 \tag{15-12}$$

The peak-inverse voltage on a tube is twice the peak voltage per phase. This assumes, as before, that the cathodes of all the tubes follow the potential of the conducting tube. Now, suppose T_4 is conducting and e_4 is at its maximum instantaneous positive value. At this instant it can be seen in Fig. 15-18, and also in Fig. 15-19, that e_1 has its maximum *negative* value. The inverse voltage on T_1 is thus twice the peak value put out by one winding. To see this in the circuit we must imagine the arrow e_1 reversed because that will be its direction at the instant e_4 is at its own positive maximum value.

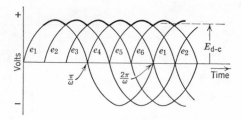

Fig. 15-19   Voltages supplied to six-phase rectifier tubes and load of Fig. 15-18.

The *rms tube current* can be shown equal to

$$I_T = I_{\text{d-c}}/\sqrt{6} \qquad (15\text{-}13)$$

15.13   Three-Phase Full-Wave Bridge Rectifier

We learned in studying the three-phase half-wave rectifier that the cathodes and the positive terminal of the load always follow *in potential* the anode which is most positive with respect to the transformer neutral. The three-phase secondary circuit can be supplied with three more tubes connected so that the *negative* terminal of the load can be made to follow in potential the *most negative* terminal of the transformer secondaries. This will result in a much larger d-c output voltage *from the same transformers.*

In Fig. 15-20, tubes 1, 2, and 3 are connected to the transformer secondaries (which we shall call *phases*) in exactly the same manner the three tubes are connected in the three-phase half-wave rectifier of Fig. 15-16a. But, instead of returning the negative load terminal to the transformer neutral (or phase neutral), we connect it to all the anodes of three more tubes. The cathode of each tube is connected to only one anode of the first group, as shown. We see that conduction will take place in the tube at the top, in the 4, 5, 6 group, which has the *most negative* cathode at the instant the anode voltage of one of the bottom tubes is *ready to send current.*

Because two tubes conduct at a time, the average tube current is *one-third* of the d-c current instead of one-sixth as in the six-phase circuit employing six tubes. This means that the d-c current is limited to three times the average-current rating of one tube, instead of six times. In other words, the three-phase bridge circuit does not utilize the tube capacity as well as the six-phase circuit does.

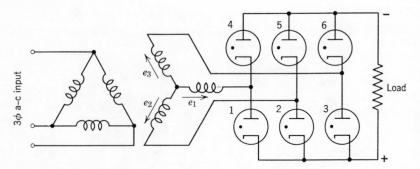

Fig. 15-20   Three-phase full-wave bridge rectifier.

There are twice as many pulses of the load current per cycle in this system as in the three-phase half-wave system of Fig. 15-16. At the middle of a bottom-tube conduction period there is a pulse when the next top tube takes over, and at the middle of a top-tube conduction period there is a pulse when the next bottom tube takes over. This means *six cycles of output ripple voltage per cycle of phase voltage.*

The d-c load voltage produced with the three-phase bridge rectifier is almost twice that of the six-phase half-wave system *for a given transformer secondary voltage.* Table 15-1 shows $1.35E$ for the value of $E_{\text{d-c}}$ with this system, but this E is twice the E per phase of the six-phase system in a given transformer.

The bridge circuit gives almost twice as much d-c load voltage for a given peak-inverse anode voltage rating on a tube as the six-phase circuit using the same number of tubes. This is a big advantage in applications where high voltage, and not large currents, is desired.

Table 15-1 shows important relationships for the three most popular polyphase rectifier systems. E is the voltage of a *phase* (or transformer secondary coil) applying positive voltage to an anode. Currents are given in terms of $I_{\text{d-c}}$, the load current. P represents the power delivered to the load in kilowatts.

TABLE 15-1   Voltage and Current Comparisons for Three Basic Rectifier Systems Supplied with Three-Phase Power

Circuit	3-Phase Half-Wave	6-Phase Half-Wave	3-Phase Bridge
Average load voltage, $E_{\text{d-c}}$	$1.17E$	$1.35E$	$1.35E$
Average anode current	$\dfrac{I_{\text{d-c}}}{3}$	$\dfrac{I_{\text{d-c}}}{6}$	$\dfrac{I_{\text{d-c}}}{3}$
Effective secondary transformer current	$\dfrac{I_{\text{d-c}}}{\sqrt{3}}$	$\dfrac{I_{\text{d-c}}}{\sqrt{6}}$	Line current $0.816I_{\text{d-c}}$
Effective primary transformer current	$\dfrac{\sqrt{2}I_{\text{d-c}}}{3}$	$\dfrac{I_{\text{d-c}}}{\sqrt{3}}$	$\dfrac{N_2}{N_1}(0.816I_{\text{d-c}})$
Peak-inverse voltage	$\sqrt{6}E$	$2\sqrt{2}E$	$\sqrt{2}E_{\text{line}}$
Fundamental frequency	$3f$	$6f$	$6f$
Fundamental rms voltage	$0.177E_{\text{d-c}}$	$0.04E_{\text{d-c}}$	$0.04E_{\text{d-c}}$
Total primary kVA rating	$1.21P$	$1.28P$	$1.05P$
Total secondary kVA rating	$1.48P$	$1.81P$	$1.05P$

Note that E is the rms voltage of one secondary winding; f is the supply frequency; P is the d-c power delivered to the load.

15.14   Large-Capacity Pool-Type Rectifiers

There are two types of large-capacity rectifiers, each using a mercury pool at the bottom of an evacuated steel tank as the negative electrode. The tank is cylindrical, surrounded by a cooling water jacket, and sealed to a very low mercury vapor pressure.

The multielectrode type has six to eighteen main anodes supported by insulating bushings in the lid of the tank. The anode is a large graphite cylinder isolated from adjacent anodes by a cylindrical insulating shield. Figure 15-21 shows the construction.

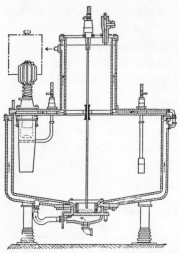

Conduction is started by a small central starting anode which can be lowered to touch the mercury surface and then automatically withdrawn. The arc thus struck is picked up by the most positive main or auxiliary anode, the succeeding anodes taking over conduction as they become more positive.

To prevent the arc from extinction in case the d-c load is removed for any reason, one or more auxiliary electrodes (shown in the right section of the rectifier in Fig. 15-21) are continuously energized to maintain ionization of the mercury vapor in the tank.

Fig. 15-21   Construction of a large mercury-arc rectifier housed in a water-cooled steel tank; approximately 6 ft in diameter and 8 ft high. (Courtesy Allis-Chalmers Manufacturing Co.)

Arc-back is a problem with the multielectrode rectifier. When an anode goes highly negative, positive ions bombard it and eject electrons. At times this action becomes sufficiently intense to cause the anode to act like a cathode, with the result that a *flashover* occurs between it and an anode that is strongly positive and near enough. Arc-back currents above the amount allowed by the protective circuits will cause shut-down of the rectifiers.

The *excitron* (Fig. 15-22) is a single-anode mercury-pool rectifier in a steel tank enclosure provided with an internal circulating cooling water system. The graphite, cylindrical, main anode is closely surrounded by a control grid which functions like the grid of a thyratron. In addition to controlling the rectifier output, the control grid reduces the deionization time and thus counteracts the cause of arc-back.

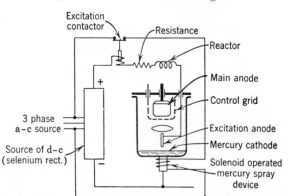

Fig. 15-22   Schematic diagram of excitron and its control circuit. (Courtesy Allis-Chalmers Manufacturing Co.)

There is only one anode per tank, and a continuous cathode spot is provided by a small excitation anode which carries direct current continuously. It is fed between 5 and 10 A through a selenium rectifier. The excitation anode is fastened securely in place and submerged in the mercury pool; just below it is a movable iron plunger of an electromagnet. The plunger floats in the mercury and makes contact with the excitation anode where the power is off, but when it is turned on the electromagnet pulls the plunger down and an arc strikes between the anode and the mercury which closes in above the plunger tip. If the arc becomes extinguished, the plunger rises and reestablishes it.

Excitrons are installed in groups of six or more tanks. They deliver large amounts of power at d-c voltages from 250 to 1,000 and more. They have the following advantages: (1) arc-back or flashover to adjacent anodes is impossible since there is only one main anode per tank; (2) deionizing effect of grid provides more stability at heavy loads; (3) they have slightly higher efficiency due to closer spacing between anode and cathode; (4) they have more dependability because a substitute unit can be switched into the system in case of trouble, which would usually start in one unit in either the multi-electrode or excitron system.

15.15   Inversion of Electric Power—Alternating to Direct Current

There are many advantages in using direct current instead of alternating current for power transmission. Direct-current transmission involves only voltage and polarity, while with alternating current there are voltage,

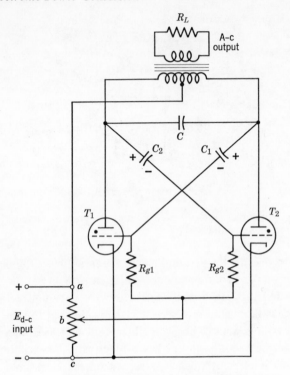

Fig. 15-23   Self-excited parallel inverter.

frequency, phase, phase sequence, and polarity. The maximum rms alternating voltage that may be used is less than the maximum direct voltage because the peak voltage determines the amount of energy loss into the air due to corona. There are no dielectric losses in d-c transmission, and line reactance is eliminated. Disadvantages are the likelihood of a power arc due to flashover of an insulator and greater difficulty of interrupting an arc. Direct voltage cannot be transformed; therefore at the receiving end it must be *inverted* to alternating voltage so it can be stepped down to safe voltages for commercial and domestic use.

There are several types of inverter circuit, the series, the parallel (each of which may be self or separately excited), the relaxation, and the grid-controlled inverter. Figure 15-23 is the circuit of a self-excited parallel inverter; using low-current thyratrons, it can be set up in the laboratory on a small scale. The action can be assumed to be in progress with T_2 just having started to conduct. C has been in a charged state with its right plate positive, and this is responsible for shutting off current in T_1 owing to its cathode being made positive by C.

Prior to the start of this explanation, T_1 was conducting, T_2 was not, and C_1 was charged (+ right, − left) to the potential between a and b plus a contribution of induced voltage in the right half of the output transformer primary. Now that T_2 is conducting, its plate potential has dropped to only about 15 V above cathode potential, and the discharge from C_1, aided by the d-c voltage from c to b, drives the grid of T_1 highly negative. From this point in the cycle, while C_1 is discharging, C is discharging through T_2, the power source, and the left half of the transformer. Soon C is discharged, then charged with reverse polarity (+ left, − right). When C_1 has discharged through R_{g1} sufficiently to allow the grid to T_1 to come up to firing potential, it fires and C puts T_2 out with its negative potential on the plate. The *commutating capacitor* C always discharges through the conducting tube, the power source, and half of the transformer primary in series. The induced voltage in half of the transformer winding, caused by the sudden increase of current going through the other half of the newly firing tube, aids in charging the small capacitors C_1 and C_2. The frequency of oscillation depends, as in multivibrator circuits and trigger circuits, on the time constant ($C_1 R_{g1} = C_2 R_{g2}$) and on the setting (b) on the voltage divider.

The circuit diagram of a six-unit large-capacity inverter installation using ignitrons with thyratron control is shown in Fig. 15-24. The power

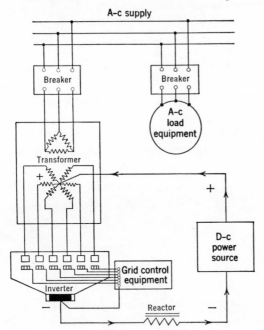

Fig. 15-24   Six-unit large-capacity inverter installation.

conversion unit receives d-c power through a series reactor which counteracts fluctuations in the instantaneous value of the direct current caused by inverter action and helps limit the input current if momentary short circuits occur in the inverter. The unit feeds a-c power into a system already getting power from one or more other sources. Alternating-current power is taken from the system by load equipment through the circuit breaker at the right.

SUGGESTED REFERENCES

1. Royce G. Kloeffler, *Industrial Electronics and Control*, 2nd ed., Wiley, New York, 1960.

2. J. M. Cage, *Theory and Application of Industrial Electronics*, McGraw-Hill, New York, 1951.

3. Westinghouse Electric Corp., *Industrial Electronics Reference Book*, Wiley, New York, 1948.

4, John Markus, *Handbook of Electronic Control Circuits*, McGraw-Hill, New York, 1959.

5. William D. Cockerell, *Industrial Electronics Handbook*, McGraw-Hill, New York, 1958.

6. Silicon Controlled Rectifier Manual, Current Edition, General Electric Co., Semiconductor Products Department, West Genessee Street, Auburn, New York.

7. B. D. Bedford and R. G. Hoft, *Principles of Inverter Circuits*, Wiley, New York, 1964.

QUESTIONS

15-1   What is the basic function of the SCR?

15-2   How is the SCR more efficient than a thyratron in controlling power in a load?

15-3   How does the gate function as the controlling element of the SCR?

15-4   What prevents the SCR from conducting internally from cathode to anode?

15-5   Sketch the schematic symbol of the SCR.

15-6   What is the significance of a "holding current" in SCR operation?

15-7   In d-c operation of the SCR, the gate loses control of conduction once it starts. Why doesn't this happen in a-c operation of the SCR?

15-8   What advantages does the TRIAC have over the SCR in controlling load power?

15-9 What are the advantages of using three-phase power instead of single-phase power?

15-10 Draw a set of delta-connected transformer secondaries supplying a balanced three-phase resistance load. Draw a phasor diagram of the voltages between the transformer terminals. On each phasor lay off a distance about half its length representing winding *current* which, in this case, is in phase with the terminal voltage. Select one of the lines going to the load and represent current flowing in it by I_L. Now use Kirchhoff's current law and determine graphically the amount and phase of I_L by *phasor addition*.

15-11 Draw a set of wye-connected transformer secondaries supplying a balanced three-phase resistance load with three wires. Draw a phasor diagram of the three winding voltages. Select two of three line wires and obtain the phasor that represents the potential difference between them. How is the line current related to the winding current?

15-12 In Fig. 15-16a the potential of point P is said to be always at the potential of the most positive anode. What assumption must be made? If the voltage drop through a conducting SCR is 1 V, how high in potential would the next positive going anode have to be before it starts conduction?

15-13 If you had an enlarged drawing, to scale, of the voltage waves in Fig. 15-16b, tell how you would proceed to evaluate $E_{\text{d-c}}$ graphically?

15-14 How does the fundamental ripple voltage in the load of Fig. 15-18 compare with that of Fig. 15-16a in frequency and in amplitude?

15-15 Compare the three-phase and six-phase half-wave rectifiers in regard to advantages and disadvantages.

15-16 Explain how conduction takes place in the tubes of a three-phase full-wave bridge rectifier. In what ways is this circuit superior to the other two?

15-17 How is a tank-type multielectrode rectifier started? What provision is made to prevent loss of an arc, with resulting shut-down?

15-18 Discuss arc-back in a mercury-pool multielectrode rectifier.

15-19 Describe the *excitron* and compare its structure with that of the multi-electrode tank rectifier.

15-20 Name the ways in which the excitron is an improvement on the multi-electrode rectifier.

15-21 Explain how the *commutating capacitor* functions in the self-excited parallel inverter circuit. How does the circuit manage to give this capacitor alternate polarities as needed?

15-22 How is a nonconducting tube in the parallel inverter kept from conducting until just the right moment?

15-23 Can you see any advantage to be gained by separating the grid circuit resistors in Fig. 15-23 and connecting each to a separate sliding contact on the voltage divider?

15-24 Name the most important advantages of transmitting power by direct current. What are some disadvantages?

PROBLEMS

15-1 A resistance load that requires 1 kW of power at 100 V is being supplied by a single SCR from a sine-wave source of 200 V rms. If half-wave phase control is being used for the SCR, what can be its maximum conduction angle without exceeding the 1 kW rating of the load?

15-2 Repeat Problem 15-1 for full-wave phase control using two SCR's.

15-3 Sketch the waveforms of voltage across the SCR and its resistive load, for half-wave phase control of a 100-V sine-wave source, for firing angles of 30°, 60°, 90°, 180°. (Assume the forward drop of the SCR is negligible.)

15-4 A TRIAC is used to proportion power to a resistive load of 1 kW rated at 100 V. The power source is a 200-V sine wave. Determine the required maximum average voltage and current ratings of the TRIAC.

15-5 A delta-connected resistance load made of three 10 Ω elements is connected to terminals 1, 2, 3 of Fig. 15-15a. The terminal voltage of each winding is 120 V. The current in each resistance element is 12 A. How much current is in each of the line wires? Calculate the delivered power, using I^2R, and check with $\sqrt{3}E_L I_L \cos\theta$, where θ is the angle between current and voltage of a load element.

15-6 The three windings of Problem 15-5, delivering 120 V each, are rearranged into a wye connection. The three 10-Ω resistors are also connected in wye, instead of delta, and then the group is connected to terminals a, b, c of Fig. 15-15b. Calculate (a) the voltage across each resistance unit, (b) the voltage between lines, (c) the total I^2R in the load, (d) $\sqrt{3}E_L I_L \cos\theta$.

15-7 (a) If each transformer winding in Fig. 15-15a has a terminal voltage of 100 V rms, what is the phasor sum of all three? (b) If these transformer windings are connected as in Fig. 15-15b, what is their phasor sum? Show by means of a phasor diagram.

15-8 Assume that winding C in the delta connection (Fig. 15-15a) has 90 V at its terminals, while windings A and B each have 100 V. The phase angles of the voltages remain at 120° with respect to adjacent voltages. What would then be their phasor sum? Show by means of a phasor diagram.

15-9 The windings described in Problem 15-8 may be connected in wye as in Fig. 15-15b. Show with a phasor diagram the sum of line-to-line voltages. (Although there is no closed path to carry a circulating current, it is important in three-phase rectifier operation that all supply voltages have the same amplitude.)

15-10 Compare the frequency of the fundamental component of load voltage supplied by the full-wave single-phase, the half-wave three-phase, the half-wave six-phase, and the three-phase bridge rectifiers.

15-11 A load of 5 Ω resistance is to receive 500 V d-c from a rectifier operating from either a single-phase or a three-phase 60 Hz supply. A series inductance of 1 H is connected with the load resistance. Calculate the rms ripple voltage at the load, when it is supplied from a single-phase full-wave rectifier.

15-12 Repeat Problem 15-11 for a three-phase half-wave rectifier.

15-13 Repeat Problem 15-11 for a three-phase bridge rectifier.

15-14 In Table 15-1 the value of average load voltage produced by the half-wave three-phase rectifier is shown to be equal to 1.17 times E, the rms value of the phase voltage in the wye-connected transformer secondary. Plot to a reasonably large scale one of the heavy loops of the instantaneous output voltage wave given in Fig. 15-16b, using 400 V for E. Determine E_{d-c} by graphical integration and compare your result with that given in the table.

15-15 (a) How much is the peak-inverse voltage on a tube of the rectifier described in Problem 15-14? (b) If the average value of tube current is 5 A, how much output power is delivered?

15-16 A six-phase, 60-Hz rectifier is supplied from transformer windings having 200 V rms at their terminals. The load is a resistance of 25 Ω. Determine: (a) average tube current, (b) peak-inverse voltage on the tubes, (c) fundamental ripple voltage and its frequency.

15-17 For the rectifier of Problem 15-16, what are the (a) d-c load voltage, (b) total primary kVA rating, (c) total secondary kVA?

15-18 A three-phase 60 Hz bridge rectifier is supplied from transformer windings having 200 V rms at their terminals. The load resistance is 25 Ω. What are the (a) average tube current, (b) peak inverse voltage on the rectifiers, (c) fundamental ripple voltage and its frequency, (d) d-c load voltage, (e) total primary kVA rating, (f) total secondary kVA rating? Compare these values with the corresponding results of Problems 15-16 and 15-17.

15-19 Calculate the power-conversion efficiency for the three-phase half-wave circuit. Neglect tube drop and transformer losses.

15-20 A three-phase half-wave rectifier uses a mercury-vapor diode rated at 1,600 A peak, 150 A average, and 800 V peak inverse. (a) Neglecting tube drop and the reactance of the transformer, find the maximum d-c current and voltage which may be obtained from such a rectifier with resistance load. (b) Determine the voltampere ratings of the transformer primary and secondary.

CHAPTER SIXTEEN

Photoelectric Devices

One of the most fascinating electronic devices in common use is the phototube, popularly called the electric eye. By means of this electron tube or of its close relative, the photocell, electric current may be controlled by a change in the intensity of light that passes through its window. Transistors with this property have also been developed.

Fundamental principles of *photoelectric emission*, which is the ejection of electrons from a surface by light energy shining upon it, and of the behavior of photoelectric devices in circuits are presented in this chapter and numerical examples are given.

An introduction to semiconductor photodevices and the laser are included.

16.1 Simple Phototube Circuit

A photograph of a high-vacuum phototube and the symbol that represents the tube in a circuit diagram are shown in Fig. 16-1. The same kind of symbol is used to represent a gas-type phototube, but a heavy dot is added in the circle to indicate that gas molecules are present. Phototubes require electric potential on their electrodes before they will operate. The anode is made positive with respect to the cathode. Phototube current is of the order of microamperes. A simple circuit containing a phototube is shown in Fig. 16-2. Consider the voltage (E) to be constant. When light shines on the cathode, electrons will be ejected from it and will pass to the anode owing to the forces exerted by the electrostatic field within the tube. The stronger the light, the more electrons per second will leave the cathode, and so the larger will be the current.

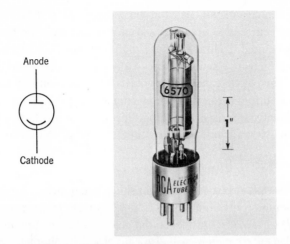

Anode

Cathode

Fig. 16-1   Symbol and photograph of a high-vacuum phototube. (Courtesy Radio Corporation of America.)

The current of a phototube is usually too small to operate a practical control circuit. Some highly sensitive relays are available which operate on approximately 0.1 mA, but they do not stay in proper adjustment so well as do more rugged types that require from 5 to 50 mA (or more) for operation.

It is readily observed that current in the circuit of Fig. 16-2 will produce a voltage drop across the series resistor (R). Suppose the current is 2 μA and the resistance is 10 MΩ. The voltage drop across R is then 20 V, with positive polarity on the right, and negative on the left. Now consider that the beam of light shining on the cathode is interrupted by some object placed in its path. The current will instantly drop to almost zero, but for this example consider that it drops to 0.1 μA. The voltage across R is then only 1 V. This change in voltage, when applied to the grid of a high-vacuum triode, can be made to result in a change of its plate current from zero to full current capacity of many

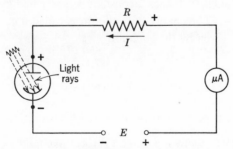

Fig. 16-2   Simple phototube circuit.

triode types. A relay coil connected in the plate circuit of the triode will then carry sufficient current to actuate its armature and close its contacts. The contacts may be used to close another circuit and thus operate a bell, a light, a counting mechanism, a small motor, or any other device or circuit that will operate with current that the relay contacts can carry without overheating. Of course, the current through the relay contacts and the device they control is supplied directly from a line or other adequate source. Examples of such a circuit, employing a triode amplifier tube, will be presented later.

16.2   Photoelectric Emission

Some metals emit electrons when light shines upon them. Common among these are sodium, potassium, and cesium. An electrode made of more rugged metal, such as steel or nickel, but coated with a compound of one of these three elementary metals, will also emit electrons photoelectrically. The compound and the parent metal are said to be *photoemissive*.

Visible light, as well as so-called black light (which cannot produce visible effects on most objects), has certain properties that indicate it must be wavelike in nature. This means it has frequency and wavelength in the sense that alternating voltage was described as having these characteristics. An important difference between black light and visible light is that the wavelength of black light is either too short (ultraviolet) or too long (infrared) to produce effects which the human eye can detect unaided.

In order to explain certain phenomena, one of which is *photoelectric emission*, it is necessary to think of light as being made up of minute specks of concentrated energy—corpuscles, Sir Isaac Newton* called them—that strike all objects in machine-gun-bullet fashion. Each speck, or bundle, of energy is called a *photon*, and it contains an amount of energy that depends on its *wavelength only*. An important physical constant relates the energy and wavelength of a photon. It is called *Planck's constant*, named after a brilliant physicist, Max Planck.** Planck's constant is designated by the letter $\hbar$ and is 6.624×10^{-34} J-s. The energy of a photon is called a *quantum* of energy, and its amount is given by the equation

$$W = \hbar f \quad (\text{J})$$
$$= \frac{\hbar f}{q_e} \quad (\text{eV}) \tag{16-1}$$

* Sir Isaac Newton (1642–1727) was a brilliant English scientist who discovered fundamental laws in physics and in mathematics.

** Max Planck (1858–1947) was a German physicist who made great contributions to the fields of mechanics, heat, and mathematics.

in which f is the frequency in cycles per second, the constant $\hbar$ has the value described in the preceding sentence, and q_e is electronic charge.

Electrons in a photoemissive material, when struck by a photon, receive the photon's energy, and some of them become able to escape from the material in spite of certain opposing forces that exist at the surface of the material.

16.3   Conditions at a Metal Surface

The surface of a metal, even when highly polished, is, from an atomic point of view, very rough. The atoms that make up metals are found by experiment and calculation to be arranged in regular geometric formations and equally spaced from each other. An atom consists of a heavy core called a *nucleus*, which is *positively* charged, and a number of electrons that move rapidly around the nucleus. In the atoms of photoemissive metals there are *free electrons* (each atom has at least one) which are farthest out from the nucleus and which readily exchange places with free electrons of other atoms of the same metal. Beneath the surface of a metal there is an abundance of electrons and nuclei both above and below each free electron, and so the electron is able to wander about. But at the surface there are forces that oppose the escape of a free electron from the metal. Because no positive charges (nuclei) are above the electron, there is a net force holding it back which is due to positive charges below it. Its escape is also hindered by opposition forces of electrons above it that decrease its energy as it tries to force its way between them to penetrate the surface.

This surface barrier is *aided* by a so-called *image force*, due to the positive charge, equal to the electron's negative charge, which the escaping electron leaves on the metal. It may be seen, therefore, that, in order to escape from a metal, an electron must receive energy from some external source. Photons supply that energy when the metal is of the type that loses its free electrons easily. Such electrons need to be supplied with only enough energy to overcome the surface barrier just described. There is a *critical frequency* of light that has just enough energy in its photons to eject electrons from a given metal. That frequency is called the *threshold frequency*.

The minimum energy required by an electron to escape from a metal is called the *work function* (or *electron affinity*) of the metal. It may be expressed in electronvolts, and its value varies between 1 and 7* depending on the

---

* See W. G. Dow, *Fundamentals of Engineering Electronics*, 2nd ed., Wiley, New York, 1952, for an excellent treatment of the escape of electrons from metals.

kind of metal and to some extent on the nature of the surface. Designating the work function as W, and remembering that the energy of a photon is hf, the *Einstein equation for photoelectric emission is*

$$hf = W + \tfrac{1}{2}mv^2 \text{ J} \qquad (16\text{-}2)$$

The energy of the electron *after escape* is $\tfrac{1}{2}mv^2$ J. If this energy is zero, the electron was *just able to escape* and had no energy left over. The threshold frequency (f_0) is then found to be

$$f_0 = \frac{W}{h} \text{ cycles/sec or hertz} \qquad (16\text{-}3)$$

in which W is expressed in joules.

If W is expressed in *electronvolts*, it may be converted to joules by the equation

$$\text{Joules} = Wq_e \qquad (16\text{-}4)$$

in which W is volts (or electronvolts), q_e is the charge on an electron, expressed in coulombs. *The number of joules in one electronvolt is* 1.602×10^{-19}.

Two important *laws of photoelectric emission* may now be readily understood. They are:

I. *The number of electrons per second ejected from a photoemissive surface is directly proportional to the intensity of the light that falls upon the surface.*

II. *The energy of electrons ejected photoelectrically from a surface depends on the frequency of the light waves striking the surface but not on the intensity (strength) of the light.*

16.4   Illumination Units

In order to understand properly the operation of phototubes and photocells, we should be familiar with some important quantities in the illumination system.

Light is radiant energy having wavelengths that produce visual sensation; i.e., light is radiant energy that makes it possible for us to see things. The range of wavelengths of the radiant energy called *light* is from 0.0000004 m to 0.00000076 m. Wavelengths of light are often expressed in *Angstrom Units* (AU) or in microns (μ). The Angstrom Unit is 10^{-10} m, and the micron is 10^{-6} m. The range of visible light, specified above in meters, is thus from 4,000 to 7,600 Angstrom units or from 0.4 to 0.76 μ. Quantity of light is expressed in *lumen-seconds* and is denoted by the letter Q.

A *lumen* is a unit of *luminous flux*. The *unit of intensity* (*I*) of a *luminous source* is the *candle*. A *source* having a luminous intensity of one candle sends out 4π lumens of flux. The symbol for luminous flux is ϕ in this book.

The number of lumens per unit area received on a surface is called the illumination on the surface. It may be denoted by *E*, and its units are lumens per square foot or foot candles.* The number of lumens per square foot (density of luminous flux) reaching the surface of a sphere of one-foot radius about the one-candle source (also called a one-candlepower source) is

$$E = \frac{\text{flux}}{\text{surface area}} = \frac{\phi}{A} = \frac{4\pi}{4\pi r^2}$$

$$= \frac{4\pi}{4\pi \times 1^2} = 1 \text{ lm/ft}^2$$

$$= 1 \text{ fc}$$

The illumination on a small surface perpendicular to a straight line from the center of a luminous source at a distance *D* is

$$E = \frac{I}{D^2}$$

in which *I* is the source intensity, in candles, and *D* is the distance, in feet. *D* must be 10 feet or larger, and the area of the source must be such that its largest dimension does not exceed about *D*/10. Otherwise the inverse-square relations do not hold. The intensity of a source may be determined by measuring *E*, and a distance *D* to the source which should be at least 10 feet away. If a shorter distance is used, or if the source area is too large, the *apparent intensity* is obtained. *E* is measured by means of a foot-candle meter (light meter).

16.5   Characteristics of a High-Vacuum Phototube

When proper voltage is applied to a high-vacuum phototube it will conduct current that varies in amount in direct proportion to the *intensity* of light striking it. This is a verification of the first law of photoelectric emission stated

* Units used in illumination studies in this book are those of the English system, which expresses lengths in feet. It should be noted that we are dealing with what may be considered *two intensity concepts*. One is the intensity of the *source*, and, while we could specify the surface receiving the light as having a certain *intensity* of illumination, we use the term *illumination* and express it in terms of *density of luminous flux* (lumens per square foot) on the surface.

in Section 16.3. In order to show this relation graphically, it is convenient to use data from a set of characteristic curves for the phototube. The curves are plotted from readings of current against voltage at several values of constant light intensity. Curves for a typical high-vacuum phototube are shown in Fig. 16-3. At the points where the curves become straight lines, the electrons are streaming to the anode as rapidly as they are being emitted. To the left of these points, the anode voltage is not high enough to cause that, and there is limitation of current by space charge.

The similarity in shape between these curves and the curves of a pentode tube is very striking. It will be recalled that each curve on the pentode-characteristic chart is identified by a value of constant negative-grid bias, which is the controlling factor that determines the current when other electrode voltages are fixed. Here only light intensity determines the current if the anode voltage is fixed. Note, however, that the current of the phototube is in *microampere units*, whereas the current of the pentode is one thousand times as great (milliamperes). This means that the high-vacuum phototube has a dynamic plate resistance of *hundreds of megohms* and an internal d-c resistance of 10 to 50 (or more) *megohms*.

Figure 16-4 shows the relation between phototube current and light intensity when the voltage across the phototube is held fixed at 70 V. Values were obtained from the curves of Fig. 16-3, at the vertical line through 70 on the horizontal scale.

This direct-proportion relation between current and light intensity is a

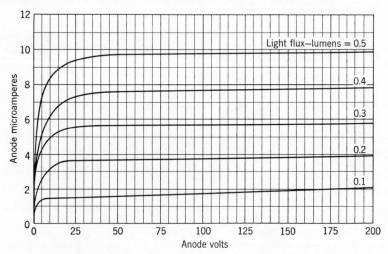

Fig. 16-3 Anode characteristics, type 925 vacuum phototube.

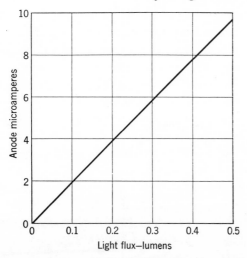

Fig. 16-4   Variation of anode current with light flux striking cathode, type 925 vacuum phototube with 70 V on the anode.

very important factor in the successful use of the high-vacuum phototube in the reproduction of sound from movie film. The "sound track" on the film is a path of varying transparency that was recorded by varying light intensities which were controlled by the original sound. In the reproduction of sound, the light striking the phototube is caused to pass first through the sound track so that the intensity of light reaching the cathode of the tube will vary in accordance with the transparency of the track. These light variations will cause proportional current variations through the phototube and its circuit resistor. The voltage variations across the resistor may then be amplified to reproduce the original sound in a loud speaker.

Practical calculations relating to the performance of a high-vacuum phototube in a circuit are relatively simple. A load line is constructed in the same manner as was done in the analysis of amplifier-tube performance. That is, the slope of the load line is $-1/R_L$, in which R_L is the total resistance in series with the anode. The load line is drawn from a point on the anode-voltage axis corresponding to the supply voltage of the circuit.

Figure 16-5 shows the effect of different values of load resistance on the variation of output voltage with light intensity. The circuit is that of Fig. 16-2 with $E_{BB} = 125$ V. When $R_L = 7.5$ MΩ, a variation of light intensity from 0.5 to 0.1 lm produces a voltage variation across R_L from 53 to 113 V, while at $R_L = 2.5$ MΩ the voltage variation is from 102 to 122 V. In certain applications the smaller voltage change may be entirely adequate.

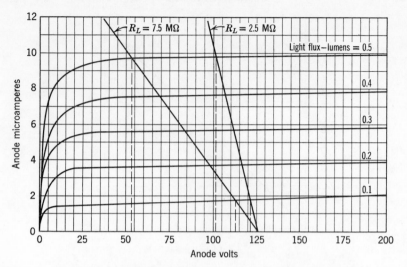

Fig. 16-5   Anode characteristics, type 925 vacuum phototube.

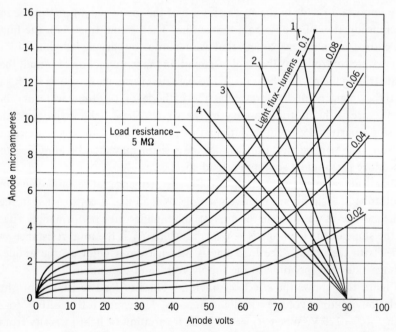

Fig. 16-6   Anode characteristics, type 930 gas phototube.

16.6 The Gas Phototube

Gas phototubes have generally the same construction as vacuum phototubes, except that a small and definite amount of inert gas is introduced to
increase the current capacity and sensitivity. The gas serves as a source of
supply of additional electrons which result from ionization of gas atoms by
electron bombardment. An electron emitted from the cathode is accelerated
toward the anode by the field and acquires sufficient velocity, and hence
kinetic energy, to drive a second electron out of its gas atom if it strikes the
atom. The two electrons gain velocity as they proceed and may release two
more electrons from gas atoms before entering the anode. It is possible in this
manner for two to a dozen or more electrons to reach the anode for each
electron that leaves the cathode. This process is called *gas amplification*.
Characteristic curves for a typical gas phototube are shown in Fig. 16-6.
Notice how much less light is needed to produce 5 μA at 80 V in this tube than
in the vacuum phototube.

Although a gas phototube has the advantage of *high current capacity*, its
curves showing current plotted against light intensity are nonlinear when
the advantage of high current is utilized (see Fig. 16-7). In the higher range of

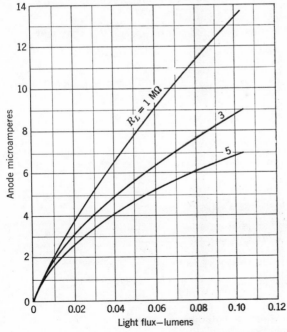

Fig. 16-7   Current-illumination characteristics, type 930 gas phototube.

anode voltages there is danger of breakdown in the tube, with resulting serious damage to the cathode. Nevertheless, gas phototubes are widely used and they perform satisfactorily in circuits properly designed and operated. A gas phototube is very sensitive to changes in light intensity. It produces larger changes in current through its load resistor, for a given change in light intensity, than does a high-vacuum phototube of comparable size.

In addition to having nonlinear current-light-intensity curves, the gas phototube is subject to slight delays between changes of light intensity on the tube and resulting changes in tube current. This *time lag* is objectionable when the tube is used in the reproduction of sound that has been recorded on film, particularly in the reproduction of music in which frequencies above 2,000 or 3,000 Hz should be heard.

16.7   Direct-Current Circuits Employing Phototubes

Phototubes are seldom used without amplifier tubes because phototube currents are so small. The change in voltage across the load resistor of a phototube, caused by a change in light intensity falling on the tube, is easily employed to vary the grid bias of a high-vacuum amplifier tube and thus materially change its plate current.

Figure 16-8 shows a simple phototube circuit that will actuate a relay when light of sufficient intensity enters the window of a phototube. The grid of the triode is biased so that its plate current is either zero or too small to operate the relay when the light, which is to actuate the circuit, is prevented

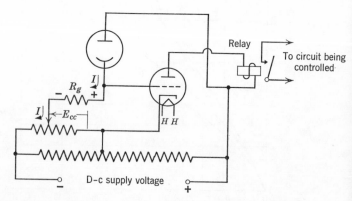

Fig. 16-8   Direct-current phototube circuit. Relay contact will be closed when light entering phototube increases.

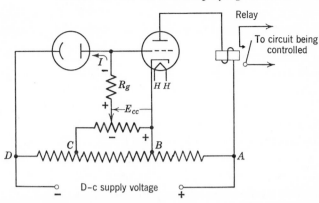

Fig. 16-9   Direct-current phototube circuit. Relay contact will be closed when the light shining on the phototube is interrupted.

from shining on the phototube. Phototube current contributes a positive amount of voltage on the grid by flowing through R_g. This positive contribution will become large enough to allow adequate plate current to flow to close the relay only when the phototube receives the required intensity of light. The grid-voltage supply E_{cc} may be adjusted to the proper value by means of the slide-wire resistor.

A circuit that will actuate the relay when the light shining on the phototube is *interrupted* is shown in Fig. 16-9. Note that the connections to the phototube are reversed, and this necessitates a change in connections to the supply. Light shining on the phototube results in the flow of phototube current (I) through R_g, and the voltage drop (IR_g) prevents the flow of enough triode plate current to actuate the relay. When the light is cut off, interrupted, or sufficiently reduced, the grid bias is diminished so that adequate plate current will flow to actuate the relay.

Some relays have a pair of contacts, so that both lines of a control circuit may be broken instead of only one. The contact part of that type of relay performs as a double-pole, single-throw switch. Some single-pole relays have two stationary contacts mounted so that the movable arm is held against one of them by a spring. When the relay coil is energized, the magnetic force overcomes the spring and pulls the arm away, opening the normally closed contact and closing the normally open contact. By means of a relay of this type, a control circuit may be either energized or deenergized when the phototube action takes place. That is, an independent circuit may either be closed or opened by phototube action. Double-pole relays with both normally open and normally closed contacts are on the market.

16.8   Principles of Design of D-C Phototube Circuits

The fundamental theory presented on d-c circuits and on the operation of high-vacuum triodes in this book is sufficient to equip the reader with adequate background to understand the basic principles used in the design and construction of a circuit like the one in Fig. 16-9. The circuit "operates" when the light to the phototube is cut off.

The first thing that should be known is how much current is required by the relay coil to actuate its armature. This information is often found stamped on a plate attached to the relay. It is available from the manufacturer at the time the relay is purchased. It it cannot be obtained in any other way, the value of current required to close the relay may be measured with a milliammeter or an ammeter.

When the required current is known, an amplifier tube must be chosen that will safely carry that current. From the plate-characteristic curves for the tube, we may determine the amount of negative grid bias required (1) to cut off the plate current, and (2) to allow the relay-closing current to flow. These bias voltages depend on the operating plate voltage that must also be chosen.

After the selection of a phototube has been made, we should determine from its characteristic-curve chart (Fig. 16-3 or Fig. 16-6) how much *change in current* will be caused by a variation of light over the operating range of the phototube. Suppose it is 5 μA. A 10-MΩ load resistor (R_g) carrying phototube current that varies in the amount of 5 μA will have a *voltage change* of 50 V across its terminals. That is ample to cut off the plate current of most triodes without any help from an E_{cc} voltage as provided in Fig. 16-9. However, the E_{cc} voltage should be present to limit the triode plate current to a safe value when the phototube is not conducting.

The value of the R_g may be chosen after the voltage between points B and D in the circuit is known. The voltage divider section furnishing E_{cc} has negligible resistance compared to R_g. The slope of the load line, $-1/R_g$, should be such that the load line intersects the characteristic curve for the largest lumen value just to the right of the knee of the curve. This will give maximum change in current for a given change in light intensity.

The resistance of the relay coil is usually so low that the load line for the triode may be taken as vertical at the value of plate voltage used on the tube. This is the voltage between points A and B on the circuit. The plate current must be allowed to rise to a value sufficiently high to operate the relay when the bias is greatly reduced owing to the loss of voltage across R_g.

The voltage-dropping resistor $ABCD$ should have sufficiently high resistance to keep the total input power to the circuit low. It could be made of three separate units, instead of having two taps on a single resistor as shown. The steady current through the section AB need not be more than $\frac{1}{3}$ A. The cathode heater coil of the triode may be inserted in series with the AB and BC sections, and thus a separate heater supply could be avoided. The relay-coil current flowing from B to C will not change the tube bias too much, because the resistance of the BC section will be low (about 20 Ω). The slide-wire resistance connected between points B and C should be at least 500 Ω in order to get fine adjustment of the bias voltage E_{cc}.

16.9   Alternating-Current Operation of Phototube Circuits

Because direct current is seldom available for phototube operation, most circuits are designed and built for use on alternating current. We may readily understand that a-c operation is possible *if the pulses of phototube current flowing through the grid-circuit resistor of the amplifier tube are phased properly with the pulses of plate voltage.*

In Fig. 16-10 the capacitor (C_g) charges with the polarity indicated, and in the time of a few cycles the grid bias is large enough to prevent the flow of triode plate current. The capacitor is small, but large enough so that it will hold the grid negative and thus prevent the flow of grid current during

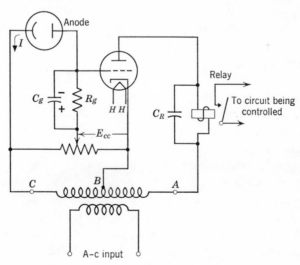

Fig. 16-10   Phototube circuit for a-c operation. Relay closes when light beam on phototube is interrupted.

the alternate half-cycles when the phototube cannot conduct. It also prevents appreciable fluctuations in negative grid voltage.

The capacitor (C_R) is necessary to prevent objectionable fluctuations in relay-coil current. Such fluctuations would cause vibration of the relay arm, with accompanying noise and burning of contacts. The general principles of operation are the same as for the d-c phototube circuit of Fig. 16-9. Soon after illumination is reduced, C_g discharges sufficiently to allow the grid potential to rise the required amount. The cathode-heater current is usually supplied by a filament winding on the transformer core or from a separate filament transformer.

The circuit of Fig. 16-10 may easily be changed to operate on increasing light on the phototube. It is only necessary to disconnect the phototube entirely, then connect its anode to point A and its cathode to the grid. The capacitor C_g will then become charged in reverse polarity to that shown in Fig. 16-10, when light shines on the phototube. E_{cc} must be made sufficiently large so that the triode current, caused by a small amount of light, will not actuate the relay. An increase in light will increase the phototube current and the contribution to grid voltage across R_g. This will add enough positive voltage to the *negative* E_{cc} voltage to permit the flow of plate current, and thus the relay will be energized. E_{cc} is present across the right-hand part of the slide-wire resistor, and it contributes negative bias, while the triode's plate is positive during positive half-cycles. Both anodes are negative on negative half-cycles, but the grid goes positive. A current limiting resistor between the grid and R_g is required.

16.10   Photomultiplier Tubes

Light intensities that are too low to affect conventional phototubes a useful amount can be handled by a *photomultiplier* tube with good efficiency. This is made possible by amplification of current from a photoemissive surface by means of secondary emission. It will be recalled that secondary emission is the ejecting of electrons from an anode by other high-energy electrons that strike it.

A series of anodes, called *dynodes*, are arranged in precise orientation in the phototube. Each successive one is higher in potential than the one preceding it. These potentials are taken from taps on a voltage divider connected across a high-voltage power supply. The steps in voltage are 75 to 150 V each, depending on the design and type of the tube.

If k_m is the ratio of the number of *secondary* electrons ejected to the number

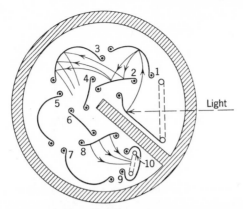

Fig. 16-11   Circular photomultiplier tube with nine dynodes.

of *primary* electrons striking the surface, after two strikings the *current* i_2, made up of the electron flow, is expressed in terms of the current i_o, representing the *input electron flow*, as follows:

$$i_2 = i_o k_m \times k_m = i_o k_m{}^2 \tag{16-5}$$

After n dynodes have been struck, the current is

$$i_n = i_o k_m{}^n \tag{16-6}$$

The RCA type 931A photomultiplier has nine dynodes which give a current amplification of 200,000. The *sensitivity* is 2 A/lm. This does not mean that a lumen of light flux is needed by the tube for operation; in fact, this would probably ruin it. It means, for example, that, when weak light is entering the tube, a *change* of 0.001 lm will cause a *change* of 2 mA in the output current. Recall that the type 925 vacuum phototube produces about 10 μA when 0.5 lm shines upon it, and this decreases to about 7.5 μA at 0.4 lm, giving a sensitivity of $2.5 \div 0.1 = 25$ μA/lm. The type 931A photomultiplier tube has the electrode structure shown in Fig. 16-11. It is about the size of a voltage amplifier tube built for an octal socket.

Another form of photomultiplier tube is shown in Fig. 16-12, which illustrates the DuMont linear photomultiplier. It has a window at the end made of a translucent cesium-antimony photoemissive surface. Light striking the outer surface ejects electrons inward, and they are focused onto the first dynode by a charged focusing shield. The dynodes are the curved surfaces of quarter-cylinders, and they are mounted so the ejected electrons leave the dynode surface approximately at right angles, heading for the next dynode. The secondary-emission surfaces are silver-magnesium for which

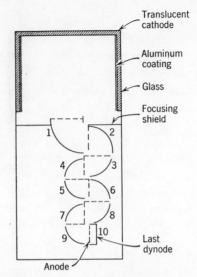

Fig. 16-12   Linear photomultiplier.

k_m is 3 to 4. Most of these tubes have ten secondary emitters and operate at voltage steps as high as 150 V. They are said to provide a current amplification of 3,000,000, and a sensitivity of 100 A/lm.

16.11   Image Orthicon and Vidicon

The image orthicon is a television camera tube that uses a five-stage photomultiplier located near the back end. The first dynode receives a return electron beam from a target near the front end of the tube. The multiplier output signals carry video information to the transmitter circuits. The image orthicon is illustrated in Fig. 16-13. It is about 14 in. long and 4 in. in diameter.

A light image is focused on the photosensitive cathode, and electrons

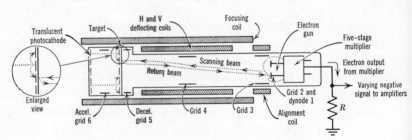

Fig. 16-13   Image orthicon (a television camera tube).

are ejected in amounts proportional to the amount of light striking each tiny patch of the surface. These electrons are accelerated to, and magnetically focused in the plane of, the target. They strike the glass target, which is only about a ten-thousandth of an inch thick, and eject secondary electrons which are collected by an adjacent mesh screen. This screen is at a small positive potential with respect to the target.

On the photocathode side of the target there is thus established a pattern of *positive* charges which is an electrical image of the light image of the scene being televised. This pattern of positive electrical charges extends through the very thin glass target, which is actually a *positive potential* pattern. The points of higher positive potentials are where more electrons are missing, and the lower positive potential points have fewer electrons missing. A scanning beam of low-velocity electrons hits this pattern of positive charges and deposits the necessary number of electrons at each point or tiny patch to neutralize the charge there. If the patch is at cathode potential, corresponding to a black picture area, no electrons are deposited.

The excess electrons from the main beam are turned back at each patch and focused into the multiplier. This electron stream evidently varies in intensity in accordance with the brightness of patches on the light image represented by positively charged patches at corresponding points on the potential image. As this electron current goes through the resistor R to ground it produces an output signal that carries the video intelligence. The sidewise resistivity of the glass target is such that the charges will neutralize each other before the next scanning cycle begins.

The *vidicon* is a small television camera tube about 6 in. long. It is used in industrial television and in studio film pick-up and also remote (outside the studio) pick-up. It does not have a return electron beam. It gets its output signal from a signal electrode consisting of a transparent conducting film on the inner surface of a face plate. The signal electrode is at about 20 V positive with respect to the grounded cathode of the electron gun. Each tiny patch of the photoconductor signal electrode can be considered as a plate of a leaky capacitor whose other plate is a "floating plate" until scanned by the electron stream.

Initially the gun side of the photoconductive surface is charged to cathode potential by the gun, thus leaving a positive charge on each elemental capacitor. During the time of one complete scanning of the scene these capacitors discharge in accordance with the value of their leakage resistance, which is determined by the amount of light falling on the patch. As a result there appears on the gun side of the photoconductive surface a positive potential pattern corresponding to the light pattern of the screen imaged on the opposite surface of the layer. As in the image orthicon, the scanning beam deposits electrons on the patches in accordance with how much positive charge is there. At this moment the electrical circuit is complete through the resistor in the signal electrode circuit to ground.

The vidicon has good definition and sensitivity. It is of much simpler construction and is less expensive than the image orthicon.

16.12   Barrier-Layer Photocells

Perhaps the most common application of self-generating photocells is their use in meters that measure intensity of illumination. Light shining on the cell surface generates a very low voltage, perhaps not more than a

Fig. 16-14   Foot-candle meter. (Courtesy Weston Instruments Division of Day-strom, Inc., Newark, N.J.)

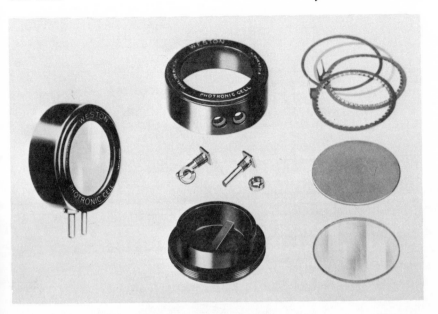

Fig. 16-15   Weston Photronic photoelectric cell and its component parts. (Courtesy Weston Instruments Division of Daystrom, Inc., Newark, N.J.)

few tenths of a volt in very bright light. Because the internal resistance of the cell is small, sufficient current will flow through the coil of a microammeter of suitable size to produce deflections over the full scale of the meter. Photographs of a foot-candle meter and a barrier-layer cell are shown in Fig. 16-14 and Fig. 16-15, respectively.

The iron-selenium cell is a popular type. It consists of only one plate, not basically different from one of the disks or plates of the selenium dry-plate rectifier described in Chapter 5. A thin layer of selenium is deposited on a chemically clean and highly polished plate of iron. A transparent film of cadmium metal is sprayed over the selenium layer, and this is protected by a coat of transparent lacquer. A ring of bronze with contact fingers along its inside circumference is clamped against the cadmium surface. Light falling within this ring passes through the transparent lacquer and releases electrons in the selenium layer. The cell is inherently a rectifier, and so the electrons pass much more readily to the cadmium layer and its contacts than to the iron plate. In fact, a barrier layer is said to form within the semiconducting selenium, through which electrons pass with difficulty.

A potential difference is thus set up across the cell with the iron plate positive and the conducting cadmium layer negative. If the external resistance

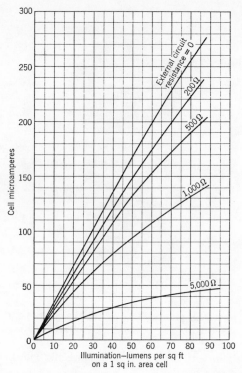

Fig. 16-16   Characteristics of an iron-selenium cell.

connected to the cell is low, the current will be almost directly proportional to the illumination intensity (see Fig. 16-16).

16.13   Photosensitive Transistor

We have found that an amplifier tube must be used with a phototube if output currents in the milliampere range are required, because phototubes have current limitations of about 10 μA. Photosensitive transistors may deliver as much as 6 mA at this writing, and we should expect a substantial increase in the current output when the demand arises. Other advantages of the photosensitive transistor are: no filament heating power is required, and the power received from bias sources is utilized with greater efficiency.

Like the first plain transistor developed, the first photosensitive transistor was the point-contact type. It was composed of a small *N*-type germanium wafer, shaped like a medicine pill, with a spherical depression ground into one face until the thickness of the wafer at the bottom of the depression was only 0.003 in. The wafer was pressed firmly in a cylindrical metal container, which made contact around its curved surface and provided one electrical connection. Then a phosphor-bronze wire with its sharp point pressing against the bottom of the depression was mounted in an insulating plug coaxially with the case.

The area directly under the wire point becomes *P*-type germanium when forming current flows, and so a *P-N* junction diode is produced. When it is biased in the reverse direction a *minute* current flows without light shining on the junction area. When light energy strikes the junction, the current increases substantially and the amount of current is proportional to the light intensity. The photons give their energy to the breaking up of covalent bonds, producing both electrons and holes. The applied electric field drives them to opposite terminals of the bias supply source, and thus current is produced.

A phototransistor made of a *P-N* junction grown in surface contact will operate with outputs comparable to those of the point-contact type. However, one made of *NPN* construction in which only the *extremely thin* central section (base) is photosensitive is most popular. No connection is made to the base, but a potential difference is applied to the ends.

Light is focused on the photosensitive base by means of a lens. The resulting current, 4 to 6 mA, is sufficient to operate reliable relays without further amplification. Furthermore, small changes in emitter-to-base potential produced by a changing light intensity will result in substantially larger changes in collector current. Thus the current amplifying properties are present. The terminal with forward bias is the emitter terminal (see Fig. 16-17). The emitter section must have low resistivity.

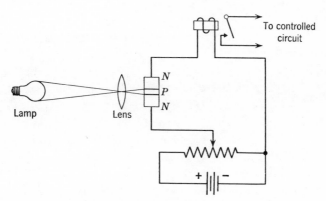

Fig. 16-17   *NPN* **phototransistor in a control circuit.**

16.14   Photovoltaic Action; the Solar Cell

Photovoltaic action occurs in semiconductors constructed to absorb incident light energy. An internal potential difference is generated across a *P-N* junction in the material making up the device. The potential difference can generate a current in an external load connected to the photovoltaic device. The most common photovoltaic energy converter is the *solar cell.* Solar cells have been used extensively as electrical power supplies in artificial earth satellites and space probes. They have been used to convert the sun's energy into useful electrical energy for powering and controlling the equipment on board.

Two similar semiconductor materials are used to form a photovoltaic solar cell. The materials differ mainly in the nature of impurity atoms present in the two materials. One of them contains impurity atoms that make it an *N*-type semiconductor and the other contains *P*-type atoms. The cell consists of a thin slice of single-crystal *P*-type silicon up to 2 cm² into which a layer (about 0.5 μ) of *N*-type material is diffused. The *N*-type material is usually phosphorus, although other materials have been used in experimental designs.

Electrical connections are made to the *P*-region and *N*-region of the cell. An external load may then be connected to the cell. An arrangement as shown in Fig. 16-18 is sometimes used. Incident light strikes the upper face of the cell and generates the internal potential difference that will force current

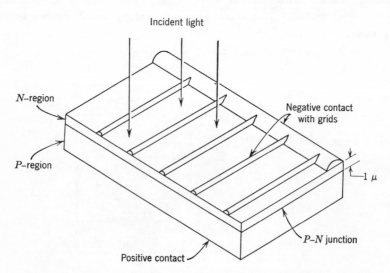

Fig. 16-18   Negative contact grid structure on exposed surface of a cell.

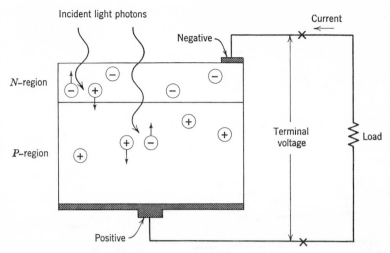

Fig. 16-19   Incident light generates "electron-hole" pairs, causing current in the load.

through an external load. The load is connected between the bottom surface, labeled positive contact, and the negative contact along one edge of the upper surface. The contact on the dark side of the cell can have a large area, covering the entire surface as a continuous layer of solder. However, the negative contact should have a small area so that it does not block the incident light. It should permit maximum exposure to incident light but at the same time make good electrical contact with the surface. To meet these conflicting requirements in a cell having a large surface, a grid electrode arrangement is sometimes used as shown in the figure. These grids are spaced so that a low-resistance connection can be made to the illuminated surface but are narrow enough to permit exposure to light. Very small cells make contact along one edge only.

Where the two doped semiconductor materials join is a *P-N* junction. It is a thin, neutral region separating the two materials. The junction region contains both types of impurities and is essentially electrically neutral. An inherent property of such a *P-N* junction is an internal electric potential that is developed across it.

Incident light photons generate electron-hole pairs in the cell, as shown in Fig. 16-19. The impinging photon can pass into the cell and react with an atom, either in the *N*-region or *P*-region of the cell. An electron may absorb the energy of the photon, increasing its own energy sufficiently to release it from its parent atom. The electron may then be conducted about in the cell

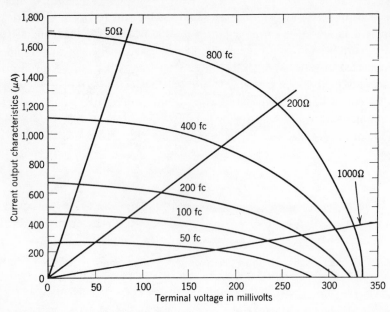

Fig. 16-20   Voltage-current characteristics of a solar cell for various external loads and illumination levels.

by diffusion and drift, leaving a hole in its atom. The electron and hole thus generated are in excess of the number needed for thermal equilibrium in the cell. In reestablishing equilibrium in the cell, they may recombine in the region of their generation or they may wander randomly about until they are forced across the junction by the electric field there. Those that cross the junction contribute to current in the cell and its external load.

The electrical behavior of a typical commercial cell is given in Fig. 16-20. The load current and terminal voltage were measured for varying values of load resistance from short-circuit to open-circuit conditions. The open-circuit voltage is less than 0.350 V and short-circuit current is only a few milli-amperes for the range of illuminations up to 800 fc. The voltage-current relations are marked for load lines representing 50, 200, and 1000 Ω.

For a given illumination there will be some load resistance that absorbs maximum power. Constant-power hyperbolas may be drawn on the figure to determine the maximum power for the selected illumination. The constant-power hyperbola that is tangent to the illumination curve will indicate the maximum power condition. Then, the load line that passes through the tangent point will represent the load resistance that will absorb the amount of power. The maximum power at 400 fc is approximately 290 μW.

In their application to furnish power from the sun to operate electrical equipment in space, a single cell such as this is certainly inadequate. Its voltage, current, and power are miniscule. However, many similar cells may be connected in series and parallel in large arrays, either on the surface of the space vehicle or on movable "paddles" that extend from it. Movable paddles covered with solar cells can be positioned by the vehicle so that maximum exposure to the sun is maintained regardless of the position and attitude of the vehicle.

16.15   Light Amplification by Laser

One of the most exciting developments in photo-technology was the *laser*. From the pioneering work of T. H. Maiman in 1960,* lasers have fired the imagination of engineers and scientists around the world. They have been used to cut holes in diamonds, weld metals almost instantly, perform delicate surgery of the human body, and carry communications channels from earth to space vehicles. Their ability to concentrate power in a narrow beam of light is an important property in many applications.

The word *laser* is an abbreviation of "*l*ight *a*mplification by *s*timulated *e*mission of *r*adiation." The laser controls a process of generation of coherent light by stimulating emission of radiation from the atoms in a solid or gaseous material. Lasers are sometimes referred to as "optical masers," where the word maser refers to the same action, but for microwaves rather than lightwaves.

Transfer of energy by radiation in an atom can be described by three different types of radiation: (1) *absorption*—occurs as an atom reacts with a photon of incident light. The photon must have sufficient energy to increase the energy of an electron and move it to a higher orbit in the atom. During the process the electron increases its energy by an amount equal to the absorbed energy of the incident photon. Any excess energy of the photon is lost in the process, either as fluorescence or as heat. The electron may acquire only discrete amounts of additional energy, even though the impinging photon may have more than is needed to raise the energy of an electron. The important result is the excited state of the atom when an electron has absorbed additional energy. As in all physical systems, the atom will tend to return to its original lower energy state; (2) *spontaneous emission*—occurs when the atom is in an excited, unstable state. An electron may spontaneously return to its original

* T. H. Maiman, "Optical Maser Action in Ruby," *Brit. Commun. Electron.*, **7,** 674 (1960).

energy level, or it may lose only part of its increased energy. In the latter case, it will take up an orbit position somewhere between that of its excited state and its original position. It is then in a *metastable state*. Whenever an electron loses some energy, a photon containing the amount of energy lost may be emitted; (3) *stimulated emission*—occurs in excited atoms that have electrons in metastable states. An electron may exist in such a "semistable" state for some period of time. An incident photon at the same energy level as the metastable electron may stimulate the electron to return to its original stable state. If the photon is not absorbed, but only passes through the atom, then *two photons* at the same energy may be emitted from the process. An "amplification" has occurred in the atom because of the *stimulated emission of radiation*. This action is the basis of laser operation.

16.16   Laser Radiation Is Coherent

In the process of stimulated radiation from an atom, the two photons that derive from a single incident photon will have the same energy. They will also have the same frequency of oscillation and the same phase relationship in time. Such waves are said to be *coherent*. If two waves have different frequencies or phase, or both, they are incoherent. An incoherent wave is difficult to modulate with information. It may be used only in on-off operation. A coherent wave, on the other hand, is easily modulated with information, either as amplitude, frequency, or phase variations as discussed in Chapter 13.

The output of a laser is coherent radiation at frequencies in the visible light region. This means that a laser has a bandwidth of about 10^{14} Hz. This bandwidth is greater than is required for all radio and television transmitting stations in the United States. A single laser could carry simultaneously all of the information being transmitted by these stations.

16.17   Types of Laser

Since the pioneering development of the optical maser by Maiman, other types of lasers have been produced. The first laser was a ruby rod doped with chromium ions as its active element. Recent developments use a gas as the active element, and semiconductors have been used in others. Basic differences in the types of lasers occur in their output power levels.

The energy level diagram for chromium in Fig. 16-21 will be used to discuss the internal operation of a ruby rod laser. An unexcited electron will be in the stable energy state, labeled level 1. During absorption of an incident photon

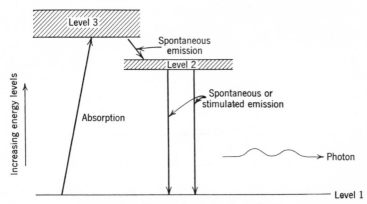

Fig. 16-21   Energy levels in chromium.

the electron may absorb the energy of the photon and take an energy level somewhere in the broad level 3, provided the photon had an energy great enough. Next, the electron may emit part of its energy and make a transition to level 2, an intermediate, metastable state of the chromiun atom. The final step in the process is a transition from level 2 to level 1. This final step can occur as either spontaneous or stimulated emission. In either case it will emit light energy equivalent to the difference in energy of the two levels.

Figure 16-22 represents a typical pattern of photon motion inside the ruby rod. The first atom undergoes spontaneous emission, yielding a photon. It strikes the partially reflective surface at the end of the rod and is reflected, stimulating a second atom to emit a photon as it travels back through the rod. Two photons are now present to continue the process. The intensity of the photon beam moving through the rod has been amplified. Further reflections and stimulations increase the number of photons moving back and forth in

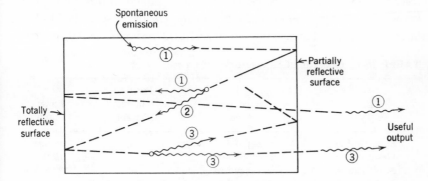

Fig. 16-22   Typical photon amplification in ruby rod.

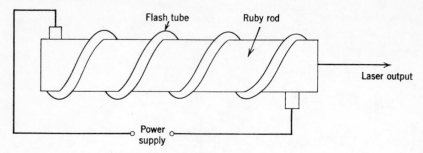

Fig. 16-23   Sketch of ruby rod laser system.

the rod. Some of them will pass through the partially reflective end of the rod and become the useful output of the laser.

Figure 16-23 is a sketch of a ruby laser system. The power supply drives a flash lamp, generating photons that enter the ruby rod to initiate laser action. The light from the flash lamp excites some of the chromium atoms. A few will emit spontaneous radiation to start the laser. Their emission triggers stimulated emission from other atoms, continuing the process. Further flashes are synchronized with the emission from the laser to continue the operation.

Similar systems are capable of stimulating coherent radiation in gaseous matter. An explanation of their internal behavior is far beyond the scope of this discussion. The same is true of semiconductor types of lasers. We will limit our study to a brief look at the comparable parameters of the types of lasers, listed in Table 16-1.

Ionized-argon lasers have been used as an "optical torch" in several experiments. They provide a continuous beam of light as a high-energy density over a small area. They have been used to mechanically trim microcircuit resistors and to make incisions in laboratory animals. Probably their best known application is in connection with the Gemini 7 space flight. Argon lasers were installed at geographically separated ground locations to serve as

TABLE 16-1   Operating Characteristics of Lasers

Laser Type	Operating Mode	Practical Efficiency (Percent)	Power Output
Solid (Ruby)	Pulsed	1	50 mW
Gas-discharge	Continuous	1	10 W
Semiconductor	Either pulsed or continuous	10	500 mW

aiming beacons or visual targets for the astronauts. Even though the power output may be small, the laser is by far the brightest of controllable, man-made light sources, because of its coherent output.

SUGGESTED REFERENCES

1. Herbert J. Reich, *Theory and Applications of Electron Tubes*, 2nd ed., McGraw-Hill, New York, 1944.

2. W. G. Dow, *Fundamentals of Engineering Electronics*, 2nd ed., Wiley, New York, 1952.

3. V. K. Zworykin and E. G. Ranberg, *Photoelectricity and Its Applications*, 4th ed., Wiley, New York, 1949.

4. *Phototubes and Photocells*, Radio Corporation of America, Lancaster, Pa., 1963.

QUESTIONS

16-1 What is photoelectric emission?

16-2 What is the order of magnitude of currents obtainable with phototubes? With phototransistors?

16-3 In what fundamental way does "black" light differ from visible light?

16-4 Define (*a*) photon, (*b*) quantum of energy, (*c*) Planck's constant, (*d*) work function of a metal.

16-5 Write Einstein's equation for photoelectric emission and discuss it.

16-6 (*a*) Define threshold frequency. (*b*) What is another name for work function?

16-7 Describe conditions at, and just under, the surface of a metal from which an electron is emerging.

16-8 State the two laws of photoelectric emission.

16-9 What is the meaning of the word *light*? What are the approximate wavelength limits of the visible part of the spectrum? What is an Angstrom unit? A micron?

16-10 Define (*a*) lumen, (*b*) footcandle, (*c*) candle-power.

16-11 How many lumens are given out by a 50-candle-power automobile lamp?

16-12 What is the approximate plate resistance of a high-vacuum phototube? The d-c resistance?

16-13 What is the reason for higher current capacity in a gas phototube?

16-14 Compare high-vacuum and gas phototubes, naming advantages and disadvantages.

16-15 What functions are performed by the two capacitors in Fig. 16-10?

16-16 Explain the operation of a phototube and amplifier circuit where interruption of the light beam causes an increase in amplifier current. Why is a fixed bias used?

16-17 Describe the principle of the photomultiplier tube. What is meant by its sensitivity?

16-18 Explain how an output signal is produced in an image orthicon tube.

16-19 What is the basic action in a phototransistor that accounts for an increase in current when light shines on it?

16-20 How does the current in a phototransistor circuit compare in amount with phototube currents caused by the same illumination intensity?

16-21 Which terminal of an *N-P-N* phototransistor is the collector? What part of the transistor is made photoemissive during manufacture? What is a requirement in the emitter section?

16-22 What do the letters of the word *laser* indicate?

16-23 How is light "amplified" by a laser?

16-24 What characteristics of a laser are of primary importance in potential applications?

PROBLEMS

16-1 In the circuit of Fig. 16-2, *E* is 100 V; choose a value for *R* which will give a maximum change in current when the light flux changes from 0.3 to 0.2 lm using the curves of Fig. 16-3.

16-2 What is the energy of a photon at a wavelength of 5,900 Angstrom units, which is in the yellow part of the visible spectrum? Calculate the number of these photons required to give enough energy to an electron so that it can escape from a metal when it needs the amount represented by the work function for cesium, which is 1.81 eV.

16-3 Calculate the threshold frequency for cesium. What is the wavelength of the radiation in Angstrom units?

16-4 The light shining on a photomissive surface ejects electrons fast enough to establish a current 1 μA. (*a*) How much current flows in the circuit when the intensity of the light is doubled? (*b*) How does the energy of an electron before the light intensity is doubled compare with the energy of an electron ejected afterward?

16-5 A point source of light has an intensity of 50 candle-power. Six feet from the source is a phototube with a window 1 in.^{2} through which light from the source enters. (*a*) Calculate the amount of luminous flux entering the phototube. (*b*) How intense must another source be to send the same amount of light flux through the window, if it is assumed that by a mirror and lens arrangement 90 per cent of its flux can be made to enter the window?

16-6 A 925 vacuum phototube has 70 V on its anode. Light flux entering it varies sinusoidally between 0.3 and 0.1 lm extremes. Determine the rms value of phototube current produced in a load resistance of 2 MΩ.

16-7 In Problem 16-6, calculate the value of a load resistor required in series with the phototube so there will be an output signal of 10 V rms. How much d-c voltage would be required at the input to the phototube circuit?

16-8 R_g in Fig. 16-10 is 20 MΩ. (*a*) What should be the capacitance of C_g if the time constant is ten times the period of a negative half-cycle on a 60-Hz line? (*b*) Assume that the capacitor voltage drops linearly from a maximum of 10 to 3.7 V (the voltage curve is a straight line). Calculate the voltage to which it would fall in the time of 1 full cycle.

16-9 The relay coil in Fig. 16-10 has a resistance R_R of 200 Ω and an inductance of 10 H. (*a*) Calculate the capacitance of C_R so that its reactance is one-tenth that of the relay coil on 60 hertz. (*b*) Calculate the resonant frequency of the parallel circuit, using the approximate formula.

16-10 In a photomultiplier tube, dynode 1 receives 0.1 μA current. There are 10 dynodes and $k_m = 3$. Calculate the output current from the last dynode.

16-11 The photomultiplier of Problem 16-10 has a sensitivity of 2 A/lm. By what fraction of a lumen must the entering light flux be changed, if the output current is to be reduced 50 per cent?

16-12 The rate of change of tube current with respect to light flux in lumens is called the dynamic *luminous sensitivity* of a phototube. (*a*) Find this sensitivity of the type 925 vacuum phototube operating as indicated by the 2.5-MΩ load line in Fig. 16-5. The *Q*-point is at 0.25 lm. (*b*) Find the sensitivity when $R_L = 7.5$ MΩ (*c*) Compare the a-c voltage drops across the load resistors when the quantity of light flux is varied sinusoidally between 0.1 and 0.5 lm in both cases.

16-13 (*a*) Find the dynamic luminous sensitivity (see Problem 16-12) of the type 930 gas phototube at 2.5 and at 5 MΩ load resistance. The *Q*-point is at 0.06 lm. Use Fig. 16-6. (*b*) Find the a-c voltage drops across the load resistors when the quantity of light is varied sinusoidally between 0.02 and 0.1 lm in both cases. How does the waveform of the a-c voltage compare with that of the voltage in Problem 16-12*c*?

16-14 Repeat Problem 16-13, using a 1-MΩ load resistance. What are the advantages and disadvantages of using a load resistance as small as this?

16-15 A type 925 vacuum phototube is used to control a relay through an amplifier circuit in Fig. P16-15. The relay resistance is 5,000 Ω, and it closes when its current

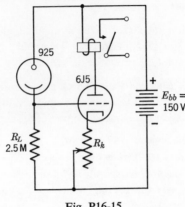

Fig. P16-15

increases to 5 mA and opens when the current decreases to 3 mA. R_k is adjusted so the relay opens when 0.1 lm of light enters the phototube. (*a*) Find the value to which R_k is adjusted. (*b*) How much light flux must fall on the phototube to close the relay? (*c*) What variation in light intensity is needed to change the relay from the just closed to the just opened conditions if the window of the phototube has an area of 0.36 in.²?

CHAPTER SEVENTEEN

Integrated Circuits and Microminiaturization in Electronics

The advent of the Space Age, with its missiles, rockets, and satellites, has demanded that electrical and electronic components *and whole systems* be built to occupy a small fraction of the space only recently required by those constructed with miniature tubes and transistors of conventional physical sizes. The needs of the aerospace industry have accelerated improvements in electronic packaging. This industry uses some of the most complex electronic systems ever developed, and reliability, size, and weight are of the utmost importance. Every extra pound of load placed in a rocket, for example, adds that many extra pounds to the weight of the rocket. Such things as additional structural material and more fuel needed for lift-off greatly increase the costs of the system.

Instead of an oscillator, an amplifier, or a whole radio transmitter-receiver system occupying several cubic inches of space, each must require only a small fraction of that amount. Furthermore, these systems must operate at extremes of temperature, acceleration, and radiant-energy fields and must possess a high degree of reliability. The failure of a single component worth only a few cents may mean that millions of dollars and months of effort have been wasted.

Microminiaturization in electronics has produced circuits and systems that combine discrete, miniature components, such as resistors and capacitors, with semiconductor "chips" on which transistors and diodes are fabricated. These combinations are referred to as "hybrid" or "multichip" circuits. The techniques used to minimize space requirements in this type of circuit construction led to the development of *whole circuits*, including passive

661

resistor and capacitor elements and active semiconductor diodes and transistors, on a common material, called a *substrate*. These circuits are called *integrated circuits*, because they include the required active and passive components as a *monolithic* structure. The fabrication of elements on a common substrate depends on *thin-film* technologies using materials such as tantalum, nichrome, and various oxides in sophisticated manufacturing processes.

This chapter presents a description of several types of microminiaturization techniques used to fabricate integrated circuits and examples of the structure of components and systems commercially available. Integrated circuits are mass produced as components of computer systems and have been announced as the working elements in consumer television sets, portable radios small enough to fit in a lady's earring, and hearing-aids that weigh only a few grams, including the battery used as a power supply.

17.1   Fabrication of Integrated Circuits

The fundamental requirement in the fabrication of an integrated circuit is that circuit components be processed simultaneously from common materials. Capacitors and resistors operating as linear networks can be effectively fabricated by thin-film technologies using nichrome, tantalum, and tin oxide. Manufacturing methods developed for cadmium sulfide are capable of producing both field-effect transistors and passive components on a common substrate.

The technology presently used to build integrated circuits that include transistors and resistive and capacitive circuit elements, however, is based on methods developed for silicon planar semiconductor components. Its popularity for fabricating integrated circuits is based on its ability to provide high-quality active devices.

Because the basic process for integrated circuit construction is similar to that used to make transistors, these components may be formed to have characteristics that are almost identical to those of discrete units. The major difference between discrete and integrated transistors is the capacitance associated with the separation of the transistor and the substrate on which it is formed.

The basic result of the silicon planar process for fabricating integrated circuits is shown in Fig. 17-1. The starting material is a polished wafer cut from a single-crystal *N*-type or *P*-type silicon, which is the starting *substrate* for the process that leads to a *monolithic integrated circuit*. The circuit

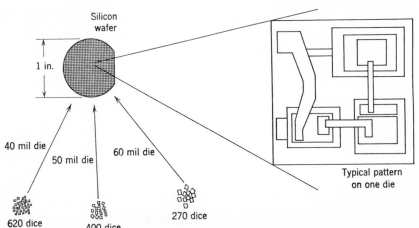

Fig. 17-1   Several identical circuits may be fabricated on a single wafer. The number of dice depends on the size of an individual die. Circuits requiring few components may use a smaller die size.

elements are built up as layers of doped semiconductor and silicon dioxide, SiO₂, (quartz).

The process illustrated here represents about 40 major manufacturing steps. Each is completed under precise control of parameters that affect the "yield," the number of good units that result. If each of the 40 steps is 97 per cent correct, on the average, the complete yield will be about 30 per cent. This figure is typical in the industry.

A normal production of small integrated circuits (IC's) might include as many as 100 to 700 circuits side by side on the starting wafer. Each die, representing a single circuit, contains 10 to 100 active and passive circuit components. The ability to fabricate simultaneously all the components of many circuits permits close matching of finished IC's.

The major steps in the fabrication of IC's, starting with the 1 in. diameter slice of silicon, are as follows.

1. Surface preparation.
2. Epitaxial growth of starting layer.
3. Diffusion of subsequent semiconductor layers.
4. Metalization and connections.

17.2   Use of Photomasks

In order to control the fabrication of the IC structure, two separate and distinct features are important. One of them, the depth or vertical dimensions

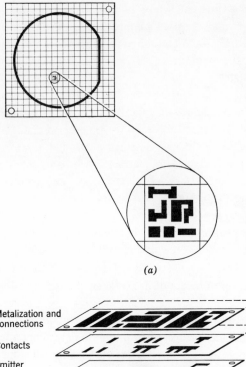

(a)

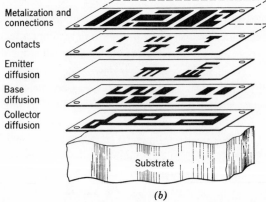

Metalization and
connections

Contacts

Emitter
diffusion

Base
diffusion

Collector
diffusion

Substrate

(b)

Fig. 17-2 Photomask patterns used to fabricate IC's. (*a*) Typical mask pattern. (*b*) Set of photomasks.

from the surface of a slice, is determined by a sequence of operations, e.g., epitaxial and diffusion processes. The second feature determines the lateral dimensions and shapes of the various layers that are built up during these processes. Photographic and chemical processes are used to fit the patterns required for the various components that are built into a circuit. Photomasks are the key to the making of high performance IC's.

A manufacturer may go from the production of one type of circuit to another by using a different set of photomasks. The same diffusion and assembly processes may be used to fabricate a different circuit. A set of photomasks is the only tooling required for a new device or circuit configuration. In essence, the photomasks restrict the processing steps to specific locations on the wafer.

In a series of repeated photographic, etching, diffusion, andother processing steps, the components of an individual IC may be formed and repeated simultaneously side by side for the 100 to 700 dice on a single wafer. To do this the processes are performed in minute selected areas over the entire wafer, while the remainder of the surface is not affected. Patterns on the photomask that is used in a particular step of the manufacture limit the process in that step to specific shapes and locations. The use of photolithography to produce the required patterns is sketched in Fig. 17-2.

Lateral sizes are set and controlled by the patterns of the photomasks. Photomasks may be made to produce 0.0004 in. line widths and a location accuracy of 0.0002 in. over a 1-in. area. The masks that are used in production start as a layout of components that will make up the final circuit. When the circuit design is complete, the pattern of the circuit is accurately drawn and photographed to make a large-scale negative of it, on which the photographer may make minor corrections to line widths and stray spots. A reduction in size and creation of a matrix of identical patterns completes the photomask.

A set of accurately registered photomasks can produce good quality IC's. A manufacturer can produce a whole family of circuits just by using a different set of photomasks for each circuit, using the same diffusion, metalization, packaging, and other processes. Changes in a specific circuit may be made by redesigning one or more of the masks of a given set. Accurate sets of photomasks are vital to making complex, yet reliable, integrated circuits.

17.3   The Epitaxial Layer

An integrated circuit comprising 10 to 100 components and their interconnections might be fabricated on a piece of silicon that is only 40 to 60 mil^2 (0.040–0.060 in.). Typically, the silicon slice would be about 5 or 6 mil thick, or slightly thicker than this page. Many of the advantages of IC's arise from the microscopic size of the silicon chips and the circuit components, but this size also creates problems in manufacture.

Figure 17-3 shows two enlarged views of the sizes and locations of various areas on a 6-mil chip. The areas labeled "N" are *N*-type semiconductor on

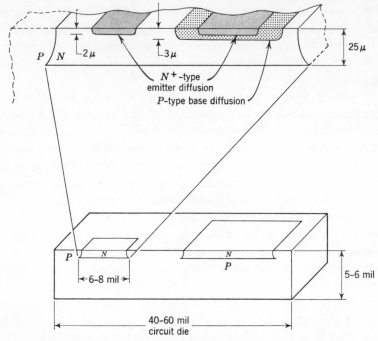

Fig. 17-3   Epitaxial layers used to fabricate circuit components.

which the circuit components are fabricated. The small area at the left corresponds to the cross section needed for a typical transistor in the IC. The magnified view directly above this section shows the depth dimensions of the transistor. The emitter and base regions have depths of only 2 to 3 μ. Other circuit components have about the same depth relationships, seldom requiring more than 1 mil (25 μ) of thickness.

Since the circuit components only require about 1 mil of thickness, it might seem desirable to use silicon slices that are not much thicker than this. However, as the silicon wafers are made thinner, they break more easily during manufacturing operations. Losses during the fabrication processes due to wafer breakage are substantial, even for 5 to 6 mil slices, but this thickness has been found to be practical for economical use of silicon crystals. The bulk of a slice is not used for circuit components but is mechanically important as a "handle" during the manufacture of IC's. It is also electrically important because of its influence on the circuit properties. The bulk material is called the substrate.

Epitaxy is the process of growing one material on the surface of another. As used in manufacturing IC's, epitaxy refers to the growth of an additional semiconductor on a wafer by deposition from a vapor phase. In this process, a thin layer of additional single-crystal silicon is deposited on the polished surface of the starting wafer. While it is being deposited, the desired resistivity is obtained by controlling the amount of *N*-type impurity present in the silicon vapor phase. A photomask is used to restrict the epitaxial layer to the desired locations and areas on the wafer, as shown in Fig. 17-3.

17.4 Diffusion of Additional Layers

Diffusion of additional layers of materials is the key process in forming the desired circuit components. *Diffusion* is the substitution of *N*- or *P*-type atoms for silicon atoms within the crystal structure of a material. It results from heating the wafer to a temperature of 900° to 1300°C in the presence of controlled amounts of impurity. The temperature required depends on the particular dopant being used, and is chosen to maintain close control over the diffusion process. The penetration of dopant increases by a factor of about 5 for every 100°C temperature rise above ambient. Typical diffusion time at 1200°C is 1 to 6 h.

The diffusion processes form *P-N* junctions in the epitaxial layers at locations permitted by placement of photomasks, as shown in Fig. 17-3. Precise locations on the surface may be held to tolerances as tight as ±15 millionths of an inch. The additional layers of materials on the IC form the various diodes, transistors, resistors, and capacitors of the circuit.

The materials used as dopants modify the silicon material by providing regions of *N*-type and *P*-type characteristics. Phosphorus (P) and arsenic (As) are widely used *N*-type dopants and boron (B) is usually used as a *P*-type dopant. The depth of penetration of a dopant depends on the temperature, length of time the slice is exposed, and properties of the doping material. Some dopants have diffusion rates that are orders of magnitude faster than others; for example, boron diffuses in silicon more than 10 times as fast as arsenic.

As a minimum, IC fabrication requires diffusion layers for an isolation from the substrate, a base for transistors, and an emitter diffusion. More complex circuit components may be made with additional diffusion processes. For example, *PNP* transistors may be formed in addition to *NPN* types, controlled layers may form Zener diodes with specific breakdown characteristics, and different values of resistor and capacitor elements may be provided.

The regions of the silicon wafer and layers into which diffusion of additional materials occurs are controlled by a combination of oxidation, photomasking, and chemical etching. Before each diffusion the slice is prepared by exposing it to oxygen at temperatures in excess of 900°C. A layer of silicon dioxide (quartz) can easily be grown in this environment and is usually allowed to form a thickness ranging from 2,000 Å to 10,000 Å. This oxide layer is not easily penetrated by the common dopants and can be used as a mask to prevent diffusion where it is not desired in further processing. Parts of the oxide coating may be selectively removed by photomasking and etching processes, to form "windows" that set the size and location of additional circuit elements. On the photomask, the areas to be etched are opaque and those to be left with an oxide coating are transparent. The patterns of the mask are transferred to the silicon slice using a thin, uniform, photosensitive film that is coated on the surface. Exposure to light through the transparent photomask exposes the photographic film in these areas. The unexposed areas are removed by a developing step. A hydrofluoric acid etch removes the oxide, leaving the silicon surface exposed for further diffusion steps. The result is an accurate pattern of etched areas through the oxide coating to permit diffusion in the selected areas.

17.5 Metalization to Form Electrical Connections

A metal coating is vacuum deposited onto the diffusion coatings to inter-connect the circuit elements and form the desired circuit. Metalization

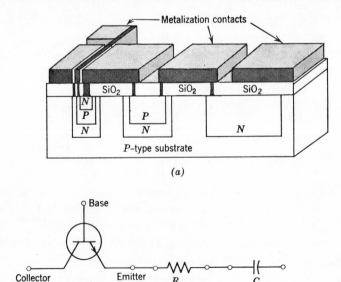

Fig. 17-4 Completed IC comprising *NPN* transistor, resistor, and capacitor. (*a*) Layer construction of circuit components. (*b*) Circuit constructed in (*a*).

provides contacts for external connections to the circuit, as well. The most common metallic film used is aluminum, because it adheres well to silicon dioxide and is easy to handle. Other metals used alone, or in combination, include gold, silver, chromium, and nickel. When resistors and capacitors are required, titanium, tantalum, and tin may be used.

Figure 17-4 illustrates the formation of circuit components by the various processes in the fabrication of IC's, concluding with the metalization connections as the outermost layer. To form a resistor, the N-type emitter diffusion is omitted and two ohmic contacts are made to the P-type region that is formed with the base diffusion. In forming a capacitor, the oxide itself may be used as the dielectric, and metalization provides the electrodes. A typical circuit and its interconnections are shown in the figure.

17.6 Cost Factors

Integrated resistors and capacitors are significantly different from their discrete counterparts. Discrete resistors and capacitors are usually made in standard physical sizes, corresponding to power and voltage ratings, and different values are obtained by varying the characteristics of the materials making up the components. A discrete resistor may be formed to have a given value by changing the resistivity of the material. The capacitance of a capacitor can be changed by using a different dielectric or by changing the surface area of its plates. However, in integrated circuits, the resistivity of the material cannot be varied to obtain different values of resistance, because it is optimized to obtain the best performance from the base diffusion of transistor structures. An integrated resistor, then, depends primarily on its geometry to set its resistance value. Its value is determined by the "square resistance," resistivity of a volume of material and the ratio of its length and width. As a result, resistors having large ohmic values are long and narrow, and small ohmic values are formed by short, squat dimensions. Similarly, the capacitance of capacitors directly depends on their surface area, because the thickness of the dielectric is a constant in well-controlled production.

Most of the cost of making monolithic IC's is associated with the processing steps in their manufacture. Because these costs are the same for any wafer, smaller circuits and more circuit dice from a single wafer lower the cost of an individual circuit. Therefore, circuit area is an important consideration in the design of an initial layout for manufacturing IC's. The *relative areas* required for integrated components are approximately 1 for a transistor or diode, 2 for a 1,000-Ω resistor, and 3 for a 10-pF capacitor. Thus, for the economical production of IC's, each individual circuit should minimize passive components and use active components whenever possible. This is exactly the opposite of the economic rule for circuits using discrete components.

In a conventional transistor circuit, the transistors are relatively expensive compared with resistors and ceramic disc capacitors. In IC construction, the least expensive components are transistors. The resistors and capacitors of the IC are much more expensive. Table 17-1 lists a cost comparison of IC components.

TABLE 17-1 Cost Comparison of Components*

Component	Unit Value	Area mil²	Cost ¢
Base-Diffused	2.5 kΩ	100	0.2
Resistor	20 kΩ	500	1.1
Metal-Oxide-	100 pF	600	1.2
Semiconductor	350 pF	1,700	3.5
(MOS) Capacitor			
Transistor	Typical	30	0.1
	Large	100	0.3

* Adapted from *The Electronic Engineer*, October 1966.

Certain familiar discrete components, such as inductors and transformers, are impossible to obtain in integrated form. Their use must be either avoided or taken into consideration in the design of the circuit so that external connections are provided for them. Novel circuit designs have evolved from many sources, based on the need to eliminate these components insofar as possible in IC applications. Cost factors associated with them, however, often prove to be prohibitive.

Resistors must be considered differently in integrated and discrete forms. One undesirable characteristic of semiconductor resistors is their temperature dependence. A relatively large variation in ohmic value with temperature makes it difficult to achieve close tolerances on absolute values. However, resistance ratio values can be closely controlled by the geometry of photomasks. In circuits where resistance ratios, rather than absolute values, are important, integrated resistors have a number of advantages over discrete units. Because they receive almost identical processing and are closely matched in characteristics, temperature effects can be minimized over a wide operating range. As a result of their proximity, temperature differences that occur between components will be small.

Cost factors may be minimized in IC design by maximizing the use of transistors and other active components, by using resistance ratios instead of absolute values, and by taking advantage of matched component characteristics. Even though the comparative costs of the circuit elements are small, integrated circuit packages are relatively expensive. A linear IC having open-loop gain of about 2,000 may cost $30.00, but future costs of a fraction of this amount are predicted by IC manufacturers.

17.7   IC Packaging

The type of package in which the IC is housed is an important consideration to system designers. Two types of package are generally available. A flat package, with the external leads extending from the ends of the package as in Fig. 17-5, permits a greater density of the IC's in a complete system. The circular package uses a TO-5 transistor configuration, with up to 10 external pins provided for connecting the IC in a conventional socket. Both types are available with a variety of pin numbers and with different dimensions. In general, the flat package is more expensive than the round can package, because of higher manufacturing costs.

Figure 17-6 illustrates the interior connections that are brought out to the pins of a flat package. The wires connecting the IC to external pins are usually gold or aluminum, and are about 1 mil in diameter.

17.8   Component Ratings and Unit Values

Because of their microscopic dimensions, integrated components have not matured to rival discrete components with respect to maximum ratings and unit values. Table 17-2 shows the ranges of values and ratings of components that are formed in the monolithic fabrication of IC's.

17.9   Thin-Film and Multichip Construction of IC's

Thin-film circuits consist of a passive substrate, such as glass or ceramic, on which resistors and capacitors are deposited as thin patterned films of conducting and nonconducting layers. Thin-film resistors may have values up to 100 kΩ or so, and thin-film capacitors, using a tantalum oxide dielectric, may have a capacitance of 2.5 pF/mil^2 of surface area. When active components (transistors and diodes) are added in discrete form to thin-film passive components, the resulting fabrication is called a *multichip circuit*.

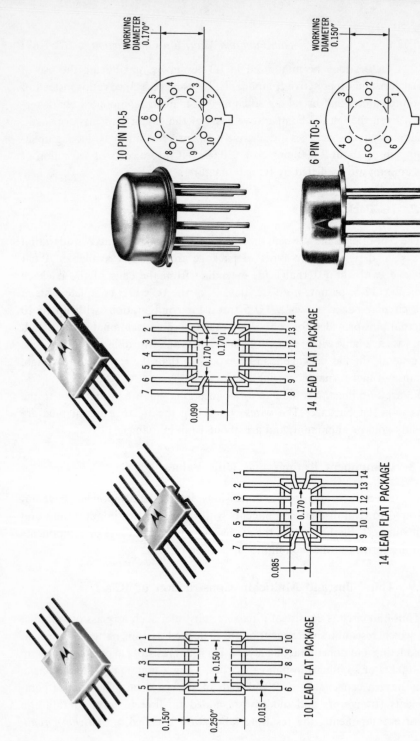

WORKING DIAMETER 0.170"

10 PIN TO-5

WORKING DIAMETER 0.150"

6 PIN TO-5

14 LEAD FLAT PACKAGE

14 LEAD FLAT PACKAGE

10 LEAD FLAT PACKAGE

Fig. 17-6 Internal connections from IC to external leads in a flat package IC (Courtesy Motorola, Inc., Semiconductor Products Div., Phoenix, Arizona.)

TABLE 17-2 Integrated Component Ratings and Values

Component	Characteristic	Ratings	Typical Values
Resistor	Ohms per square Maximum power Range of values	0.1 W	2.5 or 100–300 15 Ω–30 kΩ
Capacitor	Maximum capacitance Breakdown voltage Q (at 10 MHz) Voltage coefficient Tolerance	20 V d-c ±20%	0.2 pF/mil^2 1–10 $C = KV^{-1/2}$
Transistor	BV_{CBO} BV_{CEO} BV_{EBO} h_{FE}/I_C	35 V 7 V 7 V	 40/1 mA 60/10 mA

673

Fig. 17-7   Internal structure of a multichip circuit, using two IC's and other discrete components on individual chips. (Courtesy Motorola, Inc., Semiconductor Products Div., Phoenix, Arizona.)

Figure 17-7 shows a typical arrangement of separate components and two IC's mounted on a TO-5 base.

Such a combination of monolithic and thin-film elements permits the fabrication of more complex circuits than the monolithic approach alone. There is a limit to the maximum size of an individual monolithic chip because of cost factors of economic yield, so that circuits having many components may be built on two or more chips. Further, circuits that need complementary transistors, or transistors having widely different characteristics, can be designed on separate chips as a practical solution. Multichip circuit construction can greatly improve circuit performance and permit combinations of components that cannot be fabricated by the monolithic process.

17.10   Miniaturization by High-Density Packaging

Early approaches to miniaturization of electronic circuits included extensions of conventional packaging techniques. One method simply shrank the physical sizes of discrete components and the spacing between them in constructing the circuit. Manufacturers have reduced the size of various components, but the space required for mounting and for inter-connecting wires approaches a minimum which is many times the volume used by an integrated circuit that performs the same circuit function. High-density packaging also creates problems in heat dissipation. Essentially the same power will be dissipated by a given circuit using miniature components in closely spaced packaging, but it is generated in perhaps one-third the volume of conventional packaging. Heat dissipation is a limiting factor in the high-density packaging of electronic circuits.

Sylvanic Electric Products, Inc. developed a modular construction of circuit packaging, in which portions of a circuit are mounted on individual wafers, consisting of several layers of combined thin-film and deposited components. Figure 17-8 illustrates a circuit wafer that may be used in this method of construction. Note that the wafer has a number of stubs, or tabs, on the outside edges. These are used to stack the wafers and for making terminal connections to the completed circuit.

Low-resistance terminal connections on the tabs are produced at the time thin-film components are placed on the wafers. Similarly, the spacer frames have conductors formed on them for making the connections from one frame

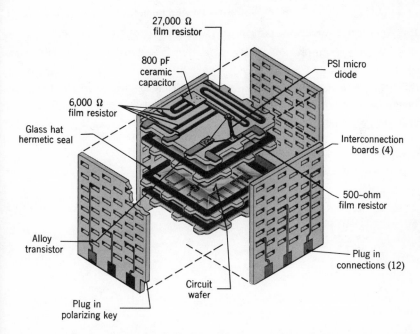

Fig. 17-8 Circuit miniaturization by modular construction of thin-film and deposited components. (Adapted from a figure by Sylvania Electric Products, Inc.)

to another and to external terminations. The spacer frames keep the wafers from touching each other and permit them to be spaced closely together and mounted firmly in the final construction.

SUGGESTED REFERENCES

1. *RCA Linear Integrated Circuit Fundamentals*, Radio Corporation of America, Harrison, N.J., 1966.
2. Glen R. Madland et al., "Integrated Circuits Course, Part I and Part II," *The Electronic Engineer*, August 1966 and October 1966.

QUESTIONS

17-1   Why is the term "monolithic" applied to the fabrication of integrated circuits on a common substrate?

17-2   How does the monolithic structure differ from multichip construction?

17-3   Describe the steps used to fabricate a monolithic integrated circuit.

17-4   What is the purpose of photomasks in the monolithic fabrication process?

17-5   Describe the epitaxial process for growing layers of semiconductors.

17-6   What is the purpose of forming oxide layers in certain regions of an integrated circuit?

17-7   Why is it desirable to use transistors and diodes whenever possible in an integrated structure, rather than passive resistors and capacitors?

17-8   Compare some of the advantages and disadvantages of discrete and integrated resistors.

CHAPTER EIGHTEEN

Application of Electronics Principles in Practical Circuits

A book on electronics principles should contain some complete practical and useful circuits in which many of the principles have application. This chapter contains a number of such examples and an explanation of the operation of each. There are hundreds of electronic devices that employ in their operation one or more of the principles studied in this text. Only a few representative types here. After becoming thoroughly acquainted with them, the reader should be able to understand the operation of many others described in scientific journals.

18.1 Complete Voltage-Regulated Power Supply

A complete power supply circuit with voltage regulation is shown in Fig. 18-1. Here the control circuit is connected to the load side of the series tube (6B4G), with the result that the control is a little more sensitive than that of Fig. 12-5 to slight variations in output voltage.

The power transformer produces 375 V to ground from either side of the center tap, and it has two filament windings. One is a 5-V winding for the rectifier tube, and the other is a 6.3-V winding that may be used to supply the heaters or filaments of tubes served by the high-voltage output of the power supply. A separate filament transformer is used for the other two tubes. After passing through the 6B4G, d-c output power flows from the center tap of the filament transformer to the + Reg. terminal.

The regulated output voltage may be set at any value between 150 and 250 V, approximately. The regulator circuit operation is identical with that of

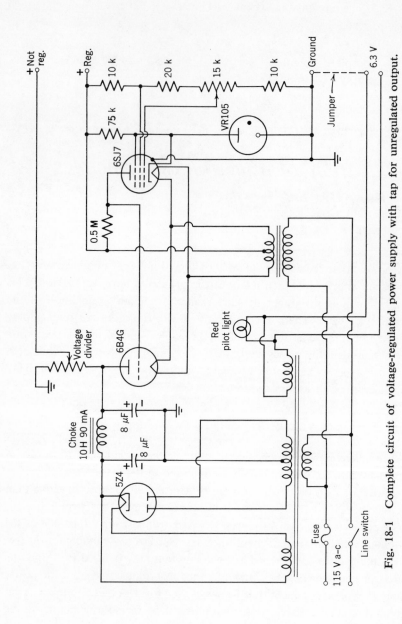

Fig. 18-1　Complete circuit of voltage-regulated power supply with tap for unregulated output.

Fig. 12-5. The unregulated output may be adjusted at any value from about 350 V to zero.

18.2 Complete Selenium-Rectifier Voltage-Doubler Power Supply (Unregulated)

The advantages of a selenium-type rectifier power supply are: no filament transformer needed; small space required; no warm-up time necessary; and very little heat given off. Many of these power supplies employ the 100-mA selenium unit in full-wave or in half-wave circuits.

A practical rectifier circuit that delivers up to 100 mA at either 325 or 250 V, approximately, is shown in Fig. 18-2. The output terminals at the extreme right may be connected to the input terminals of a voltage-regulator circuit such as that shown in Fig. 12-5, or to the voltage-regulator section of Fig. 18-1.

Because the 115-V lighting circuit (to which the a-c plug is to be attached) is usually grounded on one wire, it is important to insert the plug in the receptacle in the position that will make the bottom wire of the circuit diagram the grounded wire. This wire, marked G is negative at the output; the negative is usually grounded in electronic and radio power supply circuits. However, the positive line is usually grounded on a bias supply. Care should be taken to observe proper grounding conditions. If the wire G is connected to the grounded line of the a-c supply, and, if then the positive output terminal should be grounded, R_4 and C_6 would be short-circuited and there would, of course, be no output.

The operation with the switch S_2 in the high position is as follows. During the time the top line (containing switch S_1) is positive, the input voltage is

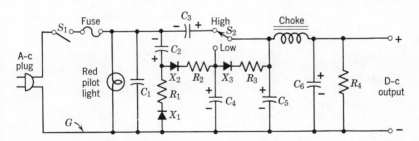

Fig. 18-2 Selenium-rectifier power supply. $C_1 = 0.05\ \mu F$, 600-V paper; C_2 to $C_6 = 40\ \mu F$, 450-V electrolytic each; $R_1, R_2, R_3 = 25\ \Omega$, 2 W each; $R_4 = 125{,}000\ \Omega$, 2 W; choke is 10 to 15 H, 110 mA; rectifiers $X_1, X_2, X_3 = 100$ mA; fuse = 2 A size; pilot light = 115 V, 6 W.

aided by the voltage across C_3 and current flows through the choke and downward through the bleeder (R_4) and also through the load circuit if one is connected to the output terminals. At the same time, current flows down through C_2 and through X_2 and X_3, charging C_4 and C_5 with the polarities shown. C_4 and C_5 send current through the choke and load on negative half-cycles while C_3 is being charged.

18.3 Bias Supply

A variable bias voltage is needed in many laboratory experiments and in some electronics and radio service work. Figure 18-3a shows a simple circuit employing a rectifier that may be connected directly to an a-c lighting circuit of 110 to 120 V. The rectifier must be connected for proper polarity in order that the variable tap on the potentiometer will be negative with respect to the bottom output terminal. The resistors have a 1-W capacity. The first capacitor should have a working voltage of not less than 250 V. The other two may have smaller voltage rating (especially the one on the right), because the maximum output voltage is about 50 or 60 V.

Care should be exercised if either of the output terminals is to be grounded. One of the a-c supply lines of a city lighting system is usually grounded. If

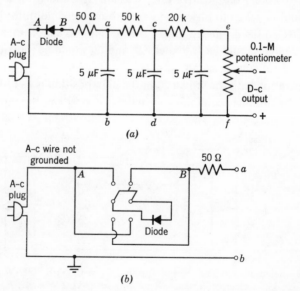

Fig. 18-3 (a) Variable supply. (b) Reversing switch to change polarity of output of (a). In down position, diode is reversed and so is the polarity of the d-c output.

the bottom line of Fig. 18-3a were connected to the *ungrounded* supply line at the left, the right end of the bottom line should not be grounded. If it were, a short circuit would result, and a fuse would blow or some damage result. Often some damage is done, such as the burning of contacts, even before the fuse or circuit breaker opens the circuit.

It is a very common practice to have the *positive terminal* of a bias supply grounded. This would require that the *bottom line* of Fig. 18-3a always be connected to the *grounded a-c line*. Which line is grounded may be determined by using an a-c voltmeter or a test lamp, which is simply a common light bulb and socket with its two wires separated and each terminating in metal prods made of stiff copper wire. If one prod is connected to "ground," which may be the iron conduit containing the a-c supply lines, and the other is touched to one of the a-c lines, the lamp will light if the a-c line touched is not the grounded one. That means that the other a-c line is grounded. If a voltmeter is used, a reading of the voltmeter corresponds to a lighted lamp. If the lamp does not light or if the voltmeter does not read when either a-c line is touched while the other test line is grounded, then neither of the a-c lines is grounded. It is then safe to ground either of the d-c output terminals.

If one of the a-c input lines is grounded and it is desired that the *negative d-c output line be grounded*, the connections to the selenium rectifier should be reversed. The a-c plug should be inserted so that the ungrounded side will be connected to point A and the bottom line to the grounded a-c input. That will reverse the output polarity and make negative the bottom line, which may then be grounded. If frequent changes from grounded positive terminal to grounded negative terminal and back are to be made, it would be convenient to have a reversing switch in the top line of Fig. 18-3a as shown in Fig. 18-3b.

There is practically no ripple in the d-c output voltage of the circuit of Fig. 18-3a, as may be verified by connecting an oscilloscope to the output. To read the output voltage, we should use a very high-resistance voltmeter, so that very little current will be drawn through the output potentiometer to cause voltage drop.

18.4 A Regulated and Filtered 6-V D-C Power Supply

A regulated and filtered 6-V d-c power supply is shown in Fig. 18-4. It is said by the designers* to be capable of supplying up to 1 A essentially free of ripple. Transistors Q_1 and Q_2 are rectifiers. C is a high-capacitance

* Sylvania Electric Products, Inc.

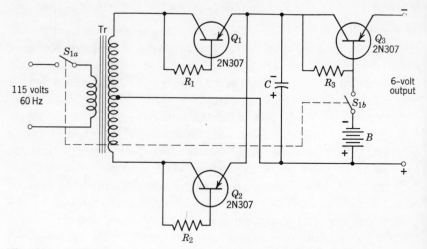

Fig. 18-4   A 6-V power supply. B, 6-V battery (lantern type); C, electrolytic capacitor, 12 V or more, 1,000 μF for loads to 0.5 A, 2,000 μF for loads to 1.0 A; Q_1, Q_2, Q_3 Sylvania 2N307 transistor; R_1, R_2, 22-Ω ½-W resistor; R_3, resistor, see text; S_1 D.P.S.T power switch; Tr, 12.6-V C.T. secondary transformer (such as Triad F26X). (Courtesy Sylvania Electric Products, Inc.)

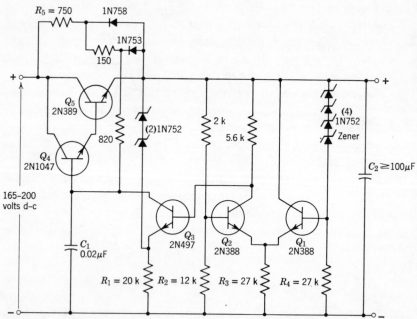

Fig. 18-5   150-V regulator circuit. For 50 to 70 V input and 30 V output use: $R_1 = 2.5$ kΩ, $R_2 = 1$ kΩ, $R_3 = 2.5$ kΩ, $R_4 = 2.5$ kΩ, $R_5 = 12$ kΩ. All other values are unchanged. (Courtesy Texas Instruments Inc., Dallas, Texas.)

682

filter element. Q_3 is an emitter follower which assures essentially ripple-free load voltage. Switch S_{1b} is part of the power switch, and it disconnects the reference-voltage battery B when the supply is not used.

With R_3 in the circuit the current in the battery may be reduced to zero, thus extending its use to "shelf life." This may introduce a slight amount of ripple, however, depending on the internal impedance of the battery. R_3 may be made up of a 47-Ω resistor and a 250-Ω potentiometer. A setting of the potentiometer may then be chosen to make the battery current zero. The ripple introduced depends on the amount of battery current to be balanced out, which is also dependent on the load current. It is of the order of $\frac{1}{10}$ mV at $\frac{1}{2}$ A load. The transistors should be mounted on a *heat sink* to keep their operating temperatures from going too high.

A 12-V power supply of the same design using four transistors in a bridge circuit may be built. The center tap on the transformer is then not used.

18.5   Transistorized Voltage-Regulator Circuit

The voltage-regulator circuit of Fig. 18-5 is capable of regulating the voltage to loads demanding up to 600 mA, according to its designers.* By changing resistors, as indicated in the illustration, the input operating-voltage range can be changed from 165–200 V to 50–70 V, giving 30 V output. Intermediate values of d-c output voltage may be obtained with other values of resistance.

The four 1N752 diodes operate in the Zener region and serve as voltage reference. They have a very small variation in voltage when their temperature changes. Transistors Q_1 and Q_2 constitute a *differential amplifier* which functions to compare the output voltage (the drop across the 2-kΩ resistor) with the reference voltage. Feedback current goes to Q_3, which amplifies it and applies it as a control to Q_4 and Q_5 transistors (a Darlington pair**) which serves like the series tube of a regulator circuit such as that in Fig. 12-5.* The operating voltages for the control stages (Q_4, Q_5) are stabilized by two 1N752 Zener diodes.

18.6   Power Zener-Diode Current Limiters

Regulation of heater supplies for vacuum tubes may be accomplished with constant-voltage transformers, ballast tubes, and rather complicated

---

* Texas Instruments, Inc., Semiconductor Components Division, Dallas, Texas.   ·
** Sorab K. Ghandi, "Darlington's Compound Connection for Transistors," *I.R.E. Transactions*, **CT-4**, 291–292 (1957).

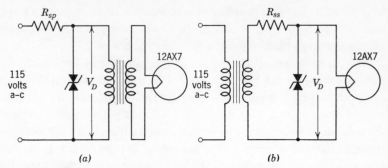

Fig. 18-6   Power Zener-diode regulator for vacuum-tube heater supply. (*a*) Primary-side regulator; (*b*) secondary-side regulator.

regulating circuits. The use of power diodes as clippers in regulating circuits provides advantages of light weight, small space requirements, and simplified circuits.

Diode shunt regulators may be used on either the primary or secondary side of the heater transformer. The two circuits are shown in Fig. 18-6 with a heater transformer and a 12AX7 twin triode tube. The resistances R_{sp} and R_{ss} prevent excessive current to the Zener diode and the heater.

On each half-cycle the reverse-connected diode operates in the Zener region and clips the voltage, limiting it to the value required for proper current to the tube heater. Fluctuations in input voltage may be quite extensive without affecting the heater current. Neither a rise nor a fall will change the voltage value at which Zener operation occurs. The Zener diode will accommodate substantial current changes with practically no voltage change across its terminals. This means very little, if any, change in V_D and, therefore, in the heater current. Tests have shown that with a line voltage change from 100 to 130 V rms, the following changes in 12AX7 *plate current* occurred as the result of changes in the heater current: heater current unregulated 1.5 to 2.2 mA (46.6 per cent increase); regulated 1.8 to 1.95 mA (8.3 per cent increase), both with 100 V d-c plate to cathode.

18.7   Transistor Bias Supply

A regulated bias voltage of 25 volts at 150 mA is provided by the circuit of Fig. 18-7. No transformers or filter chokes are needed. The ripple voltage is very low.

A bridge rectifier has a 10-Ω surge-limiting resistor and a 100 μF capacitor

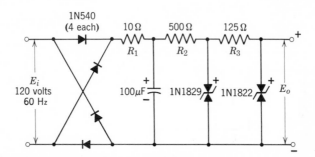

Fig. 18-7 Shunt-regulated power supply. Peak diode current 1.14 A (1N540); output 25.2 V at 150 mA, 25.7 V at 0 mA; ripple 4 mV peak-to-peak at 150 mA; at 150 mA load current: $E_o = 25.16$ V at 100 V rms input, $E_o = 25.22$ V at 140 V rms input. (Courtesy Texas Instruments, Inc., Dallas, Texas.)

at its output, followed by two pairs of Zener diodes and two filter resistors. The designers* report the following test data, in addition to the results given in the illustration.

Voltage at the 100 μF capacitor: 150 V d-c, 11.5 V peak-to-peak ripple.

R_2 and first Zener regulator reduce voltage to 47 V and ripple by a factor of 180.

R_2 was calculated to allow 50 mA to pass through the first diode at a nominal input of 120 V rms.

Second regulator further reduces ripple by 16:1 at full load.

Regulation of output voltage is effective with load-current variations from 0 to 150 mA, the output voltage dropping 2 per cent, giving an output resistance of approximately 3.3 Ω.

The circuit was tested to take input-voltage fluctuations from 100 to 140 V with little effect on the output.

18.8   Timing Circuits

A D-C Operated Timer

It was shown in Chapter 1 that, when a charged capacitor having a capacitance of C F is allowed to discharge through a resistor of R Ω, the voltage of the capacitor decreases exponentially, reaching a value that is 36.7 per cent of its original value in a time interval of RC seconds. For example, if a 4-μF capacitor, charged to 100 V, is in series with a 2-MΩ resistor and an

* Courtesy Texas Instruments, Inc.

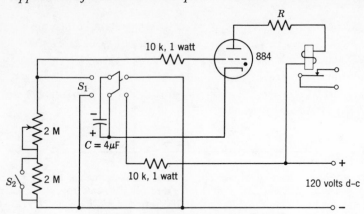

Fig. 18-8   Direct-current timing circuit.

open switch, the capacitor voltage will begin to decrease at the instant the switch is closed and 8 s later the voltage will be down to 36.7 V.

By letting the voltage bias a thyratron tube, use may be made of the manner in which the capacitor voltage decreases. The conduction of the thyratron may thus be delayed for the desired time interval by selecting the value of the time constant RC.

Figure 18-8 shows a timing circuit which employs that principle. When the switch S_1 is closed to the right, the capacitor is charged to 100 V almost instantaneously. At the instant the switch is moved to the left, the capacitor voltage makes the grid of the tube 120 V negative with respect to the cathode. This bias decreases as the capacitor discharges through the two 2-MΩ resistor units. With maximum resistance in the circuit (S_2 open), the time constant is a maximum of 16 s. Using switch S_2 and the slider, the rate of discharge of the capacitor and therefore the delay in tube conduction may be varied from about 12 s to zero when the d-c supply voltage is 120. The value of plate resistor R must be chosen to limit the relay current to the desired value and to protect the tube. For trial purposes, a 5,000-Ω slide wire may be used and also a milliammeter of 50 to 100 mA capacity may be substituted for the relay to indicate that the tube has started to conduct.

This rise in voltage across a capacitor that is being charged may be used to provide time delay. Or the fall in voltage across a resistor connected in series with the charging capacitor may be used. It is only necessary to provide the grid of the tube with the *proper polarity* of one of these voltages. If the voltage of the charging capacitor is used, it must be in series with a negative voltage that will hold the tube in the nonconducting state until the capacitor voltage has risen to the required value to bring the tube bias up to allow conduction.

High-vacuum tubes may be used in many types of timing circuits. A 6K6 or a 6V6 may be used in place of the 884 of Fig. 18-8, but a self-biasing resistor must be connected in the cathode line to limit the plate current of the tube. That would make the plate resistor R unnecessary.

An A-C Operated Timing Circuit

Figure 18-9 shows a timing circuit that will operate on a 115-V a-c line. When the switch is open, the capacitor charges because electrons in the tube pass to the grid and to the top plates during the half-cycles when input terminal a is positive and b is negative. Owing to the small time constant of the grid resistor and the capacitor ($5 \times 10^3 \times 10^{-7} = 0.0005$ s), the capacitor will quickly charge up to about 11 V. It is aided in preventing the tube from firing by the voltage across the 1-kΩ resistor, because this voltage contributes additional negative potential each time the plate of the tube goes positive.

The delay-time interval is started by closing the switch. This not only removes the 11 V of the 1-kΩ resistor from the biasing circuit of the tube, but also stops the action that kept the capacitor charged. Thus the capacitor immediately starts discharging through the variable resistor R, and soon the tube bias is reduced to the value at which the tube will fire. It is obvious that the amount of delay is determined by the setting of R, which is a 2-MΩ potentiometer.

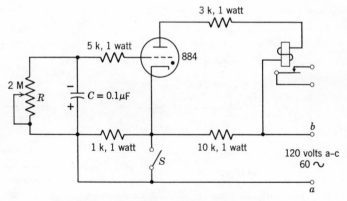

Fig. 18-9 Alternating-current timing circuit.

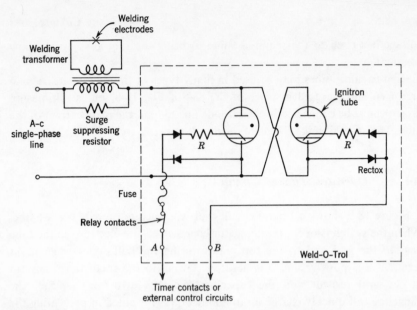

Fig. 18-10 Schematic diagram of Weld-O-Trol. (Courtesy Westinghouse Electric Corporation.)

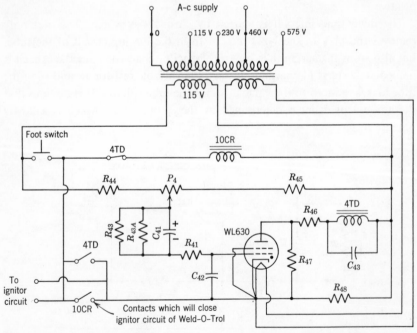

Fig. 18-11 Schematic diagram of 1A timer. (Courtesy Westinghouse Electric Corporation.)

18.9   Ignitron Firing Circuit

It was learned in Chapter 14 that the firing of an ignitron tube is initiated by sending current through its igniter rod. Figure 18-10 is a circuit that controls current to a welding transformer.

The igniter-rod current is supplied by the a-c single-phase line, as is the welding current. When the top supply line is positive, a small current flows through the transformer primary, down between the ignitron tubes, and to the right through the copper oxide rectifier marked Rectox. It then goes from point *B* through a pair of timer contacts, or perhaps a pair of foot-switch contacts, back to point *A*. Thence it passes through another rectifier unit, a resistance, and into the igniter rod to the mercury-pool cathode of the ignitron on the left, which is connected to the bottom supply line. This fires that ignitron, and thus power current flows toward the right through the welding transformer's primary.

During the next half-cycle when the bottom line of the a-c input is positive, current flows to the left through the rectifier on the cathode of the ignitron on the left and on through the contacts from *A* to *B*. It then flows through the igniter rod of the ignitron on the right and fires it, with the result that a wave of current is sent through the welding transformer's primary from right to left. The flow of alternating current in the primary will continue as long as the contacts between *A* and *B* are closed, but it will be stopped when they are opened. An automatic timer is usually employed to control the opening of the contacts and in some applications their closing also.

It is also practicable to delay the firing of the ignitrons for a variable time interval so that the *average value* of transformer current that flows each cycle may be varied. This controls the *amount of heat* in the welded joint. Such heat control is an advantage when various types of metal must be welded and when various thicknesses are used.

18.10   Automatic Timer for Welder

The circuit of a timer (NEMA type 1A) that controls the flow of the igniter-rod current in Fig. 18-10 is shown in Fig. 18-11. It contains only one tube, a small inert-gas thyratron (WL630). The following description of operation and the circuit diagrams have been supplied through the courtesy of the Westinghouse Electric Corporation:

A set of contacts on the 1A timer is in series with the ignitron circuit of the Weld-O-Trol as shown at *A* and *B* [in Fig. 18-10.] If these contacts are closed, welding current flows. If the foot switch is closed, the coil of the 10CR relay [Fig. 18-11] is energized and its contacts close. This starts the welding operation, which continues until the normally closed contacts of relay 4TD are opened; this de-energizes the 10CR relay coil. Contacts of the 10CR relay then open and the Weld-O-Trol ignitron circuit is interrupted for lack of igniter current. Consider the conditions in the circuit before the foot switch is closed. A voltage-divider circuit, consisting of R_{44}, P_4, R_{45} in series, is connected across the transformer secondary winding. Resistors R_{43} and R_{43A} and capacitor C_{41} are connected in parallel, and a common lead is connected to the moving point of potentiometer P_4. The other side of these paralleled elements is connected through R_{41} to the grid of the tube. Contacts 4TD and 10CR are open and the grid of the tube is connected to the cathode through capacitor C_{42}. An a-c voltage is thus applied between grid and cathode.

Whenever the grid of the WL630 thyratron is at a higher voltage than the cathode, a small current flows from grid to cathode; but the applied voltage is alternating, so that, if the grid is at a lower potential than the cathode, no appreciable current flows from cathode to grid. Therefore *the tube acts like a rectifier*, and only "positive" half-cycle, unidirectional current flows through the grid to the cathode. Under these conditions the voltage drop across resistors R_{43} and R_{43A} charges capacitor C_{41}, with polarity as shown in Fig. 18-11.

When the foot switch is open, the anode and cathode of the tube are at virtually the same potential, so that no anode current flows through the tube and the relay. Therefore relay 4TD remains de-energized.

When the foot switch is closed, the 10CR relay is energized and its contacts close, connecting the cathode of the tube through the foot switch to the junction of R_{44} and the transformer secondary winding. At the same time another set of contacts on 10CR close the Weld-O-Trol ignitron circuit.

Because the cathode and grid now become positive at the same time, the grid-cathode rectifying action of the tube ceases, but that portion of the alternating voltage that appears from the foot switch to the slide on P_4 is now applied across the grid and cathode of the tube. Because the charging action of the tube has stopped, C_{41} begins to discharge through R_{43} and R_{43A}, thereby permitting the lower (or negative) voltage on the grid to rise toward that of the cathode. When the actual voltage (the sum of the decaying capacitor voltage and the alternating voltage) on the grid is approximately equal to the voltage of the cathode, the thyratron tube fires each time the anode potential becomes higher than that of the cathode. [For a graphical representation refer to Fig. 18-12.]

When the tube fires, relay 4TD is energized. Its normally closed contacts open and de-energize 10CR, *thus interrupting welding current*.

At the same time, the normally open contacts of relay 4TD close. Because these contacts are in parallel with one set of the 10CR contacts, the tube continues to fire as long as the foot switch is closed. As long as the tube conducts, the normally closed 4TD contacts in series with the 10CR coil are *open*. The coil remains

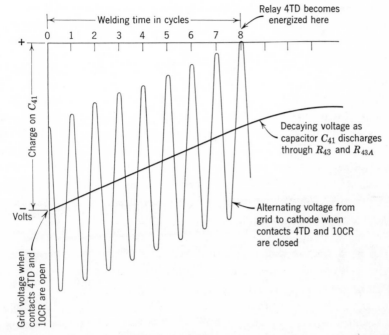

Fig. 18-12 Graphical illustration of the operating principle of the 1 A timer. (Courtesy Westinghouse Electric Corporation.)

de-energized and the normally open 10CR contacts in the igniter circuit are *open*, thus keeping the welding current shut off. To start another welding cycle, the foot switch is opened long enough for capacitor C_{41} to become fully charged again, a matter of a few cycles.

Because the tube anode-cathode circuit also acts like a rectifier, only pulsating, positive half-cycle, unidirectional current is conducted to relay coil 4TD. Capacitor C_{43} is charged during this positive half-cycle, so that it can transfer its energy to coil 4TD during the off, or negative, half-cycles when the rectifying action of the tube prevents current flow. This prevents the fast-acting relay 4TD from chattering or attempting to drop out during the half-cycle when no current flows. Because a discharged capacitor is, theoretically, a short circuit when voltage is suddenly applied, charging current to C_{43} is limited to a safe value by resistor R_{46}. Resistor R_{47} permits a small alternating current to flow through 4TD when the cathode is connected to R_{44}. This reduces the tendency of relay 4TD to saturate, thereby preventing it from becoming sluggish in operation.

If the movable point of P_4 is moved to the left, the increased alternating voltage charges C_{44} to a higher value because of the grid-rectification action of the tube. Then, as the foot switch is closed, it takes a longer time for the charge to leak off

C_{41}, and the tube remains non-conducting for a longer time; hence relay 4TD remains de-energized longer. To adjust the time during which 4TD remains de-energized requires adjustment of P_4.

A timer (NEMA type 3B) which will control four important steps in the welding process is available for use with this welding equipment. The operator can start the operation by simply stepping on a foot switch. If he holds it closed, the type 3B timer controls will (1) cause the welding electrodes to come together under pressure on the pieces to be welded and hold them for a desired time interval; (2) send a desired number of cycles of welding current through the work; (3) hold the electrodes together on the weld a desired length of time after the welding current has been shut off so that the weld can "set;" (4) separate the electrodes to allow movement of the work to another welding position.

The timer will automatically put the welder through another series of the four steps if the foot switch is kept closed, and it will repeat the series as long as it is closed. However, a nonrepeat switch may be thrown and this will make it necessary to release the foot switch in order to start another series of the welding operations. Each of the steps is controlled by one of four sections of the 3B timer, each of which is identical with the 1A timer of Fig. 18-11. It is interesting to note that the firing of the thyratron *stops* the flow of welding current. The firing of the thyratron of any section of the 3B timer *stops the action which that section controls, and* it *simultaneously starts the action which the next section controls.*

18.11 Variation of Heat in Weld by Thyratron Control

It has been shown that the timer which controls welding current in the circuit of Fig. 18-10 selects the number of cycles of flow of the current. It is important to be able to vary the portion of each cycle during which current flows, i.e., to *delay the start of ignitron conduction* each cycle so that the *average value of the welding current* may be adjusted to give the best operating conditions for each welding job. This process, as applied to the welding apparatus described, is called *heat control* and the piece of equipment that does the job is called the *heat-control attachment.*

By means of the heat-control attachment, the firing of the ignitron tubes in each half-cycle of their conduction may be delayed by any desired amount. This means that, although the timer is set for a certain number of cycles of current flow, the heat in the weld may be varied from the maximum caused by the flow of current throughout practically complete half-cycles of conduction of each tube to the minimum caused by the greatest delay in firing the ignitrons. The igniter-rod current flows through thyratrons whose conduction is timed by phase control.

The basic circuit of the heat control attachment is shown in Fig. 18-13. It is seen that the voltage between point a on potentiometer P_2 and point b on the center tap of the supply transformer secondary supplies the primaries of the grid transformers. The a-c supply is taken from the lines that supply current through the ignitron tubes to the welding transformer primary of Fig. 18-10. Therefore a delay in phase of the voltage a-b, which means a delay each cycle in the firing of thyratrons T_1 and T_2, will delay the flow of igniter-rod current and cause the ignitron tubes to fire at a later instant each cycle.

The inductance (L) has a reactance (X_L) such that the voltage between the 70-tap and point b will lag the transformer secondary voltage c-b by the angle whose cosine is 0.70 (i.e., an angle of 45°), when the slider of P_1 is at the extreme left. By moving the slider to the right, the phase of the voltage supplied to the grids of the thyratrons may be advanced so that the thyratrons will fire at an earlier point in each half-cycle, and thus the average welding current will be increased, thus increasing the heat in the weld. Points A and B of Fig. 18-13 connect to points A and B of Fig. 18-10.

18.12   Transistor Timer Circuit

Figure 18-14 shows a timer circuit, using transistors, which will energize a relay coil and keep it energized for a chosen period of time. The circuit designers, Sylvania Electric Products, Inc., describe the circuit and its operation as follows.

The timer circuit is a one-shot multivibrator. It holds the following advantages over simpler single-transistor relay circuits: (1) a high sensitivity relay is not required, (2) more stable calibration, (3) less sensitivity to temperature variation, and (4) relatively insensitive to supply voltage variation. The combination of an additional transistor and an inexpensive relay is quite likely to be less expensive than the sensitive relay required in single transistor timer circuits.

Transistor Q_1 is normally conducting and Q_2 is cut off. When S_1 is closed, the base of Q_1 is brought to ground potential momentarily, as C_2 has no voltage across it, causing the collector current of Q_1 to be transferred to the base of Q_2. Q_2 conducts closing the relay, and the base of Q_1 is driven negative by approximately 6 V, which is the initial voltage across C_1. Thus Q_2 is conducting and Q_1 is cut off. During the time Q_2 is on, C_1 is discharging through R_2 and R_3 toward +6 V. When the

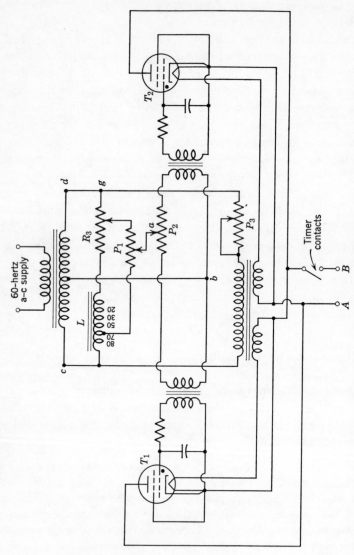

Fig. 18-13 A simplified circuit diagram of heat-control attachment. (Courtesy Westinghouse Electric Corporation.)

694

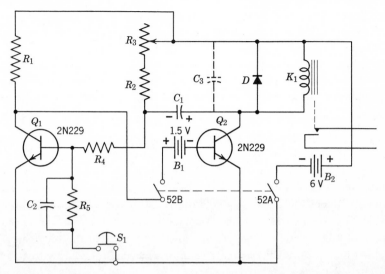

Fig. 18-14 Transistor timer circuit. B_1, penlite battery or mercury cell. B_2, latern-type 6-V battery. C_1, electrolytic; value controlled by R_3: 1000 μF, 60 s to 5 s; 200 μF, 6 s to 1 s; 100 μF, 3 s to 0.5 s. C_2, 0.5 to 1.0 μF. C_3, 0.05 μF. D 1N34A diode. K_1, 6-V, 300-Ω (or more) relay. Q_1, Q_2, Sylvania 2N229 transistor. R_1, 3.3 kΩ; R_2, 4.7 kΩ; R_4, 150 Ω; R_5, MΩ (all $\frac{1}{2}$ W), R_3, 25-kΩ potentiometer. S_1, normally open push button; S_2, DPDT switch. (Courtesy Sylvania Electric Products, Inc.)

base voltage of Q_1 has become slightly positive, Q_1 conducts and turns off Q_2, opening the relay. This completes the cycle. In this arrangement the operation of the relay is positive; the relay current being higher when Q_2 conducts and negligible when Q_2 is off.

The operating time of the relay is controlled by C_1 and the resistance R_2 and R_3 which is shown adjustable over the range of about 30 to 4.7 kΩ, a 6 to 1 range. The operating time is given by

$$T = 0.7RC_1$$

where T is in seconds, R is the sum of R_2 and R_3 in ohms, and C_1 is in microfarads. Because of high values of C_1 required to obtain operating times in seconds, C_1 must of necessity be an electrolytic capacitor. For this reason the expression given above is to be regarded as approximate as the actual capacitance value of electrolytic capacitors may depart considerably from the labelled value.

Several ranges of operating time may be obtained by switching different values of C_1 into the circuit. The circuit may be calibrated with a stop watch or if values of C_1 are accurately known, the expression for T becomes quite accurate.

In the circuit, R_4 protects Q_1 from excessive emitter current surges; D damps the positive overshoot appearing when the relay opens; R_5 keeps C_2 discharged until

S_1 is closed; and battery B_1 insures Q_2 being off when Q_1 is conducting. B_1 need not be disconnected from the circuit as only a few microamperes are drawn when the 6 V supply is disconnected.

Condenser C_3 may be necessary with some types of relay to suppress a negative jog in the relay transient appearing when the armature moves away from the pole piece and possibly retriggering the circuit. A 0.05 μF value is likely to be sufficient.

18.13 Photoelectric Light Relay

An operating circuit which counts the number of times a light beam is interrupted is shown in Fig. 18-15. Each time a person passes through a doorway, where the device is mounted, the decrease in phototube current caused by the large decrease in illumination allows the bias voltage on a thyratron tube to rise to conduction value. The firing of the tube energizes a relay which connects the actuating coil of a mechanical counter to a 115-V

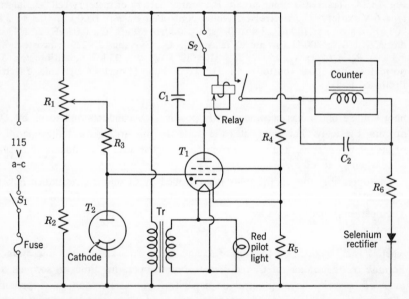

Fig. 18-15   Photoelectric counter circuit. C_1, 2 μF, 250 V; C_2, 8 μF, 250 V. R_1, 5,000 Ω, 5-W wire-wound pot; R_2, 1,000-Ω carbon, 1 W; R_3, 20 MΩ, 1-W carbon; R_4, 1,000 Ω, 10-W wire wound; R_5, 500 Ω, 5-W wire wound; R_6, 750 Ω, 1-W carbon. T_1, type 2050 thyratron; T_2, type 925 phototube. Relay, 70 V d-c 7.5 mA; counter, 115-V 60-hertz 45-mA coil, electromechanical (Central Scientific Co.). Rectifier, 120-V, 100-mA half-wave selenium. S_1, SPST, 2-A, 125-V toggle; S_2, SPST, 2-A, 125-V toggle. Tr, filament transformer, 115-V to 6.3 V, 1-A secondary. Pilot light, 6-8 V, 0.25-A S-47 bayonet base. Fuse, $\frac{1}{2}$ A, 250-V Littelfuse.

a-c line and trips the register, increasing its reading by one. When the light beam is again allowed to reach the phototube, its current increases sufficiently to reduce the grid bias on the thyratron a few volts below the critical value that prevents firing. This means that, as the thyratron goes out on the next half-cycle of a-c line voltage when its plate goes negative, the grid will regain control and keep the tube from firing during the following cycles when its plate goes positive. The high negative bias will remain until the light beam is again interrupted.

Electrons flow upward from the phototube cathode to its anode, and none can flow from the anode downward to the cathode because the anode cannot give off electrons. This means that phototube current flows only during the half-cycles when the top wire of the circuit is positive with respect to the bottom wire. Tube T_1, the thyratron (type 2050), can conduct current during only those half-cycles also.

The resistor (R_5) causes the cathode of T_1 to be a potential of about 39 V above that of the bottom line. The voltage drop through R_3 and the upper part of R_1, through which the phototube current flows, is about 42 V, so that the grid is more negative than the cathode by about $42 - 39 = 3$ V. When the light is cut off from the phototube, the voltage drop through R_3 is reduced practically to zero, the 3 V bias on tube T_1 is reduced to only a small fraction of a volt, and the tube can fire.

The capacitor (C_1) is placed across the relay coil to eliminate much of the ripple that otherwise would be in the coil current because the current flows in half-cycle pulses. This prevents hum due to vibration of the relay contacts.

The counter was at first connected directly across the 115-V a-c lines through the relay contacts, but each time it operated it made too much noise (a-c hum). Rectification of the counter's current by the selenium rectifier eliminated the hum entirely. Capacitor C_2 serves the same purpose for the counter as C_1 does for the relay. The resistor (R_6) was necessary because the counter's coil, built for a-c operation, took too much current when operated with the rectifier.

18.14   Sensitive Photorelay Circuit Using Transistors

The relay in the circuit of Fig. 18-16 will operate when the light intensity on the photocell PC decreases below a value determined by the adjustment of R_1. The circuit constants are such that the relay may be energized in bright daylight or held open in complete darkness.

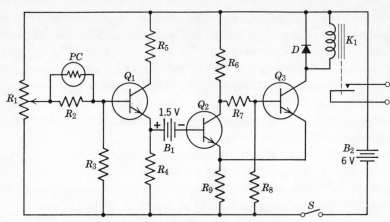

Fig. 18-16   Sensitive photorelay circuit. R_1, 10-kΩ potentiometer; R_2, R_3, 100 kΩ, ½ W; R_4, 4.7 to 5.6 kΩ, ½ W; R_5, 200 Ω to 1 kΩ, ½ W; R_6, 1.2 kΩ, ½ W; R_7, R_8, 3.3 kΩ, ½ W; R_9, 27 Ω, ½ W. Q_1, Q_2, Q_3, Sylvania 2N229 transistor. D, 1N34A diode. B_1, penlite battery; B_2, lantern type. K_1, 6 V, 300 Ω (or more) relay. *PC*, Clairex CL-3 photocell. (Courtesy Sylvania Electric Products, Inc.)

The first amplifier stage (Q_1) offers high impedance to the photocell and low output impedance to Q_2 through the emitter-follower connection. Q_2 and Q_3 form a trigger circuit which has the characteristic that Q_3 is cut off when the base voltage of Q_2 is above a certain value, but Q_3 conducts heavily and the relay is energized if the base voltage of Q_2 is lowered a few tenths of a volt below that certain value.

The operating point of the circuit is selected by adjustment of R_1. With light entering the photocell, R_1 is set a little beyond the point where the relay just opens. A small decrease in light intensity will then cause the relay to close. The battery B_1 provides bias and, inasmuch as the drain is only a few microamperes, it need not be disconnected when the 6-V power (B_2) is shut off. However, the power switch could be of the double-pole type with the second blade arranged to disconnect B_1. The circuit diagram and other information were furnished by Sylvania Electric Products, Inc.

18.15   Other Applications of Photodevices

The early phototube and photocell found their original uses in simple alarms and object counters. Later applications included reproduction of sound from film, door openers, measurement of light intensity and color, and various forms of industrial controls.

New applications and improvements in old ones have rapidly followed the developments of solid-state photodevices. Photocells, photodiodes, and light-sensitive switches and transistors have almost supplanted the phototube, particularly in applications where the mounting space requires small components. In many instances, the need for a power supply is eliminated by using photovoltaic components, which generate a potential related to the light intensity falling on them. Modern photodevices act as sensors in punched-tape and punched-card readers in various computer applications.

As an example of an application of a phototransistor to read punched cards, consider the operation of the circuit shown in Fig. 18-17. As the card passes between the light source and phototransistor, a hole in the card will pass light to the base of the transistor. When the crystal in the phototransistor is not illuminated, the collector current will be the normal small value for collector cutoff, but when light is focused on the base, collector current increases by an amount equal to beta times the change in base current carriers. A large collector current, proportional to the light intensity, then flows. A typical silicon planar phototransistor can operate at 40 V d-c and provide 10 mA collector current at 1,000 fc illumination. An advantage of the phototransistor is its amplification; relatively low-level illumination can generate a comparatively large collector current. Transistors Q_2 and Q_3 provide further amplification to drive a load that responds when light strikes the phototransistor.

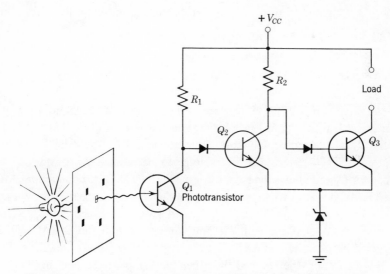

Fig. 18-17   Application of a phototransistor to read punched cards.

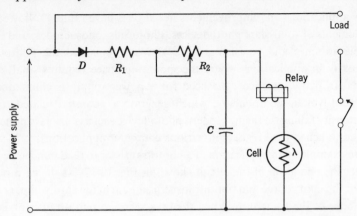

Fig. 18-18   Photoconductive cell and relay control a load in accordance with light intensity on the cell.

As an example of the application of the photoconductive cell to control lighting circuits, consider the circuit of Fig. 18-18. The photoconductive cell varies its resistance in accordance with the intensity of light falling on it. The relay will actuate its load circuits whenever the resistance of the photoconductive cell permits sufficient current to flow. Conversely, the relay will become inoperative whenever cell current drops below the amount necessary to hold the relay closed. This circuit may be used to control lighting circuits where it is desired to turn on and turn off lights automatically at dusk and dawn, for example.

18.16   Use of TRIAC as a Motor Control

The circuit of Fig. 18-19 can be used to replace a magnetic a-c motor control. When the "stop" switch is closed, the gate will lose control of the TRIAC and the motor will stop. To start the motor, the "start" switch is depressed. This permits the line voltage to be applied to resistor R_1 and capacitor C_1 and allows the gate to trigger the TRIAC into conduction. After the TRIAC has been triggered, the line voltage appears across the motor and across the combination of the gate, R_1, and C_1. Gate current through R_1 and C_1 will be out of phase with the line voltage, so that the TRIAC will be triggered into conduction each time the load voltage goes through zero.

The motor may be stopped by either the "stop" switch or the thermal-overload thermostat that is associated with the motor. Action of either will

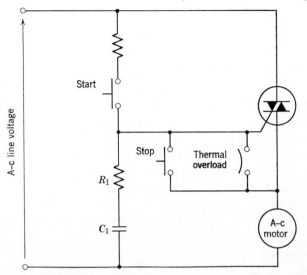

Fig. 18-19   Use of TRIAC as a motor control. (Courtesy General Electric Co.)

bypass the triggering network and unlatch the motor control, causing the motor to stop until the "start" switch is again actuated.

18.17   Automatic Dimmer for Automobile Headlights

An interesting practical application of the phototransistor is shown in Fig. 18-20. The *P-N-P* transistor is biased by adjustment of the variable

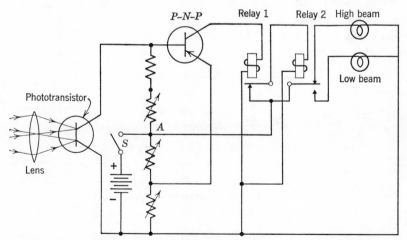

Fig. 18-20   Phototransistor-controlled automobile headlight dimmer.

resistors until its base current is practically zero. Although the emitter is negative with respect to point A, the base is still more negative so that the emitter is positive with respect to the base, as it should be.

In the absence of approaching lights, relay 1 is not energized and its contact is held closed by a spring. Relay 2 is thus energized and the high-beam lamps are on. When approaching lights shine on the phototransistor, the base current and collector current of the P-N-P transistor increase substantially and relay 1 is energized. This deenergizes relay 2, which is a power relay fed directly from the car battery, and a spring pulls its arm downward, shutting off the high-beam and turning on the low-beam lamps. Because the manual on-off light switch, S, on the dash of the car is off during the day, sunlight entering the lens and phototransistor presents no problem.

18.18　Blinker Light

A warning signal in the form of a light that continuously goes on and off, for use where electric power service is not available, is produced by the circuit of Fig. 18-21. The 2N307 power transistor should be mounted on a heat sink. The battery must supply not only transistor bias current, but lamp current also. This totals about 0.25 to 0.4 A. The following description is reproduced from the booklet "Performance-tested Transistor Circuits" with permission from Sylvania Electric Products, Inc., Woburn, Mass.

Transistors Q_1 and Q_2 are part of a direct-coupled amplifier circuit with a lamp as a load in the collector circuit of Q_2, a 2N307 power transistor. The output voltage across the lamp is of the same polarity as a signal applied to the base of Q_1, i.e., the voltage across the lamp is in phase with the voltage applied to

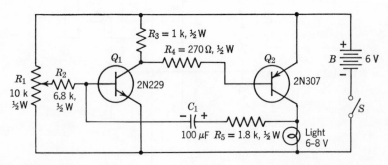

Fig. 18-21　Blinker light. On-off rate changed conveniently by changing C_1. Lamp may be No. 40, 44 or 46. (Courtesy Sylvania Electric Products, Inc.)

the base of Q_1. Thus, if the signal voltage across the lamp is applied to the base of Q_1 through C_1 and R_5, as shown in the diagram, the circuit becomes regenerative as the voltage gain from the base of Q_1 to the collector of Q_1 is greater than unity.

The operation of the circuit is as follows: Suppose the lamp to be initially off and the potentiometer advanced toward the positive voltage to cause Q_1 to conduct slightly; the voltage across the lamp increases slightly and the base of Q_1 is made more positive as the voltage across the lamp is coupled to the base of Q_1 through C_1 and R_5. This causes Q_1 to conduct more heavily and likewise Q_2 until Q_1 and Q_2 are both saturated and nearly all the battery voltage is across the lamp causing it to light.

While the lamp is on, C_1 is becoming charged and the current applied to the base of Q_1 through C_1 and R_5 is diminishing. Finally C_1 becomes charged to the point where the current supplied to the base of Q_1 is insufficient to hold Q_2 in saturation and the voltage across the lamp begins to drop. When this happens the base current of Q_1 decreases causing the voltage across the lamp to decrease further. This results in the lamp being rapidly extinguished and the base of Q_1 driven negative due to the charge accumulated on C_1 during the period that the lamp was on. Now both transistors are cut-off. The charge on C_1 is now leaked off through R_1, R_2, R_5 and the lamp and the base voltage of Q_1 rises toward a positive value determined by the setting of R_1. When the base of Q_1 becomes slightly positive, Q_1 commences to conduct and the cycle of operation is repeated.

The circuit is adjusted with R_1. If R_1 is set too near the positive end, the lamp will remain on continuously and if too near the low end, it will not light. Between these extremes, the circuit will function with the setting of R_1 determining to a limited degree the on and off times of the lamp. With the values shown in the circuit diagram, the on time of the lamp is about 2 or 3 s and the off time correspondingly about 4 to 5 s. If it is desired to increase or decrease the on and off times, this may be most conveniently accomplished by changing the value of C_1. Doubling C_1 will double the period of the cycle and reducing C_1 by 2 will reduce the cycling period by 2, etc.

The time constant may also be changed by changing the value of R_5 but this should be done with care. For 6 V operation, R_5 should not be smaller than about 250 Ω in order to protect Q_1 from excessively large emitter current. On the other hand, R_5 should not be so large as to stop the oscillations.

If it is desired to operate the circuit with a 12 V supply using a 12 V, 0.15 to 0.25 A lamp (or two 6 V lamps in series), resistors R_3 and R_4 should be made twice the value shown and R_5 should not be made less than about 500 Ω. The voltage rating of C_1 should then be greater than 12 V.

Because transistor Q_2 acts as a switch, its power dissipation is low. However, it is suggested that the practice of mounting power transistors on a metal plate be followed here.

18.19   Metronome

A steady sound beat, of controllable repetition rate, will be produced by the multivibrator circuit of Fig. 18-22. Its principle of operation is briefly described.

Assume that the collector current of Q_2 is increasing. This lowers the potential of the collector and also the base bias on Q_1. The collector current of Q_1 thus decreases and its collector voltage goes up. This lowers the base bias of Q_2, which quickly takes saturation current. Meanwhile the bias on Q_1 has started to increase, and it begins taking more collector current as the reverse action accelerates, shifting all the current to Q_1.

Adjustment of the potentiometer R_3 varies the repetition rate of the tones in the loud speaker by changing the bias on C_1 and thus controlling the discharge rate of C_1. Switch S_2 provides for the selection of two volume levels. The designers suggest that this circuit has many applications, one of which is temperature measurement, as follows. Q_1 could be placed in a container kept in ice water and Q_2 could be located in a warm place. The resulting repetition rates of the speaker tone at various temperatures at the warm location could be plotted against temperature to obtain a calibration curve.

Another suggestion is that the cap of one of the transistors be removed to make it light-sensitive. A match held over the transistor would then change the multivibrator frequency. Frequencies denoting the presence of infrared and ultraviolet would be observed.

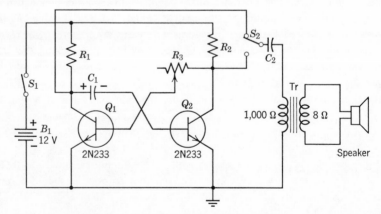

Fig. 18-22   Metronome application of multivibrator circuit. B_1, two 6-V lantern batteries in series. C_1, 50 μF electrolytic; C_2, 0.1 μF paper. R_1, R_2, 4.7 kΩ, $\frac{1}{2}$ W; R_3, 50-kΩ wire-wound potentiometer. Tr, output transformer, primary 1,000 Ω, secondary 8 Ω. (Courtesy Sylvania Electric Products, Inc.)

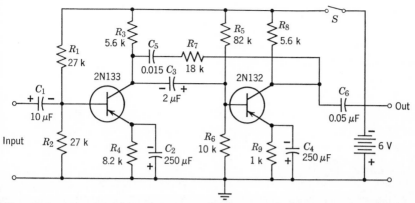

Fig. 18-23   Low-noise, stabilized hi-fi preamplifier for variable-reluctance pick-ups. C_1, C_3, 25-V electrolytic; C_2, C_4, 6-V electrolytic; C_5, C_6 200-V paper or ceramic. All resistors $\frac{1}{2}$ W carbon. Battery, four penlite cells in series. (Courtesy Raytheon Manufacturing Co.)

18.20   Hi-Fi Preamplifier

The first large-scale use of transistors was in audio amplifiers, and in the subminiature component field they have replaced vacuum tubes entirely. The circuit of Fig. 18-23, designed by Raytheon Manufacturing Company, is said to have low noise level. Both stages are d-c stabilized by emitter-circuit capacitors C_2 and C_4. Four penlite, or flashlight, cells may be used in series to provide 6 V of battery power.

Two common-emitter circuits are *RC* coupled. The circuit is *designed for use with variable-reluctance-type pick-up cartridges.* Voltage divider (R_1, R_2) provides bias for the first stage, in addition to the d-c drop across R_4. C_2 provides stabilization. The amplified audio signal across collector load resistor R_3 is coupled by C_3 and R_6 to the base-emitter input of the second stage. The output signal across R_8 is to be coupled by C_6 to the *high-impedance input* of a power amplifier.

C_5, R_7 provide negative feedback to reduce distortion and increase the frequency range. If the preamplifier is mounted in the base of the record player or changer, it should be located as far away from the motor and a-c wiring as possible. Similar circuits are available with input impedances to match crystal pick-ups, crystal microphones, or high-impedance tuners.

18.21   Audio Power Amplifier, Class B

An 8-W audio power amplifier using power transistors, with a push-pull output stage, is shown in Fig. 18-24. The input transformer must be selected

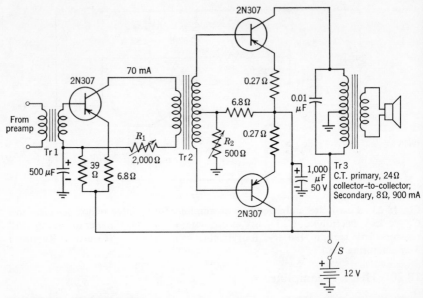

Fig. 18-24 Class B, 8-W audio amplifier. Adjust: R_1 to 70 mA in collector, R_2 to 50 mA per collector. Mount the push-pull transistors on heat sinks. (Courtesy Sylvania Electric Products, Inc.)

to match the output stage of the preamplifier, which may be high-impedance if it is the collector circuit of a transistor. The driver transformer of the push-pull stage and the output transformer must be selected to match the low impedance of the Sylvania 2N307 power transistor.

The designers state that less than $\frac{1}{2}$ W of input power from the preamplifier is needed and that the collector current of Q_1 is constant after having been set at 70 mA by using about 1,000 Ω of the 2,000-Ω base resistor R_1. The zero-signal total collector current in the Class B output stage is given as 100 mA and the maximum total collector current as 900 mA. These require that transformer Tr 2 be able to handle a peak current of 150 mA, and Tr 3, 1 A. It must be understood that, for a given power output, higher *currents* are needed with transistors than with vacuum tubes because the voltages are so low.

The designers recommend the following:

Because of the heat dissipation required, both transistors in the output stage must be mounted, utilizing the heat sink, and insulated from each other as well as the chassis. At least a $3'' \times 5'' \times \frac{1}{16}''$ aluminum surface should be used, and preferably larger. These two dissipating surfaces can be mounted either on top or underneath the chassis whichever is convenient to the builder. Although the driver need not be mounted with heat sink, care must be exercised to insulate the metal container from the chassis. Some power transistors have the collector terminal connected to the case. A mica washer is usually used to insulate the transistor and case from the chassis.

18.22   Public-Address System

A public-address system employs many of the principles studied in this book. A small system that operates with a crystal microphone and delivers 5 W is shown in Fig. 18-25. Its construction would serve as an excellent project for an extended laboratory exercise.

The full-wave rectifier power supply is well filtered with a 10-H choke (L), and two 40-μF, 450-V electrolytic capacitors (C_9 and C_{10}). The power transformer (Tr 1) delivers 300 V between each plate of the rectifier tube and ground. The 6.3-V winding of T_1 supplies the heaters of the other three tubes, which require 1.5 A.

A crystal microphone is used. The ungrounded terminal of the jack (J) is connected to the grid of T_1 by a shielded cable having a grounded sheath. This preamplifier drives the second voltage-amplifier stage, which employs T_2, through the volume control R_8. This stage drives the grid of the power-output tube (T_3) through capacitor C_4. The audio-frequency output transformer (Tr 2) should have the proper turns ratio to provide the correct load resistance for the 6L6 tube, which should be 2,500 Ω. If N represents the ratio of *secondary* turns to *primary* turns of the output transformer and if R_L represents the resistance of the connected load (loud speaker or long line going to the speaker), then

$$\frac{R_L}{2,500} = N^2 \quad \text{and} \quad N = \frac{\sqrt{R_L}}{50}$$

Thus a loud speaker with a 4-Ω voice coil would require a transformer turns ratio $N = \frac{1}{25}$. Resistor R_{11} is used to apply inverse feedback to the operation of T_3.

18.23   Intercommunication System Using Vacuum Tubes

Private communication systems that provide talking facilities between offices, and to remotely located points, have come into general use. Many of them are electronically operated and consist of an audio-frequency amplifier with a loud speaker and a suitable switching arrangement. The speaker,

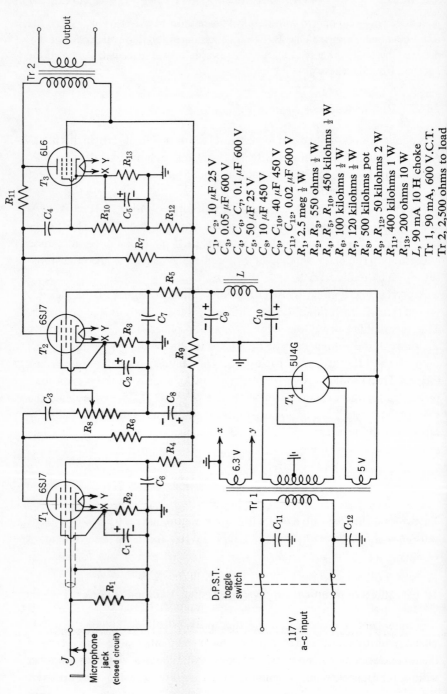

C_1, C_2, 10 μF 25 V
C_3, 0.05 μF 600 V
C_4, C_6, C_7, 0.1 μF 600 V
C_5, 50 μF 25 V
C_8, 10 μF 450 V
C_9, C_{10}, 40 μF 450 V
C_{11}, C_{12}, 0.02 μF 600 V
R_1, 2.5 meg $\frac{1}{2}$ W
R_2, R_3, 550 ohms $\frac{1}{2}$ W
R_4, R_5, R_{10}, 450 kilohms $\frac{1}{2}$ W
R_6, 100 kilohms $\frac{1}{2}$ W
R_7, 120 kilohms $\frac{1}{2}$ W
R_8, 500 kilohms pot
R_9, R_{12}, 50 kilohms 2 W
R_{11}, 400 kilohms 1 W
R_{13}, 200 ohms 10 W
L, 90 mA 10 H choke
Tr 1, 90 mA, 600 V.C.T.
Tr 2, 2,500 ohms to load

708

usually a permanent magnet type, serves as a microphone as well as a receiving device. The units are usually provided with a talk-listen switch that must be depressed when the operator is speaking, and released when he wishes to listen. However, some master stations are provided with a privacy earphone that may be connected into the circuit by raising the switch to a stable position. When this is done the conversation is permitted in the same manner as with a telephone.

Figure 18-26 is a schematic diagram of a simple master unit, without the privacy earphone feature, that may be used with from one to five sub-stations. If the substation shown were called No. 1, it would be connected X to X, R to R, and No. to 1 of the master circuit. Then a second substation could be called No. 2 and it would be connected X to X, R to R, and No. to 2 of the master circuit. A three-conductor cable would be used for making the connections. A description of the operation will now be given.

Suppose that the master station is to call a substation, in this case No. 1. The selector switch is turned to 1 and the talk-listen switch is depressed. The sound waves of the voice strike the diaphragm of the loud speaker, which vibrates and *moves the attached voice coil in the magnetic field of its permanent magnet*. The induced voltage sends very weak current through the primary of the input transformer. Thus a voltage is applied, by transformer action, to the grid of T_1. It is amplified and then applied to the grid of T_2, its strength being regulated by the volume control.

The *output-transformer secondary winding* delivers the strong audio power to the second section of the talk-listen switch (2 in dotted position) where it is passed on to the selector switch and contact No. 1. From there it goes to the substation speaker *No.* terminal, through the voice coil, and then back on line R to ground, thence to the output-transformer secondary, completing the circuit.

If points A and B are connected by the nonprivate jumper, as indicated by the dash line, the person at the substation need only answer without pushing the switch in line X. Tracing his reply circuit from the grounded R terminal, we go through the substation speaker to the *No.* terminal, to selector switch 1, through the second and third sections of the talk-listen switch (which is in the up position for listening), and through the primary of the input transformer, and also through the 47-Ω resistor, to ground. The amplifier thus receives the voice signal and delivers it from the output transformer to the master speaker.

If private operation is desired (this means that the master station amplifier cannot reproduce sounds originating at any substation unless the substation

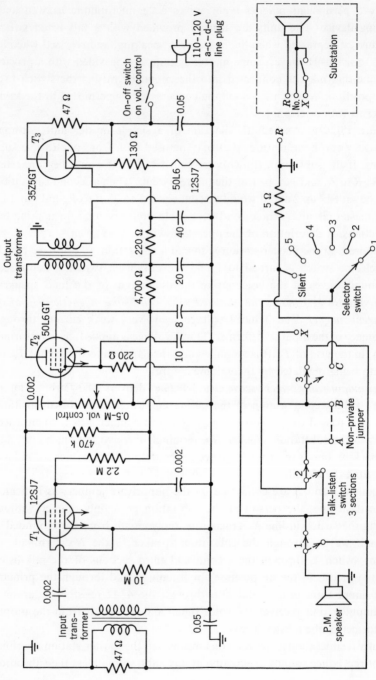

Fig. 18-26 Schematic diagram of intercommunication system master station. All capacitor ratings are in microfarads.

talk switch is depressed), the jumper between points A and B is removed and connected between points B and C instead. With this arrangement, calling by the master is accomplished over the same circuits as described above. The substation's answer comes through the substation switch on line X and goes through the input-transformer primary to ground and back to the substation on R. The 5-Ω resistor prevents the B–C jumper from short-circuiting the substation speaker.

It is important that the person at the master station not be disturbed by sounds from any substation when communication is not desired. This is automatically achieved with the connection arranged for private operation, because the substation switch must be closed before sound is produced in the master station. But, in *nonprivate operation*, the master operator will hear sound from the substation to which the selector switch is connected. To eliminate all sound, the selector switch must be placed in the *silent* position, which is the contact beyond No. 5 in the diagram. With this setting the master station will receive from any substation whose talk switch is depressed. In fact, it is always necessary to depress any substation switch to which the selector switch is not connected if the master station is to be called.

There is a provision for the master station to call all stations at once. The selector-switch position for this feature is not shown in the diagram. Lines running from all five terminals on the selector switch come to separate, closely spaced points but remain separate until the switch is moved to that position, in which event they are all tied together and receive power in parallel.

One advantage of the nonprivate connection is that we may answer at a substation from a great distance away from the unit. This means the operator need not leave his working place to come to the station to press the talk switch.

18.24 Intercommunication System Using Transistors

Figure 18-27 is a complete circuit of an intercommunication system, designed by Radio Corporation of America, using transistors and operating from a 12-V d-c power supply. For a-c operation a filtered power supply operating on 115 V, 60 cycles, and using crystal diodes or selenium rectifiers is recommended.

When switch S_1 is pressed to *talk*, Sp 1 serves as a microphone and sends the signal through capacitor C_1 to the first amplifier stage. The output stage delivers audio power through capacitor C_9 and switch S_2 to the speaker Sp 2. The person at the substation presses the switch to the *talk* position

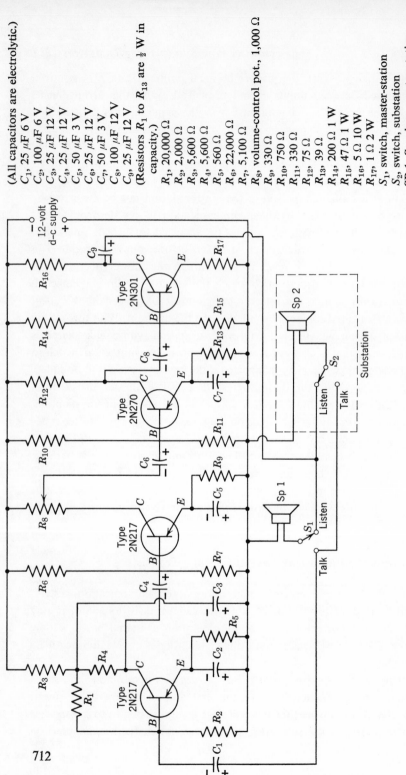

(All capacitors are electrolytic.)

C_1, 25 μF 6 V
C_2, 100 μF 6 V
C_3, 25 μF 12 V
C_4, 25 μF 12 V
C_5, 50 μF 3 V
C_6, 25 μF 12 V
C_7, 50 μF 3 V
C_8, 100 μF 12 V
C_9, 25 μF 12 V

(Resistors R_1 to R_{13} are $\frac{1}{2}$ W in capacity.)

R_1, 20,000 Ω
R_2, 2,000 Ω
R_3, 5,600 Ω
R_4, 5,600 Ω
R_5, 560 Ω
R_6, 22,000 Ω
R_7, 5,100 Ω
R_8, volume-control pot, 1,000 Ω
R_9, 330 Ω
R_{10}, 750 Ω
R_{11}, 330 Ω
R_{12}, 75 Ω
R_{13}, 39 Ω
R_{14}, 200 Ω 1 W
R_{15}, 47 Ω 1 W
R_{16}, 5 Ω 10 W
R_{17}, 1 Ω 2 W
S_1, switch, master-station
S_2, switch, substation
SP 1, Speaker, master-station, 12 Ω 1 W
SP 2, Speaker, substation, 12 Ω 1 W

712

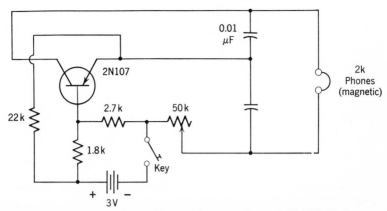

Fig. 18-28   Code practice oscillator. (Courtesy General Electric Co.)

to answer, and uses the same circuit to deliver audio power to Sp 1, which then serves as a speaker.

By means of a multiple-contact switch with suitable connections, the audio power output through C_9 can be directed to other substations, as is done in Fig. 18-26.

18.25   Code Practice Oscillator

Radio amateurs who desire to obtain an operator's license from the Federal Communications Commission usually use a *code practice oscillator* in acquiring the ability to send and receive messages in *radiotelegraph code* with the necessary speed. *Practice* is necessary, and an oscillator built with the circuit of Fig. 18-28, designed by Raytheon Manufacturing Company, should serve the purpose very well.

Feedback from the output circuit containing the headphones is effected through the C_1-C_2 voltage divider, and it is adequate to start and sustain oscillations. Base-bias current is fed through R_2, R_1 serves as the emitter "load," and the magnetic headphones are the collector load. C_3 bypasses R_2 to prevent fluctuations of bias. A hand key turns operating power on and off.

The layout and wiring are not critical. Current drain is not too much for penlite or flashlight cells—four to six in series.

18.26   Transistor in Automobile Ignition System

After fifty years of use of the conventional d-c ignition system for gasoline engines in automobiles and other vehicles, a major step was taken by

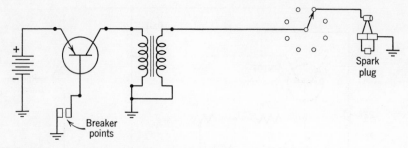

Fig. 18-29 Automobile engine ignition circuit using power transistor. Large reduction in current interrupted by breaker points.

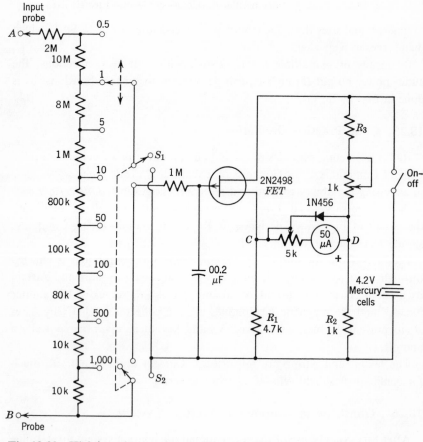

Fig. 18-30 High impedance voltmeter using the *FET*. (Courtesy Texas Instruments Incorporated, Dallas, Texas.)

Autolite Electric Corp. in applying the transistor to the conventional d-c system. The circuit is shown in Fig. 18-29.

A power transistor with a large heat sink and heat-radiating fins has the breaker contacts (points) in its base circuit where the maximum current is about 250 mA. The collector current flows through the primary of the induction coil, and its rapidly changing flux induces a high voltage in the secondary. This is fed to the spark plugs in the standard manner.

Tests show that in modern high-speed driving a current of 5 to 8 A is required in the primary to produce sufficient induced voltage in the secondary. In the conventional circuit the breaker points must carry and interrupt this much current. In the transistor circuit where the points carry only 3 to 5 per cent of that amount of current, they last about ten times as long as do those in the conventional circuit, the predicted life being 100,000 miles. A more important result lies in the fact that the spark is much hotter at high engine speeds. This is because the voltage in the secondary circuit has been found to drop from 28 kV at slow speeds to about only 26 kV at 5,000 rpm. In contrast, the voltage in the conventional circuit drops to about 13 kV at that speed. The transistor circuit is said to produce perfect timing at all engine speeds and at weather temperatures from subzero to over 100°F.

18.27 High Impedance Voltmeter Using *FET**

The field-effect-transistor (*FET*) has an extremely high input impedance and other characteristics usually attributed to a pentode vacuum tube. The voltmeter, whose circuit is shown in Fig. 18-30, is similar in many respects to a VTVM, but it uses a single *FET* instead of vacuum tubes in its circuitry.

The input impedance of the voltmeter is determined by the string of resistors between the probe terminals. Since the overall accuracy is largely determined by these resistors, they should be stable and preferably have an accuracy of 1 per cent. To provide protection from transient overloads and stray a-c voltages, the filter consisting of the 1 MΩ resistor and 0.02 μF capacitor is introduced at the input to the *FET*.

To analyze the circuit operation, assume that zero potential exists between points A and B, and that current is flowing in the *FET* and resistor R_1. Point C will then be negative with respect to point B. Adjustment of resistors R_2 and R_3 can make the potential at point D equal to that at point C, and the

---

* Adapted from *Popular Electronics*, July 1964, and Texas Instruments Application Note, *Field-Effect Transistor Voltmeter*, August 1965.

meter will indicate zero. When point C becomes more negative than point D, as when an input potential is connected between the probes at A and B, the meter will indicate the difference in potentials at C and D. Calibration of the meter according to the current change in the *FET* and resistor R_1 will permit the meter to read directly in volts to indicate the input potential applied to points A and B.

Switches S_1 and S_2 are ganged together to permit the selection of either positive or negative input potentials so that the meter will read upscale for either. The 5-kΩ potentiometer and diode associated with the meter are used to adjust the linearity of the calibration and to provide a shunt to limit the meter current to a value slightly greater than its full-scale rating.

SUGGESTED REFERENCES

1. Royce G. Kloeffler, *Industrial Electronics and Control*, 2nd ed., Wiley, New York, 1960.

2. M. G. Young and H. S. Bueche, *Fundamentals of Electronics and Control*, Harper and Brothers, New York, 1952.

3. Westinghouse Electric Corp., *Industrial Electronics Reference Book*, Wiley, New York, 1948.

4. George M. Chute, *Electronics in Industry*, 2nd ed., McGraw-Hill, New York, 1956.

5. William D. Corkerell, *Industrial Electronics Handbook*, McGraw-Hill, New York, 1958.

QUESTIONS

18-1 In the voltage-regulated power-supply circuit of Fig. 18-1: (*a*) What is the advantage of using a gas diode VR105, instead of a resistor, to connect the 6SJ7 cathode to the negative line? (*b*) Explain the action of the 6SJ7 tube when the resistance of the load connected to the output is increased. (*c*) Assume that there is no load and that the sliding contact on the 15-kΩ resistor is at its middle point. What is the effect when the slider is moved to the top? To the bottom? Justify your answers by explanations. (*d*) Assume that the plate of the 6B4G tube is at 350 V potential above ground. An unregulated output of 25 mA at 120 V is desired. Determine the total resistance and watts capacity that the voltage divider should have. The voltage divider itself should not draw excessive current, probably not more than 10 mA.

18-2 In Fig. 18-2 it is observed that R_1, R_2, and R_3 are only 25 Ω each. Why are they so small and why should they not be omitted?

18-3 Discuss the subject of grounding a bias-voltage supply.

18-4 In the transistor power supply circuit of Fig. 18-4, account for the fixed polarity of C during both half-cycles of the 12.6-V center-tapped secondary voltage. Why is the third transistor Q_3 used instead of merely connecting the battery to negative output terminal?

18-5 Explain the action of both circuits of Fig. 18-6. Could the same two diodes be used in either circuit? On what do the relative values of R_{sp} and R_{ss} depend?

18-6 Refer to the shunt-regulated power supply of Fig. 18-7 and explain what the first pair of diodes do to a ripple current that passes through the 500-Ω resistor. Considering the action of the second pair of diodes, do you see how large changes in voltage across the capacitor are negligibly small at the output (in E_o)? Suppose E_i fluctuates from the assumed-constant 120 V rms value. How does the diode regulator prevent this disturbance from reaching the output?

18-7 Trace the flow of igniter-rod current and of welding current in both cycles in the circuit of Fig. 18-10.

18-8 (a) How does C_{41} become charged in Fig. 18-11? (b) Explain what happens after the foot switch is closed. (c) The *normally closed* 4TD contact in series with 10CR is *opened* as soon as the thyratron conducts. What is the purpose of this? (d) How is the length of the time interval varied, and what is the principle involved?

18-9 (a) Why does controlling the phase of the grid voltages of T_1 and T_2 in Fig. 18-13 alter the amount of heat in the weld? (b) How is phase control accomplished?

18-10 Explain in detail the action of the light-relay circuit of Fig. 18-15 when the light beam is interrupted. What would be the effect of making R_3 (a) twice as large as it is? (b) one-half as large as it is?

18-11 In the blinker-light circuit of Fig. 18-21, what is the main function of the capacitor?

18-12 Trace the charge and discharge paths of C_1 in the metronome circuit of Fig. 18-22. What is the effect of varying R_3?

18-13 In the circuit of Fig. 18-25 (a) what is the function of capacitors C_6 and C_7? (b) what provision is made to reduce distortion?

18-14 Trace the circuit of the intercommunication system of Fig. 18-26 to show how (a) the master calls a substation; (b) a substation initiates a call to the master station when the private condition exists (nonprivate jumper disconnected).

18-15 (a) What kind of rectifier and filter is used in the power supply of the intercommunication system of Fig. 18-26? (b) Why is it not necessary to watch for grounds when the line plug is connected to an a-c line?

PROBLEMS

18-1   In the circuit of Fig. 18-3a, assume a 60-Hz ripple voltage. E_{cd} rms exists across points c and d. Calculate the fraction of E_{cd} that appears across the output at e-f.

18-2   (a) In the circuit of Problem 18-1, how many times larger than E_{cd} is the ripple voltage E_{ab}? (b) What fraction of the ripple voltage between points a and b reaches points e and f?

18-3   In the circuit of Fig. 18-3a, what is the rms value of the fundamental component of ripple voltage at a and b, if the a-c supply is 115 V, 60 Hz? How much of this ripple voltage appears across the output potentiometer? (Neglect the drop across the 50-Ω resistor.)

18-4   In Fig. 18-8, a negative bias of 12.75 V will just keep the 884 thyratron from firing when the plate is $+120$ V with respect to the cathode. Assume that the fully charged capacitor must discharge through 4 MΩ when switch S is connected to the left terminals. How much delay time will elapse after S_1 is closed before the thyratron fires?

18-5   Compute values and plot a curve for the circuit of Problem 18-4, showing delay time versus resistance from 0.25 to 4 MΩ.

18-6   In Fig. 18-8, how long will it take a completely discharged capacitor, C, to charge to 119 V when switch S_1 is connected to the right?

18-7   Assume the maximum safe value of average current in the 884 thyratron is 75 mA. (a) Determine the minimum safe value of R in series with the plate, in Fig. 18-8. Assume a 15 V drop between plate and cathode. (b) Is there a practical situation that would limit the maximum value of R?

18-8   Explain fully the action of the circuit in Fig. 18-9 both before and after the switch is closed.

18-9   Calculate approximately how much voltage drop, in per cent, occurs on the capacitor in Fig. 18-9 during negative half-cycle (switch S open) when resistance R is set at 1 MΩ.

Appendix

A.1 Proof of Maximum Power Transfer Theorem

Figure A-1 represents a generator with internal impedance $R_G + j0$ and its load $R_L + j0$. The generated voltage is E_G(a-c or d-c). The load current is

$$I_L = E_G/(R_G + R_L) \qquad \text{(A-1)}$$

The power delivered to the load is

$$P_L = E_G^2 R_L/(R_G + R_L)^2 \qquad \text{(A-2)}$$

Since R_L can be varied until P_L is a maximum, the rate of change of P_L with respect to R_L may be expressed mathematically and set equal to zero, thus satisfying conditions for maximum P_L.

Fig. A-1

$$\frac{dP_L}{dR_L} = \frac{E_G^2(R_G + R_L)^2 - 2E_G^2 R_L(R_G + R_L)}{(R_G + R_L)^4} \qquad \text{(A-3)}$$

Therefore

$$E_G^2(R_G + R_L)^2 = 2E_G^2 R_L(R_G + R_L)$$

$$R_G + R_L = 2R_L$$

$$R_G = R_L \qquad \text{(A-4)}$$

In Section 1.28, the case where reactance is present is treated. The equation for power in the receiver circuit (1-82) reduces to Equation (A-2) when X_G and X_R are equal but opposite in sign. R_R there and R_L here represent the same quantity, i.e., load resistance.

719

A.2   Derivation of Charging Current in a Capacitor

Consider an uncharged capacitance C to which a constant d-c voltage E is suddenly applied. There will be some resistance R in series, even if it is only in the connection wires.

By Kirchhoff's voltage law,

$$E = iR + \frac{1}{C} \int i \, dt \qquad (A\text{-}5)$$

in which i is the current at any instant. Differentiating with respect to t,

$$0 = R\frac{di}{dt} + \frac{i}{C} \qquad (A\text{-}6)$$

separating the variables

$$\frac{di}{i} = -\frac{1}{RC} dt \qquad (A\text{-}7)$$

$$\ln i = -\frac{1}{RC} t + k_1 \qquad (A\text{-}8)$$

By the *definition of logarithm*,

$$i = \epsilon^{[-(1/RC)t + k_1]} \qquad (A\text{-}9)$$

$$i = \epsilon^{-t/RC} \epsilon^{k_1} = k_2 \epsilon^{-t/RC} \qquad (A\text{-}10)$$

At the instant E is applied, $t = 0$ and $i = E/R$,

$$\frac{E}{R} = k_2$$

Therefore, at any time t after $t = 0$,

$$i = \frac{E}{R} \epsilon^{-t/RC} \qquad (A\text{-}11)$$

A.3   Discharge Current in a Capacitor

When a capacitor, charged to a voltage E, is suddenly (at $t = 0$) connected to a resistance R, discharge current immediately starts to flow. *At any instant* the current is i, the charge on the capacitor is q, and the Kirchhoff voltage equation is

$$e_C + e_R = 0 \qquad (A\text{-}12)$$

Denoting the capacitance by C, this equation becomes

$$\frac{q}{C} + iR = 0 \tag{A-13}$$

Differentiating and transposing,

$$R\frac{di}{dt} = -\frac{1}{C}\frac{dq}{dt} = -\frac{1}{C}i \tag{A-14}$$

Separating the variables,

$$\frac{di}{i} = -\frac{1}{RC}dt \tag{A-15}$$

This is Equation (A-7), and the solution

$$i = \frac{E}{R}\epsilon^{-t/RC} \tag{A-16}$$

is obtained exactly as before.

A.4   Voltage Applied to Inductor

When a coil having inductance L and resistance R has a constant voltage E suddenly applied, the Kirchhoff voltage equation, in which i is the instantaneous value of current, is

$$E = iR + L\frac{di}{dt} \tag{A-17}$$

Transposing and separating the variables,

$$L\frac{di}{dt} = E - iR$$

$$\frac{di}{E - iR} = \frac{dt}{L}$$

Multiplying both sides by $-R$,

$$\frac{-R\,di}{E - iR} = -\frac{R}{L}dt$$

Integrating,

$$\ln(E - iR) = \frac{-Rt}{L} + k_1 \tag{A-18}$$

From the definition of logarithm,

$$E - iR = \epsilon^{[-(R/L)t+k_1]}$$

$$E - iR = k_2\epsilon^{-(R/L)t} \tag{A-19}$$

At $t = 0$, $i = 0$ because current in an inductance cannot change in zero time, so that k_2 is found:

$$k_2 = E$$

Substituting into Equation (A-19),

$$E - iR = E\epsilon^{-(R/L)t} \tag{A-20}$$

from which

$$i = \frac{E}{R}(1 - \epsilon^{-(R/L)t}) \tag{A-21}$$

A.5   Current Decay in an Inductor

When an inductance L carrying current i is suddenly connected* across a resistance R, the instantaneous current i and the initial voltage ($E = L\, di/dt$) are related by Kirchhoff's voltage law:

$$iR + L\frac{di}{dt} = 0 \tag{A-22}$$

Transposing and separating the variables,

$$\frac{di}{i} = -\frac{R}{L}dt$$

$$\ln i = -\frac{R}{L}t + k_1$$

$$i = \epsilon^{-(R/L)t}\epsilon^{k_1}$$

$$i = k_2\epsilon^{-(R/L)t} \tag{A-23}$$

At $t = 0$, $i = E/R$, so that

$$\frac{E}{R} = k_2$$

and

$$i = \frac{E}{R}\epsilon^{-(R/L)t} \tag{A-24}$$

---

* These conditions are electrically identical with *short circuiting*, with a zero-resistance switch, a coil of inductance L and resistance R while it is carrying current. Draw the circuit, showing the coil in series with another resistance R_2 and a battery. E is the *EMF of self-induction* of the coil ($L\, di/dt$) at the instant the switch is closed.

A.6 Required Gain of Phase-Shift Oscillator

Proof that a two-stage phase-shift oscillator must have a gain of at least 3 to sustain oscillations, when the feedback circuit shown in Fig. A-2 is used.

$$Z_1 = R - jX_c, \; Z_2 = \frac{-jRX_c}{R - jX_c}$$

where

$$X_c = \frac{1}{\omega C}$$

From voltage division,

$$\frac{e_2}{e_1} = \frac{Z_2}{Z_1 + Z_2} = \frac{-jRX_c}{R - jX_c} \div \left(R - jX_c - \frac{jRX_c}{R - jX_c} \right)$$

$$\frac{e_2}{e_1} = \frac{-jRX_c}{R - jX_c} \times \frac{R - jX_c}{(R - jX_c)^2 - jRX_c} = \frac{-jRX_c}{R^2 - X_c{}^2 - 3jRX_c}$$

$$\frac{e_2}{e_1} = \frac{1}{3 + j(R^2 - X_c{}^2)/RX_c} = \frac{1}{3 + j(R/X_c - X_c/R)}$$

When $R = X_c$,

$$\frac{e_2}{e_1} = \frac{1}{3} \quad \text{or} \quad e_2 = \frac{e_1}{3}$$

The two-stage amplifier must therefore have a voltage gain of 3 so that the a-c voltage between the plate and cathode of the second stage will be adequate to provide enough input voltage to the first stage to sustain oscillations. The operating frequency is such that $X_c = R$.

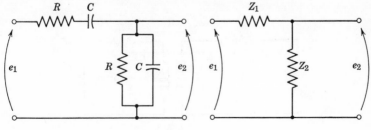

Fig. A-2

A.7   Reflected Impedance Across a Transformer

Figure A-3 represents the general case of a transformer with a closed-circuit secondary.

$$Z_p = R_p + j\left(\omega L_p - \frac{1}{\omega C_p}\right)$$

$$Z_s = R_s + j\left(\omega L_s - \frac{1}{\omega C_s}\right)$$

The circuit equations are

$$E_1 = I_p Z_p + j\omega M I_s$$

Induced voltage in the secondary is

$$-j\omega M I_p = I_s Z_s$$

Solving these equations to eliminate I_s gives

$$E_1 = I_p\left[Z_p + \frac{\omega^2 M^2}{Z_s}\right]$$

This equation shows that the effective primary impedance with a closed-circuit secondary *so that load current can flow* is $Z_p + \omega^2 M^2/Z_s$, of which the second term represents the coupled impedance arising from the presence of the closed secondary. The impedance to be reflected is represented by Z_s here.

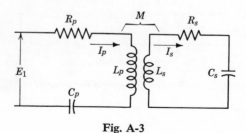

Fig. A-3

A.8   General-Purpose Curves

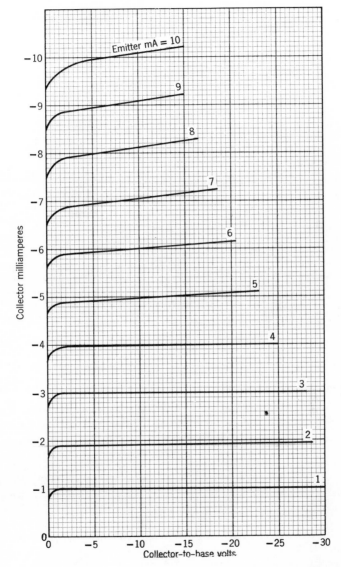

Fig. A-4   Average collector characteristics, common-base circuit, RCA transistor 2N105.

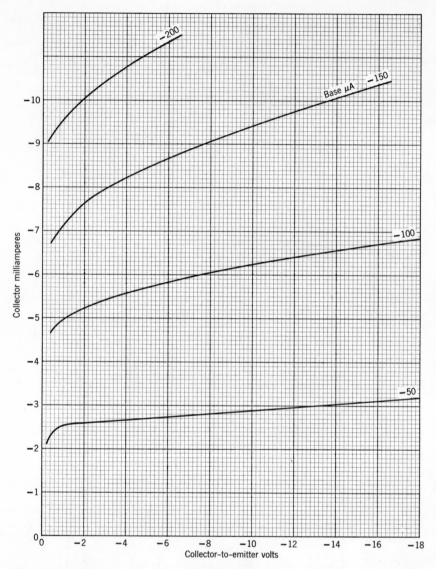

Fig. A-5 Average collector characteristics, common-emitter circuit, RCA transistor 2N105.

PNP Alloy-Junction Germanium Transistor
Type 2N404
(Courtesy Texas Instruments Incorporated)

Maximum Ratings (25°C):

Collector-Base Voltage	-25 V
Collector-Emitter Voltage	-24 V
Emitter-Base Voltage	-12 V
Collector Current	-100 mA
Total Device Dissipation (see note 1)	150 mW

Note 1: Derate linearly to 85°C free-air temperature: 2.5 mW/°C.

Characteristics (25°C):

Collector-Base Breakdown Voltage, BV_{CBO} -25 V

Common-Emitter *h*-parameters:

(taken at $V_{CE} = -6$ V,	h_{fe}	135
$I_C = -1$ mA)	h_{ie}	4 kΩ
	h_{oe}	50 μmho
	h_{re}	7×10^{-4}

The common-emitter characteristic curves are shown in Fig. A-6.

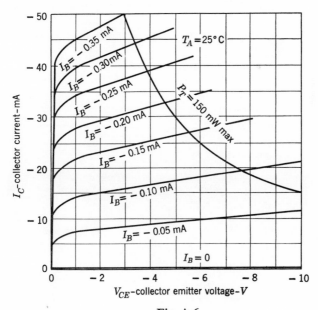

Fig. A-6

NPN Grown Junction Silicon Transistor
Type 2N334
(Courtesy Texas Instruments Incorporated)

Maximum Ratings:

Collector-Base Voltage	45 V
Emitter-Base Voltage	1 V
Collector Current	25 mA
Emitter Current	−25 mA
Device Dissipation (25°C)	150 mW
(100°C)	100 mW
(150°C)	50 mW

Characteristics—Common Base:

Collector Breakdown Voltage BV_{CBO}	45 V
h_{ib}, Input Impedance	30–80 Ω
h_{ob}, Output Admittance	0–1.2 μmho
h_{rb}, Feedback Voltage Ratio	0–1,000 × 10^{-6}
h_{fb}, Current Transfer Ratio	−0.948 to −0.989

The characteristic curves (operation at 25°C) are shown in Fig. A-7.

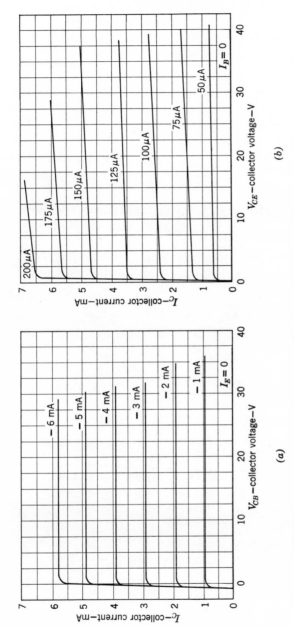

Fig. A-7 2N334 transistor (*a*) Common base output characteristics. (*b*) Common emitter output characteristics.

729

Complementary Alloy-Junction Germanium Transistors
NPN Types 2N1302, 1304, 1306, 1308
PNP Types 2N1303, 1305, 1307, 1309
(Courtesy Texas Instruments Incorporated)

Maximum Ratings (25°C):	NPN Types	PNP Types
Collector-Base Voltage	25 V	30 V
Emitter-Base Voltage	25 V	25 V
Collector Current	300 mA	300 mA
Total Device Dissipation (see note 1)	150 mW	150 mW

Note 1: Derate linearly to 85°C free-air temperature: 2.5 mW/°C.

Characteristics:

(*NPN* Types)	2N1302	2N1304	2N1306	2N1308
Common-base h-parameters:				
(taken at $V_{CB} = 5$ V				
$I_E = -1$ mA)				
h_{ib}	28	28	28	28 Ω
h_{rb}	5×10^{-4}	5×10^{-4}	5×10^{-4}	5×10^{-4}
h_{ob}	0.34	0.34	0.34	0.34 μmho
Common-emitter forward current transfer ratio, h_{fe}.	105	120	135	170

Characteristics:

(*PNP* Types)	2N1303	2N1305	2N1307	2N1309
Common-base h-parameters:				
(taken at $V_{CB} = -5$ V,				
$I_E =$ mA)				
h_{ib}	29	29	29	29 Ω
h_{rb}	7×10^{-4}	7×10^{-4}	7×10^{-4}	7×10^{-4}
h_{ob}	0.40	0.40	0.40	0.40 μmho
Common-emitter forward current transfer ratio, h_{fe}	115	130	150	190

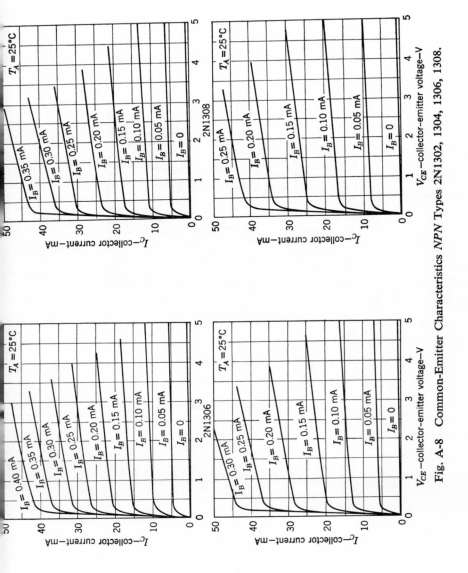

Fig. A-8   Common-Emitter Characteristics *NPN* Types 2N1302, 1304, 1306, 1308.

731

Normalized *h*-parameters, types 2N1302–1309:

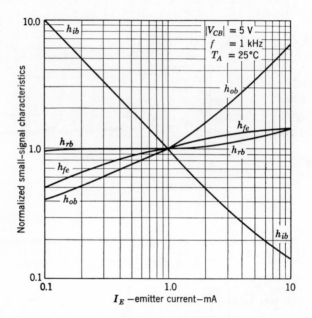

Fig. A-8 (continued)

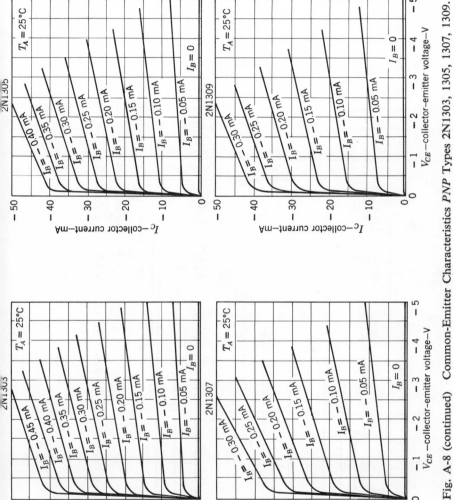

Fig. A-8 (continued) Common-Emitter Characteristics *PNP* Types 2N1303, 1305, 1307, 1309.

733

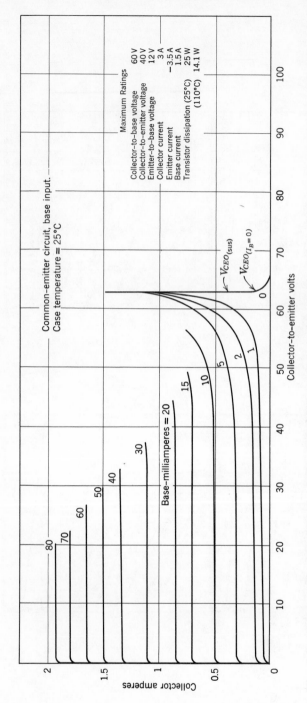

Fig. A-9 Collector characteristics for Silcon *NPN* Diffused-Junction Transistor, Type 2N1485 (courtesy Radio Corporation of America).

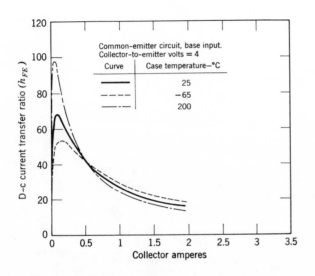

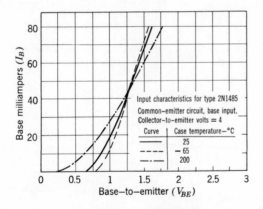

Fig. A-9 (continued)

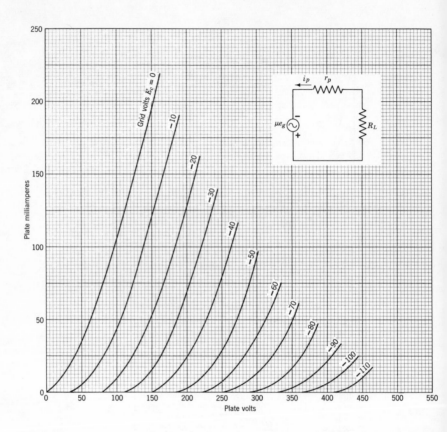

Fig. A-10 Average plate characteristics, 2A3 power triode. $E_{bo} = 250$ V, $I_{bo} = 60$ mA, $E_c = -43.5$ V, $r_p = 800\ \Omega$, $\mu = 4.2$, $g_m = 5{,}250\ \mu$mho.

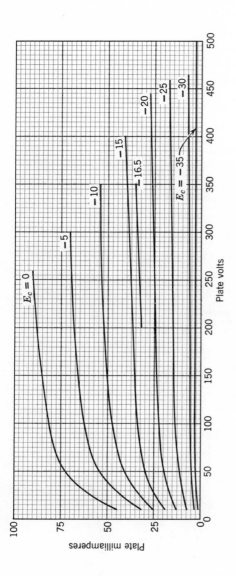

Fig. A-11 Average plate characteristics, 6F6 pentode. Class A power amplifier, $E_{bo} = 250$ V, $E_c = -16.5$ V, $I_{bo} = 34$ mA, $r_p = 80,000\ \Omega$, $g_m = 2,500\ \mu$mho.

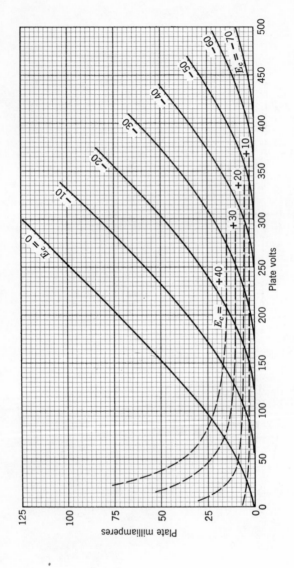

Fig. A-12 Average plate characteristics, 6F6 triode (screen connected to plate). Class A power amplifier. $E_{bo} = 250$ V, $E_c = -20$ V, $I_{bo} = 31$ mA, $\mu = 6.8$, $r_p = 2,600\ \Omega$, $g_m = 2,600\ \mu$mho.

738

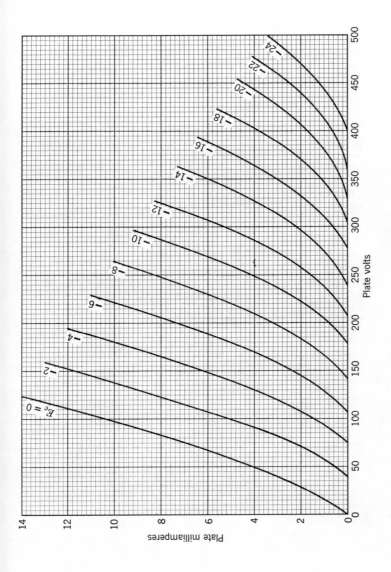

Fig. A-13 Average plate characteristics, 6J5 triode. $E_{bo} = 250$ V, $I_{bo} = 9$ mA, $E_c = -8$ V, $r_p = 7,700\,\Omega$, $\mu = 20$, $g_m = 2,600\,\mu$mho.

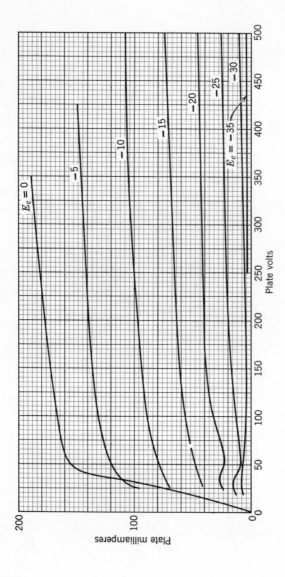

Fig. A-14 Average plate characteristics, 6L6 beam-power tube. Class A amplifier. $E_{bo} = 350$ V, $I_{bo} = 54$ mA, $r_p = 33,000$ Ω, $g_m = 5,200$ μmho, $E_{c2} = 100$ V.

740

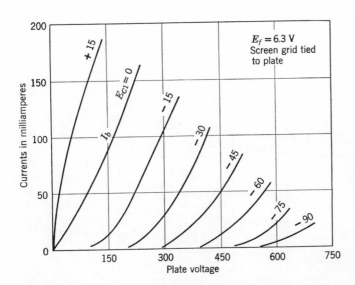

Fig. A-15 Average characteristics, 6L6GB triode connected. E_f = rated value, E_{bo} = 250 V, E_c = −20 V, I_{bo} = 40 mA, r_p = 1,700 Ω, g_m = 4,700 μmho. (Courtesy Sylvania Electric Products, Inc.)

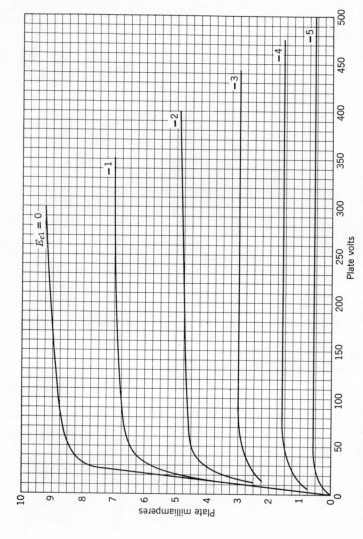

Fig. A-16 Average plate characteristics, 6SJ7 pentode. $E_{bo} = 250$ V, $E_{c1} = 3$ V, $I_{bo} = 3$ mA, $r_p = 1$ MΩ (approx.), $g_m = 1,650$ μmho, $E_{c2} = 100$ V, $E_{c3} = 0$ V.

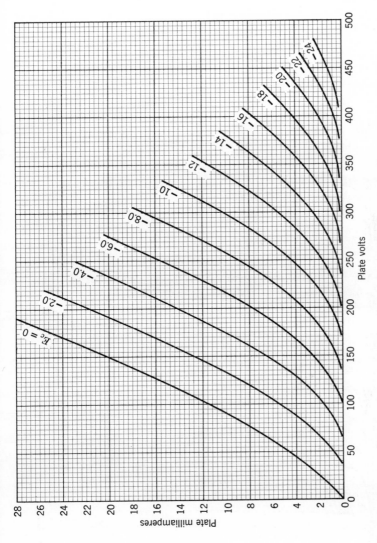

Fig. A-17 Average plate characteristics, 6SN7GT. E_f = rated value, E_{bo} = 250 V, E_c = −8 V, I_{bo} = 9 mA, r_p = 7,700 Ω, g_m = 2,600 μmho, μ = 20.

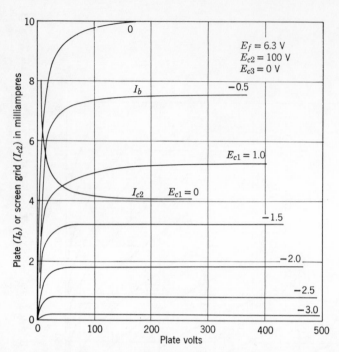

Fig. A-18 Average plate characteristics, 6AU6. E_f = rated value, E_{bo} = 100 V, E_{c1} = −1 V, I_{bo} = 5 mA, r_p = 0.5 MΩ, g_m = 3,900 μmho. (Courtesy Sylvania Electric Products, Inc.)

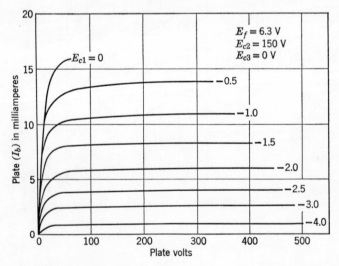

Fig. A-19 Average plate characteristics, 6AU6. E_f = rated value; E_{bo} = 250 V, E_c = −1 V, I_{bo} = 10.6 mA, r_p = 1 MΩ, g_m = 5,200 μmho. (Courtesy Sylvania Electric Products, Inc.)

744

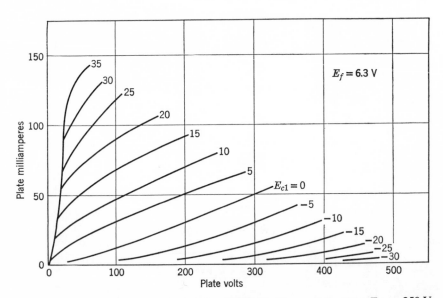

Fig. A-20 Average plate characteristics, 12AU7A. E_f = rated value; E_{bo} = 250 V, E_c = −8.5 V, I_{bo} = 10.5 mA, r_p = 7,700 Ω, g_m = 2,200 μmho, μ = 17; E_{bo} = 100 V, E_c = 0 V, I_{bo} = 11.8 mA, r_p = 6,500 Ω, g_m = 3,100 μmho, μ = 20. (Courtesy Sylvania Electric Products, Inc.)

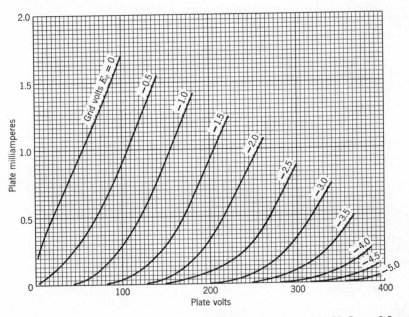

Fig. A-21 Average plate characteristics, 6SF5 triode. E_{bo} = 250 V, I_{bo} = 0.9 mA, E_c = 2 V, r_p = 66,000 Ω, μ = 100, g_m = 100 μmho.

A.9 Bessel Function Coefficients Selected Values of F-M Coefficients

m_f	$J_0(m_f)$	$J_1(m_f)$	$J_2(m_f)$	$J_3(m_f)$	$J_4(m_f)$	$J_5(m_f)$	$J_6(m_f)$	$J_7(m_f)$	$J_8(m_f)$	$J_9(m_f)$	$J_{10}(m_f)$
0	1.0000	0	0	0	0	0	0	0	0	0	0
1	0.7652	0.4401	0.1149	0.0196	0.0025	0.0003	0	0	0	0	0
2	0.2239	0.5767	0.3528	0.1289	0.0340	0.0070	0.0012	0.0002	0	0	0
3	−0.2601	0.3391	0.4861	0.3091	0.1320	0.0430	0.0114	0.0026	0.0005	0	0
4	−0.3971	−0.0660	0.3641	0.4302	0.2811	0.1321	0.0491	0.0152	0.0040	0.0009	0.0002
5	−0.1776	−0.3276	0.0466	0.3648	0.3912	0.2611	0.1310	0.0534	0.0184	0.0055	0.0015
6	0.1506	−0.2767	−0.2429	0.1148	0.3576	0.3621	0.2458	0.1296	0.0565	0.0212	0.0070
7	0.3001	−0.0047	−0.3014	−0.1676	0.1578	0.3479	0.3392	0.2336	0.1280	0.0589	0.0235
8	0.1717	0.2346	−0.1130	−0.2911	−0.1054	0.1858	0.3376	0.3206	0.2235	0.1263	0.0608
9	−0.0903	0.2453	0.1448	−0.1809	−0.2655	−0.0550	0.2043	0.3275	0.3051	0.2149	0.1247
10	−0.2459	0.0453	0.2546	−0.0584	−0.2196	−0.2341	−0.0145	0.2167	0.3179	0.2919	0.2075
11	−0.1712	−0.1768	0.1390	0.2273	−0.0150	−0.2383	−0.2016	0.0184	0.2250	0.3089	0.2804
12	0.0477	−0.2234	−0.0849	0.1951	0.1825	−0.0735	−0.2437	−0.1703	0.0451	0.2304	0.3005
13	0.2069	−0.0703	−0.2177	0.0023	0.2193	0.1316	−0.1180	−0.2406	−0.1410	0.0670	0.2338
14	0.1711	0.1334	−0.1520	−0.1768	0.0762	0.2204	0.0812	−0.1508	−0.2320	−0.1143	0.0850

A.10   Temperature Conversion Factors

The various temperature scales used in the text are based on the freezing point and boiling point of water. On the Celsius scale (often called Centigrade), the freezing point of water is taken as a starting point—0°C. The boiling point is then set 100 degrees higher—100°C. The Fahrenheit scale (the common temperature scale in use in the United States) uses 32°F for the freezing point and 212°F for the boiling point of water. This represents a total of $212 - 32 = 180$ degrees between the two extremes, while the Celsius scale uses only 100 degrees for the same temperature range. Then, for each °C there are 1.8°F. In addition, when converting from °C to °F, there must be added another 32 degrees because of the different starting points on the separate scales.

$$°F = 1.8°C + 32°$$
$$= \tfrac{9}{5}°C + 32°$$

Similarly, temperature given in °F may be converted to °C

$$°C = \tfrac{5}{9}(°F - 32°)$$

The Kelvin scale of temperatures uses 0° for the *absolute zero of temperature.* Each degree Kelvin (°K) represents the same temperature change as each degree on the Celsius scale (°C). However, absolute zero occurs at about −273°C (0°K), making 0°C = 273°K. To convert temperature given in °C to °K, simply add 273° to the temperature in °C. For example, 27°C = (27 + 273) = 300°K.

A.11   Sine Wave Template

The REPCO Sine Wave Template (shown in Fig. A-22) is designed to permit rapid sketching of sinusoidal waveforms. In many electronic circuits, the accurate sketching of waveforms helps to show important changes that occur. It is in use by many students and practicing engineers, for drawing accurate waveforms for examinations and technical reports. The transparent vinyl construction permits easy placement over prior work. The larger profile is based on a 1-in. amplitude and 2 in./cycle, and the smaller profile is half the amplitude on the same angular scale.

The REPCO Sine Wave Template is available at most bookstores serving colleges and technical institutes. It may be obtained at nominal cost from REPCO, Inc., P.O. Box 4002, Lexington, Ky. 40504.

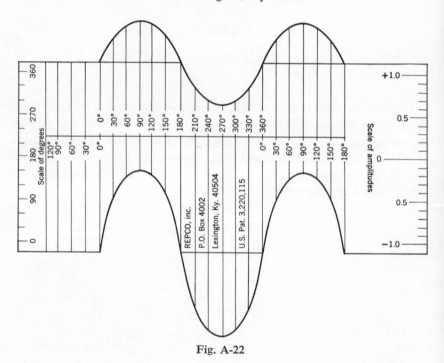

Fig. A-22

Answers to Drill Problems and Odd-Numbered Problems

CHAPTER 1

1-1 (a) 1 A; (b) 1 A. **1-3** (a) 15 Ω, 8 A; (b) G_{AB} 1/12 mho, G_{BC} 1/12 mho, G_{AC} 1/24 mho, G_T 8/120 mho; 8 A; (c) 1 A. **1-5** 274 Ω. **1-7** 60 V. **1-9** (a) 80 V; (b) 66 V; (c) 79.16 V; (d) 14 V diff., 17.5%; 0.84 V diff., 1.05%. **1-11** I_1 $4\frac{2}{9}$ A; I_2 $1\frac{1}{9}$ A. **1-13** 11/400 A. **1-15** Exercise. **1-17** 0.00316 A. **1-19** 0.00316 A. **1-21** 9.23 W. **1-23** Z_1 4 kΩ, Z_2 3 kΩ, Z_3 6 kΩ. **1-25** (a) 13 A; (b) 0.3846. **1-27** Sum, $141.4\underline{/-15°}$; difference, $141.4\underline{/75°}$. **1-29** (c) 50 W; 120 Hz. **1-31** 14.55 A. **1-33** 34.95 A. **1-35** 50 mH. **1-37** (a) 0 A, 0.1 A; 100 V; 0 V; (b) 2500 A/sec; (c) 40 μs; 63.3 mA, 63.3 V, 36.7 V. **1-39** Sketches. **1-41** Sketches. **1-43** 0.75. **1-45** (a) $1.33\underline{/56.5°}$ A; (b) 88.9 W; (c) 0.552; (d) 66.2 V, 100 V; (e) 0.735 A, 1.11 A; (f) 90.2 Ω. **1-47** (a) 50 mH; (b) 0.1905 A, 1.16 W. **1-49** 0.75. **1-51** 0.147. **1-53** (a) 50 − j31.6; (b) 333 − j31.6 = $334\underline{/-5.4°}$; $333\underline{/0°}$. **1-55** (a) 10^6 Hz; (b) 100. **1-57** (a) 0.25 J; (b) 25 mA; (c) 0.08 s, 9.175 mA; (d) 0.0152 A, 152 V; (e) −0.313 A/s; (f) 0.0554 s; (g) 0.4416 s; (h) 0.25 J. **1-59** $995\underline{/-90°}$ Ω. **1-61** 7920 Hz, 8080 Hz; 100. **1-63** (a) 177,800 Ω; (b) 188.3; (c) 5.62 μA. **1-65** (a) 0.01 s; (b) 0.02 A; (c) 0 A. **1-67** 0.09 J, 6 × 10⁻⁴ C, 40 W in resistor, 60 W to capacitor, 100 W from source. **1-69** $6.7\,\text{A}\underline{/-11.9°}$. **1-71** 68.2°. **1-73** 89.8 W. **1-75** I $2.91\underline{/14°}$ A, I_L $4.615\underline{/-57.5°}$ A, I_C $4.615\underline{/85.6°}$ A. **1-77** (a) I_1 13.52 − j27.92 mA, I_2 50.9 + j65.9 mA, I 64.4 + j38.0 mA; 0.647 W; (b) $8.39\underline{/-7.9°}$ V; in 20 Ω, 0.112 W; in 120 Ω, 0.115 W; in 60 Ω, 0.420 W; (c) 0.861. **1-79** (a) 50.3 Hz; (b) 20 mA, 63.2 V, 63.2 V; (c) 46.325 Hz, 54.275 Hz. **1-81** Plot. **1-83** C, 833 pF. **1-85** 62.5 mW. **1-87** (a) $6.57\underline{/90°}$ mA; (b) 4.32 mW; (c) $100\underline{/0°}$ V; (d) $111.75\,\text{V}\underline{/0°}$. **1-89** 120 − j50; C 7.95 μF. **1-91** 100 pF. **1-93** 0.175 V. **1-95** 46 dB. **1-97** I_1 3 A, I_2 5 A, I_3 8 A. **1-99** 8.185 A.

CHAPTER 2

D2-1 1.602×10^{-19} J, 16.02×10^{-19} J. **D2-2** 1.1214×10^{-19} J, 1.7622×10^{-19} J. **D2-3** 1.69×10^{14} Hz; yes. **D2-4** 2.66×10^{14} Hz; yes. **D2-5** -0.117 eV; 0.665 eV. **D2-6** -0.084 eV; 1.121 eV. **D2-7** $600,000$ V/m. **D2-8** 9.6×10^{-14} N. **D2-9** 0.133 Ω-m. **2-1** 1.1214×10^{-19} J. **2-3** 1.602×10^{-19} J. **2-5** 1.69×10^{14} Hz. **2-7** 2.66×10^{14} Hz. **2-9** -0.084 eV; 1.121 eV. **2-11** Germanium. **2-13** 1.5 A; 0.8 mA. **2-15** (*a*) 0.6×10^6 V/m; (*b*) 9.6×10^{14} N. **2-17** $-1.1 - j58.1 \, \Omega$. **2-19** $-333 \, \Omega$.

CHAPTER 3

D3-1 Apparently silicon, from V_F and temperature. **D3-2** 1.25 A. **D3-3** 77, 212, 302, $392°$F. **D3-4** 0.7 V, $0.75 \, \Omega$. **D3-5** When applied voltage is much greater than 0.7 V; when external series resistance is much greater than $0.75 \, \Omega$. **D3-6** (*a*) 1.414 kΩ; (*b*) 45 V, 100 mA. **D3-7** $0.214 \, \Omega$. **D3-8** (*a*) 10 V; (*b*) 20 V; (*c*) $1314 \, \Omega$. **D3-9** 83.4 V; 86.5 V. **D3-10** 50 V, 18.9 V. **D3-11** (*a*) reverse current, reverse voltage polarity; (*b*) double time constant, less voltage variation. **D3-12** 50 Hz period is 0.02 s; sketches; (*a*) e_L decreases to 37% of peak after $\frac{1}{2}$ cycle; (*b*) after 5 cycles; (*c*) after 50 cycles. **D3-13** None; -10 V; 10 mA. **D3-14** None; -10 V. **3-1** Apparently silicon, from temperature. **3-3** 0.75 V, $0.625 \, \Omega$. **3-5** (*b*) 7.225 W; (*c*) current rating exceeded at high temperatures. **3-7** Sketch. **3-9** 10.45 V maximum, sinusoidal. **3-11** 10 V. **3-13** 12.06 V. **3-15** Current reverses, load voltage reverses; double time constant, less voltage variation; no filtering. **3-17** 10.48 mA, 8 V. **3-19** 11 V, 28 V, 17 V. **3-21** 5 V, positive. **3-23** 1Z8.2 Zener, R_s 0, R $12.5 \, \Omega$. **3-25** 5. **3-27** (*a*) 0.75 V; (*b*) $214 \, \Omega$ vs. 50 kΩ.

CHAPTER 4

D4-1 0.782×10^6 mA/cm^2, 0.543×10^5 mA/cm^2, 89.8 mA/cm^2. **D4-2** Ratio of areas 1: 14.4: 8,720. **D4-3** Plot on log-log scales; $n = 1.51$, $K = 0.03$. **4-1** 5.93×10^5 m/s; $1,325,000$ mi/h. **4-3** 4.77×10^{-10} s. **4-5** Plot I_b (A) $= 51.8 \times 10^{-6} E_b^{\frac{3}{2}}$. **4-7** Design. **4-9** 12.1×10^5 m/s; no, right angle component too small; $1.414 \times 12.1 \times 10^5$ m/s. **4-11** (*a*) $9,700 \, \Omega$, maximum $1,257 \, \Omega$, minimum $297 \, \Omega$; (*b*) 150-Ω, 39 W; R_1, 2.39 W, R_L, 315 W. **4-13** (*a*) 4×10^{-17} J; (*b*) 12.5×10^{17} s; (*c*) 50 W or 50 J/s. **4-15** 104 mA. **4-17** 90 mA. **4-19** $555 \, \Omega$.

CHAPTER 5

D5-1 Graph. **D5-2** (*a*) both 68 mA; (*b*) 0.935 W; (*c*) 34 mA; (*d*) 32.5%; (*e*) 170 V. **D5-3** (*a*) both 68 mA; (*b*) 3.74 W; (*c*) 47 mA; (*d*) 65%; (*e*) 340 V. **D5-4** Diode conducts from $27°$ to $117°$. **D5-5** Sketch. **D5-6** Sketches. **D5-7** Sketch. **5-1** 50 mA, 3.75 W. **5-3** Minimum $100 \, \Omega$. **5-5** (*a*) both 32.7 mA; (*b*) 2.08 W; (*c*) 23.1 mA; (*d*) 67.8%; (*e*) 333.5 V. **5-7** 215 V, 239 V. **5-9** (*a*) 0.314 A; (*b*) 237 V; (*c*) 114.6; (*d*) 1.1 V; (*e*) 0.00464. **5-11** 0.0332. **5-13** (*a*) 106.5 VA; (*b*) 0.927 A; (*c*) $1,630$ center-tapped. **5-15** $540 \, \Omega$; 0.1 A. **5-17** and **5-19** Refer to text. **5-21** Design. **5-23** Exercise. **5-25** Reverse polarity load voltage. **5-27** (*a*) 987 μA; (*b*) sketch; (*c*)

100 μA. **5-29** Sketch. **5-31** (*a*) 159 mA average, 500 mA peak; (*b*) 132.5 μF; (*c*) yes, 412 mA peak current.

CHAPTER 6

D6-1 21.25, 9 kΩ, 2,400 μmho; 15.75, 20 kΩ, 1,100 μmho. **D6-2** 4.5 mA. **D6-3** 1,600 μmho, 1 MΩ, 1,600; 2,150 μmho, 5 kΩ, 10.75. **D6-4** E_{bb} greater than 150 V; R_b greater than 0. **D6-5** 31.8 Hz, C_k 50 μF; product fC_k greater than 1/628. **D6-6** 150 kΩ, 0.133 μF. **6-1** 10.3 $\times$ 10^5 m/s. **6-3** (*a*) 21.21 V; (*b*) 120 V; (*c*) 4.24 mA; (*d*) 26 mA. **6-5** Curves like Fig. 6-8. **6-7** 2,400 μmho. **6-9** 4.8 V; 12.5 μF minimum. **6-11** 4, 880 Ω, 4,500 μmho. **6-13** Sketch. **6-15** 5 mA. **6-17** 10. **6-19** 180 V, -6 V, 5 mA; 1 kΩ. **6-21** 4 mA. **6-23** 3.3 kΩ. **6-25** (*a*) 16 mA; (*b*) 192 V increase. **6-27** Exercise. **6-29** 16 V increase. **6-31** 2.6 mA, 152 V, -6 V; 0, 0, 0.

CHAPTER 7

D7-1 and **D7-2** Student exercises. **D7-3** Depends on value of V_{CC}. **D7-4** Sketch. **D7-5** 100 μA. **D7-6** 6.67 kΩ, -25 V, 1 V. **D7-7** 1.5 V. **D7-8** 11 V. **D7-9** 667 Ω, 120 μF. **7-1** 3.56 $\times$ 10^5 m/s. **7-3** No, 0.25 $\times$ 10^{-19} J, 0.153 eV. **7-5** Variable. **7-7** (*a*) Sketch; (*b*) 5 mA, 10 V; 200 kΩ. **7-9** Sketch. **7-11** $1/(1 - \alpha)$; $1 - \beta$. **7-13** 50 to 150 μA; about 100 μA. **7-15** 150 kΩ. **7-17** 0.218 V. **7-19** 400 kΩ. **7-21** (*a*) 4 mA; (*b*) 40. **7-23** 6 kΩ, 670 Ω, 15 μF. **7-25** 66.7 kΩ. **7-27** Graph. **7-29** Not linear.

CHAPTER 8

D8-1 (*a*) 320 V, 16 mA; (*b*) $-1/20,000$; (*c*) same; independent of E_{bb}; (*d*) through Q-point with slope $-1.02/20,000$. **D8-2** (*a*) One point 225 V, -15 V; (*b*) sketch; (*c*) 140 to 285 V. **D8-3** Sketch. **D8-4** No; depends on plate current. **D8-5** 19.5. **D8-6** 725 Ω, 69.5 μmho, 10.9 $\times$ 10^{-4}, 45. **8-1** (*a*) Sketch; (*b*) 190 V, 110 V; (*c*) 28.3 V rms, 32.6 dB; (*d*) 159 Ω. **8-3** (*a*) Sketch; (*b*) 175 V, 155 V; (*c*) 7.07 V rms, 20 dB; (*d*) 159 Ω. **8-5** Slightly less than 1; about 32. **8-7** (*a*) 40; (*b*) 40; (*c*) 1,600. **8-9** 0.375 mA. **8-11** $(\mu + 1)Z_L/(r_p + Z_L)$. **8-13** 30 V p-p. **8-15** (*a*) R_L, 5 kΩ; R_B, 660 kΩ; (*b*) I_C, 2.28 mA and V_C, 4.6 V; (*c*) I_C, 2.214 mA and V_C, 4.93 V; (*d*) I_C, 2.256 mA and V_C, 4.72 V. **8-17** R_2/R_1. **8-19** h_{fe}, 285; h_{ib}, 4.35 Ω; h_{ob}, 24.8 $\times$ 10^{-6} mho; h_{rb}, 10.5 $\times$ 10^{-4}. **8-21** Sketches. **8-23** 341 kΩ; 230 kΩ.

CHAPTER 9

D9-1 Frequency distortion. **D9-2** Exercise. **D9-3** Phase and frequency. **D9-4** Exercise. **D9-5** (*a*) Sketch; (*b*) 62.5; (*c*) 41.7. **D9-6** Sketch. **D9-7** Exercise. **9-1** (*a*) 19.1 at 180°; (*b*) 16.2 at 212°; from curve, 16.05 at 213°. **9-3** 62.7 Hz, 18.06 kHz. **9-5** (*a*) For 180 V: 57.8; 0.674 V; (*b*) for 300 V: 63.5; 0.755 V. **9-7** 19.8 at 131.8°, 75 Hz. **9-9** (*a*) -444; (*b*) 117 Hz, 9 kHz; (*c*) 0.225 V. **9-11** 29 kHz. **9-13** (*a*) 37.3; (*b*) R_{i2}, 2,075 Ω; R_{i1}, 1,990 Ω; R_{o1}, 33,900; R_{o2}, 32,500 Ω; (*c*) 0.895 at 26.4°; (*d*) approx. 1 at 0°. **9-15** 7,420$/$$-8.5°$ Ω. **9-17** $(1.58 + j1.91) \times 10^{-3}$ mho. **9-19** 61 μH. **9-21** $\mu/(\mu + 1 + r_p/R_L)$. **9-23** 60 V. **9-25** Sketch. **9-27** Asymptotes 4,000 and 2,110 Ω. **9-29** (a) 1,145 Ω, 3.78 × 10^{-5} mho, 5.18 × 10^{-4}, 38.1; (b) −150 31.8. **9-31** 125 μF.

CHAPTER 10

D10-1 11/10 W, 1/10 W. **D10-2** 5.55%. **D10-3** 100 Ω. **10-1** (a) 0.87 W; (b) 2.38%, (c) 12.6%; (d) 6.01 W. **10-3** (a) 1.34 W; (b) 66.2 μF or larger. **10-5** (a) 2.52 W; 12.42 W, 16.9%; (b) 240 V. **10-7** 2,500 Ω, 15.37 W. **10-9** 6.1 W, 30.3%. **10-11** 1,380 Ω, 0.282 W. **10-13** 6.1%, 10.2 W. **10-15** (a) Sketch; (b) 6,060 Ω; (c) 5.35 W; (d) 73.8%; (e) 5.5. **10-17** 58.4 mA, 50.3 V. **10-19** Design. **10-21** 138 V. **10-23** (a) Sketch; (b) 92 mA, 10 mA; 427 V, 32 V; (c) 3.9 W; (d) 7.3%; (e) 11.22 W, 7.32 W. **10-25** (a) 38.5 W; (b) 38.5/2 W; 0 W. **10-27** (a) Sketch; (b) center of load line and on maximum dissipation hyperbola; (c) 3 W, 3 W. **10-29** 40 V, 6.32, 20 V and 25 mA for Q-point. **10-31** Exercise. **10-33** 27.2% vs. 78.5%. **10-35** Exercise.

CHAPTER 11

D11-1 (0.796 − j23.9) × 10^5 Ω. **D11-2** (a) 31.8 Hz; (b) depends on frequency; 3 dB error at f_1; (c) plot. **D11-3** 12.25, 1.63 kΩ **11-1** R_1 750 Ω, 12 mW; R_2 63.4 kΩ, 0.57 W; R_3 750, 12 mW; R_4 16 kΩ, 0.576 W; R_5 20 kΩ, 0.5 W; 100 kΩ, 0.1 W; 10 kΩ, 1 mW. **11-3** 1/2.74, 2.74. **11-5** (a) 307 Hz; (b) 0.6%; (c) 30.6 Hz. **11-7** Curve. **11-9** (a) 108 kΩ at −90°, 207 Ω; (b) 0.92. **11-11** 12.25 H. **11-13** 12.5 MΩ, 354 Ω. **11-15** Sketch. **11-17** Sketch. **11-19** Smaller C, 20 kHz B; larger C, 6.66 kHz B; from 1.42 MHz to 0.821 MHz. **11-21** Q_1: 22.7 V, 3.915 mA; Q_2: 23.4 V, 0.127 mA. **11-23** 182 Ω.

CHAPTER 12

D12-1 $\beta = (\mu + 1)R_k/\mu R_b$; negative. **D12-2** −0.0133. **D12-3** −0.04 **D12-4** (a) 0.429 Z_o; (b) 2.33 Z_{in}. **D12-5** −24.6, 0.04. **D12-6** 20 kΩ change. **D12-7** 1.6 to 8.0 μF. **D12-8** (a) yes; (b) h_{fe} greater than 0.95 permits oscillation. **D12-9** $R_1 = R_f$: R_{20}, 3.95 MΩ; R_{200}, 0.395 MΩ; R_{2000}, 39.5 kΩ. **12-1** (a) −30; (b) −26.1, −23.1, −20.6, −18.8, −12.0, −7.5, −4.3; (c) plot. **12-3** −12.85. **12-5** From 75 kΩ to 800 kΩ. **12-7** A_{v1}, 238; A_{v2}, 27.8; current feedback: $A_{v1}{}'$, 113.5; $A_{v2}{}'$, 18.5; voltage feedback: A_v, 118. **12-9** 124 pF. **12-11** 90 mV; 3.21 mV. **12-13** 0.5; 62.5 V. **12-15** 20/3; 4 kΩ. **12-17** (a) −0.196; 910 Ω; (b) a = 0.162. **12-19** gain −4.88. **12-21** Exercise.

CHAPTER 13

D13-1 3,100 mi, 7.5 MHz. **D13-2** Sketches. **D13-3** Sketch. **D13-4** 5 kHz. **D13-5** (a) 92.2 MHz; (b) 200 kHz; (c) 10; (d) 10; (e) plot. **13-1** (a) fundamental and second harmonic; (b) high-Q tuned load; (c) 0.05. **13-3** (a) radian frequencies 100,000; 101,000; 99,000; (b) 0.5; (c) sideband power is 0.25/4 × carrier power. **13-5** 3 × 10^6 m, 300 m. **13-7** less than 15 cm. **13-9** 18 times. **13-11** Sketch. **13-13** Exercise. **13-15** (a) 10 kHz; (b) 16 kHz; (c) 20 kHz. **13-17** 10 kHz 5 times larger than 2 kHz. **13-19** No. **13-21** Choose 4 pulses.

CHAPTER 14

D14-1 Variable for a-c; constant for d-c. **D14-2** Only at 50°C. **D14-3** 2,075 Ω.
14-1 1 MΩ. **14-3** −70 V. **14-5** From 90° to approx 0°. **14-7** (*a*) 154.6°; (*b*) increases
θ, delays conduction, reduces average current. **14-9** 122.9°. **14-11** 57.9°, grid leads
plate. **14-13** 75.4 Ω. **14-15** 2,315 Ω. **14-17** (*a*) average current, 4.63 A. (*b*) 25.9 h.

CHAPTER 15

15-1 90° to 180° region. **15-3** Sketch. **15-5** 20.8 A; 4,320 W. **15-7** (*a*) zero; (*b*) zero.
15-9 Sum is 5 + *j*8.66 V. **15-11** −*j*1.565 V. **15-13** −*j*0.0443 V. **15-15** (*a*) 980 V;
(*b*) 2,340 W. **15-17** (*a*) 270 V d-c; (*b*) 3.73 kVA; (*c*) 5.28 kVA. **15-19** 0.675 or
67.5%.

CHAPTER 16

16-1 (*a*) 8 μA; (*b*) 42 V. **16-3** 9 MΩ. **16-5** Sufficient energy. **16-7** 2 μA; same. **16-9**
1.22. **16-11** 1.41 rms μA. **16-13** 0.0042 μF; *E* = 0.74 V. **16-15** 5.905 mA. **16-17** (*a*)
19.5 μA/lm; (*b*) 19.6 μA/lm. **16-19** (*a*) 75 and 60 μA/lm; (*b*) 5.44 V rms and 7.42 V
rms; some distortion. **16-21** 840 Ω.

CHAPTER 18

18-1 0.02655. **18-3** 81.3 V; 0.0572 V. **18-5** *R* = 0.25 MΩ, *t* = 2.24 s. **18-7** (*a*)
1,400 Ω; (*b*) *R* must allow current for relay. **18-9** 8%.

Index

755